ANTENNAS

ANTENNAS
FROM THEORY TO PRACTICE

Second Edition

Yi Huang
Professor of Wireless Engineering
The University of Liverpool, UK

WILEY

Registered Offices

John Wiley & Sons, Inc., 111 River Street, Hoboken, NJ 07030, USA

John Wiley & Sons Ltd, The Atrium, Southern Gate, Chichester, West Sussex, PO19 8SQ, UK

Editorial Office

The Atrium, Southern Gate, Chichester, West Sussex, PO19 8SQ, UK

For details of our global editorial offices, customer services, and more information about Wiley products, visit us at www.wiley.com.

Wiley also publishes its books in a variety of electronic formats and by print-on-demand. Some content that appears in standard print versions of this book may not be available in other formats.

Library of Congress Cataloging-in-Publication Data

Names: Huang, Yi, 1964– author.
Title: Antennas : from theory to practice / Yi Huang, Professor of Wireless
    Engineering, The University of Liverpool, UK.
Description: Second edition. | Hoboken, NJ, USA : Wiley, 2022. | Includes
    bibliographical references and index.
Identifiers: LCCN 2021007217 (print) | LCCN 2021007218 (ebook) | ISBN
    9781119092322 (cloth) | ISBN 9781119092339 (adobe pdf) | ISBN
    9781119092346 (epub)
Subjects: LCSH: Antennas (Electronics)
Classification: LCC TK7871.6 .H79 2021 (print) | LCC TK7871.6 (ebook) |
    DDC 621.382/4–dc23
LC record available at https://lccn.loc.gov/2021007217
LC ebook record available at https://lccn.loc.gov/2021007218

Cover Design: Wiley
Cover Image: © bluebay2014/Getty Images

Set in 10/12pt TimesLTStd by Straive, Pondicherry, India
Printed and bound by CPI Group (UK) Ltd, Croydon, CR0 4YY

C9781119092322_190821

Contents

Preface to the Second Edition

Since the publication of the first edition of this book in 2008, it has attracted a lot of attention from university students, professors, researchers as well as engineers in the industry. The book has been adopted by many universities around the world as a textbook for undergraduate and postgraduate teaching and used as a reference book for antenna research and designs. It was translated to Korean as a major antenna book in 2014. The main attraction and success of the book are its distinct and unique feature from other well-known antenna books: it has integrated the antenna theory with real-world examples from the past to the present into one volume without excessive mathematics. This feature has been maintained in this new edition.

Based on the success of the first edition, this edition has provided a better coverage in many important areas and added some recent and latest developments in antenna theory and practice. In particular, the theory of characteristic modes (Section 6.5) for antenna design and analysis, new materials (e.g. composite materials, metamaterials, and meta-surfaces in Section 7.1), fabrication processes (Section 7.2), and advanced measurement methods for antennas (Section 7.3). MIMO antennas, automotive antennas, and reflector antennas have been added to Special Topics in Chapter 8 to reflect some of the current hot topics. Every chapter has been updated with the latest information. Of course, the mistakes and errors identified in the first edition have also been corrected. As a result, the number of pages is increased significantly from 360 to over 500.

As an essential element of a radio system, the antenna has always been an interesting but difficult subject for radio-frequency (RF) engineering students and engineers. Many good books on antennas have been published over the years, and some of them were used as our major references.

This book is different from other antenna books. It is especially designed for the people who are relatively new to antennas but would like to learn this subject from the basics to advanced antenna analyses, designs, and measurements within a relatively short period of time. In order to gain a comprehensive understanding of antennas, one must also know transmission lines and radio propagation. At the moment, people often have to read a number of different books, which may not be well correlated. Thus, it is not the most efficient way to study this subject. In this book, we put all the necessary information about antennas into a single manuscript and try to examine the antenna from both the circuit point of view and the field point of view. The book covers the basic transmission line and radio propagation theories, which are then used to gain a good understanding of antenna basics and theory. Various antennas are examined, and design examples are presented. Particular attention is given to modern computer-aided antenna designs. Both basic and advanced computer software packages are used as examples to illustrate how they can be used for antenna analysis and design. Antenna materials, fabrication processes, measurement theory, and techniques are also addressed. Some special topics on the latest antenna development are covered in the final chapter.

The book material is mainly based on a successful short course on antennas for practicing professionals at the University of Oxford and the Antennas module for the students at the University of Liverpool. The book covers almost all aspects about antennas, especially important and timely issues involving modern practical antenna design and theory. Many examples and questions are given in each chapter. It is an ideal textbook for university antenna course, professional training course, and self-study. It is also a valuable reference for engineers and researchers who work with RF engineering, radar and radio communications. The organization of this book is as follows:

Chapter 1: Introduction. The objective of this chapter is to introduce the concept of antenna and review essential mathematics and electromagnetics, especially Maxwell's equations. Material properties (permittivity, permeability, and conductivity) are discussed, and some common ones are presented in tables.

Chapter 2: Circuit Concepts and Transmission Lines. The concepts of lumped and distributed systems are established. The focus is placed on the fundamentals and characteristics of transmission lines. A comprehensive coverage on Smith chart, bandwidth, and impedance match techniques is provided in this edition. A wide range of conventional and new transmission lines and connectors are introduced and compared.

Chapter 3: Field Concepts and Radiowaves. Field concepts, including plane wave, intrinsic impedance, and polarization, are introduced and followed by a discussion on radio propagation mechanisms and radiowave propagation characteristics in various media. Some basic radio propagation models are introduced, and circuit concepts and field concepts are compared at the end of this chapter.

Chapter 4: Antenna Basics. The essential and important parameters of an antenna (such as the radiation pattern, gain, and input impedance) are addressed from both the circuit point of view and field point of view. Through this chapter, you will become familiar with antenna language, understand how antennas work, and know what the main design considerations are.

Chapter 5: Popular Antennas. In this long chapter, some of the most popular antennas (wire-type, aperture-type, and array antennas) are introduced and analyzed using relevant antenna theories. The aim is to see why they have become popular, what their major features and properties are (including advantages and disadvantages), and how they should be designed.

Chapter 6: Computer-Aided Antenna Design and Analysis. The aim of this special and unique chapter is to give a good review of antenna modeling methods and software development, introduce the basic theory behind computer simulation tools, and demonstrate how to use industry standard software to analyze and design antennas. Two software packages (one is simple and free) are presented with step-by-step illustrations. The theory of characteristic modes (TCM) is introduced, and a patch antenna is employed as an example to illustrate how to use TCM for antenna design and analysis.

Chapter 7: Antenna Materials, Fabrication, and Measurements. This is another practical chapter to address three important issues: what materials are suitable for antennas, how to make an antenna, and then how to conduct antenna measurements accurately and efficiently. Some popular measurement equipment and facilities are introduced and discussed. A good overview of antenna measurement systems is provided with real-world examples. Some latest measurement techniques and problems are also presented and discussed.

Chapter 8: Special Topics. This final chapter presents some of the latest important developments in antennas. It covers electrically small antennas, mobile terminal and base-station antennas, diversity and MIMO antennas, RFID antennas, multiband and broadband antennas, reconfigurable antennas, automotive antennas, and reflector antennas. Relevant theory, design techniques, and practical examples are provided for in-depth understanding.

On completion of this book, the reader should be ready to conduct advanced antenna design, analysis, and measurements and ready to become an antenna researcher and engineer.

I am very grateful to the many individuals who have provided great contributions, comments, suggestions, and assistance to make this much improved second edition a reality. In particular, I would like to extend my sincere appreciation to:

1. Dr Kevin Boyle from Qualcomm, who has contributed significantly to the first edition as coauthor and provided many good suggestions, corrections, and references to the second edition. It is a pity that his new job has made it impossible for him to directly contribute to the second edition;
2. Mr Lars Foged, Scientific Director of MVG, who is a well-known measurement expert and has contributed to various parts and a section on antenna measurements in Chapter 7;
3. Prof Tony Brown, a leading antenna expert from Manchester University, who has contributed a new section on reflector antennas in Chapter 8;
4. Prof Tae-Hoon Yoo at Dongyang Mirae University for having translated the book into Korean and provided many great suggestions and feedbacks on how to make the second edition better;
5. Dr Hanyang Wang, Chief Antenna Expert of Huawei, for inspiring discussions on industrial practices in antenna designs, fabrication, and measurements.
6. My students: Ahmed Alieldin, Lyuwei Chen, Qian Xu, Sheng Yuan, Chaoyun Song, Jiayou Wang, Shahzad Maqbool, Barry Cheeseman, Yang Lu, and many other current and past students at the University of Liverpool for constructive feedbacks and production of figures;
7. The individuals and organizations who have provided us with their figures or allowed us to reproduce their figures;
8. A special one who has kindly provided a list of corrections of the first edition via John Wiley, but I was not given his/her name;
9. The team at John Wiley who have provided the guidance and great support throughout the process.

Prof Yi Huang
Chair in Wireless Engineering
The University of Liverpool, UK
Yi.Huang@IEEE.org
January 2021

Preface to the First Edition

As an essential element of a radio system, the antenna has always been an interesting but difficult subject for radio-frequency (RF) engineering students and engineers. Many good books on antennas have been published over the years, and some of them were used as our major references.

This book is different from other antenna books. It is especially designed for people who know little about antennas but would like to learn this subject from the very basics to practical antenna analysis, design, and measurement within a relatively short period of time. In order to gain a comprehensive understanding of antennas, one must know about transmission lines and radio propagation. At the moment, people often have to read a number of different books, which may not be well correlated. Thus, it is not the most efficient way to study the subject. In this book, we put all the necessary information about antennas into a single volume and try to examine antennas from both the circuit point of view and the field point of view. The book covers the basic transmission line and radio propagation theories, which are then used to gain a good understanding of antenna basics and theory. Various antennas are examined and design examples are presented. Particular attention is given to modern computer-aided antenna design. Both basic and advanced computer software packages are used in examples to illustrate how they can be used for antenna analysis and design. Antenna measurement theory and techniques are also addressed. Some special topics on the latest antenna development are covered in the final chapter.

The material covered in the book is mainly based on a successful short course on antennas for practicing professionals at the University of Oxford and the Antennas module for students at the University of Liverpool. The book covers important and timely issues involving modern practical antenna design and theory. Many examples and questions are given in each chapter. It is an ideal textbook for university antenna courses, professional training courses, and self-study. It is also a valuable reference for engineers and designers who work with RF engineering, radar and radio communications.

The book is organized as follows:

Chapter 1: Introduction. The objective of this chapter is to introduce the concept of antennas and review essential mathematics and electromagnetics, especially Maxwell's equations. Material properties (permittivity, permeability, and conductivity) are discussed, and some common ones are tabulated.

Chapter 2: Circuit Concepts and Transmission Lines. The concepts of lumped and distributed systems are established. The focus is placed on the fundamentals and characteristics of transmission lines. A comparison of various transmission lines and connectors is presented. The Smith Chart, impedance matching, and bandwidth are also addressed in this chapter.

Chapter 3: Field Concepts and Radio Waves. Field concepts, including the plane wave, intrinsic impedance, and polarization, are introduced and followed by a discussion on radio propagation mechanisms and radio wave propagation characteristics in various media. Some basic radio propagation models are introduced, and circuit concepts and field concepts are compared at the end of this chapter.

Chapter 4: Antenna Basics. The essential and important parameters of an antenna (such as the radiation pattern, gain, and input impedance) are addressed from both the circuit point of view and field point of view. Through this chapter, you will become familiar with antenna language, understand how antennas work, and know what design considerations are.

Chapter 5: Popular Antennas. In this long chapter, some of the most popular antennas (wire-type, aperture-type, and array antennas) are examined and analyzed using relevant antenna theories. The aim is to see why they have become popular, what their major features and properties are (including advantages and disadvantages), and how they should be designed.

Chapter 6: Computer-Aided Antenna Design and Analysis. The aim of this special and unique chapter is to give a brief review of antenna-modeling methods and software development, introduce the basic theory behind computer simulation tools, and demonstrate how to use industry standard software to analyze and design antennas. Two software packages (one is simple and free) are presented with step-by-step illustrations.

Chapter 7: Antenna Manufacturing and Measurements. This is another practical chapter to address two important issues: how to make an antenna and how to conduct antenna measurement, with a focus placed on the measurement. It introduces S-parameters and equipment. A good overview of the possible measurement systems is provided with an in-depth example. Some measurement techniques and problems are also presented.

Chapter 8: Special Topics. This final chapter presents some of the latest important developments in antennas. It covers mobile antennas and antenna diversity, RFID antennas, multiband and broadband antennas, reconfigurable antennas, and electrically small antennas. Both the theory and practical examples are given.

The authors are indebted to the many individuals who provided useful comments, suggestions, and assistance to make this book a reality. In particular, we would like to thank Shahzad Maqbool, Barry Cheeseman, and Yang Lu at the University of Liverpool for constructive feedback and producing figures, Staff at Wiley for their help and critical review of the book, Lars Foged at SATIMO and Mike Hillbun at Diamond Engineering for their contribution to Chapter 7, and the individuals and organizations who have provided us with their figures or allowed us to reproduce their figures.

Yi Huang and Kevin Boyle

Acronyms and Constants

ε_0	8.85419×10^{-12} F/m
μ_0	$4\mu \times 10^{-7}$ H/m
η_0	$\approx 377\ \Omega$
h	*Planck's constant* $= 6.63 \times 10^{-34}$ Js
j	$\sqrt{-1}$
2D	Two-dimensional
3D	Three-dimensional
2G	Second generation (mobile system)
3G	Third generation (mobile system)
3GPP	The 3rd Generation Partnership Project (for mobile)
4G	Fourth generation (mobile system)
5G	Fifth generation (mobile system)
AC	Alternating current
ACC	Automatic cruise control
ADAS	Advanced driver assistance systems
AF	Antenna factor
AiP	Antenna in package
AM	Amplitude modulation
AMC	Artificial magnetic conductor
AR	Axial ratio
AUT	Antenna under test
BER	Bit error rate
BNC	Baby N connector
BPR	Branch power ratio
CA	Characteristic angle
CAD	Computer aided design
CAM	Computer-aided manufacturing
CATR	Compact antenna test range
CDF	Cumulative distribution function
CEM	Computational electromagnetics
CFC	Carbon-fiber composite
CM	Common mode
CNT	Carbon nanotube
CP	Circular polarization

CPU	Central processing unit
CPW	Co-planar waveguide
CSRR	Complementary split ring resonator
CST	Computer simulation technology (a simulation tool)
CTIA	Cellular Telecommunications Industry Association
DAB	Digital audio broadcasting
dB	Decibel
DC	Direct current
DCS	Digital cellular system
DECT	Digital enhanced cordless telecommunications
DG	Diversity gain
DM	Differential mode
DNG	Double negative (material)
DPS	Double positive (material)
DRA	Dielectric resonant antenna
DUT	Device under test
EBG	Electromagnetic bandgap (material)
ECC	Envelope correlation coefficient
EGC	Equal gain combining
EIRP	Effective isotropic radiated power
EIS	Effective isotropic sensitivity
EM	Electromagnetic
EMC	Electromagnetic compatibility
EMI	Electromagnetic interference
ENG	Epsilon negative (material)
ENZ	Epsilon near zero (material)
EQC	Equivalent current
ERP	Effective radiated power
FCC	Federal Communications Commission
FDTD	Finite-difference time domain
FEM	Finite element method
FET	Field effect transistor
FM	Frequency modulation
FMCW	Frequency modulated continuous wave
FNBW	First null beamwidth
FoM	Figure of merit
FPC	Flexible printed circuit
FSS	Frequency selective surface
GaAs	Gallium Arsenide
GO	Geometrical optics
GPS	Global positioning system
GSM	Global system for mobile communications
GTD	Geometrical theory of diffraction
HF	High frequency
HFSS	High-frequency structure simulator (a simulation tool)

HIS	High impedance surface
HPBW	Half-power beamwidth
HPBW	Half-power bandwidth
HW	Hansen–Woodyard (condition)
IEEE	Institute of Electrical and Electronics Engineers
IFA	Inverted F antenna
InP	Indium Phosphide
IoT	Internet of Things
ISI	Inter symbol interference
ISM	Industrial, scientific and medical (frequency band)
LCP	Left-hand circular polarization
LCP	Liquid crystal polymer
LDS	Laser direct structuring
LF	Low frequency
LHM	Left-handed materials
LNA	Low-noise amplifier
LPDA	Log-periodic dipole antenna
LRR	Long-range radar
LTCC	Low-temperature co-fired ceramic
LTE	Long-term evolution (4G mobile system)
LUF	Lowest usable frequency
MCX	Miniature coaxial (connector)
MEG	Mean effective gain
MF	Medium frequency
MEMS	Microelectromechanical systems
MID	Moulded interconnect devices
MIMO	Multiple input and multiple output
MMIC	Monolithic microwave integrated circuits
mm-Wave	Millimeter wave
MNG	Mu-negative
MoM	Method of moments
MRC	Maximal ratio combining
MRR	Medium range radar
MS	Modal significance
MSTL	Mode-selective transmission line
MVG	Microwave Vision Group
NEC	Numerical electromagnetic code
NFC	Near-field coupling
NZI	Near-zero refractive index
OATS	Open area test site
OTA	Over-the-air (test/measurement)
PCB	Printed circuit board
PCS	Personal communications system
PDF	Power density function
PDF	Probability density function

PET	Polyethylene terephthalate
PIFA	Planar inverted F antenna
PIM	Passive inter-modulation
PTD	Physical theory of diffraction
PTFE	Polytetrafluoroethylene
PO	Physical optics
PVC	PolyVinyl Chloride
PWG	Plane wave generator
QZ	Quiet zone
RAM	Radio absorbing material
RC	Reverberation chamber
RCP	Right-hand circular polarisation
RCS	Radar cross section
RF	Radio frequency
RFID	Radio-frequency identification
RMS	Root mean square
SAR	Specific absorption rate
SC	Selection combining
SDR	Software defined radio
SI	International system of units (metric system)
SiP	System in package
SISO	Single-input single-output
SIW	Substrate integrated waveguide
SLL	Side lobe level
SMA	Sub-miniature version A (connector)
SNR	Signal-to-noise ratio
SoC	System on chip
SRR	Split ring resonator
SRR	Short-range radar
SSC	Source stirred chamber (or cap or cavity)
SWC	Switch combing
SWR	Standing wave ratio
TCM	Theory of characteristic modes
TDR	Time-domain reflectometer
TE	Transverse electric (mode/field)
TEM	Transverse electro-magnetic (mode/field)
THz	Terahertz, 10^{12} Hz
TIS	Total isotropic sensitivity
TLM	transmission line modelling/matrix (method)
TM	Transverse magnetic (mode/field)
TMM	Thermoset microwave material
TPMS	Tire pressure monitor system
TRP	Total radiated power
TV	Television
UE	User equipment

UHF	Ultrahigh frequency
UMTS	Universal mobile telecommunications system (3G mobile system)
UTD	Uniform theory of diffraction
UWB	Ultrawide band
VHF	Very high frequency
VNA	Vector network analyzer
VSWR	Voltage standing wave ratio
Wi-Fi	Wireless fidelity, a WLAN
WLAN	Wireless local area network

About the Author

Prof Yi Huang received BSc in Physics (Wuhan, China) in 1984, MSc (Eng) in Microwave Engineering (Nanjing, China) in 1987, and DPhil in Communications from the University of Oxford, United Kingdom, in 1994. He has been conducting research in the areas of antennas, wireless communications, applied electromagnetics, and radar since 1987. More recently, he is focused on mobile antennas, wireless energy harvesting, and power transfer. His experience includes three years spent with NRIET (China) as a Radar Engineer and various periods with the Universities of Birmingham, Oxford, and Essex in the United Kingdom as a member of research staff. He worked as a Research Fellow at British Telecom Labs in 1994 and then joined the Department of Electrical Engineering & Electronics, the University of Liverpool, United Kingdom, as a Faculty in 1995, where he is now a Chair Professor in Wireless Engineering, the Head of High Frequency Engineering Group.

Dr Huang has published over 400 refereed papers in leading international journals and conference proceedings and authored three books. He has received many patents, research grants from research councils, government agencies, charity, EU, and industry and is a recipient of over 10 awards (e.g. BAE Systems Chairman's Award 2017 for Innovation for Next Generation GNSS Antenna, Highly Recommended IET Innovation Award 2018, and Best Paper Awards). He has served on a number of national and international technical committees and been an Editor, Associate Editor, or Guest Editor of five international journals. In addition, he has been a keynote/invited speaker and organizer of many conferences and workshops (e.g. IEEE iWAT2010, LAPC2012, and EuCAP2018). He is at present the Editor-in-Chief of Wireless Engineering and Technology, Associate Editor of IEEE Antennas and Wireless

Propagation Letters, United Kingdom, and Ireland Rep to European Association of Antenna and Propagation (EurAAP), a Fellow of IET and IEEE, and Senior Fellow of Higher Education Academy.

More information about him can be found from:

https://www.liverpool.ac.uk/electrical-engineering-and-electronics/staff/yi-huang/

About the Companion Website

Antennas: From Theory to Practice, Second Edition is accompanied by a companion website:

www.wiley.com/go/huang_antennas2e

The website includes:

- Lecture PowerPoint Slides
- Answers to questions

1

Introduction

1.1 A Brief History of Antennas

Work on antennas started many years ago. The first well-known satisfactory antenna experiment was conducted by the German physicist Heinrich Rudolf Hertz (1857–1894), pictured in Figure 1.1. The SI (International Standard) frequency unit, Hertz, is named after him. In 1888, he built a system, as shown in Figure 1.2, to produce and detect radio waves. The original intention of his experiment was to demonstrate the existence of electromagnetic radiation. In the transmitter, a variable voltage source was connected to a dipole (a pair of 1 m wires) with two conducting balls (capacity spheres) at the ends.

The gap between the balls could be adjusted for circuit resonance as well as for the generation of sparks. When the voltage was increased to a certain value, a spark or break-down discharge was produced. The receiver was a simple loop with two identical conducting balls. The gap between the balls was carefully tuned to receive the spark effectively. He placed the apparatus in a darkened box to see the spark clearly. In his experiment, when a spark was generated at the transmitter, he also observed a spark at the receiver gap at almost the same time. This proved that the information from location A (the transmitter) was transmitted to location B (the receiver) in a wireless manner – electromagnetic (EM) waves! The information in his experiment was actually in binary digital form by tuning the spark on and off. This could be considered as the very first digital wireless system that consisted of two of the best-known antennas: the dipole and the loop. For this reason, the dipole antenna is also called Hertz (dipole) antenna (Figure 1.2).

While Heinrich Hertz conducted his experiments in a laboratory and did not quite know what radio waves might be used for in practice, Guglielmo Marconi (1874–1937, pictured in

Antennas: From Theory to Practice, Second Edition. Yi Huang.
© 2022 John Wiley & Sons Ltd. Published 2022 by John Wiley & Sons Ltd.
Companion website: www.wiley.com/go/huang_antennas2e

Figure 1.1   Heinrich Rudolf Hertz

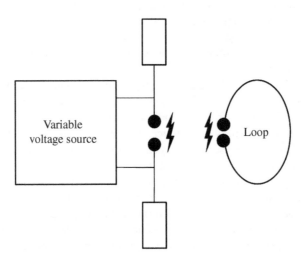

Figure 1.2   1887 experimental setup of Hertz's apparatus

Figure 1.3), an Italian inventor, was the man who developed and commercialized wireless technology by introducing a radiotelegraph system, which served as the foundation for the establishment of numerous affiliated companies worldwide. His most famous experiment was the transatlantic transmission from Poldhu, UK, to St Johns, Newfoundland, in the USA in 1901 employing un-tuned systems. He shared the 1909 Nobel Prize in Physics with Karl Ferdinand Braun 'in recognition of their contributions to the development of wireless

Figure 1.3 Guglielmo Marconi. Source: https://commons.wikimedia.org/wiki/File:Marconi_1909.jpg#/media/File:Marconi_1909.jpg

telegraphy'. Monopole antennas (near quarter wavelength) were widely used in his experiments, thus vertical monopole antennas are also called Marconi antennas.

During World War II, battles were won by the side that was first to spot enemy airplanes, ships, or submarines. To give the Allies an edge, British and American scientists developed radar technology to 'see' targets from hundreds of miles away, even at night. The research resulted in the rapid development of high-frequency radar antennas, which are no longer just wire-type antennas. Some aperture-type antennas such as reflector and horn antennas were developed, an example is shown in Figure 1.4.

Broadband, circularly polarized antennas, as well as many other types, were subsequently developed for various applications. Since an antenna is an essential device for any radio broadcasting, communication, and radar systems, there has always been a requirement for better or new antennas to meet existing and emerging applications.

For example, the cellular radio communication system is moving to its 5th generation (5G), the operational frequencies are extended from sub-6 GHz (e.g. 698–960, 1710–2690, and 3300–3800 MHz) to millimeter waves. The number of antennas in both the mobile portable and the base station is increased significantly to form massive multiple input and multiple output (MIMO) system to dramatically increase the communication data rate and capacity. This means massive new challenges to antenna designers: the antennas are to be placed in a relatively small device, such as a smartphone, and need to perform well at different frequencies (including 3G and 4G mobile frequencies) at the presence of other electronic systems (e.g. Wi-Fi, GPS, cameras, and a large display) and human body/hands. At millimeter waves, the antennas are also expected to produce beaming forming and steering functionalities to combat increased path loss which poses one of the main challenges for 5G mobile antenna design and measurement. The ultrawide band (UWB) wireless system is another example of

Figure 1.4   A WWII radar. Source: From ATNF, used with permission

recent broadband radio communication and positioning systems. The allocated frequency band is from 3.1 to 10.6 GHz. The beauty of UWB system is that the spectrum, which is normally very expensive, can be used free of charge but the power spectrum density is limited to −41.3 dBm/MHz. Thus, it is only suitable for short-distance applications (like Bluetooth but with a much larger bandwidth). The antenna design for these systems faces many challenging issues.

The role of antennas is becoming increasingly important. In some systems, the antenna is now no longer just a simple transmitting/receiving device, but a device which is integrated with other parts of the system to achieve better performance. For example, the MIMO antenna system has been introduced as an effective means to combat the multipath effects in the radio propagation channel and increase the channel capacity, where several co-ordinated antennas are required.

Things have been changing quickly in the wireless world. But one thing has never been changed since the very first antenna was made, that is, that the antenna is a practical engineering subject! It will remain as an engineering subject. Once an antenna is designed and made, it must be tested. How well it works is not just determined by the antenna itself, it also depends on the other parts of the system and the environment. The standalone antenna performance can be very different from that of an installed antenna. For example, when a mobile phone antenna is designed, we must take the case and other parts of the phone, even our hands, into account to ensure that it will work well in the real world. The antenna is an essential device of a radio system, but not an isolated device! This makes it an interesting and challenging subject.

1.2   Radio Systems and Antennas

A radio system is generally considered as an electronic system that employs radio waves, a type of EM wave up to GHz frequencies. An *antenna*, as an essential part of a radio system, is defined as a device that can radiate and receive EM energy in an efficient and desired manner. It is normally made of metal, but other materials may also be used. For example, ceramic materials have been employed to make dielectric resonator antennas (DRAs). There are many things in our lives, such as a power leads that can radiate and receive EM energy but cannot be viewed as antennas because the EM energy is not transmitted or received in an efficient and desired manner or because they are not a part of a radio system, thus they cannot be called antennas.

Since radio systems possess some unique and attractive advantages over wired systems, numerous radio systems have been developed. TV, radar, and mobile radio communication systems are just some examples. The advantages include at least:

- *Mobility*: it is essential for mobile communications;
- *Good coverage*: the radiation from an antenna can cover a very large area that is good for TV and radio broadcasting and mobile communications;
- *Low pathloss*: this is distance (and frequency) dependent. Since the loss of a transmission line is an exponential function of the distance (the loss in dB = distance × per unit loss in dB) and the loss of a radio wave is proportional to the distance square (the loss in dB = $20 \log_{10}$ (distance)), thus the pathloss of radio waves can be much smaller than that of a cable link. For example, assume that the loss is 10 dB for both a transmission line and a radio wave over 100 m, if the distance is increased 10 times to 1000 m, the loss for the transmission line becomes $10 \times 10 = 100$ dB but the loss for the radio link is just $10 + 20 = 30$ dB, which is much smaller than 100 dB! Therefore, the radio communication system is extremely attractive for long-distance communication. It should be pointed out that optic fibers are also employed for long-distance communications since they are of very low loss and UWB – but it is for point-to-point communications and fibers/cables normally need to be buried in subsurface, which could be costly in practice.

Figure 1.5 illustrates a typical radio communication system. The source information is normally modulated and amplified in the transmitter and then passed on to the transmit antenna via a transmission line, which has a typical characteristic impedance (which will be explained in the Chapter 2) of 50 Ω. The antenna radiates the information in the form of an EM wave in an efficient and desired manner to the destination, where the information is picked up by the receiver antenna and passed on to the receiver via another transmission line. The signal is demodulated, and the original message is then recovered at the receiver.

Thus, the antenna is actually a transformer that transfers electrical signals (voltages and currents from a transmission line) into EM waves (electric and magnetic fields) or vice versa. For example, a satellite dish antenna receives the radio wave from a satellite and transfers it into electrical signals which are output to a cable to be further processed. Our eyes may be viewed as another example of antennas. In this case, the wave is not a radio wave but an optical wave, another form of EM wave that has much higher frequencies.

Now it is clear that the antenna is actually a transformer of voltage/current to electric/magnetic field; it can also be considered as a bridge to link the radio wave and transmission line. An *antenna system* is defined as the combination of the antenna and its feed line. As an antenna is

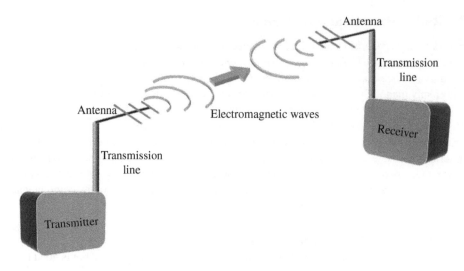

Figure 1.5   A typical radio system

usually connected to a transmission line, how to make this connection is a subject of interest since the signal from the feed line should be radiated into the space in an efficient and desired way. Transmission lines and radio waves are in fact two different subjects in engineering. To understand the antenna theory, one has to understand transmission lines and radio waves, which will be discussed in detail in Chapters 2 and 3, respectively. Thus, there is no need for the reader to study these subjects using other books.

In some applications where space is very limited (such as hand portables and aircrafts), it is desirable to integrate the antenna and its feed line. In other applications (such as the reception of TV broadcasting), the antenna is far away from the receiver and a long transmission line has to be used.

Unlike other devices in a radio system (such as filters and amplifiers), the antenna is a very special device; it deals with electrical signals (voltages and currents) as well as EM waves (electric fields and magnetic fields) that have made antenna design an interesting and difficult subject. For different applications, the requirements on the antenna could be very different, even for the same frequency band.

In conclusion, the subject of antennas is about how to design a suitable device that will be well matched with its feed line and radiate/receive the radio wave in an efficient and desired manner.

1.3   Necessary Mathematics

To thoroughly understand antenna theory requires a considerable amount of mathematics. However, the intention of this book is to provide the reader with a solid foundation of antenna theory and apply the theory to practical antenna design. Here, we are just going to introduce and review the essential and important mathematics required for this book. More in-depth study materials can be obtained from other references [1, 2].

1.3.1   Complex Numbers

In mathematics, a complex number, Z, consists of real and imaginary parts, that is

$$Z = R + jX \tag{1.1}$$

where R is called the real part of the complex number Z, i.e. Re(Z), and X is defined as the imaginary part of Z, i.e. Im(Z). Both R and X are real numbers, and j (not the traditional notion i in mathematics to avoid confusion with a changing current in electrical engineering) is the imaginary unit and defined by

$$j = \sqrt{-1} \tag{1.2}$$

Thus,

$$j^2 = -1 \tag{1.3}$$

Geometrically, a complex number can be presented in a two-dimensional plane where the imaginary part is found on the vertical axis while the real part is presented by the horizontal axis as shown in Figure 1.6.

In this model, multiplication by −1 corresponds to a rotation of 180 degrees about the origin. Multiplication by j corresponds to a 90-degree rotation anti-clockwise, and the equation $j^2 = -1$ is interpreted as saying that if we apply two 90-degree rotations about the origin, the net result is a single 180-degree rotation. Note that a 90-degree rotation clockwise also satisfies this interpretation.

Another representation of a complex number Z is to use the amplitude and phase form:

$$Z = Ae^{j\varphi} \tag{1.4}$$

where A is the amplitude and φ is the phase of the complex number Z, which are also shown in Figure 1.6. The two different representations are linked by the following equations:

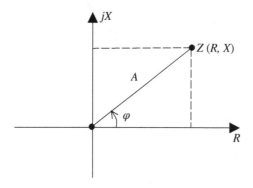

Figure 1.6   Complex plane

$$Z = R + jX = Ae^{j\varphi};$$

$$A = \sqrt{R^2 + X^2}, \qquad \varphi = \tan^{-1}(X/R) \tag{1.5}$$

$$R = A \cos\varphi, \qquad X = A \sin(\varphi)$$

1.3.2   Vectors and Vector Operation

A scalar is a one-dimensional quantity that has magnitude only, whereas a complex number is a two-dimensional quantity. A vector can be viewed as a three-dimensional (3D) quantity, and a special one – it has both a magnitude and a direction. For example, force and velocity are vectors. A position in space is a 3D quantity, but it does not have a direction, thus it is not a vector. Figure 1.7 is an illustration of vector A in Cartesian coordinates. It has three orthogonal components (A_x, A_y, A_z) along the x, y, and z directions, respectively. To distinguish vectors from scalars, the letter representing the vector is printed in bold, as A or a, and a unit vector is printed in bold with a hat over the letter as $\hat{x}$ or $\hat{n}$.

The magnitude of vector A is given by

$$|A| = A = \sqrt{A_x^2 + A_y^2 + A_z^2} \tag{1.6}$$

Now let us consider two vectors A and B:

$$A = A_x\hat{x} + A_y\hat{y} + A_z\hat{z}$$
$$B = B_x\hat{x} + B_y\hat{y} + B_z\hat{z}$$

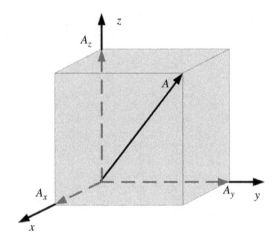

Figure 1.7   Vector A in Cartesian coordinates

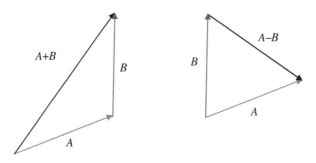

Figure 1.8 Vector addition and subtraction

The addition and subtraction of vectors can be expressed as

$$A + B = (A_x + B_x)\hat{x} + (A_y + B_y)\hat{y} + (A_z + B_z)\hat{z}$$
$$A - B = (A_x - B_x)\hat{x} + (A_y - B_y)\hat{y} + (A_z - B_z)\hat{z}$$

(1.7)

Obviously, the addition obeys the *commutative law*, that is $A + B = B + A$.

Figure 1.8 shows what the addition and subtraction mean geometrically. A vector may be multiplied or divided by a scalar. The magnitude changes but its direction remains the same. However, the multiplication of two vectors is complicated. There are two types of multiplication: dot product and cross product.

The *dot product* of two vectors is defined as

$$A \bullet B = |A||B| \cos \theta = A_x B_x + A_y B_y + A_z B_z$$

(1.8)

where θ is the angle between vector A and vector B and $\cos \theta$ is also called the direction cosine. The dot $\bullet$ between A and B indicates the dot product that results in a scalar, thus it is also called a *scalar product*. If the angle θ is zero, A and B are in parallel – the dot product maximized, whereas for an angle of 90 degrees, i.e. when A and B are orthogonal, the dot product is zero.

It is worth noting that the dot product obeys the *commutative law*, that is, $A \bullet B = B \bullet A$.

The *cross product* of two vectors is defined as

$$A \times B = \hat{n}|A||B| \sin \theta = C$$
$$= \hat{x}(A_y B_z - A_z B_y) + \hat{y}(A_z B_x - A_x B_z) + \hat{z}(A_x B_y - A_y B_x)$$

(1.9)

where $\hat{n}$ is a unit vector normal to the plane containing A and B. The cross $\times$ between A and B indicates the cross product that results in a vector C, thus, it is also called a *vector product*. The vector C is orthogonal to both A and B, and the direction of C follows a so-called right-hand rule as shown in Figure 1.9. If the angle θ is zero or 180 degrees, that is, A and B are in parallel, the cross product is zero, whereas for an angle of 90 degrees, i.e. A and B are orthogonal, the cross

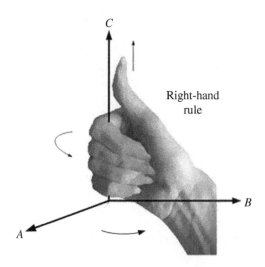

Figure 1.9 The cross product of vectors A and B

product of these two vectors reaches a maximum. Unlike the dot product, the cross product does not obey the commutative law.

The cross product may be expressed in determinant form as follows, which is the same as Equation (1.9) but it may be easier for some people to memorize:

$$A \times B = \begin{vmatrix} \hat{x} & \hat{y} & \hat{z} \\ A_x & A_y & A_z \\ B_x & B_y & B_z \end{vmatrix} \qquad (1.10)$$

Another important thing about vectors is that any vector can be decomposed into three orthogonal components (such as x, y, and z components) in 3D or two orthogonal components in a 2D plane.

Example 1.1   Vector operation
Vectors $A = 10\hat{x} + 5\hat{y} + 1\hat{z}$ and $B = 2\hat{y}$. Find:

$$A + B; \quad A - B; \quad A \bullet B; \quad \text{and } A \times B$$

Solution

$$A + B = 10\hat{x} + (5 + 2)\hat{y} + 1\hat{z} = 10\hat{x} + 7\hat{y} + 1\hat{z};$$
$$A - B = 10\hat{x} + (5 - 2)\hat{y} + 1\hat{z} = 10\hat{x} + 3\hat{y} + 1\hat{z};$$
$$A \bullet B = 0 + (5 \times 2) + 0 = 10;$$
$$A \times B = 10 \times 2\hat{z} + 1 \times 2\hat{x} = 20\hat{z} + 2\hat{x}$$

1.3.3   Coordinates

In addition to the well-known Cartesian coordinates, the spherical coordinates (r, θ, φ), as shown in Figure 1.10, will also be used frequently throughout this book. These two coordinate systems have the following relations:

$$
\begin{aligned}
x &= r \sin \theta \cos \varphi \\
y &= r \sin \theta \sin \varphi \\
z &= r \cos \theta
\end{aligned}
\tag{1.11}
$$

and

$$
\begin{aligned}
r &= \sqrt{x^2 + y^2 + z^2} \\
\theta &= \cos^{-1} \frac{z}{\sqrt{x^2 + y^2 + z^2}}; \quad 0 \le \theta \le \pi \\
\varphi &= \tan^{-1} \frac{y}{x}; \quad\quad\quad\quad\quad 0 \le \varphi \le 2\pi
\end{aligned}
\tag{1.12}
$$

The dot products of unit vectors in these two coordinator systems are

$$
\begin{aligned}
\hat{x} \bullet \hat{r} &= \sin \theta \cos \varphi; & \hat{y} \bullet \hat{r} &= \sin \theta \sin \varphi; & \hat{z} \bullet \hat{r} &= \cos \theta \\
\hat{x} \bullet \hat{\theta} &= \cos \theta \cos \varphi; & \hat{y} \bullet \hat{\theta} &= \cos \theta \sin \varphi; & \hat{z} \bullet \hat{\theta} &= -\sin \theta \\
\hat{x} \bullet \hat{\varphi} &= -\sin \varphi; & \hat{y} \bullet \hat{\varphi} &= \cos \varphi; & \hat{z} \bullet \hat{\varphi} &= 0
\end{aligned}
\tag{1.13}
$$

Thus, we can express a quantity in one coordinate system using the known parameters in the other coordinate system. For example, if A_r, A_θ, A_φ are known, we can find

$$
A_x = A \bullet \hat{x} = A_r \sin \theta \cos \varphi + A_\theta \cos \theta \cos \varphi - A_\varphi \sin \varphi
$$

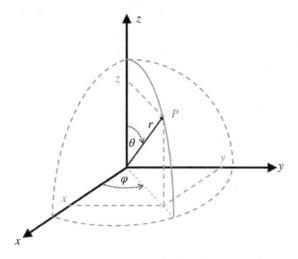

Figure 1.10   Cartesian and spherical coordinates

1.4   Basics of EMs

Now let us use basic mathematics to deal with antennas, or precisely, EM problems in this section.

EM waves cover the whole spectrum; radio waves and optical waves are just two examples of EM waves. We can see light but cannot see radio waves. The whole spectrum is divided into many frequency bands. Some EM bands and their applications are listed in Table 1.1. There are

Table 1.1   EM frequency bands and applications

Frequency	Band	Wavelength	Applications
3–30 kHz	VLF	100–10 km	Navigation, sonar, fax
30–300 kHz	LF	10–1 km	Navigation
0.3–3 MHz	MF	1–0.1 km	AM broadcasting
3–30 MHz	HF	100–10 m	Tel, Fax, CB, ship communications
30–300 MHz	VHF	10–1 m	TV, FM broadcasting
0.3–3 GHz	UHF	1–0.1 m	TV, mobile, radar
3–30 GHz	SHF	100–10 mm	Radar, satellite, mobile, microwave links
30–300 GHz	EHF	10–1 mm	Radar, wireless communications
0.3–3 THz	THz	1–0.1 mm	THz imaging
3–430 THz	Infrared	0.1 mm–700 nm	Heating, communications, camera
430–770 THz	Light	700–400 nm	Lighting, camera
Radar frequency bands according to IEEE standard			
1–2 GHz	L	0.3–0.15 m	Long wave, mobile radio
2–4 GHz	S	0.15–0.075 m	Short wave, mobile radio
4–8 GHz	C	7.5–3.75 cm	Compromise between S and X, radar
8–12 GHz	X	3.75–2.5 cm	Radar, satellite
12–18 GHz	Ku	2.5–1.7 cm	Satellite and radar
18–27 GHz	K	1.7–1.1 cm	Satellite and radar
27–40 GHz	Ka	11–7.5 mm	Communications and radar
40–75 GHz	V	7.5–4.0 mm	Communications and radar
75–110 GHz	W	4.0–2.7 mm	Communications and radar
Frequencies of some popular wireless systems			
535–1605 kHz	AM radio broadcast band		
3–30 MHz	Short-wave radio broadcast band		
13.56 MHz	NFC		
88–108 MHz	FM radio broadcast band		
175–240 MHz	DAB radio broadcast band		
470–890 MHz	UHF TV (14-83)		
698–960 MHz	Cellular mobile radio (2/4G)		
1710–2690 MHz	Cellular mobile radio (2/3/4G)		
3.3–3.8 GHz	Cellular mobile radio (5G)		
1.227 GHz	GPS L2 band		
1.575 GHz	GPS L1 band		
2.45 GHz	Microwave, Bluetooth, Wi-Fi		
3.1–10.6 GHz	UWB band		
5.180–5.825	Wi-Fi bands		

other letter band designations from organizations such as NATO. Here, we have used the IEEE standard.

Although the whole spectrum is infinite, the useful spectrum is limited and some frequency bands, such as the UHF, are already very congested. Normally significant license fees have to be paid to use the spectrum, although there are some license-free bands: the most well-known ones are the industrial, science, and medical (ISM) bands. The 433 MHz and 2.45 GHz bands are just two examples. Cable operators do not need to pay the spectrum license fees, but they have to pay other fees for things such as digging out the roads to bury the cables.

The *wave velocity v* is linked to the frequency *f* and wavelength λ by this simple equation:

$$v = f\lambda \tag{1.14}$$

It is well known that the speed of light (an EM wave) is about 3×10^8 m/s in free space. The higher the frequency, the shorter the wavelength. An illustration of how the frequency is linked to the wavelength is given in Figure 1.11, where both the frequency and wavelength are plotted on a logarithmic scale. The advantage, by doing so, is that we can see clearly how the function is changed even over a very large scale.

Radio waves, lights, and X-ray ($f = 10^{16}$ to 10^{19} Hz) are EM waves at different frequencies although they seem to be very different. One thing that all the forms of EM waves have in common is that they can travel through empty space (vacuum). This is not true for other kinds of waves; sound waves, for example, need some kind of material, such as air or water, in which to move. EM energy is carried by photons, the energy of a photon (also called quantum energy) is *hf*, where *h* is Planck's constant = 6.63×10^{-34} Js, and *f* is frequency in Hz. The higher the frequency, the more the energy of a photon. X-Ray has been used for imaging just because

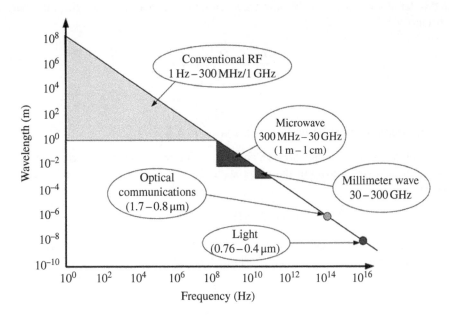

Figure 1.11 Frequency vs wavelength

of its high frequency: it carries very high energy and can penetrate through most objects. Also due to this high energy, X-ray can kill our cells and cause ionizing radiation that is not safe for our health. However, lights and radio waves operate at lower frequencies and do not have such a problem.

Logarithmic scales are widely used in RF (radio frequency) engineering and antennas community since the signals we are dealing with change significantly (over 1000 times in many cases) in terms of the magnitude. The signal power is normally expressed in dB (decibel), which is defined as

$$P\,(\text{dBW}) = 10\log_{10}\frac{P\,(\text{W})}{1\,\text{W}}; \quad P\,(\text{dBm}) = 10\log_{10}\frac{P\,(\text{W})}{1\,\text{mW}} \tag{1.15}$$

Thus, 100 W is 20 dBW, or just expressed as 20 dB in most cases; 1 W is 0 dB or 30 dBm; and 0.5 W is −3 dB or 27 dBm. Based on this definition, we can also express other parameters in dB. For example, since the power is linked to voltage V by $P = V^2/R$ (so $P \propto V^2$), the voltage can be converted to dBV by

$$V\,(\text{dBV}) = 20\log_{10}\left(\frac{V\,(\text{V})}{1\,\text{V}}\right) \tag{1.16}$$

Thus, 300 kVolts is 70 dBV, and 0.5 V is −6 dBV (not −3 dBV) or 54 dBmV.

1.4.1 Electric Field

The *electric field* (in V/m) is defined as the force (in Newtons) per unit charge (in Coulombs). From this definition and Coulomb's law, the electric field E created by a single point charge Q at a distance r is

$$E = \frac{F}{Q} = \frac{Q}{4\pi\varepsilon r^2}\hat{r}\,(\text{V/m}) \tag{1.17}$$

where

F is the electric force given by Coulomb's law $\left(F = \dfrac{Q_1 Q_2}{4\pi\varepsilon r^2}\hat{r}\right)$;

$\hat{r}$ is a unit vector along r direction which is also the direction of the electric field E.

ε is the electric *permittivity* of the material. Its SI unit is Farads/m. In free space, it is a constant:

$$\varepsilon_0 = 8.85419 \times 10^{-12}\,\text{F/m} \tag{1.18}$$

The product of permittivity and the electric field is called the *electric flex density, D*, which is a measure of how much electric flux passes through a unit area, i.e.

$$D = \varepsilon E = \varepsilon_r \varepsilon_0 E\,\left(\text{C/m}^2\right) \tag{1.19}$$

where $\varepsilon_r = \varepsilon/\varepsilon_0$ is called the *relative permittivity* (also called *dielectric constant*, but it is normally a function of frequency, not really a constant, thus relative permittivity is preferred in this

Table 1.2 Relative permittivity of some common materials at 100 MHz

Material	Relative permittivity	Material	Relative permittivity
ABS (plastic)	2.4–3.8	Polypropylene	2.2
Air	1	Polyvinylchloride (PVC)	3
Alumina	9.8	Porcelain	5.1–5.9
Aluminum silicate	5.3–5.5	PTFE-teflon	2.1
Balsa wood	1.37 @ 1 MHz	PTFE-ceramic	10.2
	1.22 @ 3 GHz	PTFE-glass	2.1–2.55
Concrete	~8	RT/Duroid 5870	2.33
Copper	1	RT/Duroid 6006	6.15 @ 3 GHz
Diamond	5.5–10	Rubber	3.0–4.0
Epoxy (FR4)	4.4	Sapphire	9.4
Epoxy glass PCB	5.2	Sea water	80
Ethyl alcohol (absolute)	24.5 @ 1 MHz	Silicon	11.7–12.9
	6.5 @ 3 GHz	Soil	~10
FR-4(G-10)		Soil (dry sandy)	2.59 @ 1 MHz
– low resin	4.9	Water (32 °F)	88.0
– high resin	4.2	(68 °F)	80.4
GaAs	13.0	(212 °F)	55.3
Glass	~4	Wood	~2
Gold	1		
Ice (pure distilled water)	4.15 @ 1 MHz		
	3.2 @ 3 GHz		

book). The relative permittivities of some common materials are listed in Table 1.2. Note that they are functions of frequency and temperature. Normally, the higher the frequency, the smaller the permittivity in the radio frequency band. It should also be pointed out that almost all conductors have a relative permittivity of one.

The electric flux density is also called the *electric displacement*, hence, the symbol D. It is also a vector. In an isotropic material (properties independent of direction) D and E are in the same direction and ε is a scalar quantity. In an anisotropic material, D and E may be in different directions if ε is a tensor.

If the permittivity is a complex number, it means that the material has some loss. The *complex permittivity* can be written as

$$\varepsilon = \varepsilon' - j\varepsilon'' \tag{1.20}$$

The ratio of the imaginary part to the real part is called the *loss tangent*, that is

$$\tan \delta = \frac{\varepsilon''}{\varepsilon'} \tag{1.21}$$

It has no unit and is also a function of frequency and temperature.

Table 1.3 Conductivities of some common materials at room temperature

Material	Conductivity (S/m)	Material	Conductivity (S/m)
Silver	6.3×10^7	Graphite	$\approx 10^5$
Copper	5.8×10^7	Carbon	$\approx 10^4$
Gold	4.1×10^7	Silicon	$\approx 10^3$
Aluminum	3.5×10^7	Ferrite	$\approx 10^2$
Tungsten	1.8×10^7	Sea water	≈ 5
Zinc	1.7×10^7	Germanium	≈ 2
Brass	1×10^7	Wet soil	≈ 1
Phosphor bronze	1×10^7	Animal blood	0.7
Tin	9×10^6	Animal body	0.3
Lead	5×10^6	Fresh water	$\approx 10^{-2}$
Silicon steel	2×10^6	Dry soil	$\approx 10^{-3}$
Stainless steel	1×10^6	Distilled water	$\approx 10^{-4}$
Mercury	1×10^6	Glass	$\approx 10^{-12}$
Cast iron	$\approx 10^6$	Air	0

The electric field E is related to the current density J (in A/m^2), another important parameter, by Ohm's law. The relationship between them at a point can be expressed as

$$J = \sigma E \tag{1.22}$$

where σ is the *conductivity*, which is the reciprocal of *resistivity*. It is a measure of a material's ability to conduct an electrical current and is expressed in Siemens per meter (S/m). Table 1.3 lists conductivities of some common materials linked to antenna engineering. The conductivity is also a function of temperature and frequency.

1.4.2 Magnetic Field

Whilst charges can generate an electric field, currents can generate a magnetic field. The *magnetic field*, H (in A/m), is the vector field that forms closed loops around electric currents or magnets. The magnetic field from a current vector I is given by the Biot–Savart law as

$$H = \frac{I \times \hat{r}}{4\pi r^2} \; (A/m) \tag{1.23}$$

where

$\hat{r}$ is the unit displacement vector from the current element to the field point and
r is the distance from the current element to the field point.
$I, \hat{r}$ and H follow the right-hand rule, that is H is orthogonal to both I and $\hat{r}$, as illustrated by Figure 1.12.

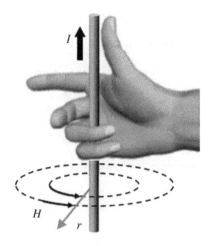

Figure 1.12   Magnetic field generated by current I

Like the electric field, the magnetic field exerts a force on electric charge. But unlike an electric field, it employs force only on a moving charge, and the direction of the force is orthogonal to both the magnetic field and the charge's velocity:

$$F = Qv \times \mu H \tag{1.24}$$

where

F is the force vector produced, measured in Newtons;
Q is the electric charge that the magnetic field is acting on, measured in Coulombs (C);
v is the velocity vector of the electric charge Q, measured in meters per second (m/s);
μ is the magnetic permeability of the material. Its unit is Henries per meter (H/m). The permeability of free space is

$$\mu_0 = 4\pi \times 10^{-7} \, \text{H/m} \tag{1.25}$$

In Equation (1.24), Qv can actually be viewed as the current vector I and the product of μH is called the *magnetic flux density* B (in Tesla), the counterpart of the electric flux density. Thus,

$$B = \mu H \tag{1.26}$$

Again, in an isotropic material (properties independent of direction), B and H are in the same direction and μ is a scalar quantity. In an anisotropic material, B and E may be in different directions and μ is a tensor.

Like the relative permittivity, the relative permeability is given as

$$\mu_r = \mu/\mu_0 \tag{1.27}$$

Table 1.4   Relative permeabilities of some common materials

Material	Relative permeability	Material	Relative permeability
Superalloy	$\approx 1 \times 10^6$	Aluminum	≈ 1
Purified iron	$\approx 2 \times 10^5$	Air	1
Silicon iron	$\approx 7 \times 10^3$	Water	≈ 1
Iron	$\approx 5 \times 10^3$	Copper	≈ 1
Mild steel	$\approx 2 \times 10^3$	Lead	≈ 1
Nickel	600	Silver	≈ 1

The relative permeabilities of some materials are given in Table 1.4. Permeability is not sensitive to frequency and temperature. Most materials, including conductors, have a relative permeability very close to one.

Combining Equations (1.17) and (1.24) yields

$$F = Q(E + v \times \mu H) \tag{1.28}$$

This is called the *Lorentz force*. The particle will experience a force due to the electric field of QE and the magnetic field $Qv \times B$.

1.4.3   Maxwell's Equations

Maxwell's equations are a set of equations first presented as a distinct group in the latter half of the nineteenth century by James Clerk Maxwell (1831–1879) pictured in Figure 1.13. Mathematically, they can be expressed in the following differential form:

$$\nabla \times E = -\frac{dB}{dt}$$
$$\nabla \times H = J + \frac{dD}{dt} \tag{1.29}$$
$$\nabla \cdot D = \rho$$
$$\nabla \cdot B = 0$$

where

ρ is the charge density.

$\nabla = \frac{\partial}{\partial x}\hat{x} + \frac{\partial}{\partial y}\hat{y} + \frac{\partial}{\partial z}\hat{z}$ is a vector operator;

$\nabla \times$ is the *curl operator* or called *rot* in some countries instead of *curl*.

$\nabla \cdot$ is the *divergence operator*.

Here we have both the vector cross product and the dot product.

Maxwell's equations describe the interrelationship between electric fields, magnetic fields, electric charge, and electric current. Although Maxwell himself was not the originator of the individual equations, he derived them again independently in conjunction with his molecular vortex model of Faraday's lines of force, and he was the person who first grouped these

Figure 1.13   James Clerk Maxwell

equations all together into a coherent set. Most importantly, he introduced an extra term to Ampere's Circuital Law, the second equation of (1.19). This extra term is the time derivative of the electric field and is known as Maxwell's displacement current. Maxwell's modified version of Ampere's Circuital Law enables the set of equations to be combined together to derive the EM wave equation, which will be further discussed in Chapter 3.

Now let us have a closer look at these mathematical equations to see what they really mean in terms of the physical explanations.

1.4.3.1   Faraday's Law of Induction

$$\nabla \times \boldsymbol{E} = -\frac{d\boldsymbol{B}}{dt} \tag{1.30}$$

This equation simply means that the induced *electromotive force* (EMF with a unit in V, it is the left-hand size of the equation expressed in the integral form as shown in Equation (1.37)) is proportional to the rate of change of the magnetic flux. In layman's terms, moving a conductor (such as a metal wire) through a magnetic field produces a voltage. The resulting voltage is directly proportional to the speed of movement. It is apparent from this equation that a time-varying magnetic field $\left(\mu\dfrac{d\boldsymbol{H}}{dt} \neq 0\right)$ will generate an electric field, i.e. $\boldsymbol{E} \neq 0$. But if the magnetic field is not time varying, it will NOT generate an electric field!

1.4.3.2   Amperes' Circuital Law

$$\nabla \times \boldsymbol{H} = \boldsymbol{J} + \frac{d\boldsymbol{D}}{dt} \tag{1.31}$$

This equation was modified by Maxwell by introducing the displacement current $\dfrac{dD}{dt}$. It means that a magnetic field appears during the charge or discharge of a capacitor. With this concept, and the Faraday's law, Maxwell was able to derive the wave equations, and by showing that the predicted wave velocity was the same as the measured velocity of light, Maxwell asserted that light waves are EM waves.

This equation shows that both the current (J) and time-varying electric field $\left(\varepsilon\dfrac{dE}{dt}\right)$ can generate a magnetic field, i.e. $H \neq 0$.

1.4.3.3   Gauss' Law for Electric Field

$$\nabla \bullet D = \rho \tag{1.32}$$

This is the electrostatic application of Gauss's generalized theorem, giving the equivalence relation between any flux, e.g. of liquids, electric or gravitational, flowing out of any closed surface and the result of inner sources and sinks, such as electric charges or masses enclosed within the closed surface. As a result, it is not possible for electric fields to form a closed loop. Since $D = \varepsilon E$, it is also clear that charges (ρ) can generate electric fields, i.e. $E \neq 0$.

1.4.3.4   Gauss' Law for Magnetic Field

$$\nabla \bullet B = 0 \tag{1.33}$$

This shows that the divergence of the magnetic field ($\nabla \bullet B$) is always zero, which means that the magnetic field lines are closed loops, thus the integral of B over a closed surface is zero.

For a time-harmonic EM field (which means the field linked to the time by factor $e^{j\omega t}$ where ω is the angular frequency and t is the time), we can use the *constitutive relations*

$$D = \varepsilon E, \quad B = \mu H, \quad J = \sigma E \tag{1.34}$$

to write Maxwell's equations into the following forms

$$\begin{aligned}
\nabla \times E &= -j\omega\mu H \\
\nabla \times H &= J + j\omega\varepsilon E = j\omega\varepsilon\left(1 - j\dfrac{\sigma}{\omega\varepsilon}\right)E \\
\nabla \bullet E &= \rho/\varepsilon \\
\nabla \bullet H &= 0
\end{aligned} \tag{1.35}$$

where B and D are replaced by the electric field E and magnetic field H to simplify the equations and they will not appear again unless necessary.

It should be pointed out that, in Equation (1.35), $\varepsilon\left(1 - j\dfrac{\sigma}{\omega\varepsilon}\right)$ can be viewed as a *complex permittivity* defined by Equation (1.20). In this case, the loss tangent is

$$\tan\delta = \frac{\varepsilon''}{\varepsilon'} = \frac{\sigma}{\omega\varepsilon} \tag{1.36}$$

It is hard to predict how the loss tangent changes with the frequency since both the permittivity and conductivity are functions of frequency as well. More discussion will be given in Chapter 3.

1.4.4 Boundary Conditions

Maxwell's equations can also be written in the integral form as

$$\oint_C \boldsymbol{E} \cdot dl = -\iint_S \frac{d\boldsymbol{B}}{dt} \cdot ds$$

$$\oint_C \boldsymbol{H} \cdot dl = \iint_S \left(\boldsymbol{J} + \frac{d\boldsymbol{D}}{dt}\right) \cdot ds$$

$$\oiint_S \boldsymbol{D} \cdot ds = \iiint_V \rho\, dv = Q \tag{1.37}$$

$$\oiint_S \boldsymbol{B} \cdot ds = 0$$

Consider the boundary between two materials shown in Figure 1.14. Using these equations, we can obtain a number of useful results. For example, if we apply the first equation of Maxwell's equations in integral form to the boundary between Medium 1 and Medium 2, it is not difficult to obtain [2]:

$$\hat{n} \times \boldsymbol{E}_1 = \hat{n} \times \boldsymbol{E}_2 \tag{1.38}$$

where $\hat{n}$ is the surface unit vector from Medium 2 to Medium 1 as shown in Figure 1.14. This condition means that the tangential components of an electric field ($\hat{n} \times \boldsymbol{E}$) are continuous across the boundary between any two media.

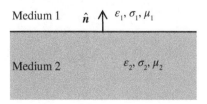

Figure 1.14 Boundary between Medium 1 and Medium 2

Similarly, we can apply other three Maxwell's equations to this boundary to obtain:

$$\hat{n} \times (\boldsymbol{H}_1 - \boldsymbol{H}_2) = \boldsymbol{J}_s$$
$$\hat{n} \cdot (\varepsilon_1 \boldsymbol{E}_1 - \varepsilon_2 \boldsymbol{E}_2) = \rho_s \qquad (1.39)$$
$$\hat{n} \cdot (\mu_1 \boldsymbol{H}_1 - \mu_2 \boldsymbol{H}_2) = 0$$

where $\boldsymbol{J}_s$ is the surface current density and ρ_s is the surface charge density. These results can be interpreted as

- The change in tangential component of the magnetic field across a boundary is equal to the surface current density on the boundary;
- The change in the normal component of the electric flux density across a boundary is equal to the surface charge density on the boundary;
- The normal component of the magnetic flux density is continuous across the boundary between two media, while the normal component of the magnetic field is not continuous unless $\mu_1 = \mu_2$.

Applying these boundary conditions on a perfect conductor (which means no electric and magnetic field inside and the conductivity $\sigma = \infty$) in the air, we have

$$\hat{n} \times \boldsymbol{E} = 0; \quad \hat{n} \times \boldsymbol{H} = \boldsymbol{J}_s; \quad \hat{n} \cdot \boldsymbol{E} = \rho_s / \varepsilon; \quad \hat{n} \cdot \boldsymbol{H} = 0 \qquad (1.40)$$

We can also use these results to illustrate, for example, the field distribution around a two-wire transmission line as shown in Figure 1.15, where the electric fields are plotted as the solid lines and the magnetic fields are shown in broken lines. As expected, the electric field is from positive charges to the negative charges, while the magnetic field forms loops around the current.

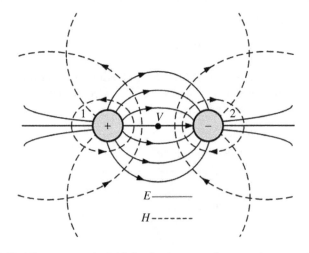

Figure 1.15 Electromagnetic field distribution around a two-wire transmission line

1.5   Summary

In this chapter, we have introduced the concept of antennas, briefly reviewed antenna history, and laid down the mathematical foundations for further study. The focus has been on the basics of EMs, which include electric and magnetic fields, EM properties of materials, Maxwell's equations, and boundary conditions. Maxwell's equations have revealed how electric fields, magnetic fields, and sources (currents and charges) are interlinked. They are the foundation of EMs and antennas.

References

1. R. E. Collin, Antennas and Radiowave Propagation, McGraw-Hill, Inc., 1985.
2. J. D. Kraus and D. A. Fleisch, Electromagnetics with Applications, 5th edition, McGraw-Hill, Inc., 1999.

Problems

Q1.1. What wireless communication experiment did H. Hertz conduct in 1887? Use a diagram to illustrate your answer.

Q1.2. Use an example to explain what a complex number means in our daily life.

Q1.3. Vector $A = 10\hat{x} + 5\hat{y} + 1\hat{z}$ and $B = 5\hat{z}$. Find
    a. the amplitude of vector A;
    b. the angle between vectors A and B;
    c. the dot product of these two vectors;
    d. a vector which is orthogonal to A and B.

Q1.4. Vector $A = 10 \sin (10t + 10z)\hat{x} + 5\hat{y}$. Find
    a. $\nabla \bullet A$;
    b. $\nabla \times A$;
    c. $(\nabla \bullet \nabla)A$;
    d. $\nabla \nabla \bullet A$

Q1.5. Vector $E = 10e^{\,j(10t - 10z)}\hat{x}$. Find
    a. The amplitude of E;
    b. Plot the real part of E as a function of t;
    c. Plot the real part of E as a function of z;
    d. What this vector means.

Q1.6. Explain why mobile phone service providers have to pay license fees to us the spectrum. Who is responsible for the spectrum allocation in your country?

Q1.7. Cellular mobile communications have become part of our daily life. Explain the major differences between the 2nd, 3rd, and 4th generations of cellular mobile systems in terms of the frequency, data rate, and bandwidth. Further explain why their operational frequencies have increased.

Q1.8. Which frequency bands have been used for radar applications? Give an example.

Q1.9. Express 1 kW in dB, 10 kV in dBV, 0.5 dB in W, and 40 dBμV/m in V/m and μV/m.

Q1.10. Explain the concepts of the electric field and magnetic field, how are they linked to the electric and magnetic flux density functions?

Q1.11. What are the material properties of interest to our electromagnetic and antenna engineers?

Q1.12. What is the Lorentz force? Name an application of the Lorentz force in our daily life.

Q1.13. If a magnetic field on a conducting surface $z = 0$ is $\boldsymbol{H} = 10\cos(10t - 5z)\hat{\boldsymbol{x}}$, find the surface current density $\boldsymbol{J}_s$.

Q1.14. Use Maxwell's equations to explain the major differences between the static EM fields and time-varying EM fields.

Q1.15. Express the boundary conditions for the electric and magnetic fields on the surface of a perfect conductor.

2

Circuit Concepts and Transmission Lines

In this chapter, we are going to review the very basics of circuit concepts and distinguish the lumped element system from the distributed element system. The focus will be placed on the fundamentals of transmission lines, including the basic model, the characteristic impedance, input impedance, reflection coefficient, return loss, and voltage standing wave ratio (VSWR) of a transmission line. The Smith Chart, impedance-matching techniques, Q-factor, and bandwidth will also be addressed. A comparison of various transmission lines and associated connectors will be made at the end of this chapter.

2.1 Circuit Concepts

Figure 2.1 shows a very basic electrical circuit where a voltage source V is connected to a load Z via conducting wires. This simple circuit can represent numerous systems in our daily life, from a simple torch – a direct current (DC) circuit, to a more complicated power supply system – an alternating current (AC) circuit. To analyze such a circuit, one has to use at least the following four quantities:

- *Electric current I* is a measure of the charge flow/movement. The SI unit of current is Ampere (A), which is equal to a flow of one coulomb of charge per second.
- *Voltage V* is the difference of electrical potential between two points of an electrical or electronic circuit. The SI unit of voltage is expressed in volts (V). It measures the potential energy of an electric field to cause an electric current in a circuit.
- *Impedance $Z = R + jX$* is a measure of opposition to an electric current. In general, the impedance is a complex number, its real part R is the electrical *resistance* (or just resistance) and reflects the ability to consume energy, while the imaginary part X is the *reactance* and indicates the ability to store energy. If the reactance is positive, it is called *inductance* since the

Antennas: From Theory to Practice, Second Edition. Yi Huang.
© 2022 John Wiley & Sons Ltd. Published 2022 by John Wiley & Sons Ltd.
Companion website: www.wiley.com/go/huang_antennas2e

Figure 2.1 A simple electrical circuit with a source and load

reactance of an inductor is positive (ωL); if the reactance is negative, it is then called as *capacitance* since the reactance of a capacitor is negative ($-1/\omega C$). The same unit, ohm (Ω), is used for impedance, resistance, and reactance. The inverses of the impedance, resistance, and reactance are called the *admittance* (Y), *conductance* (G), and *susceptance* (B), respectively. Their unit is Siemens (S). Thus, *impedance = resistance + j reactance* (inductance or capacitance) while *admittance = conductance + j susceptance*.

- *Power P* is defined as the amount of work done by an electrical current over time, or the rate at which electrical energy is transmitted/consumed. The SI unit of power is Watt (W). When an electric current flows through a device with resistance, the device converts the power into various forms, such as light (light bulbs), heat (electric cooker), motion (electric razor), sound (loudspeaker), and radiation of an antenna.

Ohm's law is the very basic and fundamental theory for electrical circuits. It reveals how the current, voltage, and resistance are linked in a DC circuit. It states that the current passing through a conductor/device from one terminal point on the conductor/device to another terminal point on the conductor/device is directly proportional to the potential difference (i.e. voltage) across the two terminal points and inversely proportional to the resistance of the conductor/device between the two terminal points. That is

$$I = \frac{V}{R} \tag{2.1}$$

In an AC circuit, Ohm's law can be generalized as:

$$I = \frac{V}{Z} \tag{2.2}$$

i.e. the resistance R is replaced by the impedance Z. Since the impedance is a complex number, both the current and voltage can be complex numbers as well, which means that they have magnitude and phase.

The average power (over time) can be obtained using

$$P = IV = V^2/R = RI^2 \qquad \text{for DC}$$
$$P_{av} = \frac{1}{2}I_0V_0 = \frac{V_0^2}{2R} = \frac{1}{2}RI_0^2 \quad \text{for AC} \tag{2.3}$$

where V_0 and I_0 are the amplitudes of voltage and current, respectively.

2.1.1   Lumped and Distributed Element Systems

In traditional circuit theory, we basically divide the circuits into DC and AC parts. The voltage, current, and impedance are real numbers in DC circuits but complex numbers in AC circuits. The effects of conducting wires can normally be neglected. For example, the current across the load Z in Figure 2.1 can be obtained using Ohm's law. It is given by Equation (2.2) and considered as the same voltage across the load.

In most countries, the electrical power supply system operates at 50 or 60 Hz, which means a wavelength of 6000 or 5000 km (which is close to the radius of the Earth: 6378 km), much longer than any transmission line in use. The current and voltage along the transmission line may be considered unchanged. The system is called a *lumped element system*. However, in some applications, the frequency of the source is significantly increased, as a result the wavelength becomes comparable with the length of the transmission line linking the source and the load. The current and voltage along the transmission line are functions of the distance from the source, thus the system is called a *distributed element system*. If Figure 2.1 is a distributed element system, Equation (2.2) is no longer valid since the voltage across the load may now be very different from the source voltage. V should therefore be replaced by the voltage across the load.

Conventional circuit theory was developed for lumped element systems whose frequency is relatively low and the wavelength is relatively large. While the frequency of a distributed system is relatively high, and the wavelength is relatively short. It is therefore important to introduce the transmission line theory which has been developed for the distributed element system and has taken the distributed nature of the parameters in the system into account.

2.2   Transmission Line Theory

A *transmission line* is the structure that forms all or part of a path from one place to another for directing the transmission of energy, such as electrical power transmission and optical waves. Examples of transmission lines include conducting wires, electrical power lines, coaxial cables, dielectric slabs, optical fibers, and waveguides. In this book, we are only interested in the transmission lines for RF engineering and antenna applications. Thus, dielectric transmission lines such as optical fibers are not considered.

2.2.1   Transmission Line Model

The simplest transmission line is a two-wire conducting transmission line as shown in Figure 2.2. It has been widely used for electrical power supply and also for radio and television systems. In the old days, the broadcasting TV signal was received by an antenna and then passed down to a TV via such a two-wire conducting wire which is now replaced by the coaxial cable. This is partially due to the fact that the antenna used now (Yagi-Uda antenna, a popular TV antenna to be discussed in Chapter 5, it has an input impedance around 75 Ω) is different from the antenna used then (folded dipole which was a popular TV antenna many years ago, it has an input impedance around 300 Ω). Also, the coaxial cable performs much better than the two-wire transmission line at the ultra-high-frequency (UHF) TV bands.

As shown in Figure 2.2, if we divide the transmission line into many (almost infinite) short segments of length Δz which is much smaller than the wavelength of interest, each segment can then be represented using a set of lumped elements. By doing so, a distributed transmission line

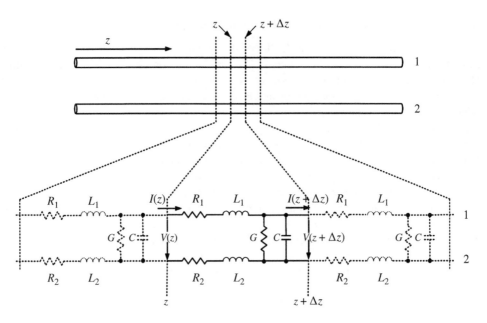

Figure 2.2   A two-wire transmission line model

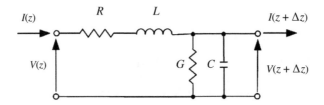

Figure 2.3   Schematic representation of the elementary component of a transmission line

is modeled as an infinite series of two-port lumped elementary components, each representing an infinitesimally short segment of the transmission line. To make the analysis easier, the equivalent circuit of the segment of the transmission line is simplified to Figure 2.3, where $R = R_1 + R_2$ and $L = L_1 + L_2$.

- The resistance R represents the conductive loss of the transmission line over a unit length, thus the unit is ohms/unit length (Ω/m).
- The inductance L is the self-inductance of the transmission line and expressed in Henries per unit length (H/m).
- The capacitance C between the two conductors is represented by a shunt capacitor with a unit of Farads per unit length (F/m).
- The conductance G of the dielectric material separating the two conductors is represented by a conductance G shunted between the two conductors. Its unit is Siemens per unit length (S/m).

It should be repeated for clarity that the model consists of an infinite series of elements shown in Figure 2.3, and that the values of the components are specified per unit length. *R, L, C,* and *G* may be functions of frequency.

Using this model, we are going to investigate how the current and voltage along the line are changed and how they are linked to *R, L, C,* and *G*.

It is reasonable to assume that the source is time-harmonic and has an angular frequency ω (= $2\pi f$, f is frequency), thus its time factor is $e^{j\omega t}$.

Using Ohm's law, we know that the voltage drop and current change over a segment of Δz can be expressed in the frequency domain as:

$$V(z + \Delta z) - V(z) = -(R + j\omega L)\Delta z \cdot I(z)$$
$$I(z + \Delta z) - I(z) = -(G + j\omega C)\Delta z \cdot V(z + \Delta z) \tag{2.4}$$

When Δz approaches zero, these two equations can be written in differential forms as:

$$\frac{dV(z)}{dz} = -(R + j\omega L) \cdot I(z)$$
$$\frac{dI(z)}{dz} = -(G + j\omega C) \cdot V(z) \tag{2.5}$$

Taking a differentiation of z on both sides of the equations and combining them gives

$$\frac{d^2 V(z)}{dz^2} = (R + j\omega L)(G + j\omega C) \cdot V(z)$$
$$\frac{d^2 I(z)}{dz^2} = (R + j\omega L)(G + j\omega C) \cdot I(z) \tag{2.6}$$

That is,

$$\frac{d^2 V(z)}{dz^2} - \gamma^2 V(z) = 0$$
$$\frac{d^2 I(z)}{dz^2} - \gamma^2 I(z) = 0 \tag{2.7}$$

where

$$\gamma = \sqrt{(R + j\omega L)(G + j\omega C)} \tag{2.8}$$

and is called the *propagation constant* which may have real and imaginary parts. Equation (2.7) is a pair of linear differential equations which describe the line voltage and current on a transmission line as a function of distance and time (the time factor $e^{j\omega t}$ is omitted here). They are called *telegraph equations,* or *transmission line equations.*

2.2.2   Solutions and Analysis

The general solution of $V(z)$ in the telegraph equations can be expressed as [1, 2]:

$$V(z) = V_+(z) + V_-(z) = A_1 e^{-\gamma z} + A_2 e^{\gamma z} \tag{2.9}$$

where A_1 and A_2 are complex coefficients to be determined by the boundary conditions, which means the voltage, current, and impedance at the input and the load of the transmission line – we need to know at least two of them in order to determine the two coefficients.

Replacing $V(z)$ in Equation (2.5) by Equation (2.9), we can find the solution of the line current as:

$$I(z) = \frac{\gamma}{R + j\omega L}\left(A_1 e^{-\gamma z} - A_2 e^{\gamma z}\right) \tag{2.10}$$

It can be written as:

$$I(z) = \frac{1}{Z_0}\left(A_1 e^{-\gamma z} - A_2 e^{\gamma z}\right) \tag{2.11}$$

where

$$Z_0 = \frac{V_+(z)}{I_+(z)} = \frac{R + j\omega L}{\gamma} = \sqrt{\frac{R + j\omega L}{G + j\omega C}} \tag{2.12}$$

and is called the *characteristic impedance* of the transmission line. Its unit is ohm (Ω). It is a function of frequency and parameters of the line. The industrial standard transmission line normally has a characteristic impedance of 50 or 75 Ω when the loss can be neglected ($R \approx 0$ and $G \approx 0$).

Since the propagation constant is complex, it can be written as:

$$\gamma = \alpha + j\beta \tag{2.13}$$

where α is called the *attenuation constant* (in Nepers/meter, or Np/m) and β is called the *phase constant*. Because $\gamma = \sqrt{(R + j\omega L)(G + j\omega C)}$, we can find that mathematically:

$$\alpha = \left[\frac{1}{2}\left(\sqrt{(R^2 + \omega^2 L^2)(G^2 + \omega^2 C^2)} + (RG - \omega^2 LC)\right)\right]^{1/2}$$

$$\beta = \left[\frac{1}{2}\left(\sqrt{(R^2 + \omega^2 L^2)(G^2 + \omega^2 C^2)} - (RG - \omega^2 LC)\right)\right]^{1/2} \tag{2.14}$$

They are functions of frequency as well as the parameters of the transmission line.

If we take the time factor into account, the complete solution of the voltage and current along a transmission line can be expressed as:

$$V(z,t) = A_1 e^{j\omega t - \gamma z} + A_2 e^{j\omega t + \gamma z} = A_1 e^{-\alpha z + j(\omega t - \beta z)} + A_2 e^{\alpha z + j(\omega t + \beta z)}$$

$$I(z,t) = \frac{1}{Z_0}\left(A_1 e^{j\omega t - \gamma z} - A_2 e^{j\omega t + \gamma z}\right) = \frac{1}{Z_0}\left(A_1 e^{-\alpha z + j(\omega t - \beta z)} - A_2 e^{\alpha z + j(\omega t + \beta z)}\right) \tag{2.15}$$

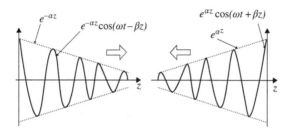

Figure 2.4   Forward and reverse traveling waves

Physically, the line voltage solution can be considered as the combination of two traveling voltage waves: the wave traveling toward the z direction (called the forward wave) has amplitude of $|V_+(z)| = |A_1 e^{-\alpha z}|$ which attenuates as z increases, whereas the wave traveling toward the $-z$ direction (called the reverse wave) has amplitude of $|A_2 e^{\alpha z}|$ as shown in Figure 2.4. The amplitudes of A_1 and A_2 are actually the voltage amplitudes of the forward and reverse waves at $z = 0$, respectively. If there is no reflection at the end of the transmission line, it means that the boundary conditions have forced A_2 to be zero, thus the reverse wave will be zero and only forward traveling voltage exist on the transmission line in this case.

Similarly, the line current can also be viewed as the combination of two traveling current waves. It is worth noting that the reverse traveling current has a minus sign with the amplitude, this means a phase change of 180° and reflects the direction change in the returned current.

The velocity of the wave is another parameter of interest and it can be determined from the phase term: $\omega t - \beta z$. At a fixed reference point, the wave moves Δz over a period of Δt, i.e. we have $\omega \Delta t - \beta \Delta z = 0$, thus the velocity:

$$v = \frac{dz}{dt} = \frac{\omega}{\beta} \tag{2.16}$$

Since the phase constant β is a function of the angular frequency as shown in Equation (2.14), the velocity is therefore a function of frequency, which is a well-known dispersion problem (change with frequency).

Using Equation (2.16), the phase constant can be expressed as:

$$\beta = \frac{\omega}{v} = \frac{2\pi f}{v} = \frac{2\pi}{\lambda} \tag{2.17}$$

where λ is the wavelength. The phase constant is also called the *wave number*. For every one wavelength, the phase is changed by 2π.

These solutions are general and can be applied to any transmission line in principle. We can see that the characteristic impedance could be complex and the attenuation constant and phase constant are complicated functions of frequency. But in practice, we always prefer something simpler and easier to use.

2.2.2.1 Lossless Transmission Lines

Since the function of the transmission line is to transmit information from one place to another with little change, the loss of the transmission line should be minimized – this is one of the requirements for transmission line manufactures. There are indeed many low-loss transmission lines available on the market.

For a lossless transmission line, elements R and G can be considered to be zero ($R \approx 0$ and $G \approx 0$). In this hypothetical case, the model depends only on elements L and C which greatly simplifies the analysis.

The characteristic impedance of the transmission line, Equation (2.12), can now be simplified as:

$$Z_0 = \sqrt{\frac{L}{C}} \tag{2.18}$$

It is just a real number (resistance) and determined only by L and C, not a function of the frequency.

Similarly, Equation (2.14) becomes

$$\begin{aligned} \alpha &= 0 \\ \beta &= \omega\sqrt{LC} \end{aligned} \tag{2.19}$$

This means that there is no attenuation and the propagation constant is now just an imaginary number:

$$\gamma = j\beta = j\omega\sqrt{LC}$$

The voltage and current along the line are

$$\begin{aligned} V(z,t) &= A_1 e^{j(\omega t - \beta z)} + A_2 e^{j(\omega t + \beta z)} \\ I(z,t) &= \frac{1}{Z_0}\left(A_1 e^{j(\omega t - \beta z)} - A_2 e^{j(\omega t + \beta z)}\right) \end{aligned} \tag{2.20}$$

Both the forward and reverse waves are not attenuated and their amplitudes are not a function of the distance.

The velocity of the waves is now

$$v = \frac{\omega}{\beta} = \frac{1}{\sqrt{LC}} \tag{2.21}$$

which is not a function of the frequency and only determined by the transmission line itself. This is an important feature required for all transmission lines.

2.2.2.2 Low-Loss Transmission Lines

In practice, most transmission lines cannot be considered as lossless structures but as low-loss transmission lines.

The definition of "low loss" is

$$R \ll \omega L, \quad G \ll \omega C$$

This seems to mean extremely high frequency. The reality is that both R and G are functions of frequency. Normally, the higher the frequency, the larger the R and G. Thus, this condition does not mean high frequency. It is for any frequency when this condition is met.

For a low-loss transmission line, the characteristic impedance is

$$Z_0 = \sqrt{\frac{R + j\omega L}{G + j\omega C}} = \sqrt{\frac{j\omega L(1 + R/j\omega L)}{j\omega C(1 + G/j\omega C)}} \approx \sqrt{\frac{j\omega L(1 + 0)}{j\omega C(1 + 0)}} = \sqrt{\frac{L}{C}} \qquad (2.22)$$

Thus, it is the same as the lossless case. The characteristic impedance is a pure resistance and determined by L and C, not a function of the frequency. This is why the characteristic impedance of industrial standard transmissions has a constant value, normally 50 or 75 Ω, over a large frequency band even when the line loss is not zero.

Similarly, the attenuation and phase constants can be approximated as:

$$\alpha \approx \frac{R}{2}\sqrt{C/L} + \frac{G}{2}\sqrt{L/C} = \frac{R}{2Z_0} + \frac{GZ_0}{2}$$

$$\beta \approx \omega\sqrt{LC} \qquad (2.23)$$

The loss (attenuation) is caused by the resistive loss R and the material loss G between the conductors. The phase constant is again the same as for lossless lines.

However, the voltage and current are attenuated as they travel along the line (due to the loss) and can be expressed as:

$$V(z, t) = A_1 e^{-\alpha z + j(\omega t - \beta z)} + A_2 e^{\alpha z + j(\omega t + \beta z)}$$

$$I(z, t) = \frac{1}{Z_0}\left(A_1 e^{-\alpha z + j(\omega t - \beta z)} - A_2 e^{\alpha z + j(\omega t + \beta z)}\right) \qquad (2.24)$$

where the attenuation constant is given by Equation (2.23).

Just like the lossless line, the velocity of the waves in a low-loss transmission line is only determined by L and C, i.e.

$$v = \frac{\omega}{\beta} = \frac{1}{\sqrt{LC}} \qquad (2.25)$$

There is no dispersion (it is not changed with frequency).

2.2.3 Terminated Transmission Line

It is now clear that the voltage and current of a transmission line are distributed quantities; they are functions of the position z. However, the characteristic impedance of a transmission line is not a distributed parameter but a constant. When the line is terminated with a load impedance Z_L as shown in Figure 2.5, what is the input impedance?

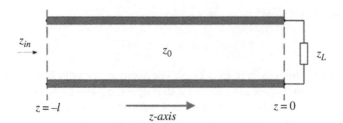

Figure 2.5 A transmission line terminated with a load

2.2.3.1 Input Impedance

The *input impedance* of a transmission line is defined as the ratio of voltage to current at the input port and is the impedance looking toward the load, i.e.

$$Z_{in}(z) = \frac{V(z)}{I(z)} = Z_0 \frac{A_1 e^{-\gamma z} + A_2 e^{\gamma z}}{A_1 e^{-\gamma z} - A_2 e^{\gamma z}} = Z_0 \frac{e^{-\gamma z} + \Gamma_0 e^{\gamma z}}{e^{-\gamma z} - \Gamma_0 e^{\gamma z}} \tag{2.26}$$

where $\Gamma_0 = A_2/A_1$ is called the *reflection coefficient* at the load and equals to the ratio of A_2 to A_1, which are the complex coefficients of forward to reverse voltage waves at $z = 0$. The input impedance at the load should be the load impedance, that is

$$Z_{in}(0) = Z_0 \frac{1 + \Gamma_0}{1 - \Gamma_0} = Z_L \tag{2.27}$$

Thus, the reflection coefficient at the load can be expressed as:

$$\Gamma_0 = \frac{Z_L - Z_0}{Z_L + Z_0} \tag{2.28}$$

A general expression of the reflection coefficient on a transmission line at reference point z is

$$\Gamma(z) = \frac{V_-(z)}{V_+(z)} = \frac{A_2 e^{\gamma z}}{A_1 e^{-\gamma z}} = \Gamma_0 e^{2\gamma z} = \frac{Z_L - Z_0}{Z_L + Z_0} e^{2\gamma z} \tag{2.29}$$

This means the reflection coefficient is a distributed parameter and is a function of the load impedance as well as the transmission line characteristic impedance.

Replacing Γ_0 in Equation (2.26) by Equation (2.28), we have

$$\begin{aligned}
Z_{in}(z) &= Z_0 \frac{(Z_L + Z_0)e^{-\gamma z} + (Z_L - Z_0)e^{\gamma z}}{(Z_L + Z_0)e^{-\gamma z} - (Z_L - Z_0)e^{\gamma z}} \\
&= Z_0 \frac{Z_L(e^{\gamma z} + e^{-\gamma z}) - Z_0(e^{\gamma z} - e^{-\gamma z})}{Z_0(e^{\gamma z} + e^{-\gamma z}) - Z_L(e^{\gamma z} - e^{-\gamma z})}
\end{aligned}$$

Thus,

$$Z_{in}(z) = Z_0 \frac{Z_L - Z_0 \tanh (\gamma z)}{Z_0 - Z_L \tanh (\gamma z)} \tag{2.30}$$

where

$$\tanh (\gamma z) = \frac{e^{\gamma z} - e^{-\gamma z}}{e^{\gamma z} + e^{-\gamma z}} \tag{2.31}$$

is the hyperbolic tangent function.

In practice, the input impedance is measured at a given distance l rather than at its z axis value as shown in Figure 2.5. Thus, the input impedance at l meters away from the load is

$$Z_{in}(l) = Z_0 \frac{Z_L + Z_0 \tanh (\gamma l)}{Z_0 + Z_L \tanh (\gamma l)} \tag{2.32}$$

Note that there is a sign change from Equation (2.30) since the distance should not be negative and we have used $l = -z$ and $\tanh(-\gamma l) = -\tanh(\gamma l)$.

If the loss of the transmission line can be neglected, that is, $\gamma \approx j\beta$, Equation (2.32) can be simplified as:

$$Z_{in}(l) = Z_0 \frac{Z_L + jZ_0 \tan (\beta l)}{Z_0 + jZ_L \tan (\beta l)} \tag{2.33}$$

This is a very useful equation and the input impedance is a periodic function of l, *the period is half of a wavelength* (not one wavelength); special attention should be paid to the following cases:

a. **Matched case:** $Z_L = Z_0$ $Z_{in}(l) = Z_0$,
   the input impedance is the same as the characteristic impedance and is not a function of the length of the line.
b. **Open circuit:** $Z_L = \infty$

$$Z_{in}(l) = Z_0 \frac{1}{j \tan (\beta l)} \tag{2.34}$$

The input impedance has no resistance just reactance (capacitive for small l).
c. **Short circuit:** $Z_L = 0$

$$Z_{in}(l) = jZ_0 \tan (\beta l) \tag{2.35}$$

Again, the input impedance has no resistance just reactance (inductive for small l).
Quarter-wavelength: $l = \lambda/4$

$$Z_{in}(l) = \frac{Z_0^2}{Z_L} \tag{2.36}$$

This special case is called the *quarter-wavelength transform* since the input impedance, load impedance, and characteristic impedance are linked by this simple equation. It is often used for impedance-matching purposes. For example, if a load impedance is 200 Ω, we can choose/make a special $l = \lambda/4$ transmission line of characteristic impedance 100 Ω to make the input impedance to become 50 Ω to match with standard line or connector.

It should be pointed out that, in calculating the wavelength λ and wave number $\beta = 2\pi/\lambda$, the wavelength inside the transmission line is generally different from that in free space. The dielectric properties of the material of the transmission line have to be taken into account when doing such a calculation. The simplest case is that the wavelength is linked to the relative permittivity ε_r (also called the relative dielectric constant) of the material by:

$$\lambda = \frac{\lambda_0}{\sqrt{\varepsilon_r}} \tag{2.37}$$

where λ_0 is the free space wavelength. More details will be given later in this chapter.

Example 2.1   Input impedance
A lossless transmission line with a characteristic impedance of 50 Ω is loaded by a 75 Ω resistor. Plot the input impedance as a function of the line length (up to two wavelengths).

Solution
Since it is a lossless transmission line, Equation (2.33) is employed to calculate the input impedance. The result is shown in Figure 2.6 where both the resistance and reactance are plotted as a function of the normalized (to wavelength) line length. It is apparent that:

The input impedance is a periodic function of the line length. The period is half of a wavelength as mentioned earlier.

The input impedance is a complex number even when the load impedance is a pure resistance. The resistance is changed between 75 and 33 Ω, while the reactance is changed between −20 and +20 Ω.

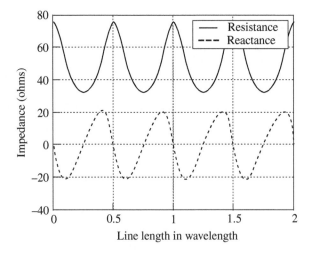

Figure 2.6   Input impedance as a function of the transmission line length for $Z_L = 75\ \Omega$ and $Z_0 = 50\ \Omega$

There are two resonant points (where the reactance is zero) over one period. That is at $l/\lambda = 0$ and 0.25 in this case.

When $0 < l/\lambda < 0.25$, the reactance is negative, i.e. capacitive. As $0.25 < l/\lambda < 0.5$, the reactance is positive, i.e. inductive.

Example 2.2   Input impedance of a low-loss transmission line
A 75 Ω resistor is now connected to a good transmission line with characteristic impedance of 50 Ω. The attenuation constant is not zero but 0.2 Np/m at 1 GHz. Plot the input impedance as a function of the line length (up to 2λ). Assume that the effective relative permittivity is 1.5.

Solution
For a low-loss transmission line, the characteristic impedance is still a constant (=50 Ω in this case). The line length changes from 0 to 2λ. Since the effective relative permittivity is 1.5 and the frequency is 1 GHz, the wavelength in the medium is

$$\lambda = \frac{c}{f\sqrt{\epsilon_r}} = \frac{3 \times 10^8}{10^9\sqrt{1.5}} \approx 0.245 \quad (m)$$

Thus, the length l is from 0 to 0.49 m.

The propagation constant is $\gamma = \alpha + j\beta = 0.2 + j2\pi/\lambda$. Using (2.32), we can plot the input impedance as shown in Figure 2.7. We can see that the only change is that the input impedance is *no longer a periodic function of the line length*. But it still exhibits a period feature if we neglect the amplitude changes. All other features remain the same as in the lossless case.

Example 2.3   Quarter-wavelength transform
A 75 Ω resistor is to be matched with a transmission line of characteristic impedance 50 Ω. If a quarter-wavelength transformer is employed, what should its characteristic impedance be?

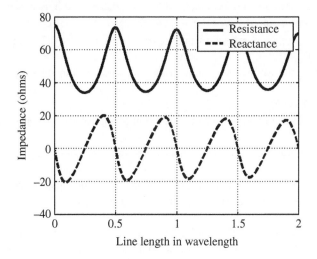

Figure 2.7   The input impedance along a low-loss transmission line for $Z_L = 75\ \Omega$ and $Z_0 = 50\ \Omega$

Solution
Using Equation (2.36), we have

$$Z_0 = \sqrt{Z_{in}Z_L} = \sqrt{75 \cdot 50} \approx 61.2 \ (\Omega)$$

Since this is not a standard characteristic impedance of a transmission line, special dimensions and/or materials will be needed to construct this line.

2.2.3.2 Reflection Coefficient and Return Loss

The reflection coefficient was defined by Equation (2.29). If we replace the z axis value by the length of the line l, it can be rewritten as:

$$\Gamma(l) = S_{11}(l) = \frac{Z_L - Z_0}{Z_L + Z_0} e^{-2\gamma l} = \Gamma_0 e^{-2\gamma l} \tag{2.38}$$

This is the voltage reflection coefficient and the notation S_{11} is widely used in practice and measurement. Since the power is proportional to the voltage square as shown in Equation (2.3), the power reflection coefficient is

$$\Gamma_P(l) = |\Gamma(l)|^2 = |\Gamma_0|^2 e^{-4\alpha l} \tag{2.39}$$

Both reflection coefficients are a good measure of how much signal/power is reflected back from the terminal. Obviously, when the load impedance is the same as the characteristic impedance, they are both zero and that is the matched case.

When the voltage reflection coefficient and power reflection coefficient are expressed in logarithmic forms, they give the same result which is called the *return loss*:

$$L_{RT}(l) = -20 \log_{10}(|\Gamma(l)|) = -10 \log_{10}(\Gamma_P(l)) \tag{2.40}$$

Since the return loss should not be smaller than zero, there is a minus sign in Equation (2.40) (but in practice some people ignore the minus sign, which is not correct). 3 dB return loss corresponds to the (voltage) reflection coefficient being $1/\sqrt{2} \approx 0.707$ or power reflection coefficient being 0.5.

It is worth noting the following special cases for a lossless transmission line:

a. Matched case: $Z_L = Z_0$
   $\Gamma(l) = \Gamma_0 = 0$ – the reflection is zero at any point of the line (even for a non-perfect transmission line).
   $L_{RT} = \infty$ dB – in practice, this means that the return loss is huge, say 50 dB.
b. Open circuit: $Z_L = \infty$
   $\Gamma_0 = 1$ and $\Gamma(l) = e^{-j2\beta l}$ – the amplitude of the reflection coefficient is 1 at any point of the line.
   $L_{RT} = 0$ dB – this means that all power is reflected back from the load.
c. Short circuit: $Z_L = 0$
   $\Gamma_0 = -1$ and $\Gamma(l) = -e^{-j2\beta l}$ – the amplitude of the reflection coefficient is 1 at any reference point of the line, there is a phase shift of 180° between the input and reflected voltages at the end of the line.
   $L_{RT} = 0$ dB – again, this means that all power is reflected back from the load.

Example 2.4   Reflection coefficient and return loss of a lossless transmission line
A 75 Ω resistor is connected to a lossless transmission line with characteristic impedance of 50 Ω.

a. What is the voltage reflection coefficient for $l = 0$ and $\lambda/4$, respectively?
b. What is the return loss for $l = 0$ and $\lambda/4$, respectively?

Solution
For a lossless transmission line, the attenuation constant α is zero.

a. Using Equation (2.38), we have

$$\Gamma(0) = \frac{Z_L - Z_0}{Z_L + Z_0} = \frac{75 - 50}{75 + 50} = 0.2$$

and

$$\Gamma\left(\frac{\lambda}{4}\right) = \frac{Z_L - Z_0}{Z_L + Z_0} e^{-2\gamma l} = 0.2 e^{-j2*2\pi/4} = 0.2 e^{-j\pi} = -0.2$$

This means that the phase of the reflection coefficient is changed 180° when the length of the transmission line is increased by a quarter-wavelength.
b. The return loss can be obtained by using Equation (2.40), i.e.

$$L_{RT}(l) = -20 \log_{10}(|\Gamma(l)|) = -10 \log_{10}(\Gamma_P(l)) = 13.98 \ \ \text{dB}$$

for $l = 0$ and $\lambda/4$, and actually any length of such a transmission line since the loss is due to mismatch, not the loss of the transmission line.

Example 2.5   Reflection coefficient and return loss of a low-loss transmission line
A 75 Ω resistor is connected to a low-loss transmission line with characteristic impedance of 50 Ω. The attenuation constant is 0.2 Np/m at 1 GHz.

a. What is the voltage reflection coefficient for $l = 0$ and $\lambda/4$, respectively?
b. Plot the return loss as a function of the line length. Assume that the effective relative permittivity is 1.5.

Solution
For this low-loss transmission line, the attenuation constant is $\alpha = 0.2$ Np/m. At 1 GHz, the wavelength in the line is

$$\lambda = \frac{c}{f\sqrt{\varepsilon_r}} = \frac{3 \times 10^8}{10^9 \sqrt{1.5}} \approx 0.245 \ \ (\text{m})$$

a. Using Equation (2.38), we have

$$\Gamma(0) = \frac{Z_L - Z_0}{Z_L + Z_0} = \frac{75 - 50}{75 + 50} = 0.2$$

and

$$\Gamma\left(\frac{\lambda}{4}\right) = \frac{Z_L - Z_0}{Z_L + Z_0} e^{-2\gamma l} = 0.2 e^{-2*0.2*0.245/4 - j2*2\pi/4} = 0.1952 e^{-j\pi} = -0.1952$$

This means that not only the phase of the reflection coefficient but also the amplitude is changed when the length of the transmission line is increased by a quarter-wavelength.

b. The return loss can be obtained by using Equation (2.40), i.e.

$$L_{RT}(0) = -20 \log_{10}(|\Gamma(0)|) = 13.98 \text{ dB}$$

and

$$L_{RT}\left(\frac{\lambda}{4}\right) = -20 \log_{10}\left(\left|\Gamma\left(\frac{\lambda}{4}\right)\right|\right) = 14.19 \text{ dB}$$

The return loss is slightly increased as expected. Over two wavelengths, the change is shown in Figure 2.8. It follows $0.2 e^{-2\alpha l}$ and is a straight line in logarithmic scale, i.e. $L_{RT}(l) = 13.98 + 3.47l$ dB.

2.2.3.3 Voltage Standing Wave Ratio (VSWR)

The VSWR (also known as the *standing wave ratio*, SWR) is defined as the magnitude ratio of the maximum voltage on the line to the minimum voltage on the line as shown in Figure 2.9. Mathematically, it can be expressed as:

$$VSWR(l) = \frac{|V|_{max}}{|V|_{min}} = \frac{|V_+| + |V_-|}{|V_+| - |V_-|} = \frac{1 + |\Gamma(l)|}{1 - |\Gamma(l)|} \tag{2.41}$$

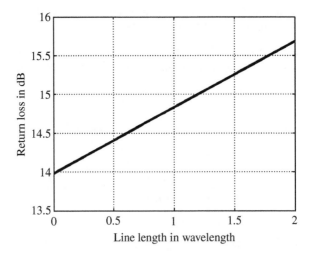

Figure 2.8   Return loss as a function of the line length

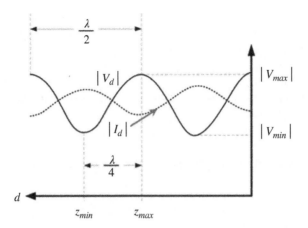

Figure 2.9   Standing waves of the voltage and current on a transmission line

Obviously, the VSWR is just another measure of how well a transmission line is matched with its load. Unlike the reflection coefficient, the VSWR is a scalar and has no phase information. For a non-perfect transmission line, the VSWR is a function of the length of the line (l) as well as the load impedance and the characteristic impedance of the line. But for a lossless transmission line, the VSWR is the same at any reference point of the line.

From Equation (2.41), we can prove that

$$|\Gamma| = \frac{VSWR - 1}{VSWR + 1} \tag{2.42}$$

This can be used to calculate the reflection coefficient once the VSWR is known.

For the following special cases:

a. Matched termination $Z_L = Z_o$:

$$VSWR = 1$$

b. Open circuit $Z_L = \infty$:

$$VSWR = \infty$$

c. Short circuit $Z_L = 0$:

$$VSWR = \infty$$

Thus, the VSWR of a line is bounded by unity and infinity:

$$1 \leq VSWR \leq \infty$$

For most applications, the VSWR is required to be smaller than 2 which is considered as a good match. But for mobile phone industry, the desired VSWR is normally less than 3, which is due to the considerable effects of human body on the performance of the phone.

Example 2.6   VSWR

A 75 Ω resistor is connected to a transmission line of characteristic impedance of 50 Ω. What is the VSWR at the termination?

Solution

The reflection coefficient at the termination is

$$\Gamma(0) = \frac{Z_L - Z_0}{Z_L + Z_0} = \frac{75 - 50}{75 + 50} = 0.2$$

Using Equation (2.41) to give

$$VSWR(0) = \frac{1 + |\Gamma(0)|}{1 - |\Gamma(0)|} = 1.5$$

It is smaller than 2, thus this can be considered as a very well-matched case.

In this section, we have discussed the reflection coefficient, return loss, and *VSWR*. All these quantities are measures of the impedance matching. Table 2.1 is a list of some typical values to show how they are interrelated. It is interesting to note that, if Z_L is a real number,

- when $Z_L/Z_0 > 1$, $Z_L/Z_0 = VSWR$;
- when $Z_L/Z_0 < 1$, $Z_L/Z_0 = 1/VSWR$.

Table 2.1   Links of normalized impedance, reflection coefficient, return loss, and *VSWR*

Z_L/Z_0	Γ	L_{RT} (dB)	VSWR	Note
∞	+1	0	∞	Open circuit
5.8470	0.7079	3	5.8470	Half power returned
3.0096	0.5012	6	3.0096	25% power returned
1.9248	0.3162	10	1.9248	10% power returned
1.2222	0.1000	20	1.2222	1% power returned
1.0653	0.0316	30	1.0653	0.1% power returned
1.0202	0.0100	40	1.0202	0.01% power returned
1	**0**	∞	**1**	**Matched**
0.9802	−0.0100	40	1.0202	
0.9387	−0.0316	30	1.0653	
0.8182	−0.1000	20	1.2222	
0.5195	−0.3162	10	1.9248	Close to $VSWR = 2$
0.3323	−0.5012	6	3.0096	Close to $VSWR = 3$
0.1710	−0.7079	3	5.8470	Half power returned
0	−1	0	∞	Short circuit

However, if Z_L is a complex number and its imaginary part is not zero, these simple relations linking Z_L/Z_0 and *VSWR* do not hold. For example: for $Z_L = j50 \, \Omega$ and $Z_0 = 50 \, \Omega$, we have $\Gamma = j$, $L_{RT} = 0$ dB and *VSWR* $= \infty \neq Z_L/Z_0$.

2.3   The Smith Chart and Impedance Matching

2.3.1   The Smith Chart

The *Smith Chart*, as shown in Figure 2.10, was invented by Phillip H. Smith (1905–1987) and is a graphical aid designed for radio frequency (RF) engineers to solve transmission lines and matching circuit problems. Although computer-aided tools for impedance matching have grown steadily over the years, the Smith Chart is still widely used today, not only as a problem-solving aid, but as a graphical demonstrator of how RF parameters behave, an alternative to using tabular information. The Smith Chart can be utilized to represent many parameters including impedances, admittances, reflection coefficients, scattering parameters, noise figure circles, constant gain contours, and regions for unconditional stability. It is most frequently used at or within the unity radius region. However, the remainder is still mathematically relevant, being used, for example, in oscillator design and stability analysis.

The Smith Chart is plotted on the complex reflection coefficient plane in two dimensions as shown in Figure 2.11. The horizontal axis is the real part of the reflection coefficient, while the vertical axis shows the imaginary part of the reflection coefficient. The origin or center is $|\Gamma| = 0$ (*VSWR* = 1). *In the standard Smith Chart, only the circle for $|\Gamma| = 1$ (VSWR = ∞) is shown and*

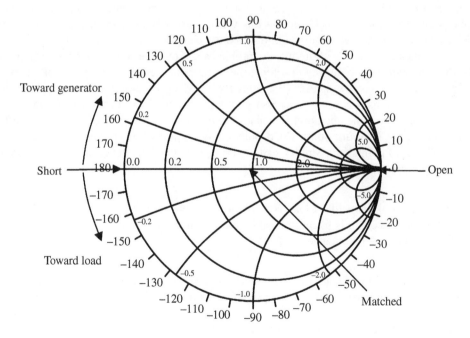

Figure 2.10   The standard Smith Chart

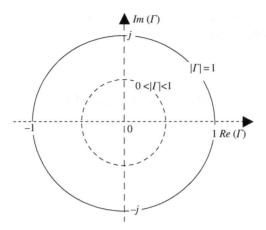

Figure 2.11   The Smith Chart showing the complex reflection coefficient

other circles of a constant reflection coefficient |Γ| *are not displayed for the reason of making the chart simple and neat.*

Most information shown on the standard Smith Chart is actually the normalized complex impedance as shown in Figure 2.12: the resistance is displayed in circles and the reactance is shown as arched lines. The upper half-space is inductive, while the lower half-space is capacitive. The middle line is pure resistance from 0 to infinity. If a reference point on a transmission line is moved away from the load (i.e. toward the source), this can be shown on the Smith Chart as the impedance point is moved on the |Γ| (or *VSWR*) circle clockwise. When the reference point is moved toward the load, it means that the impedance point is moved on the |Γ| circle anticlockwise. The distance is normalized to the wavelength.

As impedances change with frequency, problems can only be solved manually using the Smith Chart with one frequency at a time, the result being represented by a point. This is often adequate for narrow band applications (typically up to about 10% bandwidth), but for wide bandwidths it is usually necessary to apply Smith Chart techniques at more than one frequency across the operating frequency band. Provided that the frequencies are sufficiently close, the resulting Smith Chart points may be joined by straight lines to create a locus. A locus of points on a Smith Chart covering a range of frequencies can be employed to visually represent

- how capacitive or inductive a load is across the frequency range;
- how difficult matching is likely to be at various frequencies;
- how well matched a particular component is.

The accuracy of the Smith Chart is reduced for problems involving a large spread of impedances, although the scaling can be magnified for individual areas to accommodate these.

Example 2.7   Input impedance and reflection coefficient
Using a Smith Chart to redo Example 2.1 and also display the reflection coefficient on the Chart.

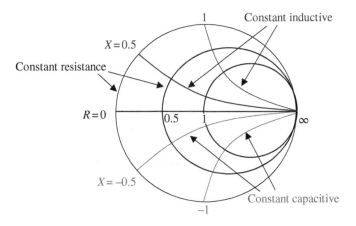

Figure 2.12   The Smith Chart showing the complex impedance

Solution

The characteristic impedance of the line is 50 Ω, thus the normalized load impedance is 75/50 = 1.5 which can be uniquely identified as point A in the Smith Chart in Figure 2.13. The distance from the origin (center) to A is 0.2, thus the reflection coefficient |Γ| = 0.2. Moving the reference plane away from this load means moving point A along the |Γ| = 0.2 circle clockwise into the capacitive half-space and then past the resonant point at R = 0.67 (i.e. 0.67 × 50 Ω = 33.5 Ω) into the inductive half-space. After 360° (half-wavelength), it is back to A. Over this period, the normalized reactance is changed between –0.4 and +0.4, i.e. –20 and +20 Ω. This periodic feature is repeated as the line length is increased. The results are therefore the same as that in Example 2.1.

The impedance Smith Chart is not great for working with parallel components (parallel inductors, capacitors, or shunt transmission lines), a more suitable form is the admittance Smith Chart which is often known as the Y Smith Chart as shown in Figure 2.14. It is formed by many circles of constant conductance and curves of constant susceptance. It is complementary to the impedance Smith Chart and very useful in the impedance-matching process as we will see soon.

The admittance is the ratio of current and voltage and has a unit of Siemens. The general definition equation is given by:

$$Y = \frac{1}{Z} = \frac{1}{R + jX} = G + jB \tag{2.43}$$

where Y is the complex admittance, G is the real-valued conductance, and B is the real-valued susceptance which can be positive (capacitive) or negative (inductive), all these parameters are measured in Siemens.

The admittance (Y) is the reciprocal of the impedance (Z). If the impedance is not zero, we can find

$$G = \frac{R}{R^2 + X^2} \tag{2.44}$$

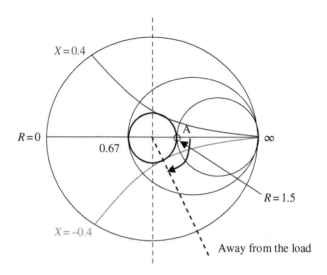

Figure 2.13   The impedance on the Smith Chart as the reference away from the load

and

$$B = \frac{-X}{R^2 + X^2} \qquad (2.45)$$

The sign of susceptance B is just opposite of reactance X which is clearly shown in Figure 2.14 and in Equation (2.45).

In this book, we are mainly interested in using impedance Smith Chart and admittance Smith Chart. Other forms which are not widely used and can be found in references (such as [1, 2]) are not discussed.

2.3.2   Impedance Matching

Impedance matching is the practice of making the output impedance of a source equal to the input impedance of the load in order to maximize the power transfer and minimize reflections from the load. Mathematically, it means the load impedance being the complex conjugate of the source impedance. That is,

$$Z_S = Z_L^* = (R_L + jX_L)^* = R_L - jX_L \qquad (2.46)$$

When the imaginary part is zero, the two impedances are the same:

$$Z_S = Z_L^* = Z_L = R_L \qquad (2.47)$$

Normally, we can use either lumped networks or distributed networks to match impedance. A lumped matching network typically consists of some lumped capacitors and/or inductors, while

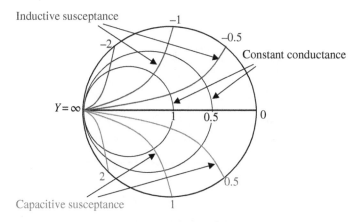

Figure 2.14   The Y Smith Chart showing the admittance

a distributed matching network consists of distributed capacitors and/or inductors. The former is mainly used at lower frequencies, while the latter is only used at higher frequencies.

2.3.2.1   Lumped Matching Networks

The lumped matching networks can be divided into three basic networks: L network (Figure 2.15), T network (Figure 2.21), and pi (π) network (Figure 2.22). The aim of matching is to make the equivalent impedance of the network to be the same as the desired resistance R_{in} – usually the characteristic impedance of a transmission line. *Generally speaking, resistors are not employed for impedance matching*, since the power could be consumed by the resistor without making useful contribution to the operation of the system.

In Figure 2.15(a). an L network is employed to match the impedance R_{in} with R_L. It can be shown that this impedance is linked to the network elements by the following equations:

$$B = \pm \frac{\sqrt{n-1}}{R_{in}}$$
$$X = \pm \frac{R_{in}\sqrt{n-1}}{n} - X_L \tag{2.48}$$

where B is susceptance (not normalized), X is reactance (not normalized), and $n = R_{in}/R_L$ should be greater than 1 (i.e. $n > 1$). The same ("+" or "−") sign is selected for both B and X. Thus, there are two sets (not four sets) of solutions. In most cases, B and X are just from a capacitor or an inductor, but in some cases, they could be from the combination of a capacitor and an inductor. If the condition for $n > 1$ cannot be met, the second lumped L network shown in Figure 2.15(b) may be used. The components are interrelated by:

$$Y_L = 1/Z_L = G_L + jB_L$$
$$B = \pm \frac{\sqrt{m-1}}{mR_{in}} - B_L \tag{2.49}$$
$$X = \pm R_{in}\sqrt{m-1}$$

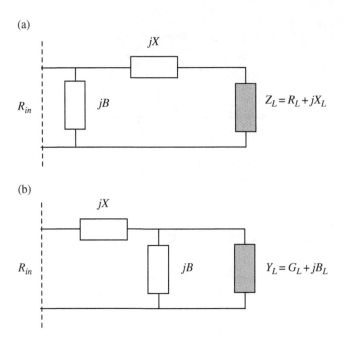

Figure 2.15   Lumped L networks. (a) for $R_{in} > R_L$. (b) for $R_{in}G_L < 1$

where $m = 1/(R_{in}G_L)$ and should be greater than 1. Again, the same ("+" or "–") sign is selected for both B and X. Thus, there are another two sets of solutions.

To understand these points better, let us walk through the following example.

Example 2.8   Impedance matching
A load with an impedance of $10 - j100\ \Omega$ is to be matched with a 50 Ω transmission line. Design a matching network and then verify your solution using the Smith Chart.

Solution
Since $Z_L = R_L + jX_L = 10 - j100$ and $n = R_{in}/R_L = 50/10 = 5 > 1$, the L network in Figure 2.15(a) is therefore a suitable matching network. Using Equation (2.48) to obtain

$$B = \pm \frac{\sqrt{n-1}}{R_{in}} = \pm \frac{2}{50} = \pm 0.04$$

$$X = \pm \frac{R_{in}\sqrt{n-1}}{n} - X_L = \pm \frac{50 \times 2}{5} + 100 = \pm 20 + 100$$

Thus, there are two sets of solutions: $(B, X) = (0.04, 120)$ and $(B, X) = (-0.04, 80)$.

Now let us examine if there is a possible solution using the L network in Figure 2.15(b). Because $Y_L = 1/(R_L + jX_L) \approx 0.001 + j0.01$ and $m = 1/(R_{in}G_L) \approx 1/0.05 = 20 > 1$, we can also use the second L network to get the impedance matched:

$$B = \pm \frac{\sqrt{m-1}}{mR_{in}} - B_L \approx \pm \frac{\sqrt{19}}{20 \times 50} - 0.01 \approx \pm 0.0044 - 0.01$$

$$X = \pm R_{in}\sqrt{m-1} \approx \pm 50\sqrt{19} \approx \pm 218$$

Here we have obtained another two sets of solutions: $(B, X) = (-0.0056, 218)$ and $(B, X) = (-0.0144, -218)$ using the second L network. Thus, there are four solutions in total.

Now we are going to verify and explain these solutions using a combined Smith Chart (also called the *immittance Smith Chart*) as shown in Figure 2.16 where the broken lines are for the admittance Smith Chart, while the solid lines are for the impedance Smith Chart. It may look a bit more complicated than the standard Smith Chart (here we have removed most circles and curves to make it simple and neat). This combined Smith Chart is extremely helpful as it can provide excellent visualization on how to get an impedance matched step by step.

For the first solution $(B, X) = (0.04, 120)$: the normalized load impedance is $Z_L/R_{in} = Z_L/50 = 0.2 - j2.0$ which is marked as point A in the Smith Chart in Figure 2.16. By adding an inductor with a normalized inductance 2.4 (=120/50) in series with the load, the impedance point A is moved

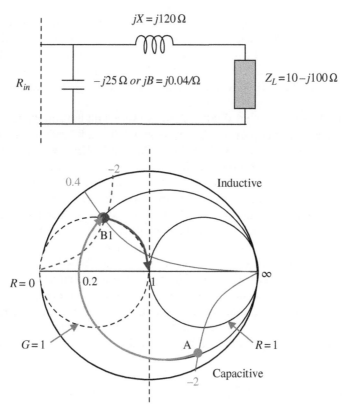

Figure 2.16   Using the Smith Chart to explain the matching process for Solution 1

along the circle of a constant resistance = 0.2 clockwise (adding series inductance) to point B1 whose normalized impedance is $0.2 + j0.4$, and the corresponding normalized admittance is $1 - j2$. Now if we add a parallel capacitor with a normalized susceptance 2 ($=0.04 \times 50$), using the admittance (Y) Smith Chart which was discussed earlier, the impedance point is now moved along the circle of a constant conductance $G = 1$ clockwise (adding capacitive susceptance in parallel) to the matching point. Thus, a perfect impedance matching is achieved. The equivalent circuit is also given in Figure 2.16.

Using the L matching network in Figure 2.16, two steps are required to achieve the matching: the first step is to move the impedance point clockwise along the circle of a contact resistance to an intersecting point with the circle of $G = 1$. The second step is to move the point clockwise to the matching point along the circle of $G = 1$. This is only possible for some, not all load impedances. In the Smith Chart, the suitable load impedance points are in the capacitive region outside the circle of $R = 1$ and all the points inside the circle of $G = 1$. The total area covers half of the Smith Chart. If the initial load impedance point is outside these suitable regions, we cannot use this L network to get the load impedance matched to the transmission line impedance (we cannot find the intersecting point B1), thus the region outside these regions is called *forbidden region* for this matching network. It is very important to know the limitations of each matching network.

Similarly, for the second solution $(B, X) = (-0.04, 80)$, we can use the same process but adding different inductance (1.6, not 2.4) and susceptance (−2, not 2), to get the circuit perfectly matched as shown in Figure 2.17. It should be pointed out that, in the second step, an inductor (not a capacitor) is added in parallel in the circuit, hence the impedance point is moved anticlockwise (not clockwise) along the circle of $G = 1$. As a result, for this matching network, the suitable load impedance points are in the capacitive region outside the circles of $R = 1$ and $G = 1$, thus, the suitable region is smaller while the forbidden region is larger than the previous case.

For the third solution $(B, X) = (-0.0056, 218)$, we start with the admittance Smith Chart. The normalized load admittance is $R_{in}/Z_L = 50/Z_L \approx 0.050 + j0.50$ which is again marked as point A in the Smith Chart in Figure 2.18. By adding an inductor with a normalized susceptance of −0.28 ($= -0.0056 \times 50$) in parallel with the load, the impedance point is moved along the circle of a constant conductance = 0.05 anticlockwise (adding inductance susceptance) to point B3 whose normalized admittance is $0.05 + j0.22$, and the corresponding normalized admittance is roughly $1 - j4.4$. Now if we add an inductor with a normalized inductance 4.4 ($\approx 218/50$) in series, using the impedance Smith Chart, the impedance point will move along the circle of $R = 1$ clockwise (adding inductance) to the matching point. The equivalent circuit for this case is also given in Figure 2.18. This is also a two-step matching process: the first step is to get the impedance to the circle of $R = 1$ by adding a parallel inductor and the second step is to move it to the matching point at the middle of the Smith Chart by adding another inductor in series. The suitable load impedance points are in the capacitive region outside the circle of $R = 1$ and outside the circle of $G = 1$. The other region is the forbidden region for this matching network. It is the same as the case of Solution 2.

For the fourth solution $(B, X) = (-0.0144, -218)$, we start with load admittance marked as point A in the Smith Chart in Figure 2.19. By adding an inductor with a normalized

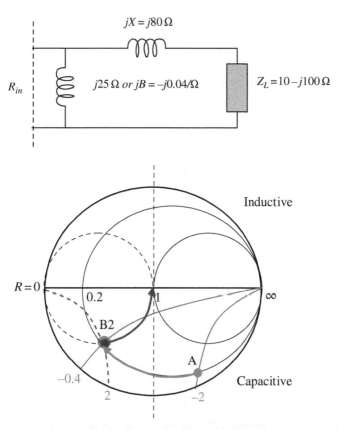

Figure 2.17   Using the Smith Chart to explain the matching process for Solution 2

susceptance -0.72 (= -0.0144×50) in parallel with the load, the impedance point is moved along the circle of $G = 0.05$ anticlockwise (adding inductance susceptance) to point B4 whose normalized admittance is $0.05 - j0.22$, and the corresponding normalized impedance is roughly $1 + j4.4$. Now if we add a capacitor with a normalized capacitance -4.4 ($\approx -218/50$) in series, using the impedance Smith Chart, the impedance point will move along the circle of $R = 1$ anticlockwise (adding capacitance) to the matching point at the center of the Smith Chart. Thus, the impedance matching is achieved. The equivalent circuit is also given in Figure 2.19.

   Again, it is a two-step matching process for this case: the first step is to get the impedance point to the circle of $R = 1$ by adding an inductor in parallel and the second step is to move it to the matching point by adding a capacitor in series. The suitable load impedance points are in the capacitive region outside the circle of $G = 1$ plus the inductive region inside the circle of $R = 1$. Compared with the case of Solution 1 in Figure 2.16, the suitable (or forbidden) region is a mirror imaging about the vertical axis (y-axis). The total suitable (or forbidden) region covers half of the Smith Chart.

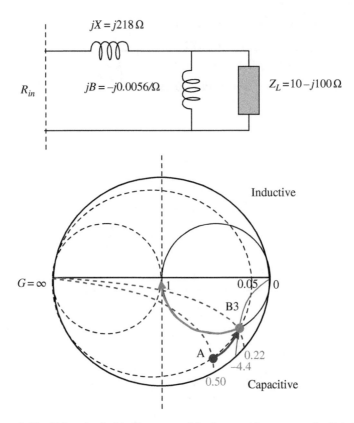

Figure 2.18   Using the Smith Chart to explain the matching process for Solution 3

There are a lot of interesting conclusions and observations can be drawn from this exercise:

1. The Smith Chart is a great design and visualization tool for impedance matching with clear physical meaning. It shows clearly how a parallel or series component will affect the impedance of a matching circuit.
2. To get a circuit matched, two steps are required for using an L matching network. The first step is to move the impedance point to the intersecting point with the circle of either $R = 1$ or $G = 1$ by adding an L or C in parallel or series. The second step is to add another L or C to move the impedance point to the matching point. The impedance Smith Chart and admittance Smith Chart are used once, respectively, in the process.
3. There are two types of L matching networks as shown in Figure 2.14. Each component can be either L or C, and there are therefore eight different combinations in total which are summarized in Figure 2.20 (four of them have been used and discussed in Example 2.8). The forbidden region in each case is also given in the figure.
4. There are four possible elements in the matching network, corresponding to four movements of the impedance point in the Smith Chart: (a) adding a series inductor, it moves clockwise

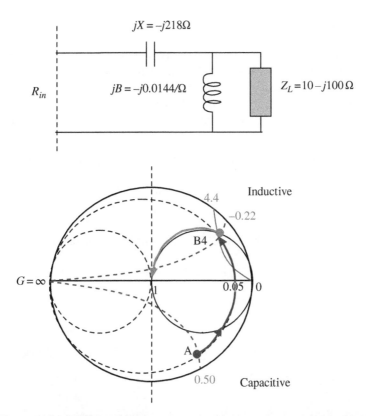

Figure 2.19 Using the Smith Chart to explain the matching process for Solution 4

along a constant R circle (such as the second step in Figure 2.20(a)); (b) adding a series capacitance, it moves anticlockwise along a constant R circle (such as the second step in Figure 2.20(b)); (c) adding a parallel inductor, it moves anticlockwise along a constant G (such as the first step in Figure 2.20(b)); (d) adding a parallel capacitor, it moves clockwise along a constant G circle (such as the first step in Figure 2.20(a)).

5. After careful examinations on the forbidden regions in Figure 2.20, we can see that there are two matching network solutions for the load impedance located inside the circles of $R = 1$ and $G = 1$ but four network solutions for the other load impedances as demonstrated in Example 2.8. For a given load impedance, the best solution depends on a number of practical considerations, such as the bandwidth, the insertion loss, and the values of the components. Some component values may not be realistic in practice.

6. Figure 2.20(a) and (c) are complementary which means, if the forbidden region of one circuit is the active region of the other circuit. Similarly, Figure 2.20(b) and (d) are also complementary. The forbidden regions for Figure 2.20(e) and (g) are the same, while the forbidden regions for Figure 2.20(f) and (h) are the same.

It should be pointed out that *lumped L networks have no degree of freedom to optimize the bandwidth, while the bandwidth is actually a very important consideration of many applications. To resolve this problem, one should consider adding further reactive elements to become T, π, or other networks to optimize the overall bandwidth.*

Figure 2.21 is a T network which may be considered as the cascade of two back-to-back L networks in Figure 2.14 by decomposing jB in the T network into two parallel components, which could be the combination of Figure 2.20(a) and (c) as a low-pass matching network (it has no forbidden region) or the combination of Figure 2.20(b) and (d) as a high-pass matching network (it has no forbidden region as well). The design process can be summarized by the following three steps:

Step 1: according to the load impedance and the desired bandwidth, choose X_1,

$$B_f = \frac{\Delta f}{f_o} = \frac{R_L}{|X_1 + X_L|} \tag{2.50}$$

Step 2: since Z_L and jX_1 are in series, the composite load impedance can be obtained as $Z_{LN} = Z_L + jX_1$.

Step 3: Use Z_{LN} and the L network design Equation (2.48) to find B and X_2.

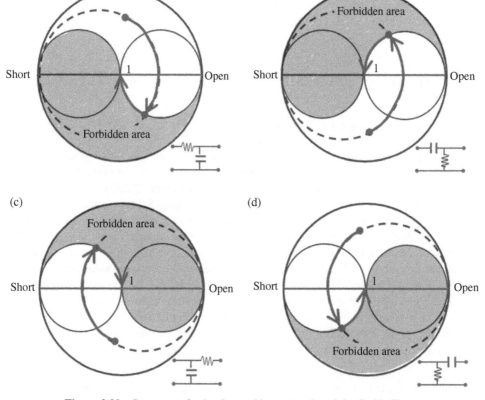

Figure 2.20   Summary of using L matching network and the Smith Chart

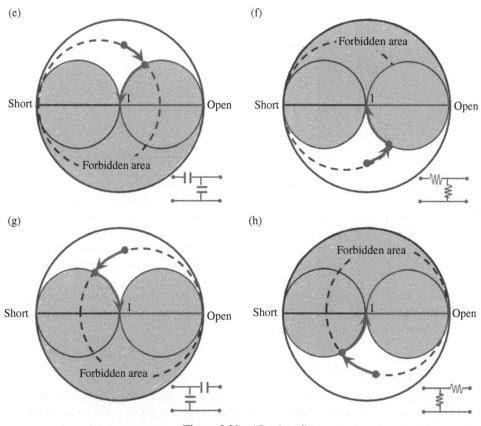

Figure 2.20 (*Continued*)

Another option for impedance matching is to use a π network as shown in Figure 2.23 which is the dual of the T network and can be seen as the cascade of two back-to-front L networks in Figure 2.14 by decomposing jX in the π network into two series components. Thus, it could be considered as the combination of Figure 2.20(a) and (c) to form a low-pass matching network (it has no forbidden region) or the combination of Figure 2.20(b) and (d) to form a high-pass matching network (it also has no forbidden region). The design process is very similar to that of the T network. That is:

Step 1: according to the load impedance and the desired bandwidth, choose B_1,

$$B_f = \frac{\Delta f}{f_o} = \frac{G_L}{|B_1 + B_L|} \tag{2.51}$$

Step 2: since Y_L and jB_1 are in parallel, the composite load admittance can be obtained as $Y_{LN} = Y_L + jB_1$.

Step 3: Use Y_{LN} and the L network design Equation (2.49) to find X and B_2.

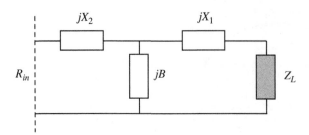

Figure 2.21   Lumped T network

Like an L network, there are also possibly two or four solutions when using a T or π network, depending on the load impedance location inside or outside the circle of $G = 1$ or $R = 1$ in the Smith Chart. Once the solutions are obtained, a further analysis should be conducted to finalize the design selection according to the desired parameters such as the realized bandwidth, insertion loss, nonlinearity, and availability of component values. To broaden the bandwidth and lower the Q-factor, a number of cascading stages may be required in the matching network.

In addition, one may wonder if a T and π network should be adopted in the design. The design freedom and complexity of both networks are the same. It would be good for the reader to make a comparison by going through the process like Example 2.8 to find out – this should be a very interesting homework for the reader to complete.

2.3.2.2   Distributed Matching Networks

At higher frequencies, the lumped elements (e.g. capacitors and inductors) may show abnormal behaviors due to parasitic element effects. Distributed elements and matching networks should be used. Distributed matching networks can be formed by a quarter-wavelength transmission line, an open-circuit transmission line (capacitor), a short-circuit transmission line (inductor), or their combinations. They can be represented mathematically by Equations (2.34)–(2.36). The process is best visualized on the Smith Chart.

The quarter-wavelength transformer mentioned earlier is a unique and popular narrowband impedance-matching technique, and the process is quite straightforward. The short- and open-circuit stub tuning is very similar to the lumped matching circuit tuning: reactance and/or susceptance is added to the matching network. The example below is a good illustration of how to realize matching using a distributed network.

Example 2.9   Impedance matching and bandwidth
A load with an impedance of $10 - j100\ \Omega$ is to be matched with a $50\ \Omega$ transmission line. Design two distributed matching networks and compare them in terms of the bandwidth performance.

Solution
The normalized load impedance is

$$z_L = Z_L/50 = 0.2 - j2$$

As shown in Figure 2.23, this corresponds to a unique point A on the Smith Chart. The reflection coefficient is

$$\Gamma = \frac{Z_L - 50}{Z_L + 50} = 0.5588 - j0.7353$$

$$|\Gamma| = 0.9235$$

The $|\Gamma| = 0.9235$ circle is shown in Figure 2.23 as a broken line. To get the impedance matched with 50 Ω, we can move this point (A) clockwise (toward the source) along this circle to points C1 (its normalized admittance $y = 1 + jb$), or C2 ($y = 1 - jb$), or C3 (its normalized impedance $z = 1 + jx$), or C4 ($z = 1 - jx$) which are the crossover points with the circle of $1/z = 1$ (for C1 and C2) and $z = 1$ (for C3 and C4). They can then be further moved either along circle $1/z = 1$ (for C1 and C2) or $z = 1$ (for C3 and C4) to the center O – the matching point. Since $|\Gamma| = 0.9235$ at these points, we can find $b = 4.8149$ and $x = 4.8149$, respectively. These four points may look similar to the four points in Example 2.8, but the impedance values are different from B1, B2, B3, and B4 since C1, C2, C3, and C4 were obtained for a constant $|\Gamma| = 0.9235$ while B1, B2, B3, and B4 were obtained through a circle of constant R or G.

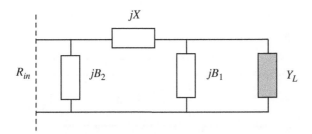

Figure 2.22 Lumped π network

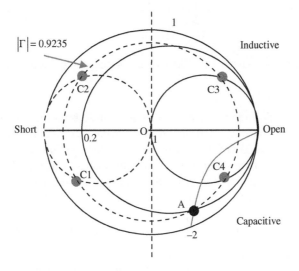

Figure 2.23 Impedance matching using Smith Chart

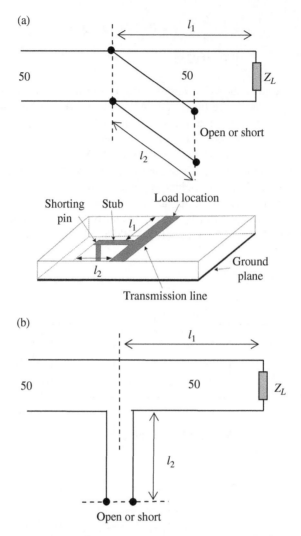

Figure 2.24   Stub matching networks. (a) Parallel stub matching using a microstrip line. (b) Series stub matching

The matching network through C1 or C2 can be realized by using the matching circuit in Figure 2.24(a) where the open- or short-circuit stub is in parallel with the load impedance, while matching network through C3 or C4 can achieved using the matching circuit in Figure 2.24(b) where the open- or short-circuit stub is in series with the load impedance. The matching process is different from the conventional lumped matching network. Here the rotational angle on the Smith Chart is required to determine the impedance and length l_1 or l_2. There are at least four possible designs.

To design a matching network through C1, we can follow the steps below:

Step 1: Since the normalized admittance at C1 is $1.0 + j4.8149$ (the normalized impedance is $0.0414 - j0.1991$), we can move the impedance point from A to C1, the rotational angle is about

0.582π (or $104.8°$) on the Smith Chart which corresponds to $l_1 = 0.1455\lambda$ and $\beta l_1 = 0.291\pi$ (i.e. the rotation angle is $2\beta l$ on the Smith Chart, since the period of the reflection coefficient is π, not 2π for a loss-free line as shown in Equation (2.38)).

Step 2: Move from point C1 to center O. This can be easily achieved using a stub connected in parallel with the line, thus it is advantageous to work in admittances. The stub in parallel with the line should produce a susceptance of -4.8149, thus the normalized total impedance is 1. This can be achieved by the following two designs:

A. a short circuit with a stub length $l_2 = 0.0325\lambda$, using Equation (2.35);
B. an open circuit with a stub length $l_2 = 0.2825\lambda$ – this is $\lambda/4$ longer than solution A.

The stub length can also be read directly from the Smith Chart, and the reading values may not be as accurate as that obtained using equations. But using the Smith Chart directly to figure out these parameters could be quicker than the approach using numerical calculation for some people, since the calculation has involved in complex numbers which could be tricky. This is probably one of the main reasons that the Smith Chart is still popular in industry even today.

Now let us examine the bandwidth of these two designs. Assuming the center frequency is 1 GHz, thus we have $l_1 = 0.1455\lambda = 4.365$ cm, and $l_2 = 0.0325\lambda = 0.975$ cm for Design A, and $l_2 = 0.2825\lambda = 8.475$ cm for Design B. Using the input impedance formula:

$$Z_{in}(l) = Z_0 \frac{Z_L + jZ_0 \tan(\beta l)}{Z_0 + jZ_L \tan(\beta l)}$$

for both l_1 and l_2, respectively, and then the following two equations:

$$\Gamma = \frac{Z_L - Z_0}{Z_L + Z_0}$$

$$VSWR = \frac{1 + |\Gamma|}{1 - |\Gamma|}$$

to obtain the VSWR as a function of frequency. The results are shown in Figure 2.25. It is apparent that:

- Both designs have an excellent impedance match at the center frequency 1 GHz;
- The stub length of Design A is shorter than that of Design B, while the bandwidth of Design A is much wider than that of Design B. The additional $\lambda/4$ transmission line in Solution B has not brought any benefit. This is a very interesting and useful result. Design A is definitely the preferred choice.

Similar conclusions can be drawn from the matching networks built at the other points (C2, C3, and C4). To employ a distributed matching network, the length should be as short as possible to minimize the size while maximizing the bandwidth.

An interesting question is: how does the lumped network solution in Example 2.8 compare with distributed network solution? As demonstrated above, the distributed network solution can easily show the matching network performance over a frequency band and the result is

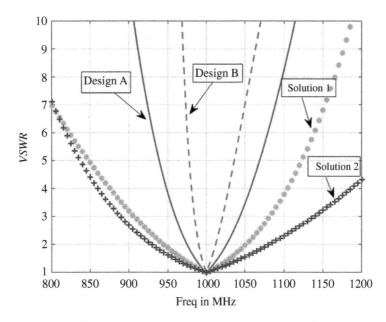

Figure 2.25   VSWR of different designs as a function of frequency

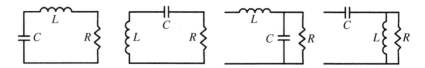

Figure 2.26   Four load impedances with LC matching networks

relatively accurate, but a lumped network solution is less straightforward: we need to find out the capacitance/inductance for each element and then sweep over a frequency band and the results may be less accurate at higher frequencies. For the case using point B2 in Figure 2.17 at 1 GHz, we can find the inductances for the two inductors are $(80/2\pi)$ nH and $(25/2\pi)$ nH, respectively, and then use the same set of equations mentioned above to obtain the VSWR as a function of frequency. The results for this case (Solution 2) and Solution 1 (Figure 2.16) are also plotted in Figure 2.25. We can see that the lumped element solutions have wider bandwidths and Solution 2 is possibly the best solution among these four designs. It is actually true in practice, around 1 GHz, most impedance-matching networks are realized using lumped elements, not distributed elements.

The frequency bandwidth limitation on matching networks has been investigated by many people. There exists a general limit on the bandwidth over which an arbitrarily good impedance match can be obtained in the case of a complex load impedance. It is related to the ratio of reactance to resistance, and to the bandwidth over which we desire to match the load.

Figure 2.26 shows four load impedances (series RL, series RC, parallel RC, and parallel RL) with matching networks which are specific examples of L matching network discussed earlier.

Taking the parallel RC load impedance as an example, Bode and Fano derived, for lumped circuits, a fundamental limitation for it and it can be expressed as [2]:

$$\int_0^\infty \ln\left(\frac{1}{|\Gamma(\omega)|}\right) d\omega \leq \frac{\pi}{RC} \tag{2.52}$$

This is known as *Bode–Fano limit* for parallel RC. Since $\ln(1) = 0$, there is no contribution to this integral over frequencies for $|\Gamma| = 1$, so it can be seen that it is desirable to have the maximum mismatch outside the band of interest if a broad bandwidth is required. If this condition is assumed, the integral is limited to the bandwidth of interest ($\Delta\omega$), and we can get an idea of how well we can match an arbitrary complex impedance over that bandwidth. For an idealized case, this equation can be simplified as:

$$\Delta\omega \ln\left(\frac{1}{|\Gamma(\omega)|}\right) \leq \frac{\pi}{RC} \tag{2.53}$$

It clearly shows how the bandwidth ($\Delta\omega$) is linked to the impedance matching (Γ) and load impedance (RC). For a given reflection coefficient (or *VSWR*) and the RC product, one can use it estimate the maximum bandwidth attainable.

Similarly, the limits for the other three matching LC networks were obtained as:

$$\text{series RL: } \Delta\omega \ln\left(\frac{1}{|\Gamma(\omega)|}\right) \leq \frac{\pi R}{L}$$

$$\text{series RC: } \Delta\omega \ln\left(\frac{1}{|\Gamma(\omega)|}\right) \leq \pi\omega_0^2 RC$$

$$\text{parallel RL: } \Delta\omega \ln\left(\frac{1}{|\Gamma(\omega)|}\right) \leq \frac{\pi\omega_0^2 L}{R}$$

There are similar limitations on other forms of complex impedance. A general implication of the Bode–Fano limit is that one should not waste any match out of band, and that the best in-band match is obtained with Chebyshev rather than maximally flat networks. The best broadband impedance-matching practice incorporates the complex load impedance into, for example, a multi-section filter structure with a design that includes the characteristics of the load which is normally a function of frequency.

It should be pointed out that there are some impedance-matching design tools available on the market. For example, Optenni Lab (https://www.optenni.com/) is a leading professional matching circuit synthesis and optimization software tool for solving impedance-matching problems. It can optimize broadband, multiband, multiport, and tunable matching circuits for antennas, filters, amplifiers, switches, tuners, and other RF/microwave devices. It has a lot of interesting features, such as estimation of obtainable bandwidth potential and components library from major manufacturers. Another important feature is that it can conduct *tolerance analysis* which is very importance in real-world applications.

2.3.2.3 Adaptive Impedance Matching

In some applications, the static impedance matching is not good enough, since the load impedance may be a variable and could change significantly. For example, the antenna impedance of a mobile phone is not a fixed value due to fluctuating body effects from such as the hands or

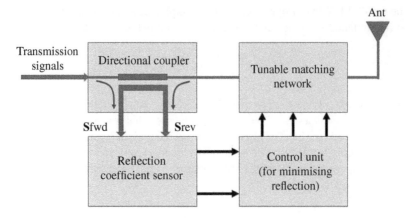

Figure 2.27   A block diagram of a typical adaptive matching network

head. This could cause a wide range of problems on antenna and mobile system performances (e.g. reduced efficiency, poor received signals, and more battery usage). Thus, techniques for adaptive control of impedance-matching networks have been developed which provide automatic compensation of antenna mismatch to ensure excellent performance in real time in a changing environment.

The block diagram of a typical adaptive impedance-matching network is shown in Figure 2.27 which consists of three major elements: mismatching detector, control unit, and tunable matching network. The mismatching detector is responsible for collecting the mismatching information through a device (such as a directional coupler to collect the forward and reverse signals) which may contain amplitude and phase information. The data are then utilized by the control unit to tune the matching network adaptively in real time. The control unit may use a simple preprepared lookup table or a more complex algorithm to control the tunable matching circuit through normally voltage-controlled capacitors (varactors) or microelectromechanical systems (MEMSs) tunable capacitors and high Q fixed inductor. A fast switch could also be used for controlling the matching network. The tunable matching network could be an L, T, or π network using both L and C. The topology could be more complex than what discussed earlier. The selection of the values of L and C is critical, and range for these components should be right for the desired application.

Let us take reference [3] as an example where adaptive impedance-matching techniques were discussed for mobile phone applications. Two L networks are employed, one for down-converting and the other for up-converting. To secure reliable convergence, a cascade of two control loops was proposed for independent control of the real and imaginary parts of impedance. A secondary feedback path is used to enforce operation into a stable region when needed. These techniques exploit the basic properties of tunable series and parallel LC networks. A generic quadrature detector that offers a power-independent orthogonal reading of the complex impedance value was presented, which was used for the direct control of variable capacitors. This approach renders calibration, elaborates software computation superfluous, and allows for autonomous operation of adaptive antenna-matching modules.

It should be pointed out that adaptive impedance matching is not limited to mobile applications. Another good example is wireless power transfer (WPT) via magnetic resonance coupling where the load impedance or the charging distance is not fixed. One of the most challenging design issues here is how to maintain a good level of power transfer efficiency, even when the distance between the transmitter and the receiver changes. When the distance varies, the power transfer efficiency drastically decreases due to the impedance mismatch between the resonator of the transmitter and that of the receiver. Reference [4] presented a serial/parallel capacitor matrix in the transmitter, where the impedance can be automatically reconfigured to track the optimum impedance-matching point in the case of varying distances. The dynamic WPT matching system was enabled by changing the combination of serial and parallel capacitors in the capacitor matrix. An interesting observation in the proposed capacitor matrix was that the resonant frequency was not shifted, even with capacitor matrix tuning. In order to quickly find the best capacitor combination that achieves maximum power transfer, a window-prediction-based search algorithm was also presented in this paper. The proposed resonance WPT system was implemented using a resonant frequency of 13.56 MHz, and the experimental results with 1 W power transfer showed that the transfer efficiency increased up to 88% when the distance changed from 0 to 1.2 m which has demonstrated the usefulness and effectiveness of the adaptive impedance-matching network.

2.3.3   Quality Factor and Bandwidth

The bandwidth is indeed a very important parameter for any electric/electronic device and system. It is closely linked to the *quality factor, Q*, which is a measure of how much lossless reactive energy is stored in a circuit compared to the average power dissipated.

Antenna bandwidth is maximized when the power dissipation is comparatively high. In other words, a low Q is required for wide bandwidth. In turn, to the extent to which this energy is associated with radiation (rather than conductor or dielectric losses) determines the antenna efficiency. For a circuit component such as an inductor or capacitor, we require the resistive losses to be low; hence, the Q is required to be high. It is often the case that antennas and circuit components have seemingly contradictory requirements: *antennas are designed to have a low Q, whereas circuit components are designed for a high Q.*

Quality factor is quoted as being either unloaded or loaded. For the latter, the losses of the external circuit – for example, the source – are included, whereas for the former they are not. The *unloaded quality factor, Q*, is defined as:

$$Q \equiv \omega \frac{\text{(total energy stored)}}{\text{(average power loss in the load)}} = \omega \frac{W_E + W_M}{P_L} \qquad (2.54)$$

where W_E is the energy stored in the electric field, W_M is the energy stored in the magnetic field, and P_L is the average power delivered to the load. The loaded quality factor, Q_L, can also be given in Equation (2.54) but with P_L replaced by the total power P_T, which is dissipated in both the external circuit and the load.

At resonance, the electric and magnetic field energies have the same magnitudes and the formulas simplify such that the unloaded quality factor at resonance Q_0 is given by:

$$Q_0 \equiv \frac{2\omega_0 \, W_E}{P_L} = \frac{2\omega_0 \, W_M}{P_L} \qquad (2.55)$$

where ω_0 is the angular resonant frequency (= $2\pi f_0$, where f_0 is the resonant frequency).

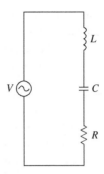

Figure 2.28   Series resonant circuit

There are many simplifications of these definitions that are widely used, but which only truly apply when they are derived directly from the above. For example, a relation that is often used is

$$Q_0 = \frac{f_0}{f_2 - f_1} = \frac{1}{B_F} \tag{2.56}$$

where f_1 and f_2 are the frequencies at which the power reduces to half of its maximum value (i.e. 3 dB) at the resonant frequency, f_0, and where B_F is the (3 dB) *fractional bandwidth*. It is important to point out that f_0 here is not necessary for the arithmetic average/mean of f_1 and f_2 (i.e. $(f_1 + f_2)/2$), but for the geometric average/mean as defined by some people. This relation only truly applies to simple circuits but is considered to be a good approximation to a wide range of unloaded configurations. It has the advantage of providing a simple relationship between Q and fractional bandwidth, but we will see later that it only accurately applies to simple, single-resonant circuits and should therefore be used with some care. The derivation of this relation is illustrated below.

Consider the series resonant circuit shown in Figure 2.28. The power dissipated in the resistance, R, is proportional to the square of the magnitude of the current, I. This is plotted in Figure 2.29 with $R = 50\ \Omega$, $L = 79.5775$ nH, and $C = 0.3183$ pF. The half-power frequencies, f_1 and f_2, are found to be 0.9513 and 1.0513 GHz, respectively. Note that the average of the two frequencies is 1.0013 GHz, not 1 GHz. As shown in Figure 2.29, the resonant frequency (when the current is maximized) is 1 GHz. Using Equation (2.56) yields a Q_0 of 10.

It is possible to find the unloaded quality factor of the circuit directly from the stored energies of the inductor and capacitor. The magnetic and electric energies are given by:

$$W_M = \frac{1}{2}\,LI^2, \quad W_E = \frac{1}{2}\,CV_C^2 \tag{2.57}$$

where V_C is the voltage across the capacitor. Writing these and the power delivered to the load (i.e. the resistor, R) in terms of the current I gives

$$W_M = \frac{1}{2}\,LI^2, \quad W_E = \frac{1}{2}\,C\frac{1}{(\omega C)^2}I^2, \quad P_L = RI^2 \tag{2.58}$$

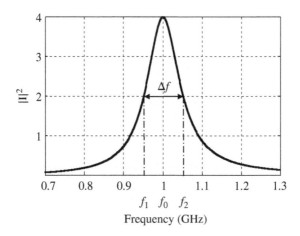

Figure 2.29   Relative power dissipated in a series resonant circuit around resonance

Further substitution in Equation (2.54) yields

$$Q = \frac{\left(\omega L + \frac{1}{\omega} C\right)}{2R} \tag{2.59}$$

At resonance, the magnitudes of the electric and magnetic energies are equal, and this relation simplifies to

$$Q_0 = \frac{\omega_0 L}{R} = \frac{1}{\omega_0 CR} \quad \text{and} \quad \omega_0 = \frac{1}{\sqrt{LC}} \tag{2.60}$$

This formula is often used, but it should be understood that it only applies at resonance. Taking $R = 50\ \Omega$, $L = 79.5775$ nH, and $C = 0.3183$ pF gives $Q_0 = 10$, as previously shown.

It is interesting to evaluate the ratio of Q at any frequency to that at resonance. From Equations (2.59) and (2.60), this is given by:

$$q = \frac{Q}{Q_0} = \frac{1}{2}\left(\frac{\omega}{\omega_0} + \frac{\omega_0}{\omega}\right) \tag{2.61}$$

This means $q \geq 1$, thus, the **unloaded quality factor is minimum at resonance**, although the variation with frequency is slow. For bandwidths of less than 20%, Q is approximately equal to Q_0 with an error of less than 0.5%, i.e. the term in the parentheses of Equation (2.61) is approximately equal to two over moderate bandwidths.

We would like to derive a relationship between the unloaded quality factor and the bandwidth of the circuit. The bandwidth is normally taken to be the range of frequencies over which the power dissipated; P_L is greater than half of the maximum, P_{L0} (at resonance). However, the sections that follow more general relations are derived based on a specified power transfer to the load.

The current in the circuit is given by:

$$I = \frac{V}{R + j\left(\omega L - 1/\omega C\right)} \tag{2.62}$$

From Equation (2.62), the ratio of the power dissipated at any frequency to the power dissipated at resonance is given by:

$$p = \frac{P_L}{P_{L0}} = \left|\frac{I}{I_0}\right|^2 = \frac{1}{1 + \left(\dfrac{\omega L}{R} - \dfrac{1}{\omega CR}\right)^2} \tag{2.63}$$

This can be written as:

$$p = \frac{1}{1 + \chi^2} \tag{2.64}$$

where, from (2.60),

$$\chi = Q_0\left(\frac{\omega}{\omega_0} - \frac{\omega_0}{\omega}\right) \tag{2.65}$$

The relation given in Equation (2.64) can be solved to give

$$Q_0\left(\frac{\omega}{\omega_0} - \frac{\omega_0}{\omega}\right) = \pm\sqrt{\frac{1-p}{p}} \tag{2.66}$$

This is a quadratic equation in ω with two positive and two negative solutions. The difference between the two positive solutions is

$$\omega_2 - \omega_1 = \frac{\omega_0}{Q_0}\sqrt{\frac{1-p}{p}} \tag{2.67}$$

This gives the fractional bandwidth as:

$$B_F = \frac{f_2 - f_1}{f_0} = \frac{1}{Q_0}\sqrt{\frac{1-p}{p}} \tag{2.68}$$

When $p = 0.5$ (i.e. 3 dB below the resonant frequency power), this simplifies to the familiar expression:

$$B_F = \frac{1}{Q_0} \tag{2.69}$$

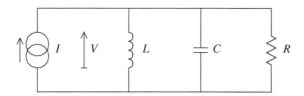

Figure 2.30 Parallel antiresonant circuit

as previously seen in Equation (2.56). This derivation does not make any assumptions about the numerical value of the quality factor or the bandwidth and is, therefore, applicable to both high and low Q systems.

The response of a parallel resonant circuit, as shown in Figure 2.30, can be found in much the same way as for series resonance.

The energies stored in the electric and magnetic fields of the capacitor and inductor, respectively, and the power dissipated in the resistance are given by:

$$W_E = \frac{1}{2} CV^2, \quad W_M = \frac{1}{2} L \frac{1}{(\omega L)^2} V^2, \quad P = GV^2 \tag{2.70}$$

where G is the conductance. Substitution in Equation (2.54) gives

$$Q = \frac{(\omega C + 1/\omega L)}{2G}. \tag{2.71}$$

At antiresonance, the magnitudes of the electric and magnetic energies are equal and this relation simplifies to

$$Q_0 = \frac{\omega_0 C}{G} = \frac{1}{\omega_0 LG} \tag{2.72}$$

The ratio between the Q and Q_0 is the same as for the series resonance, given by Equation (2.61). It can also be shown (using the same method) that the fractional bandwidth is the same as for the series resonant circuit.

The formulas for Q that have been derived so far have been for series and parallel resonant circuits, respectively. However, they are often applied (strictly incorrectly) to other circuit combinations. We will see later (in Section 8.1) that this can give large errors, so some care is required in applying these formulas.

Another important aspect which is not covered in most books is to show the Q-factor on the Smith Chart. Since any point in a Smith Chart represents a complex impedance and its corresponding Q-factor is the ratio of the imaginary part and the real part, this simple relationship means that lines of constant Q can be drawn on a Smith Chart, as shown in Figure 2.31 (for simplicity, only some selected curves are shown): the arc of a constant Q starts at $R = X = 0$ (the short-circuit point) and ends at $R = X = \infty$ (the open-circuit point).

This is very useful for bandwidth estimation of an impedance-matching network using the Smith Chart. If we take the case of Figure 2.29 as an example, we can see that the resonance

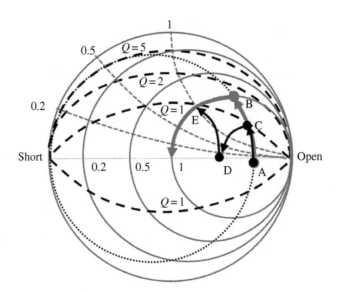

Figure 2.31   Smith Chart with constant Q lines

frequency is at 1 GHz. When the frequency is lower than 1 GHz, the lower the frequency, the larger the Q. Similarly, when the frequency is higher than 1 GHz, the higher the frequency, the larger the Q. When $Q = 1$ (i.e. the reactance is the same as the resistance on the Smith Chart), the corresponding frequencies below and above the resonant frequency are 0.95 and 1.05 GHz, respectively. Thus, the frequency bandwidth is the same as the 3-dB frequency bandwidth of Figure 2.29. It should be noted that the Q on the Smith Chart is for a single frequency, not for the whole circuit although they are linked. In this example, the Q on the Smith Chart is 1 but the Q for the circuit is 10, and they have different and have different meanings.

To achieve a particular Q, the matching elements must be chosen for the matching path to stay within the desired curves of constant Q. For example, to get impedance point A matched in Figure 2.31, we could use an L matching network whose impedance-matching path is from A to B and then to the matching point ($R = 1$) – in this case, Q would be greater than 1. Another approach is to use two L matching networks whose matching path is from A to C, to D, to E and then to the matching point ($R = 1$) – in this case, Q is kept less than or equal to 1 in the process. Thus, the second case would produce a wider bandwidth than the first case. The trade-off is that more elements are employed to keep it under the lower Q curve in order to provide a greater bandwidth. This is a general rule of thumb and widely used in practice. For example, a two-section quart-wave transformer has a wider bandwidth than a single-section quart-wave transformer.

The theoretical best bandwidth of a matching network is a function of the starting point on the Smith Chart. Real-world lumped element components have losses and tolerances, thus

- When trying to achieve a high Q solution, the losses in real components changes the trajectory of the matching path (lowering the Q of the match).
- Loss in matching elements reduces the practicality of many broadband solutions.
- Cascading the tolerances of parts rapidly expands the bounds over which the manufactured solution will fall.

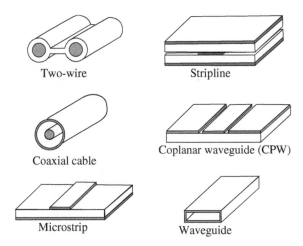

Figure 2.32   Various popular transmission lines

More discussion on impedance matching, bandwidth, and Q-factor is provided in Section 8.1, where the bandwidth broadening techniques and limitations are also introduced.

2.4   Various Transmission Lines

There are many transmission lines developed for various RF and microwave applications. The most popular ones are shown in Figure 2.32. They are two-wire transmission line, coaxial cable, microstrip, strip line, coplanar waveguide (CPW), and waveguide. We are going to examine these transmission lines in terms of their characteristic impedance, basic mode, frequency bandwidth, loss characteristic, and costs. We will also introduce and discuss some recently developed transmission lines which are particularly good for higher-frequency applications.

2.4.1   Two-wire Transmission Line

This is the simplest transmission line and its cross-sectional view is given in Figure 2.33. The separation of the wires is D and the diameter of the wires is d. The medium between the wires has a permittivity of ε.

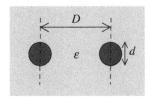

Figure 2.33   Two-wire transmission line

It can be shown that the per unit length inductance and capacitance of the transmission line are [2]

$$L = \frac{\mu}{\pi} \ln \frac{D + \sqrt{D^2 - d^2}}{d}, \quad C = \frac{\pi \varepsilon}{\ln \dfrac{D + \sqrt{D^2 - d^2}}{d}} \tag{2.73}$$

If the medium between the wires has a conductivity of σ_1, and the conductivity of the wire is σ_2, we can obtain that the resistance and conductance of a unit length line are

$$R = \frac{2}{\pi d} \sqrt{\frac{\omega \mu}{2 \sigma_2}}, \quad G = \frac{\pi \sigma_1}{\ln \dfrac{D + \sqrt{D^2 - d^2}}{d}} \tag{2.74}$$

respectively.

2.4.1.1   Characteristic Impedance

If the loss of the line could be considered very small, the characteristic impedance is therefore given by Equation (2.22), i.e.

$$\begin{aligned} Z_0 &= \sqrt{\frac{L}{C}} = \sqrt{\frac{\mu}{\pi^2 \varepsilon}} \ln \frac{D + \sqrt{D^2 - d^2}}{d} \\ &\approx \frac{120}{\sqrt{\varepsilon_r}} \ln \frac{D + \sqrt{D^2 - d^2}}{d} \end{aligned} \tag{2.75}$$

The typical value of industrial standard lines is 300 Ω. This type of transmission line was commonly used to connect a television receiving antenna (usually a folded dipole with an impedance around 280 Ω) to a home television set many years ago. The uniform spacing is assured by embedding the two wires in a low-loss dielectric, usually polyethylene. Since the wires are embedded in the thin ribbon of polyethylene, the dielectric space is partly air and partly polyethylene.

2.4.1.2   Fundamental Mode

The electromagnetic field distribution around the two-wire transmission line is illustrated by Figure 1.14. Both the electric field and magnetic field are within the transverse (to the propagation direction) plane; thus this mode is called the *TEM* (transverse electromagnetic) *mode*. This means that it is nondispersive, and the velocity is not changed with the frequency. The plane wave can also be considered as a TEM wave.

2.4.1.3   Loss

Since the lumped parameters of a transmission line are given by Equations (2.73) and (2.74), the attenuation constant α can be calculated using Equation (2.14). However, the principal loss of the two-wire transmission line is actually due to radiation, especially at higher frequencies. Thus, this type of transmission line is not suitable for higher-frequency applications. The typical usable frequency is less than 300 MHz.

Some people may be familiar with the *twisted-pair transmission line*. As the name implies, the line consists of two insulated wires twisted together to form a flexible line without the use of spacers. It has relatively good electromagnetic compatibility (EMC) performance – the twisted configuration has canceled out the radiation from both wires and resulted in a small and symmetrical total field around the line, but it is not suitable for high frequencies because of the high dielectric losses that occur in the rubber insulation (low costs) as well as the radiation. When the line is wet, the losses increase significantly.

2.4.2 Coaxial Cable

The coaxial cable consists of a central, insulated wire (inner conductor) mounted inside a tubular outer conductor as shown in Figure 2.34. In some applications, the inner conductor is also tubular. The inner conductor is insulated from the outer conductor by insulating materials which are made of Pyrex, polystyrene, polyethylene plastic, or some other material that has good insulating characteristics and low dielectric losses at high frequencies over a wide range of temperatures. In order to ensure good EMC performance, shielded and double-shielded coaxial cables have been developed and available on the market. This type of transmission line is widely used for RF engineering and antenna measurements, and the connection between the antenna and transceiver.

Coaxial cables come in three basic types: flexible, semi-rigid, and rigid. The rigid cable gives the best performance and is normally for high-performance and phase-sensitive applications, while the flexible cables are cheap and flexible. The semi-rigid is a compromise. As shown in Figure 2.34, the diameters of the inner and outer conductors of a cable are denoted as a and b, respectively, and the relative permittivity of the insulating material is ε_r. The dielectric material reduces the velocity of the wave inside the cable to $c/\sqrt{\varepsilon_r}$. Some common loading materials and corresponding velocities are shown in Table 2.2 [5].

If the conductivity between the inner and outer conductors is σ_1, and the conductivity of the conductors is σ_2, it can be shown that the per unit length parameters of the coaxial line are

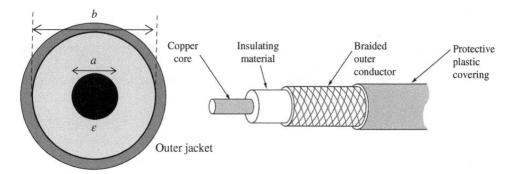

Figure 2.34   The configuration of a coaxial line

Table 2.2 Coaxial cable material and velocity

Dielectric type	Time delay (ns/m)	Propagationvelocity (% of c)
Solid polyethylene (PE)	5.05	65.9
Foam polyethylene (FE)	4.17	80.0
Foam polystyrene (FS)	3.67	91.0
Air space polyethylene (ASP)	3.77–3.97	84–88
Solid teflon (ST)	4.79	69.4
Air space teflon (AST)	3.71–3.94	85–90

Source: Modified from RF Café [5], Kirt Blattenberger, 1964–2024.

$$L = \frac{\mu}{2\pi} \ln \frac{b}{a} \qquad R = \sqrt{\frac{f\mu}{\pi\sigma_2}} \left(\frac{1}{a} + \frac{1}{b} \right)$$
$$C = 2\pi\varepsilon / \ln \frac{b}{a} \qquad G = 2\pi\sigma_1 / \ln \frac{b}{a} \tag{2.76}$$

The velocity as shown in Equation (2.25) is

$$v = \frac{1}{\sqrt{LC}} = \frac{1}{\sqrt{\varepsilon\mu}} = \frac{c}{\sqrt{\varepsilon_r}}$$

2.4.2.1 Characteristic Impedance

Normally, the loss of the line can be considered very small, and the characteristic impedance is given by Equation (2.22), i.e.

$$Z_0 = \sqrt{\frac{L}{C}} = \frac{\sqrt{\mu/\varepsilon}}{2\pi} \ln \frac{b}{a} = \frac{60}{\sqrt{\varepsilon_r}} \ln \frac{b}{a} \tag{2.77}$$

The typical value for industrial standard lines is 50 or 75 Ω.

2.4.2.2 Fundamental Mode

The electromagnetic field distribution around the coaxial cable is illustrated by Figure 2.35. Again, both the electric field and magnetic field are within the transverse (to the propagation direction) plane, thus this field is *TEM mode*.

However, this is only true for the frequencies below the *cutoff frequency* which is [2]

$$f_c = \frac{v}{\pi(a + b)} \tag{2.78}$$

where v is the velocity of the wave in the cable, and the cutoff wavelength is therefore

$$\lambda_c = \pi(a + b) \tag{2.79}$$

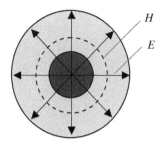

Figure 2.35   Field distribution within a coaxial line

If the operational frequency is above this cutoff frequency (or say the operational wavelength is below the cutoff wavelength), the field within the coaxial cable may no longer be TEM mode, and some higher modes such as TE_{11} (a transverse electric field, the magnetic field has non-transverse component) mode may exist which is not a desirable situation since the loss could be significantly increased.

2.4.2.3   Loss

Since the transmission line lumped parameters are given by Equation (2.76), the attenuation constant α can be calculated using Equation (2.14) or (2.23) for a low-loss line, that is

$$\alpha \approx \frac{R}{2Z_0} + \frac{GZ_0}{2} = \frac{\sqrt{f\mu\varepsilon_r}(1/a + 1/b)}{120\sqrt{\pi\sigma_2}\ln{(b/a)}} + \frac{60\pi\sigma_1}{\sqrt{\varepsilon_r}} \tag{2.80}$$

which is a function of the coax dimensions a and b as well as the conductivity and permittivity of the materials. When $b/a \approx 3.592$ (which means that the typical characteristic impedance should be around 77 Ω), the attenuation reaches the minimum. This is one of the most important considerations when people make the cable.

In addition to the characteristic impedance, mode, and loss of a cable, there are some other considerations when choosing a transmission line. Power-handling capacity is one of them since it is very important for radar and high-power applications. The breakdown electric field strength in air is about 30 kV/cm (this means the best characteristic impedance should be close to 30 Ω). A list of some commercial cables with some important specifications is given in Table 2.3, where O.D. stands for the outer diameter and V_{max} is the maximum voltage which may be applied to the cable in Volts. More information can be found from reference [5]. It is clear that there are over 100 industry standard cables on the market with various specifications. The one with the smallest loss at 400 MHz is RG-211A, about 2.3 dB/100 ft (or 7.5 dB/100 m). It is also one of the most expensive cables – normally the cost is inversely proportional to the loss of the cable.

Here, we have discussed two very important issues: one is to minimize the cable loss and the other is to maximise its power-handling capacity. The characteristic impedance of a standard cable is set to be 50 Ω because it is a good compromise by considering these two aspects. The standard cable with the characteristic impedance of 75 Ω is aimed at minimizing the loss without bothering the power-handling capacity, i.e. it is not aimed for high-power applications.

Table 2.3 Some commercial coaxial cables and their specifications

Type (/U)	MIL-W-17	Z0 (Ω)	Dielectric Type	Capacitance (pF/ft)	O. D.(in.)	dB/100 ft@400 MHz	V_{max}(rms)	Shield
RG-6A	/2-RG6	75.0	PE	20.6	0.332	6.5	2,700	Braid
RG-8		52.0	PE	29.6	0.405	6.0	4,000	Braid
RG-8A		52.0	PE	29.6	0.405	6.0	5,000	Braid
RG-9		51.0	PE	30.2	0.420	5.9	4,000	Braid
RG-11A	/6-RG11	75.0	PE	20.6	0.405	5.2	5,000	Braid
RG-55B		53.5	PE	28.8	0.200	11.7	1,900	Braid
RG-58A	/28-RG58	52.0	PE	29.6	0.195	13.2	1,900	Braid
RG-58C	/28-RG58	50.0	PE	30.8	0.195	14.0	1,900	Braid
RG-59/A	/29-RG59	73.0	PE	21.1	0.242	10.5	2,300	Braid
RG-59B	/29-RG59	75.0	PE	20.6	0.242	9.0	2,300	Braid
RG-141/A		50.0	ST	29.4	0.190	9.0	1,900	Braid
RG-142/A/B	/60-RG142	50.0	ST	29.4	0.195	9.0	1,900	Braid
RG-164	/64-RG164	75.0	PE	20.6	0.870	2.8	10,000	Braid
RG-174		50.0	ST		0.100	17.3	1,200	Braid
RG-177	/67-RG177	50.0	PE	30.8	0.895	2.8	11,000	Braid
RG-178/A/B	/93-RG178	50.0	ST	29.4	0.072	29.0	1,000	Braid
RG-180A/B	/95-RG180	95.0	ST	15.4	0.140	17.0	1,500	Braid
RG-188		50.0	ST		0.050	17.5	700	Braid
RG-211/A	/72-RG211	50.0	ST	29.4	0.730	2.3	7,000	Braid
RG-223	/84-RG223	50.0	PE	19.8	0.211	8.8	1,900	Dbl Braid
RG-316	/113-RG316	50.0	ST	29.4	0.102	20.0	1,200	Braid
RG-393	/127-RG393	50.0	ST	29.4	0.390	5.0	5,000	Braid
RG-400	/128-RG400	50.0	ST	29.4	0.195	9.6	1,900	Braid
RG-401	/129-RG401	50.0	ST	29.4	0.250	4.6	3,000	Cu. S-R
RG-402	/130-RG402	50.0	ST	29.4	0.141	7.2	2,500	Cu. S-R
RG-403	/131-RG403	50.0	ST	29.4	0.116	29.0	2,500	Braid
RG-405	/133-RG405	50.0	ST	29.4	0.086	13.0	1,500	Cu. S-R

Note: PE: Polyethylene, ST: Solid Teflon.
Source: From RF Café [5], Kirt Blattenberger, 2008.

2.4.3 Microstrip Line

As shown in Figure 2.36, a microstrip line may be viewed as a derivative of a two-wire transmission line and is perhaps the most widely used form of planar transmission line. One side of the structure is freely accessible for the mounting of packaged devices and the geometry lends itself extremely suitable to printed circuit board (PCB) patterning techniques to define the circuit. It has been used extensively in microwave and millimeter circuits and systems.

Due to the complexity of the structure, the analytical expressions of per unit length parameters are difficult to obtain. The effective relative permittivity is approximated as:

$$\varepsilon_{re} \approx \frac{\varepsilon_r + 1}{2} + \frac{\varepsilon_r - 1}{2\sqrt{1 + 12d/W}} \tag{2.81}$$

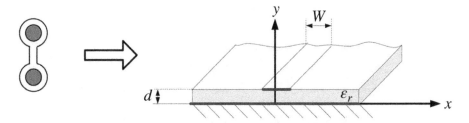

Figure 2.36 Microstrip line

This is an empirical expression and is a function of the material property and the ratio W/d. W is the width of the strip and d is the thickness of the substrate which has a relative permittivity ε_r.

2.4.3.1 Characteristic Impedance

The calculation of the characteristic impedance is not an easy task. From the transmission line theory, the relation between the velocity and per unit length inductance and capacitance is

$$v = \frac{1}{\sqrt{LC}} = \frac{c}{\sqrt{\varepsilon_{re}}}$$

Using Equation (2.18), the characteristic impedance can be expressed as:

$$Z_0 = \sqrt{\frac{L}{C}} = \frac{1}{vC} = \frac{\sqrt{\varepsilon_{re}}}{cC} \tag{2.82}$$

Thus, to compute the characteristic impedance, we just need to obtain the per unit length capacitance C once the effective permittivity is known. This approach has made a difficult task slightly easier. When the thickness of the metal strip can be neglected, it has been found that [2]:

a. when $W/d < 1$, the characteristic impedance of the line is

$$Z_0 = \frac{60}{\sqrt{\varepsilon_r}} \ln \left(\frac{8d}{W} + \frac{W}{4d} \right) > \frac{126}{\sqrt{\varepsilon_r}} \tag{2.83}$$

It decreases monotonically to $126/\sqrt{\varepsilon_r}$ as W/d increases to 1.

b. when $W/d > 1$, the characteristic impedance of the line is

$$Z_0 = \frac{120\pi}{\sqrt{\varepsilon_r}(W/d + 1.393 + 0.667 \ln (W/d + 1.44))} < \frac{126}{\sqrt{\varepsilon_r}} \tag{2.84}$$

It also decreases monotonically from $126/\sqrt{\varepsilon_r}$ as W/d increases. That is, the larger the ratio W/d, the smaller the characteristic impedance; also, the larger the permittivity, the smaller the characteristic impedance. Practical limitations exist on the range of impedances that can be manufactured. These limits depend on factors such as the dielectric constant, substrate height, and manufacturing capability. In general, the thinnest line that can be routinely etched with a

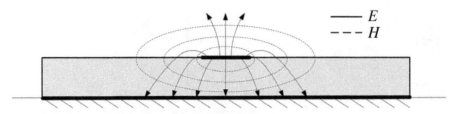

Figure 2.37 The field distribution of a microstrip

good photolithographic process is of the order of 0.1 mm. This then puts the upper bound of the impedance to 90–120 Ω. The lower bound is determined by the line width which should not be comparable to a wavelength. The typical value of the characteristic impedance for industrial standard lines is 50 or 75 Ω.

2.4.3.2 Fundamental Mode

The electromagnetic field distribution around the microstrip line is illustrated by Figure 2.37. Both the electric field and magnetic field are seen to be within the transverse plane and orthogonal to each other. But half of the wave is traveling in free space which is faster than the other half-wave traveling in the substrate, thus this field is called *quasi-TEM mode*, a sort of TEM mode.

A result of a microstrip line being an open structure is that circuits are subject to radiation. This does not mean that they are dangerous to get close to, but the performance of a device or circuit may be affected. This is a direct consequence of the "unterminated" field lines illustrated in Figure 2.37. In reality, the field lines do not just hang in free space but terminate on whatever is close to the line. The exact relations concerning radiation from a microstrip are complicated but in general narrow lines radiate less.

The first higher mode in a microstrip line is the transverse electric TE_{10} mode, and its cutoff wavelength is twice of the strip width. After taking the material and fringing effects into account, the cutoff frequency can be expressed approximately as:

$$\lambda_c \approx \sqrt{\varepsilon_r}(2W + 0.8d) \tag{2.85}$$

However, the mode analysis of a microstrip is actually more complicated than this. In addition to the conventional higher-order modes, surface modes may exist. The surface mode does not need the metal strip; it only needs the ground plane and the substrate. The lowest mode transverse electric mode is TE_1 mode and its cutoff frequency is

$$(f_c)_{TE_1} = \frac{3c\sqrt{2}}{8d\sqrt{\varepsilon_r - 1}} \tag{2.86}$$

The lowest mode transverse magnetic mode is TM_0 mode and its cutoff frequency is

$$(f_c)_{TM_0} = \frac{c\sqrt{2}}{4d\sqrt{\varepsilon_r - 1}} \tag{2.87}$$

Obviously, $(f_c)_{TE_1} = 1.5 \cdot (f_c)_{TM_0} > (f_c)_{TM_0}$, thus, in order to keep the quasi-TEM mode propagation, the operational frequency of a microstrip line should be smaller than the cutoff frequency of TE_{10} model in (2.85) and the cutoff frequency of TM_0 mode in (2.87). Higher-order modes will cause significant power loss via conductive loss and radiation loss. Surface mode may transmit the power to any direction which is of course not desirable.

2.4.3.3 Loss

The loss of a microstrip line comes from the conductor loss and dielectric substrate loss. The radiation loss is negligible at low frequencies. For most microstrip lines, conductor loss is much more significant than dielectric loss. The attenuation constant can be calculated approximately by:

$$\alpha_c = \frac{R_s}{Z_0 W} \qquad (2.88)$$

where $R_s = \sqrt{\omega\mu/2\sigma}$ is the *surface resistance* of the conductor.

A summary of some of the common substrates is given in Table 2.4. The first five are hard substrates and the rest are considered as soft substrates.

Table 2.4 Some common substrates for microstrip at 10 GHz

Substrate	ε_r	Loss tangent	Comments
Alumina (Al_2O_3)	9.8	0.0004	Low loss and cost, stable, and difficult to machine, but very hard wearing
$LaAlO_3$	24	0.0001	Low loss, but very expensive
MgO	9.8	0.00001	Very low loss, but expensive and fragile
Quartz (SiO_2)	3.8	0.0004	Low loss and low permittivity, good for mm range; fragile
Sapphire (Al_2O_3)	9.4 and 10.8	0.00002	Very low loss, single crystal and anisotropic material, and expensive
Epoxy (FR4)	4.43@ 1 GHz	0.01	Relatively high loss and low cost, popular PCB, up to ~2 GHz
FR2 (flame resistant 2)	4.5@ 1 MHz	0.025	Similar to FR4, cheap, but recommend above ~1 GHz
GaAs	13.0	0.0006	Low loss, not cheap, widely used for MMICs
LCP	3.1	0.002	Medium loss, low permittivity, up to 40 GHz. Cheap
PTFE (Teflon)	2.1	0.0004	Low loss, medium cost, and low permittivity
PTFE-glass	2.1–2.55	~ 0.001	Medium loss and cost, low permittivity
PTFE-ceramic	10.2	0.002	Medium loss, high permittivity soft substrate. But not cheap
RT/Duroid 5870	2.33	0.0012	Medium loss, low cost, and low permittivity, up to 40 GHz
RT/Duroid 5880	2.22	0.0009	Low loss, low cost, and low permittivity, for up to 77 GHz
RT/Duroid 6002, 6202	2.94	0.0012	Medium loss, low cost, and low permittivity
RT/Duroid 6006,	6.15	0.0019	Medium loss, low cost, and medium permittivity
RT/Duroid 6010,	10.2	0.0023	Medium loss, low cost, and medium permittivity

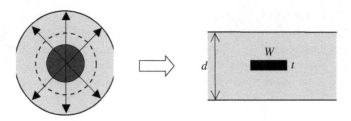

Figure 2.38   From a coaxial cable to a stripline

2.4.4   Stripline

A stripline is a conductor sandwiched by a dielectric between a pair of ground planes. It may be viewed as an evolved structure from a coaxial cable as shown in Figure 2.38. In practice, stripline is usually made by etching circuitry on a substrate that has a ground plane on the opposite face, then adding a second substrate (which is metallized on only one surface) on top to achieve the second ground plane. The stripline is often considered as a "soft-board" technology, but using low-temperature co-fired ceramics (LTCCs), ceramic stripline circuits are also possible.

Unlike a microstrip line, the stripline is basically an enclosed structure; the field is not affected by nearby components. The effective permittivity is the same as the substrate permittivity.

There are many advantages of using striplines. Whatever circuits are on a microstrip (which is a quasi-TEM mode structure), you can do better using a stripline, unless you run into fabrication or size constraints. Stripline filters and couplers always offer better bandwidth than their counterparts in a microstrip. Another advantage of the stripline is that fantastic isolation between adjacent traces can be achieved (as opposed to microstrip). The best isolation results when a picket fence of vias surrounds each transmission line, spaced at less than 1/4 wavelength. The stripline can be used to route RF signals across each other quite easily, when an offset stripline (i.e. the center conductor is not right at the middle between the two ground planes) is used.

There are two major disadvantages of a stripline:

- It is much harder and more expensive to fabricate than the microstrip. Lumped elements and active components either have to be buried between the ground planes (not as convenient as a microstrip), or transitions to the microstrip must be employed as needed to get the components onto the top of the board.
- Because of the second ground plane, the strip width is much narrower for given impedance (such as 50 Ω) and the board is thicker than that for a microstrip. A common reaction to problems with microstrip circuits is to convert them to a stripline; this may result in a much larger thickness for the same loss of the transmission line.

2.4.4.1   Characteristic Impedance

The characteristic impedance can be calculated approximately by:

$$Z_0 = \frac{30\pi}{\sqrt{\varepsilon_r}\left[\dfrac{W}{d-t} + A\right]} \qquad (2.89)$$

where

$$A = \left(2B\ln\left(B+1\right) - \left(B-1\right)\ln\left(B^2-1\right)\right)/\pi$$
$$B = 1/\sqrt{1-t/d}$$

The impedance is sensitive to the thickness of the center conductor. The typical value for industrial standard lines is 50 or 75 Ω.

2.4.4.2 Fundamental Mode

Just as the field within a coaxial cable, the field in a stripline is *TEM mode*. This means that it is non-dispersive, and the velocity is not changed with frequency.

It is also possible to generate higher-order modes if the operational frequency is above the lowest *cutoff frequency*. The smallest wavelength should meet the following condition to avoid higher-order modes:

$$\lambda_{\min} > 2d\sqrt{\varepsilon_r}$$
$$\lambda_{\min} > 2W\sqrt{\varepsilon_r}$$

$$(2.90)$$

2.4.4.3 Loss

The loss characteristics of the stripline are similar to the microstrip but have little loss due to radiation, as the structure is almost screened.

2.4.5 *Coplanar Waveguide (CPW)*

The CPW is another popular planar transmission line. Just like a stripline, it may be considered as a structure evolved from a coaxial cable as shown in Figure 2.39. This structure can also be viewed as a co-planar strip line. The center conductor is separated from a pair of ground planes. They all sit on a substrate with a dielectric permittivity of ε. In the ideal case, the thickness of the dielectric is infinite; in practice, it is just thick enough so that EM fields die out before they get out of the substrate. A variant of CPW is formed when a ground plane is provided on the opposite side of the dielectric, which is called grounded coplanar waveguide (GCPW) and it was originally developed to counter the power dissipation problems of CPW.

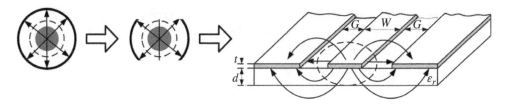

Figure 2.39 Evolution from a coaxial cable to CPW (G for gap, W for width, and d for substrate height)

The CPW offers many advantages which include at least the following four:

- It is easy for fabrication and circuit integration. Circuit components can be easily mounted on top of the line (even easier than on microstrip) due to the fact that both the conductor and ground plane are on the same side of the substrate. Unlike the microstrip and stripline, no vias are required.
- It can work to extremely high frequencies (100 GHz). Connecting to a CPW does not entail any parasitic discontinuities in the ground plane.
- Good circuit isolation can be achieved using a CPW, because there are always RF grounds between traces. Many examples of high-isolation RF switches have used a grounded CPW to get 60 dB isolation or more.
- The characteristic impedance can be kept as a constant as the signal conductor's width is tapered down/up to meet a pin. This is perfect for matching to a component pin width without changing the substrate thickness!

One disadvantage is potentially lousy heat dissipation – this depends on the thickness of the line and whether it makes contact to a heat sink. In addition, in terms of the circuit size, the CPW is at a disadvantage versus a stripline or microstrip circuit, because its effective dielectric constant is lower (half of the fields are in air). CPW circuits can be lossier than comparable microstrip circuits, if a compact layout is required.

2.4.5.1 Characteristic Impedance

The design formulas for a CPW are very complicated. There are four geometric parameters: gap G, conductor width W and thickness t, and substrate thickness d. It is not possible to obtain an accurate analytical expression of the characteristic impedance. Some approximations have to be made. If the conductor thickness is neglected, the effective permittivity is given approximately by [2]:

$$\varepsilon_{re} = \frac{\varepsilon_r + 1}{2}\{\tanh[0.775\ln(d/G) + 1.75] + \frac{kG}{d}[0.04 - 0.7k + 0.01(1 - 0.1\varepsilon_r)(0.25 + k)]\}$$

(2.91)

where

$$k = \frac{W}{W + 2G}$$

(2.92)

The effective dielectric constant of a CPW is very close to the average dielectric constant of the substrate and free space. One way to think about this is that half of the electric field lines are in free space, and half are in the dielectric.

The characteristic impedance is

$$Z_0 = \frac{30\pi}{\sqrt{\varepsilon_{re}}}\frac{K'(k)}{K(k)}$$

(2.93)

where $K(k)$ is a complete elliptic function of the first kind. We have

$$k' = \sqrt{1 - k^2}; \quad K'(k) = K(k') \tag{2.94}$$

and

$$\frac{K'(k)}{K(k)} = \begin{cases} \left[\dfrac{1}{\pi} \ln \left(2 \dfrac{1 + \sqrt{k'}}{1 - \sqrt{k'}} \right) \right] & \text{if} \quad 0 < k < 0.707 \\[4mm] \left[\dfrac{1}{\pi} \ln \left(2 \dfrac{1 + \sqrt{k}}{1 - \sqrt{k}} \right) \right]^{-1} & \text{if} \quad 0.707 < k < 1 \end{cases} \tag{2.95}$$

Again, the typical impedance value for industrial standard lines is 50 or 75 Ω.

2.4.5.2 Fundamental Mode

The EM field distribution around the CPW is illustrated in Figure 2.39 which is similar to that of a microstrip. The wave velocity in the air is faster than that in the substrate, thus the fundamental field of CPW is *quasi-TEM mode*.

Higher-order modes and surface modes may be generated in a CPW just as in a microstrip line. Thus, ground straps (bounding wires) are normally needed to tie the two grounds together in a CPW. These are especially important around any discontinuity, such as a tee junction. Care has to be taken since the bounding wires themselves could be the cause of discontinuity!

2.4.5.3 Loss

The current on the CPW is concentrated around the signal conductor. The current on the ground planes is also very focused in a small area, which results in a relatively high conductor loss as well as a heat dissipation problem. Generally speaking, the CPW exhibits a higher loss than its microstrip counterpart.

2.4.6 *Waveguide*

This is a very special and unique electromagnetic transmission line. Unlike any other transmission lines, a waveguide consists of just one piece of metal which is tubular, usually with a circular or rectangular cross section. A rectangular waveguide is shown in Figure 2.40. Due to the boundary conditions that the electric and magnetic fields have to satisfy, there are many possible wave patterns which are called transverse electric (m, n) modes (TE_{mn} modes) and transverse magnetic (m, n) modes (TM_{mn} modes). m and n represent the number of peaks (half-wavelength) along the x and y axes, respectively. Some selected mode patterns are shown in Figure 2.41 where the solid curves are for the electric fields, while broken lines represent the magnetic fields (magnetic field lines always form loops in 3D as we discussed earlier). For example, TE_{10} mode means that the electric field is within the transverse plane and there is no electric field component along the propagation direction, while the magnetic field is not confined to the transverse plane, and the electric field changes along the x-axis having one peak

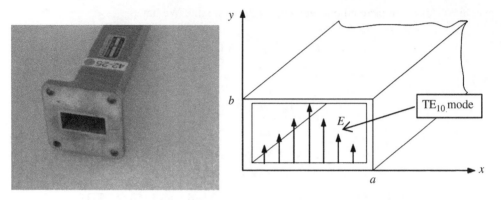

Figure 2.40   Rectangular waveguide. Source: Modified from Pozar [2]

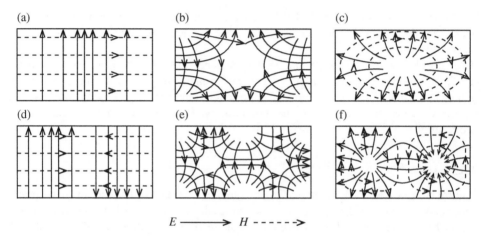

Figure 2.41   Rectangular waveguide mode patterns. (a) TE_{01}, (b) TE_{11}, (c) TM_{11}, (d) TE_{02}, (e) TE_{12}, and (f) TM_{12}

but it has no changes along the y-axis. Which modes will actually be generated inside a waveguide depend on the frequency and excitation. To generate TE_{10} mode, a probe along y-axis is normally placed in the middle of the waveguide.

If the frequency is below the cutoff frequency of the fundamental mode (which is TE_{10} mode for standard rectangular waveguides), no propagation mode can be generated. Thus, the operational frequency should be greater than the cutoff frequency which means that the waveguide can be considered as a high-pass filter. Any signals below the cutoff frequency will be filtered out by the waveguide. The larger the waveguide, the lower the cutoff frequency. Since large waveguides are heavy and expensive, they are not attractive for applications. Thus, waveguides are only used for microwave and millimeter wave frequency bands.

The main advantages of the waveguide are low loss and high power-handling capacities which are very important for high-power applications such as radars.

2.4.6.1 Fundamental Mode

The fundamental mode of a standard rectangular waveguide is TE_{10} mode. The field pattern is illustrated in Figure 2.40 (the field patterns for some higher modes can be found from references [2, 6]). The width and height of the waveguide are a and b, respectively. The electric field can be expressed as:

$$E_y = E_0 \sin\left(\frac{\pi}{a}x\right)e^{\,j(\omega t - \beta z)}$$
$$E_x = E_z = 0$$
(2.96)

and the magnetic field is given by:

$$H_x = H_1 \sin\left(\frac{\pi}{a}x\right)e^{\,j(\omega t - \beta z)}$$
$$H_y = 0$$
$$H_z = H_2 \cos\left(\frac{\pi}{a}x\right)e^{\,j(\omega t - \beta z)}$$
(2.97)

The electric field is indeed within the transverse plane and has one maximum field at $x = a/2$ (that means the mode index $m = 1$) while the magnetic field has two components, one is along the propagation direction z and the other is along direction x. They are orthogonal to the electric field. Both the electric and magnetic fields are not functions of y, i.e. the mode index $n = 0$.

2.4.6.2 Cutoff Frequency, Waveguide Wavelength, and Characteristic Impedance

The cutoff wavelength for TE_{mn} and TM_{mn} modes is given by:

$$\lambda_c = \frac{2}{\sqrt{\left(\frac{m}{a}\right)^2 + \left(\frac{n}{b}\right)^2}}$$
(2.98)

and its corresponding cutoff frequency is

$$f_c = \frac{1}{2\sqrt{\varepsilon\mu}}\sqrt{\left(\frac{m}{a}\right)^2 + \left(\frac{n}{b}\right)^2}$$
(2.99)

Thus, for TE_{10} mode, the cutoff wavelength is $\lambda_c = 2a$. This means that the waveguide works for the operational wavelength $\lambda < 2a$.

Since the next highest mode to TE_{10} is TE_{20} (for a standard waveguide, $2b < a$, i.e. TE_{01} is a higher mode than TE_{20}) and its cutoff wavelength is $\lambda_c = a$, this waveguide is therefore only suitable for the operational wavelength between these two cutoff wavelengths, i.e. $a < \lambda < 2a$. Outside the bounds, the frequency is either too low for transmission or too high to keep single-mode transmission. Higher modes are not desirable since they have higher loss and the field pattern may be changed over the transmission. Thus, each industrial standard waveguide is only suitable for a certain frequency range. A list of some selected standard waveguides with their

Table 2.5 Standard waveguides [7]

Waveguide	Freq. (GHz)	ID of a (mm)	ID of b (mm)	Freq. band
WR-137	5.85–8.2	34.85	15.80	C band
WR-112	7.05–10.0	28.50	12.60	H band
WR-90	8.2–12.4	22.86	10.16	X band
WR-62	12.4–18.0	15.80	7.90	Ku band
WR-51	15.0–22.0	12.96	6.48	K band
WR-42	18.0–26.5	10.67	4.32	K band
WR-28	26.5–40.0	7.11	3.36	Ka band
WR-19	40.0–60.0	4.78	2.39	U band
WR-12	60.0–90.0	3.10	1.55	E band
WR-8	90.0–140.0	2.032	1.016	F band

Source: Modified from Waveguide Frequency Bands with Interior Dimensions, Millimeter Wave Products Inc., 2021.

suitable frequency range, inside dimensions (ID), and frequency band is given in Table 2.5 where the frequency band designation for the waveguide used in the industry is not quite the same as the IEEE frequency band in Table 1.1 as mentioned earlier.

The field inside a waveguide exhibits periodic feature and the period is one *waveguide wavelength* which is actually different from (longer than) the free space wavelength. The waveguide wavelength can be calculated using:

$$\lambda_g = \lambda / \sqrt{1 - \left(\frac{\lambda}{\lambda_c}\right)^2} \tag{2.100}$$

It is determined by the free space wavelength λ and the cutoff frequency λ_c.

The characteristic impedance is also mode dependant. For TE_{10} mode, it is

$$Z_{TE10} = 120\pi / \sqrt{1 - \left(\frac{\lambda}{2a}\right)} \tag{2.101}$$

which is not a constant but a function of frequency. This is one of the reasons that the single mode is preferred for waveguide applications.

2.4.7 New Transmission Lines (SIW, Gap Waveguide, and MSTL) and Comparisons

Six most popular transmission lines have been introduced and discussed above. The characteristic impedance, fundamental mode, and loss characteristics have been presented. A brief summary of these established transmission lines is provided in Table 2.6. We can see that most of them are suitable up to 10s GHz. Waveguides can be used for higher frequencies but are not suitable for integration. Generally speaking, at higher frequencies, the attenuation or higher modes may become a problem. The current trend of wireless communications is that the

Table 2.6 Summary of various popular transmission lines

	Two-wire	Coax	Micro-strip	Stripline	CPW	Waveguide
Basic mode	TEM	TEM	Quasi-TEM	TEM	Quasi-TEM	TE_{10}
Frequencies	DC – 100 MHz	DC – 10s GHz	DC – up to 100 GHz	DC – 10s GHz	DC – up to 100 GHz	>Cutoff freq.
Loss	High	Med.	Med.	Med.	Med.	Low
Cost	Low	Low/med.	Med.	Med.	Med.	High
Ease of integration	Med.	Hard	Easy	Med.	Easy	Hard
Application	Low freq. short distance	General purpose	PCB circuit and MMIC	RF circuit and MMIC	RF circuit and MMIC	High power and high freq.

operational frequency is going up to millimeter wave (mm-wave) and even terahertz (THz) frequencies in order to meet higher data rate and capacity requirements. These transmission lines are not ideal for higher frequencies, thus alternative transmission lines are required.

Optical fiber is a good for much higher frequencies (optical frequencies) but not suitable for mm-wave and THz frequencies. Over the past decade or so, some new transmission lines have been proposed, studied, and developed. The most noticeable are *substrate-integrated waveguide, gap waveguide,* and *mode-selective transmission line.*

2.4.7.1 Substrate-Integrated Waveguide

A substrate-integrated waveguide (SIW) is fabricated using two rows of conducting cylinders (vias) or slots embedded in a dielectric substrate that electrically connect two parallel metal plates [8, 9] as shown in Figure 2.42. In this way, the nonplanar rectangular waveguide can be made in planar form, compatible with existing planar processing techniques (e.g. standard PCB or LTCC technology). For such a structure, its propagation characteristics are similar to a classical rectangular waveguide, that is, it has low loss, high quality factor, and high power-handling capability. However, unlike a conventional waveguide, an SIW is easy to

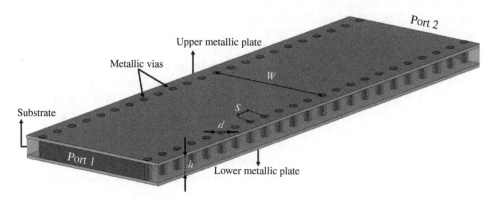

Figure 2.42 Substrate-integrated waveguide. Source: Modified from Bozzi et al. [9]

integrate components (including passive and active components, and even antennas) on the same substrate. It has therefore attracted a lot of attention since its introduction in the late 1990s. Many researchers have conducted original and innovative research in this area.

Experiments and simulations have proven that dispersion characteristics of the SIW are the same as those of its equivalent rectangular waveguide. The equivalent width of the SIW, W_e, is between W and $(W - d)$ and with a very good approximate equation [8]:

$$W_e = W - 1.08 \frac{d^2}{s} + 0.1 \frac{d^2}{W} \tag{2.102}$$

where d is the diameter of the metal vias, W represents their transverse spacing, and s is their longitudinal spacing.

Waveguide theory can be applied to the SIW, but it only supports TE_{m0} modes, not TM modes. Thus, its fundamental mode is TE_{10} mode and the desired operational frequency f should be greater than the cutoff frequency of TE_{10} mode but smaller than the cutoff frequency of TE_{20} mode. Using Equation (2.99), we have

$$f_{c\_TE10} = \frac{1}{2W_e\sqrt{\varepsilon\mu}} < f < f_{c\_TE20} = \frac{1}{W_e\sqrt{\varepsilon\mu}} \tag{2.103}$$

This means that an SIW is a band-limited transmission line whose fractional bandwidth is typically less than 66%. Furthermore, it is a periodic structure which is subject to electromagnetic bandgaps or stopband effects. Thus, one must ensure that there is no bandgap over the waveguide bandwidth of interest.

A key issue in the design of SIW structures is how to minimize the loss, which is particularly critical when operating at higher frequencies. Three major mechanisms of loss need to be considered: conductor losses (due to the finite conductivity of metal walls), dielectric losses (due to the lossy dielectric material), and possible radiation losses (due to the energy leakage through the gaps). The behavior of conductor and dielectric losses is similar to the corresponding losses in rectangular waveguides filled with a dielectric medium, and the classical equations can be effectively applied. It transpires that conductor losses can be significantly reduced by increasing the substrate thickness, being the corresponding attenuation constant almost proportional to the inverse of substrate thickness h. The other geometrical dimensions of the SIW exhibit a negligible effect on conductor losses. Conversely, dielectric losses depend only on the dielectric material and not on the geometry of the SIW, and therefore they can be reduced only using a better dielectric substrate. Finally, radiation losses can be kept reasonably small if $s/d < 2.5$, with $s/d = 2$ being the recommended value [9]. In fact, when the spacing s is small and the diameter d of the metal vias is large, the gap between the metal vias is small, thus approaching the condition of continuous metal wall and minimizing the radiation leakage. Generally speaking, the contribution of dielectric losses is predominant at mm-wave frequencies, when using a typical substrate thickness and commercial dielectric material.

A systematic comparison of SIW and microstrip components is not easy, because SIW circuits are usually implemented on a thick substrate with low dielectric constant (which is not suitable for the implementation of microstrip circuits), with the aim of minimizing conductor losses. In principle, microstrip component losses could also be mitigated by increasing the

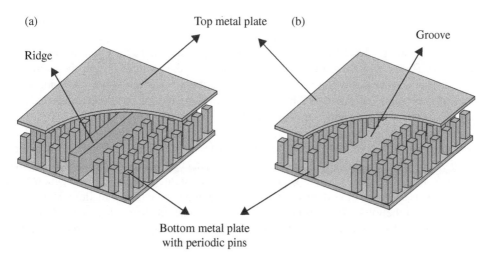

(a)   Top metal plate   (b)

Ridge

Groove

Bottom metal plate
with periodic pins

Figure 2.43 Two basic gap waveguides. Source: Based on Kildal et al. [10]

substrate thickness; in practice, however, this cannot be exploited due to the unacceptable increase in radiation loss and excitation of surface waves. It was reported that SIW structures can guarantee comparable or lower losses, compared to traditional planar transmission lines. There are many SIW-based passive and active devices and antennas developed. The typical frequency is from around 30 to 60 GHz. The highest frequency reported was 180 GHz [9]. It is envisaged that the SIW will become a standard transmission line for mm-wave applications.

2.4.7.2 Gap Waveguide

The gap waveguide was introduced to be an alternative to conventional hollow waveguide for high-frequency applications [10]. In Figure 2.43, two basic gap waveguides are presented: one has a ridge and the other has a groove to guide the wave propagation. Generally speaking, a gap waveguide is formed by two parallel metal plates separated by a small gap. There is a texture with periodic pins (called a bed of nails) or multilayer structure on one of the plates to act as a high-impedance surface (magnetic wall or perfect magnetic conductor), so as to create a cutoff condition. For that, the gap between the two plates must be less than a quarter-wavelength, while the height of the pins/nails is a quarter-wavelength at the central frequency. Unlike a microstrip or SIW whose dielectric loss at higher frequencies could be significant, the gap waveguide is made of all-metallic structure and can exhibit smaller insertion loss. Furthermore, it is considered easier to fabricate than the conventional hollow waveguide. Thus, the gap waveguide is suitable for applications above 30 GHz that include slot antennas, feeding networks for slot arrays, standard components like waveguides, filters or couplers, and cavity-mode suppression for microstrip circuits (for solving packaging problems).

It is important to note that, like the conventional waveguide, the gap waveguide is also just suitable for a frequency band determined by the fundamental mode. The two parallel surfaces forming the gap can have metal connection to each other at some distance from the gap waveguide without affecting its performance significantly. This is a vital mechanical advantage, as one of the surfaces can be made with a solid metal wall that provides support for the other surface to ensure a well-defined gap height everywhere. Thereby, the whole gap waveguide may be completely encapsulated by metal, providing strong shielding to the exterior (as a waveguide) [10]. However, unlike a conventional waveguide, the requirement on the contact of the two metal surfaces of a gap waveguide is less critical since waves are not going to be radiated/leaked from the gap. The fabrication of gap waveguide is therefore less demanding than the conventional waveguide in this regard.

There was a comparison of insertion losses of different transmission lines in reference [11]. It showed that the gap waveguide had a similar loss as the conventional rectangular waveguide but indicated a slightly smaller loss than SIW and much smaller loss than microstrip around 50 GHz. Some researchers have shown that it was easier to integrate the gap waveguide with many antennas and mm-wave devices than a conventional waveguide. This flexible mechanical feature and the low-loss performance make gap waveguide technology one of the very good candidates for future antenna systems and RF front ends in the millimeter-wave frequency range.

2.4.7.3 Mode-Selective Transmission Line

A mode-selective transmission line (MSTL) is one of the latest developments in finding a better transmission line [12]. It has been demonstrated that this transmission line has low loss and low dispersion from DC to THz which make it an excellent candidate for low-loss, ultra-high-speed, and ultra-wideband electric signal transmission/propagation in modern digital systems and inter-chip connections.

An MSTL is shown in Figure 2.44 with a 3D view and two cross-sectional views of the electric field at lower and higher frequencies, respectively. This structure consists of two conductors (separated by two slots with a width s) and a dielectric layer (with permittivity ε, width d, and thickness h), and it could be considered as a combination of a microstrip line and a waveguide. At lower frequencies, as seen in Figure 2.44(b), it acts as a microstrip line and the dominant mode is a quasi-TEM mode, thus the MSTL can easily transmit DC and low-frequency signals. While at higher frequency, it acts as a rectangular waveguide as shown in Figure 2.44 (c), and the fundamental mode is TE_{10} mode. The "mode selection" is referred to the ability of the transmission line to select either quasi-TEM or TE_{10} mode according to the operational frequency. The cutoff frequency for the TE_{10} mode is

$$f_{c\_TE10} = \frac{1}{2d\sqrt{\varepsilon\mu}} \tag{2.104}$$

This cutoff frequency is determined by the material properties and the width of the line and can be considered as the transition frequency from quasi-TEM to TE_{10} mode. If the operational frequency is below this cutoff frequency, the MSTL runs quasi-TEM mode. If the operational frequency is above this cutoff frequency, the MSTL operates at TE_{10} mode.

To design the MSTL for an ultra-broadband interconnect application, the design should be made in such a way that f_c resides around the frequency where loss and dispersion of quasi-TEM mode of operation are not negligible and may deteriorate the circuit performance. Thus, other parameters, such as the height h and line width W, should be carefully chosen. In-depth study on MSTL can be found from reference [12].

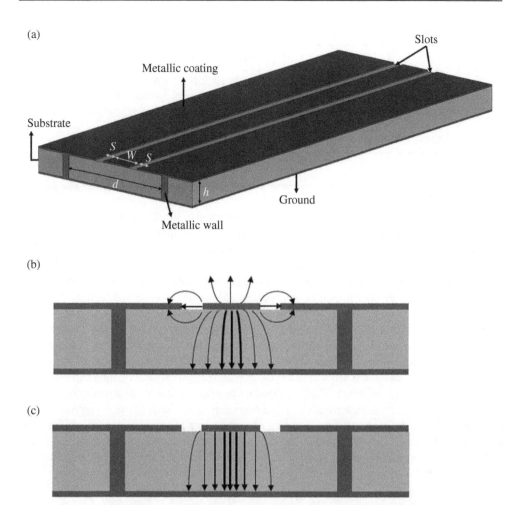

Figure 2.44   Mode-selective transmission line: (a) the 3D configuration; (b) cross-sectional view of the electric field at lower frequencies; and (c) cross-sectional view of the electric field at higher frequencies

A comparison of the loss characteristics of some selective transmission lines for higher frequencies is given in Table 2.7. The signal attenuation of 0.7 dB/mm is obtained for microstrip line and 0.1 dB/mm for CPW at 100 GHz, respectively; these values increase to 2.2 and 2.5 dB/mm at 500 GHz. The hollow metallic waveguides have been traditionally fabricated from metals with rectangular, square, and circular cross sections. TE_{01} wave mode in circular waveguide yields a transmission loss of about 0.4 dB/mm at 120 GHz and 1 dB/mm at 500 GHz. For SIW, attenuation close to 0.04 dB/mm was reported at 94 GHz. The SIW was designed using a 76-µm-thick quartz substrate with permittivity of 3.8 and loss tangent of 0.003 at 0.9 THz. The MSTL showed a lower level of signal attenuation (0.07 dB/mm at 100 GHz and 0.3 dB/mm at 500 GHz), and its loss variation is negligible over all the frequency range covering DC unless the length increases significantly, which is not usually required in short-reach chip-to-module or medium reach chip-to-chip interface [12].

Table 2.7   Comparison of the loss of some transmission lines

Line	Attenuation dB/ mm
Microstrip	0.7 @ 100 GHz, 2.2 @500 GHz
CPW	0.1 @ 120 GHz, 2.5 @500 GHz
Circular waveguide	0.378 @ 120 GHz, 1.084 @500 GHz
SIW	0.04 @100 GHz, 0.3@900 GHz
MSTL	0.07 @100 GHz, 0.3@500 GHz

2.5   Connectors

In practice, almost all transmission lines have to be terminated to suitable connectors which make the device interconnection much easy. There are many types of industrial standard connectors. Sometimes, more than one connector is available. For example, RG58 cable can be assembled to sub-miniature version A (SMA), sub-miniature version B (SMB), baby N connector (BNC), and Type N connectors. Making the right choice can be a problem and there is a lack of information in this practical subject in other books; hence, we will address this issue in this section.

Connectors are developed as a pair: a male and a female (also called plug and jack in practice, an example is shown in Figure 2.45), although some RF connectors are sexless (such as the APC-7 and the General Radio GR874). It is very important to choose the right connector for the application, since the effects of the connector (which is an additional element and may have not been taken into account in the design) on the system performance and measurements could be quite significant. This is especially true in antenna measurements.

Choosing a proper connector requires an understanding of its characteristics – both electrical and mechanical. Obviously, the physical size of a connector is an important aspect. Furthermore, multiple performance characteristics, such as frequency range and power-handling capability, must be examined when making a selection. Other important parameters include

Figure 2.45   Male/plug (left) and female/jack (right) N-type connectors

insertion loss and VSWR. Furthermore, environmental characteristics, such as operating temperature, vibration, and shock, help determine if a connector is suitable for a given application. Of course, cost plays a key role, too.

Generally speaking, coaxial connectors consist of an outer conductor contact and an inner conductor contact. These devices also must have the means to mechanically couple to another connector. Some connectors are named after the inside diameter of their outer conductor, with smaller diameters yielding higher usable frequencies. Connectors are designed with either an air or a solid dielectric. Air–dielectric types include 3.5, 2.92, and 2.4-mm connectors, among others. A good example of a connector that employs a solid dielectric is the widely used SMA connector. A solid dielectric can be implemented with either a flush or overlapping configuration; overlapping configurations are employed to prevent voltage breakdown and to handle higher power levels. Connectors can be mounted by various means. Those intended to mount onto PCBs are manufactured with either a straight or right-angle orientation. Cable-mount connectors can be attached to cables by means of either crimping or clamping. And panel-mount connectors include those with flanges, which typically have either two or four holes.

Manufacture of connectors involves various materials, each having their own set of advantages and disadvantages. Such materials are evaluated by their electrical, mechanical, and environmental properties, and they weigh heavily in terms of connector performance and reliability. Two of the more commonly used materials for building connector bodies are stainless steel and brass. Stainless steel is more durable than brass, but it comes at a higher cost. The highest-quality connectors are often manufactured with stainless-steel bodies. Connector contacts, on the other hand, are typically not made of stainless steel, which has a relatively low electrical conductivity. Rather they often consist of brass or beryllium copper.

A good summary of all RF/microwave connectors can be found in references [13, 14]. Table 2.8 is a selection of some popular connectors for antenna systems and measurements. These connectors are mainly for coaxial cables, but some of them can be used for other transmission lines. For example, SMA straight PCB-mount jack (female), tab terminal panel jack, and stub terminal panel jack are available and widely used in antenna community. For all these connectors, they have a limit on the frequency range. It is very important to select a suitable connector for your application.

The SMA connector is widely used throughout the RF/microwave industry. It originated in the late 1950s. The connector was then designated the SMA, connector. Like the Type-N connector, it has an impedance of 50 Ω and employs threaded coupling and a solid dielectric. Furthermore, while standard SMA connectors perform from DC to 18 GHz, some suppliers provide versions that can perform up to 26.5 GHz. Although they are common and inexpensive, SMA connectors have their limitations – they are rated for a very limited number of connection cycles.

The 3.5-mm connector (named after the inside diameter of its outer conductor), which can achieve mode-free performance of up to 34 GHz, first appeared in the 1970s. These connectors are known for their durability, as they were designed to allow thousands of repeatable connections.

Introduced by Wiltron (now Anritsu) as the K-connector in 1983, the 2.92-mm connector (also named after the inside diameter of its outer conductor) is available today from a wide range of suppliers. The 2.92-mm connectors can be used at higher frequencies than 3.5-mm connectors, as they offer performance of up to 40 GHz. Measurement systems and high-performance components, for example, sometimes implement 2.92-mm connectors.

The SMA connector is mechanically compatible with 3.5- and 2.92-mm connectors. Both of these employ an air dielectric and can perform at higher frequencies than their SMA counterpart.

Table 2.8 Some industry standard connectors

BNC

Baby N connector (BNC). Bayonet-style coupling for quick connect and disconnect. Available in 50, 75, and 50 Ω reverse polarity. DC – 4 GHz

K-connector

It is Anritsu's version of the 2.92-mm connector. It is compatible with SMA and 3.5-mm connectors. It is well suited to applications in components, systems, or instrumentation. DC – 40 GHz

MCX

Snap-on miniature coaxial (MCX) connector that conforms to the European CECC 22220. Since the MCX has identical inner contact and insulator dimensions as the SMB while being 30% smaller, it provides designers with options where weight and physical space are limited. DC – 6 GHz.

Mini BNC

A new generation of miniature BNC connectors that maintain the positive characteristics of our full-size BNCs for 75 Ω systems while allowing 40% more interconnects in the same area. DC – 11 GHz.

MMCX

Micro-miniature coaxial (MMCX) connector with a lock-snap mechanism allowing for 360° rotation on a printed circuit board. Conforms to the European CECC 22000 specification and comes in surface mount, edge card, and cable connectors. DC – 6 GHz.

SMA

Sub-miniature version A (SMA) connectors with a threaded coupling mechanism. Available in standard, phase adjustable, and reverse polarity. Built in accordance with MIL-C-39012 and CECC 22110/111, SMA connectors can be mated with all connectors that meet these specs regardless of manufacturer. Widely used with RG-55, 58, 141, 142, 223, 303; 122, 174, 188, 316. DC – 18 GHz.

SMB

Sub-miniature version B (SMB) connectors. Developed in the 1960s as a smaller alternative to the SMA, the SMB line features a snap-on coupling mechanism. Available in 50 Ω, 75 Ω, and miniature 75 Ω. Often used with RG-188, 196. DC – 4 GHz (usable to 10 GHz).

SMC

Sub-miniature version C (SMC) connectors. Medium-sized 50 Ω threaded connectors designed to meet MIL-C-39012 category D as generated by the US Air Force. Often used with RG-188, 196. DC – 4 GHz (usable to 10 GHz).

SMP

Used in miniaturized applications and features both push-on and snap-on mating styles. Sub-miniature connector with a frequency range of up to 40 GHz. Its micro-miniature is called **SMPM** to addresses small package design needs. It is known for its superior performance in a wide range of applications and suitable for a frequency range of DC to 65 GHz.

Table 2.8   (*Continued*)

SSMB
Scaled SMA (SSMA). Microminiature connectors with snap-on mating interface allowing quick installation in small spaces with excellent performance in devices of up to 4 GHz.

TNC
Features screw threads for mating and serves as a threaded version of the BNC connector. The TNC is a 50 Ω connector available in both standard and reverse polarity. DC – 11 GHz.

Type N
Available in standard N (coaxial cable) and corrugated N (helical and annular cable), the Type N is a durable, weatherproof, medium-sized connector. Used with RG-8, RG-58, RG-141, and RG-225. DC – 11 GHz.

UHF
Invented for use in the radio industry, UHF stands for ultra-high frequency. While at the time 300 MHz was considered as high frequency, these are now general-purpose connectors for low-frequency systems. DC – 1 GHz.

1.85 mm (V-connector)
Named after the inside diameter of its outer conductor. The outer thread size is bigger than SMA, 3.5, and 2.92. This makes the area of the outer conductor mating surface look very large compared to the relatively small air dielectric. It mates with 2.4 connectors and is suitable for up to 70 GHz

2.4 mm
Named after the inside diameter of its outer conductor. The outer thread size is bigger than SMA, 3.5, and 2.92. This makes the area of the outer conductor mating surface look very large compared to the relatively small air dielectric. It mates with 1.85 and is suitable for up to 50 GHz

2.92 mm (K-connector)
Named after the inside diameter of its outer conductor. The K-connector is Anritsu's version of the 2.92-mm connector. It mates with 3.5 and SMA and is suitable for up to 40 GHz

3.5 mm
Named after the inside diameter of its outer conductor. The 3.5-mm connector is the next upgrade from using SMA, and it mates with 2.92 and SMA and performs well up to 26 GHz.

7/16 DIN
A 50 Ω threaded RF connector used to join coaxial cable. It is among the most widely used high-power RF connectors in cellular network antenna systems. It outperforms other non-flange options, such as BNC or N connectors, when it comes to interference and intermodulation rejection or higher power handling at RF frequencies, suitable for up to 6 GHz.

There are other connectors in the same category, such as **4.3/10, 2.2/5,** and **1.0/2.3 DIN** connectors, and they are engineered for the wireless market and ideal for applications requiring low passive intermodulation (PIM).

Antenna 1: CPW-fed   Antenna 2: microstrip-fed

Figure 2.46   Wideband antennas fed by CPW and microstrip which are directly connected/soldered to SMA connectors

Even higher-frequency performance can be achieved by 2.4-, 1.85-mm (also known as the V-connector), and 1.0-mm connectors. Like 3.5- and 2.92-mm connectors, 2.4-, 1.85-, and 1.0-mm connectors employ an air dielectric. They also derive their names from the inside diameter of their respective outer conductors.

Since all RF test equipments come with coaxial connectors (Type N and SMA are popular connectors), direct connection with other forms of transmission lines (such as microstrip and CPW) would be tricky. Some adapters have been developed. For example, the industrial standard coax-to-waveguide adaptors are now widely available on the market. Figure 2.46 shows how to directly connect an SMA connector to a microstrip and CPW (feed line to an antenna) in practice – no standard adaptor is available.

2.6   Summary

This long chapter has provided a comprehensive coverage on circuit concepts and transmission lines – they are essential knowledge for antenna feeding, matching, and characterization. In summary:

- An introduction to the lumped element systems and distributed element systems has been given right at the beginning; the main idea is that the current, voltage, and impedance are all functions of the frequency and the reference position at the transmission line.
- A transmission line model has been developed to obtain the important parameters of a transmission line, which include the characteristic impedance, input impedance, attenuation constant, phase constant, and velocity. An extensive study on terminated transmission lines

has been carried out. The reflection coefficient, return loss, and VSWR have been introduced to evaluate the line impedance matching.

- The Smith Chart has been introduced as a very useful tool to analyze impedance matching. Lumped and distributed matching networks and impedance-matching techniques have also been thoroughly addressed. Furthermore, adaptive impedance matching was introduced for two practical applications.
- The bandwidth and quality factor (Q-factor) have been discussed in depth.
- Six popular transmission lines have been examined and compared in terms of their characteristic impedance, fundamental mode, loss characteristics, and frequency bandwidth. Another three of the latest transmission lines (SIW, gap waveguides, and SMTL) have been introduced for higher-frequency applications.
- Various RF/microwave cables and connectors have been presented at the end of this chapter along with their typical specifications and frequency bandwidth.

References

1. J. D. Kraus and D. A Fleisch, *Electromagnetics with Applications*, 5th edition, McGraw-Hill, Inc., 1999.
2. D. M. Pozar, *Microwave Engineering*, 4th edition, John Wiley, 2013.
3. A. van Bezooijen, M. A. de Jongh, F. van Straten, R. Mahmoudi, A. H. M. van Roe, "Adaptive impedance-matching techniques for Controlling L Networks," *IEEE Trans. Circuits Systems*, 57(2), 495–505, 2010.
4. Y. Lim, H. Tang, S. Lim, J. Park, "An adaptive impedance-matching network based on a novel capacitor matrix for wireless power," transfer. *IEEE Trans. Power Electron.* 29(8), 4403–4413, 2014.
5. RF Cafe. Coaxial cable specifications. http://www.rfcafe.com/references/electrical/coax-chart.htm
6. C. S. Lee, S. W. Lee and L. L. Chuang, "Plot of modal field distribution in rectangular and circular waveguides," *IEEE Trans. Microwave Theory Tech.*, 33(3), 271–274, 1985.
7. Mi-Wave. http://miwv.com/images/Waveguide-Chart.pdf
8. F. Xu and K. Wu, "Guided-wave and leakage characteristics of substrate integrated waveguide," *IEEE Trans. Microwave Theory Tech.*, 53(1), 66–73, 2005.
9. M. Bozzi, A. Georgiadis and K. Wu, "Review of substrate integrated waveguide circuits and antennas", *IET Microwaves Antennas Propagat.*, 5(8), 909–920, 2011.
10. P. S. Kildal, E. Alfonso, A. Valero-Nogueira and E. Rajo-Iglesias, "Local metamaterial-based waveguides in gaps between parallel metal plates," *IEEE Antennas Wireless Propagat. Lett.*, 8, 84–87, 2009.
11. A. Valero-Nogueira, M. Ferrando-Rocher and A. Zaman, "Gap waveguide technology for millimeter-wave antenna systems," *IEEE Commun. Mag.*, 56, 14–20, 2018.
12. F. Fesharaki, T. Djerafi, M. Chaker and K, Wu, "Mode-selective transmission line for chip-to-chip terabit-per-second data transmission," *IEEE Trans. Components Packaging Manufacturing Tech.*, 8(7), 1272–1281, 2018.
13. RF Cafe. http://www.rfcafe.com/references/electrical/coax-connector-usage.htm
14. https://www.amphenolrf.com/connectors.html

Problems

Q2.1. Explain the concept of the characteristic impedance of a transmission line.

Q2.2. For a low-loss transmission line, find its characteristic impedance, attenuation constant, and phase constant (or wave number). How does the frequency affect these parameters?

Q2.3. A uniform transmission line has constants $R = 500$ μΩ/m, $G = 1.5$ mS/m, $L = 0.5$ μH/m, and $C = 10$ nF/m. Find the characteristic impedance and the attenuation constant of the line at the following frequencies
   a. 50 Hz;
   b. 30 MHz;
   c. 1 GHz;
   d. 10 GHz.
   and comment on the results.

Q2.4. A coaxial transmission line has $a = 4$ mm and $b = 12$ mm. Find the characteristic impedance of the line if the dielectric is
   a. Air space polyethylene (ASP);
   b. Foam polyethylene (FE);
   c. Solid Teflon (ST).
   Hint: Use Table 2.2 for permittivity.

Q2.5. A 100 Ω resistor is connected to a good cable with characteristic impedance of 50 Ω. The attenuation constant is not zero but 0.2 Np/m at 1 GHz, and the relative permittivity of the cable dielectric is 1.5. If the cable length is 10 m. Find
   a. the reflection coefficient and return loss at the termination;
   b. the reflection coefficient and return loss at the input of the cable;
   c. VSWR at both the terminal and the input of the cable;
   d. the input impedance at the input of the cable;
   e. a method to improve the matching of the system.

Q2.6. Obtain the theoretical characteristic impedances for the lowest loss and for maximum power-handling capacity of a coaxial cable, respectively, and then use the results to justify why the most common coaxial cable impedances are 50 and 75 Ω.

Q2.7. RG-59U, a popular cable for microwave applications, has an open-circuit impedance of $130 + j75$ Ω and short-circuit impedance of $30.3 - j21.2$ Ω. Find the characteristic impedance of the line.

Q2.8. A quality transmission line is terminated in $100 + j50$ Ω. Find
   a. The voltage reflection coefficient;
   b. VSWR;
   c. The shortest length of line required to transform the impedance to a purely resistive. If 220 V is applied to the line, find the maximum and minimum line voltages.

Q2.9. Explain the concept of impedance matching and then compare lumped matching network and distributed matching network.

Q2.10. Explain what the Smith Chart is and its application.

Q2.11. A load with an impedance of $100 - j100$ Ω is to be matched with a 50 Ω transmission line. Obtain all possible solutions using L matching networks. If the operational frequency is at 1 GHz, compare the frequency bandwidth of these solutions.

Q2.12. A load with an impedance of $100 - j100$ Ω is to be matched with a 50 Ω transmission line. Design two stub matching networks and then compare their bandwidth performance.

Q2.13. Explain what Bode–Fano limits are and how they may be applied to matching network.

Q2.14. Discuss the relationship between bandwidth and quality factor. What is the major difference between the loaded Q-factor and unloaded Q-factor?

Q2.15. Design a 50 Ω microstrip line using a PCB board with PTFE (Teflon) substrate of 1-mm thickness. Find the cutoff frequency for the first higher mode in the line.

Q2.16. Design a 50-Ω CPW using a PCB board with PTFE (Teflon) substrate of 1-mm thickness.

Q2.17. Rectangular waveguides are widely used for radar applications. WR-90 standard waveguide (see Table 2.5) is mainly used for X band. Find

    a. The cutoff frequency for TE_{10} mode;

    b. The cutoff frequency for TE_{01} mode;

    c. The cutoff frequency for TE_{20} mode.

    Hence, identify the most suitable frequency range for this waveguide.

Q2.18. Compare substrate-integrated waveguide (SIW) and mode-selective transmission line (MSTL) in terms of the fundamental mode and suitable frequency band.

3

Field Concepts and Radiowaves

In this chapter, we will first see how Maxwell's equations can be used to obtain wave solutions. The concepts of the plane wave, intrinsic impedance, and polarization will then be introduced and followed by the discussion on radio propagation mechanisms and radiowave propagation characteristics in various media. A few basic radio propagation models will be introduced, and circuit concepts and field concepts will be compared at the end of this chapter. The concept of skin depth will be looked into from both the field and circuit points of views. Although the issues addressed in this chapter may not be used directly for antenna design, the knowledge will be extremely useful for gaining a better understanding of the antenna radiation characteristics as well as radiowaves – generated/received by antennas. Because antennas and radio propagation are so closely linked, some countries and universities treat antennas and radio propagation as a single subject.

3.1 Wave Equation and Solutions

As it was mentioned in Chapter 2, Maxwell's modified version of Ampere's Circuital Law enables a set of equations to be combined together to derive the electromagnetic wave equation. The derivation is relatively straightforward.

Now let us discuss a time-harmonic case with the time factor $e^{j\omega t}$ which means a single frequency and is the most common form of a wave in real life (according to Fourier's theory, more complicated cases may be decomposed to a linear combination of harmonic waves). From Maxwell's Equations (1.29) or (1.35), we have

$$\nabla \times \boldsymbol{E} = -j\omega\mu\boldsymbol{H}$$
$$\nabla \times \boldsymbol{H} = (\sigma + j\omega\varepsilon)\boldsymbol{E}$$
$$\nabla \cdot \boldsymbol{E} = \rho/\varepsilon \tag{3.1}$$
$$\nabla \cdot \boldsymbol{H} = 0$$

Take a curl operation on the first equation to yield

$$\nabla \times \nabla \times \boldsymbol{E} = \nabla(\nabla \cdot \boldsymbol{E}) - \nabla^2\boldsymbol{E} = -j\omega\mu\nabla \times \boldsymbol{H}$$

where $\nabla = \dfrac{\partial}{\partial x}\hat{\boldsymbol{x}} + \dfrac{\partial}{\partial y}\hat{\boldsymbol{y}} + \dfrac{\partial}{\partial z}\hat{\boldsymbol{z}}$, $\nabla^2 = \nabla \cdot \nabla = \dfrac{\partial^2}{\partial x^2} + \dfrac{\partial^2}{\partial y^2} + \dfrac{\partial^2}{\partial z^2}$.

Combine it with the 2nd and 3rd equations in Equation (3.1) to obtain

$$\nabla^2\boldsymbol{E} - j\omega\mu(\sigma + j\omega\varepsilon)\boldsymbol{E} = \nabla(\rho/\varepsilon) \tag{3.2}$$

Now let

$$\gamma = \sqrt{j\omega\mu(\sigma + j\omega\varepsilon)} = \alpha + j\beta \tag{3.3}$$

where α and β are the *attenuation constant* and *phase constant*, respectively. Similar definitions were introduced for a transmission line in Chapter 2. From Equation (3.3), we can represent these constants by the material properties and frequency as:

$$\alpha = \omega\sqrt{\mu\varepsilon}\left[\frac{1}{2}\left(\sqrt{1 + \frac{\sigma^2}{\varepsilon^2\omega^2}} - 1\right)\right]^{1/2}$$
$$\beta = \omega\sqrt{\mu\varepsilon}\left[\frac{1}{2}\left(\sqrt{1 + \frac{\sigma^2}{\varepsilon^2\omega^2}} + 1\right)\right]^{1/2} \tag{3.4}$$

Equation (3.2) can now be rewritten as

$$\nabla^2\boldsymbol{E} - \gamma^2\boldsymbol{E} = \nabla(\rho/\varepsilon) \tag{3.5}$$

In the source free region ($\rho = 0$), we have

$$\nabla^2\boldsymbol{E} - \gamma^2\boldsymbol{E} = 0 \tag{3.6}$$

This is called the *wave equation*. There are many possible solutions to this equation. Boundary conditions and sources are required to obtain the specific solutions. In free space, one of the solutions is

$$\boldsymbol{E} = \hat{\boldsymbol{x}}E_0 e^{j\omega t - \gamma z} = \hat{\boldsymbol{x}}E_0 e^{-\alpha z + j(\omega t - \beta z)} \tag{3.7}$$

It can be easily validated by using this representation in the wave equation. Other possible solutions include, for example

$$\boldsymbol{E} = \hat{x}E_0 e^{j\omega t + \gamma z}; \quad \boldsymbol{E} = \hat{y}E_0 e^{j\omega t + \gamma z}; \quad \boldsymbol{E} = \hat{z}E_0 e^{j\omega t \pm \gamma x}; \ldots$$

Using the electric field $\boldsymbol{E}$ in Equation (3.7) and Equation (3.1), the magnetic field $\boldsymbol{H}$ is yielded as

$$\boldsymbol{H} = \frac{j}{\omega\mu} \nabla \times \boldsymbol{E} = -\hat{y}\frac{j\gamma}{\omega\mu}E_0 e^{-\alpha z + j(\omega t - \beta z)} \tag{3.8}$$

Thus, the magnetic field has only a y component in this case, which is orthogonal to the electric field. There is a phase difference between the electric and magnetic fields if the attenuation constant is not zero.

3.1.1 Discussion on Wave Solutions

Equation (3.7) can be illustrated by Figure 3.1; it is evident that

- the wave solution is a vector. In this case, it has only an x component.
- Its amplitude is decreased exponentially as a function of the propagation distance (z, in this case). The attenuation constant α, given by Equation (3.4), is determined by the material properties and frequency. When the conductivity σ is zero, the wave amplitude is a constant.
- Its phase φ is of the form ($\omega t - \beta z$), which is a function of time, frequency, and propagation distance.

For the loss free case, $\sigma = 0$, Equation (3.4) can be simplified as

$$\alpha = 0$$
$$\beta = \omega\sqrt{\mu\varepsilon} \tag{3.9}$$

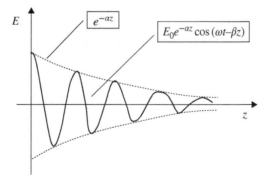

Figure 3.1 A traveling wave in a medium with loss

If we fix the phase and let the wave travel a distance of Δz over a period of time Δt, mathematically this is

$$\Delta\varphi = \omega\cdot\Delta t - \beta\cdot\Delta z = 0 \tag{3.10}$$

Thus, the velocity of the wave can be obtained as

$$v = \frac{\Delta z}{\Delta t} = \frac{\omega}{\beta} \tag{3.11}$$

Replace β by Equation (3.9) to give

$$v = \frac{1}{\sqrt{\mu\varepsilon}} \tag{3.12}$$

This means that the wave velocity is determined by the permittivity and permeability of the medium in which the wave is traveling. In free space, this velocity is

$$v = \frac{1}{\sqrt{\mu_0\varepsilon_0}} \approx 3 \times 10^8 \text{ m/s}$$

We can therefore conclude that the velocity of an electromagnetic wave (including light) in free space is about 3×10^8 m/s – this was what Maxwell obtained more than 130 years ago when he formulated the four equations, but at that time, nobody could validate this important result. This is a good example of how mathematics can be used to solve real-world engineering problems.

In addition, from Equation (3.11), we can see that

$$\beta = \frac{\omega}{v} = \frac{2\pi f}{v} = \frac{2\pi}{\lambda} \tag{3.13}$$

Thus, the phase constant is also called the *wave number* (for every one wavelength, the phase is changed by 2π), which is the same as what we obtained for a transmission line in Chapter 2. In fact, the transmission line Equation (2.7) is just a special case of the wave Equation (3.6) when x and y are fixed. The free space could be viewed as an open transmission line where the information is carried by electromagnetic (EM) waves.

3.2 Plane Wave, Intrinsic Impedance, and Polarization

3.2.1 *Plane Wave and Intrinsic Impedance*

When the conductivity of the medium is zero, the electric field in Equation (3.7) can be simplified as

$$E = \hat{x}E_0 e^{\,j(\omega t - \beta z)} \tag{3.14}$$

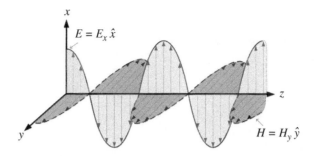

Figure 3.2 A plane wave traveling at z-direction

and the corresponding magnetic field is

$$\boldsymbol{H} = \hat{y}\,\frac{\beta}{\omega\mu}\,E_0 e^{\,j(\omega t - \beta z)} = \hat{y}\,\sqrt{\frac{\varepsilon}{\mu}}E_0 e^{\,j(\omega t - \beta z)} \tag{3.15}$$

Thus, the electric and magnetic fields are in phase, of constant amplitude and orthogonal to each other, as well as orthogonal to the propagation direction z. This EM wave is called the *plane wave* and is illustrated by Figure 3.2. It is a special but common form of EM wave whose amplitude is a constant (in theory, not in reality). The spherical wave is another common and more realistic wave form. The plane wave can only exist (approximately) far away from the source, and it is simple and suitable for wave study in a small region.

The *power flow density* of the EM wave, also known as the *Poynting vector*, is defined as the cross product of the electric and magnetic fields, i.e.

$$\boldsymbol{S} = \boldsymbol{E} \times \boldsymbol{H}^* \ \left(\text{W/m}^2\right) \tag{3.16}$$

where $*$ denotes the complex conjugate, i.e., $(R + jX)^* = R - jX$. Poynting vector describes the amplitude and direction of the flow of power density in EM waves. It is named after the English physicist John Henry Poynting, who introduced it in 1884. The power flow direction is orthogonal to $\boldsymbol{E}$ and $\boldsymbol{H}$. Equation (3.16) gives the instantaneous Poynting vector. The averaged Poynting vector is obtained by integrating the instantaneous Poynting vector over one period and dividing by one period (i.e., we take the time-varying feature into account). Thus, the time-averaged power density of an EM wave is

$$\boldsymbol{S}_{av} = \frac{1}{2}\,\text{Re}\left(\boldsymbol{E} \times \boldsymbol{H}^*\right) = \hat{z}\,\frac{1}{2}\sqrt{\frac{\varepsilon}{\mu}}E_0^2 \tag{3.17}$$

and it can be obtained by its electric field amplitude and material properties $\sqrt{\varepsilon/\mu}$ – this is actually the ratio of the electric field to magnetic field:

$$\eta = \frac{E}{H} = \sqrt{\frac{\mu}{\varepsilon}} = 120\pi\sqrt{\frac{\mu_r}{\varepsilon_r}}\ (\Omega) \tag{3.18}$$

It is called the *intrinsic impedance* of the material and determined by the ratio of the permittivity to the permeability of the medium for a loss free medium. If the medium is lossy (conductivity is not negligible, and/or the permittivity is a complex), the intrinsic impedance is then a complex number:

$$\eta = \frac{E}{H} = -\frac{\omega\mu}{j\gamma} = \sqrt{\frac{j\omega\mu}{\sigma + j\omega\varepsilon}} \; (\Omega) \tag{3.19}$$

In free space, it is

$$\eta_0 = \sqrt{\frac{\mu_0}{\varepsilon_0}} = 120\pi \approx 377 \; (\Omega) \tag{3.20}$$

and the time-averaged power density is

$$S_{av} = \hat{z}\frac{1}{2\eta_0}E_0^2 = \hat{z}\frac{1}{240\pi}E_0^2 \tag{3.21}$$

3.2.2   Polarization

A very important feature of the EM wave is the *polarization*, which is described by the locus of the tip of the E vector as the time progresses. If we use a trigonometric form (we can also use exponential form, just for a change this time), a wave propagating toward z direction can be expressed as

$$E = \hat{x}A \cos(\omega t - \beta z) + \hat{y}B \sin(\omega t - \beta z) \tag{3.22}$$

where A and B, also shown in Figure 3.3, are the amplitudes of the field components in x and y directions, respectively. It is not difficult to verify that this E field is also a solution of the wave Equation (3.6).

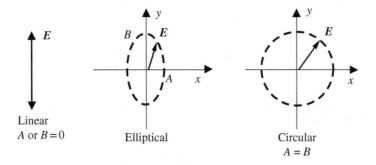

Figure 3.3   Wave polarizations

If A or $B = 0$, this expression represents a *linearly polarized* wave; if $A \neq B \neq 0$, it is an *elliptically polarized* wave; if $A = B$, it then represents a *circularly polarized* wave, which is widely employed in satellite communications. Because the ionosphere causes *Faraday rotation* to an EM wave, which means that a linearly polarized EM wave may be rotated by an unknown amount (depending on the thickness and temperature of the ionosphere, as well as the frequency – the rotation is high at lower frequencies but small at higher frequencies) and makes the linearly polarized wave hard to be matched after passing through the ionosphere. However, there is no problem for circularly polarized waves: a circularly polarized wave is still a circularly polarized wave after passing through the ionosphere: this is why most satellite systems like GPS (global positioning system) have employed circular polarization, not linear polarization, for transmission.

The circularly polarized wave may be considered as a combination of two linearly polarized waves. There are two types of circular polarization: *right-hand circular polarization* (RCP) and *left-hand circular polarization* (LCP) – one linearly polarized wave is ahead or behind the other one by 90°. When the thumb points to the propagation direction, if the tip of the E vector follows the right-hand fingers as time progresses, it is RCP. Otherwise, it is LCP. Equation (3.22) represents a right-hand circular polarized wave if $A = B > 0$. Its corresponding left-hand polarized wave can be expressed as

$$E = \hat{x}A \cos (\omega t - \beta z) - \hat{y}B \sin (\omega t - \beta z) \qquad (3.23)$$

There is just a sign change: "+" is changed to "−" for the y component. The ratio of amplitudes A to B is called the *axial ratio*:

$$AR = \frac{A}{B} \qquad (3.24)$$

For a circularly polarized wave, AR is one ($AR = 1$). For a linearly polarized wave, it is infinite or zero; thus $0 \leq AR \leq +\infty$.

It should be pointed out that a plane wave can be linearly polarized, circularly polarized, or elliptically polarized. Equation (3.14) represents a linearly polarized plane wave, whilst Equations (3.22) and (3.23) are circularly polarized plane waves.

3.3 Radiowave Propagation Mechanisms

Radiowave propagation is a special subject. A radiowave is considered a general term in this book for EM waves up to about 100 GHz. In this section, we are going to briefly review wave propagation mechanisms, which include wave reflection, transmission, diffraction, and scattering.

3.3.1 Reflection and Transmission

As we understand now, an EM wave far away from its source may be considered as a local plane wave. Let a linearly polarized plane wave incident on the surface between Medium 1

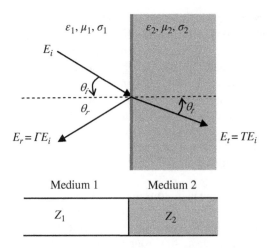

Figure 3.4 Plane wave reflection and transmission, and its analogous transmission line

and Medium 2 as shown in Figure 3.4, what is going to happen at the boundary? The wave will be partially reflected back to Medium 1 and partially transmitted (more precisely refracted; the *refraction* is the change in direction of a wave due to a change in velocity from one medium to another) into Medium 2. If the loss can be neglected, there are a few important points to note:

- The incident angle θ_i is the same as the reflected angle θ_r, that is $\theta_i = \theta_r$;
- The incident angle θ_i is linked to the transmitted angle θ_t by *Snell's law*:

$$\frac{\sin \theta_t}{\sin \theta_i} = \frac{\gamma_1}{\gamma_2} = \frac{\sqrt{\varepsilon_1 \mu_1}}{\sqrt{\varepsilon_2 \mu_2}} \tag{3.25}$$

The *reflection coefficient* is defined as the ratio of the reflected wave to the incident wave, i.e.

$$\Gamma = \frac{E_r}{E_i} \tag{3.26}$$

and the *transmission coefficient* is defined as the ratio of the transmitted wave to the incident wave, i.e.

$$T = \frac{E_t}{E_i} \tag{3.27}$$

Both coefficients are linked to the wave polarization. There are basically two orthogonal polarizations, that is the parallel polarization (E is parallel to the incident plane formed by the incident and reflected waves) and perpendicular polarization (E is perpendicular to the incident plane) as shown in Figure 3.5. Any other polarizations can be considered as a combination of these two principal polarizations.

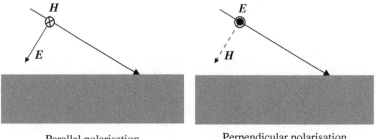

Parallel polarisation   Perpendicular polarisation

Figure 3.5   Two principal polarizations

We can employ either the field approach or the circuit approach to obtain the reflection and transmission coefficients. Using the field concepts, we need to employ the boundary conditions, which is relatively complicated. Thus, we are going to use the circuit approach to obtain these coefficients.

From Figure 3.4, we can see that the two media can be replaced by two analogous transmission lines with characteristic impedances of Z_1 and Z_2, respectively. They are determined by wave polarization, incident angle and material properties:

$$
\begin{aligned}
Z_1 &= \begin{cases} \eta_1 \cdot \cos\theta_i; & \text{for parallel pol.} \\ \eta_1 / \cos\theta_i; & \text{for perpendicular pol.} \end{cases} \\
Z_2 &= \begin{cases} \eta_2 \cdot \cos\theta_t; & \text{for parallel pol.} \\ \eta_2 / \cos\theta_t; & \text{for perpendicular pol.} \end{cases}
\end{aligned}
\tag{3.28}
$$

where the intrinsic impedance η is determined by material properties as shown in Equation (3.19). The reflection coefficient between these two transmission lines can be easily obtained by Equation (2.28), i.e.:

$$
\Gamma = \frac{Z_2 - Z_1}{Z_2 + Z_1}
\tag{3.29}
$$

which is also the reflection coefficient at the boundary between Medium 1 and Medium 2. The transmission coefficient is

$$
T = 1 + \Gamma = \frac{2Z_2}{Z_2 + Z_1}
\tag{3.30}
$$

The reflection and transmission coefficients are ratios of electric field strengths. Because the incident power equals to the sum of the reflected and transmitted powers, we have:

$$
\begin{aligned}
\sqrt{\frac{\varepsilon_1}{\mu_1}} E_i^2 &= \sqrt{\frac{\varepsilon_1}{\mu_1}} E_r^2 + \sqrt{\frac{\varepsilon_2}{\mu_2}} E_t^2 \\
|\Gamma|^2 &+ \sqrt{\frac{\varepsilon_2 \mu_1}{\varepsilon_1 \mu_2}} |T|^2 = 1
\end{aligned}
\tag{3.31}
$$

For most materials, the permeability is the same; thus, the equation can be further simplified to

$$|\Gamma|^2 + \sqrt{\frac{\varepsilon_2}{\varepsilon_1}}|T|^2 = 1$$

But note, this is not

$$|\Gamma|^2 + |T|^2 = 1$$

unless the two media are the same. This is a common mistake easy to make.

Example 3.1   Reflection on a perfect conductor

Obtain the reflection and transmission coefficients between air and a perfect conductor.

Solution

The conductivity of a perfect conductor is infinite. Using Equations (3.19) and (3.28), we know that the characteristic impedance of its equivalent transmission line is zero for any polarization and incident angle, i.e. $Z_2 = 0$, thus

$$\Gamma = -1 \quad \text{and} \quad T = 0$$

This means that all signals are reflected back and the phase is changed by 180°. There is no signal transmitted into the conductor.

Example 3.2   Reflection on a ground

If the relative permittivity of a ground is nine and the conductivity is very small and negligible, plot the reflection coefficient as a function of the incident angle for both parallel and perpendicular polarizations.

Solution

$\varepsilon_1 = 1$ and $\varepsilon_2 = 9$. To obtain the reflection coefficient, we first use Equation (3.25) to yield

$$\frac{\sin \theta_t}{\sin \theta_i} = \frac{\sqrt{\varepsilon_1 \mu_1}}{\sqrt{\varepsilon_2 \mu_2}} = \frac{1}{3}, \quad \text{thus} \quad \cos \theta_t = \sqrt{1 - \sin^2 \theta_t} = \sqrt{1 - \frac{1}{9} \sin^2 \theta_i};$$

and then use Equation (3.28) to give

$$Z_1 = \begin{cases} 120\pi \cdot \cos \theta_i; & \text{for parallel pol.} \\ 120\pi / \cos \theta_i; & \text{for perpendicular pol.} \end{cases}$$

$$Z_2 = \begin{cases} 40\pi \cdot \sqrt{1 - \sin^2 \theta_i / 9}; & \text{for parallel pol.} \\ 40\pi / \sqrt{1 - \sin^2 \theta_i / 9}; & \text{for perpendicular pol.} \end{cases}$$

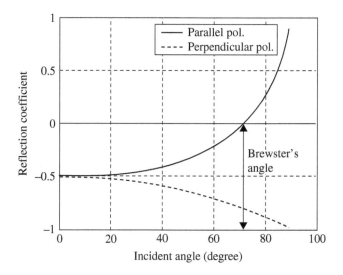

Figure 3.6   Reflection coefficient as a function of incident angle

Thus, the reflection coefficient for parallel polarization is

$$\Gamma_{//} = \frac{Z_2 - Z_1}{Z_2 + Z_1} = \frac{\sqrt{1 - \sin^2\theta_i/9} - 3\cos\theta_i}{\sqrt{1 - \sin^2\theta_i/9} + 3\cos\theta_i}$$

and the reflection coefficient for perpendicular polarization is

$$\Gamma_{\perp} = \frac{Z_2 - Z_1}{Z_2 + Z_1} = \frac{\cos\theta_i - 3\sqrt{1 - \sin^2\theta_i/9}}{\cos\theta_i + 3\sqrt{1 - \sin^2\theta_i/9}}$$

The results are plotted as a function of the incident angle in Figure 3.6. Obviously, the reflection coefficients are very different for different polarizations. There are some important observations:

- For parallel polarization, the reflection coefficient vanishes (= 0) at a particular incident angle, this angle is called *Brewster's angle:*

$$\theta_B = \sin^{-1}\sqrt{\frac{\varepsilon_r}{\varepsilon_r + 1}} \tag{3.32}$$

- For perpendicular polarization, the larger the incident angle, the larger the reflection coefficient (magnitude).
- At normal incidence, the reflection coefficient is the same for both polarizations.
- At 90°, the reflection coefficient is 1 for the parallel polarization and −1 for the perpendicular polarization.

Special Case 1: **when the incident angle is greater than the Brewster's angle** for parallel polarization, $\Gamma > 0$ as shown in Figure 3.6. Can we still employ Equation (3.30) to calculate the transmission coefficient?

The answer is no, otherwise $|T|$ would be greater than one and fail to meet the condition set by Equation (3.31). Thus, Equation (3.30) is only valid for $\text{Re}(\Gamma) < 0$. If $\text{Re}(\Gamma) > 0$, Equation (3.30) for calculating the transmission coefficient should be replaced by

$$T = 1 - \Gamma = \frac{2Z_1}{Z_2 + Z_1} \tag{3.33}$$

Special Case 2: **when the incident wave travels from a dense medium into a less dense medium at an angle exceeding the critical angle**, what will happen?

The *critical angle* is the incident angle that gives a transmitted angle of 90° when the wave is from a dense medium to a less dense medium, such as from water into air. From Snell's law, let $\sin\theta_t = 1$ to obtain this special angle

$$\theta_{ic} = \sin^{-1} \frac{\sqrt{\varepsilon_2 \mu_2}}{\sqrt{\varepsilon_1 \mu_1}} \tag{3.34}$$

If the relative permittivity of the water is 80, the critical angle is therefore 6.42°.

When the incident angle is greater than the critical angle, for whatever polarization, the wave will be totally internally reflected and will also be accompanied by a surface wave in the less dense medium. This surface wave decays exponentially away from the surface but propagates without loss along the surface [1]. People have developed communication systems for submarines utilizing this phenomenon.

3.3.1.1 Effects of Reflection and Transmission on Wave Polarization

Example 3.2 has clearly shown that, for non-normal incidence, the reflection coefficients are different for the two principal polarizations. As a result, if an incident wave is a combination of these two orthogonal waves, the combined signal after the reflection will be changed. For example, if the incident wave is a circular polarization which means that the parallel component and the perpendicular component are of the same amplitudes but 90° out of phase; after the reflection, the wave is no longer a circularly polarized wave but an elliptically polarized wave – because the wave components in these two orthogonal planes are no longer the same! Similarly, if it is a linearly polarized wave, after the reflection, this linear polarization may be rotated.

The same conclusion can be drawn to the transmitted wave, since the transmission coefficients are also different for the two different polarizations if the incident angle is not zero degree (normal incidence).

An important special case is when the reflector is a perfect conductor; the reflection coefficient is −1. For a linearly polarized wave, this means a phase change of 180°. But for a circularly polarized wave, an RCP wave becomes an LCP wave and an LCP wave becomes an RCP after the reflection. This results in polarization mismatch. Thus, circular polarization is not recommended for indoor radio communications where plenty of reflections occur for a radio path.

3.3.1.2 Radiowave Through a Wall

A very common scenario is a radiowave passing through a wall to reach a television, mobile phone, or other radio devices (such as a laptop computer). How can we find out the transmission coefficient? and the attenuation?

Again, we can use circuit concepts to deal with this problem which should be much simpler than using field concepts. As shown in Figure 3.7, there are now three media. Their equivalent transmission lines have the characteristic impedances Z_1, Z_2 and Z_3, respectively. They are defined by Equation (3.28). The thickness of the wall is d. There are multiple reflections and transmissions at the boundaries. Thus, the reflected wave is now the summation of all the reflected waves, and the transmitted wave is also the summation of all the waves transmitted. They could be combined constructively (larger) or destructively (smaller). Using the transmission line model, the calculation of the reflection and transmission coefficients is much easier than the field approach (but it is still not easy!). Z_3 may be considered as the load of the second transmission line. Using Equation (2.32), the input impedance at the interface between Medium 1 and Medium 2 can be written as

$$Z_{in} = Z_2 \frac{Z_3 + Z_2 \tanh(\gamma_2 d_e)}{Z_2 + Z_3 \tanh(\gamma_2 d_e)} \tag{3.35}$$

where $\gamma_2 = \sqrt{j\omega\mu_2(\sigma_2 + j\omega\varepsilon_2)}$ and $d_e = d/\cos\theta_t$ is the effective thickness. If the loss is negligible, we have

$$Z_{in} = Z_2 \frac{Z_3 + jZ_2 \tan(\beta_2 d_e)}{Z_2 + jZ_3 \tan(\beta_2 d_e)} \tag{3.36}$$

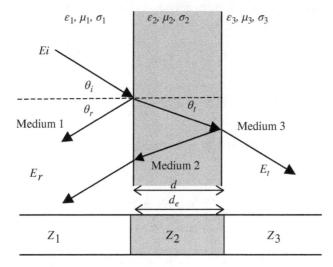

Figure 3.7 Reflection and transmission of a wall and its analogous transmission line

Thus, the reflection coefficient is:

$$\Gamma = \frac{Z_{in} - Z_1}{Z_{in} + Z_1}$$

From power point of view, the power transmitted through the wall is

$$P_T = (P_{in} - P_R) \cdot A \tag{3.37}$$

where

P_T = power transmitted through the wall;
P_{in} = incident power;
P_R = reflected power = $P_{in} \cdot |\Gamma|^2$
A = attenuation of the wall = $e^{-2\alpha d_e}$, i.e. determined by the attenuation constant and the effective thickness of the wall.

Example 3.3   Reflection of a wall
A brick wall has a relative permittivity of 4 and thickness of 20 cm, the loss is negligible.

a. If the operational frequency is 2.45 GHz for wireless applications (such as Bluetooth), plot the reflection coefficient as a function of the incident angle for both parallel and perpendicular polarizations.
b. If the incident angle is 45°, plot the reflection coefficient as a function of the frequency for both parallel and perpendicular polarizations..

Solution
We know $\varepsilon_1 = 1$, $\varepsilon_2 = 4$, and $\varepsilon_3 = 1$, and all conductivities are zero. In Medium 2, the wavelength $\lambda_2 = \dfrac{\lambda_0}{\sqrt{4}} = \dfrac{\lambda_0}{2}$, and $\beta_2 = \dfrac{2\pi}{\lambda_2}$.

To obtain the reflection coefficient, we first use Equation (3.25) to yield

$$\frac{\sin \theta_t}{\sin \theta_i} = \frac{\sqrt{\varepsilon_1 \mu_1}}{\sqrt{\varepsilon_2 \mu_2}} = \frac{1}{2}, \quad \text{thus} \quad \cos \theta_t = \sqrt{1 - \sin^2 \theta_t} = \sqrt{1 - \frac{1}{4} \sin^2 \theta_i};$$

and then use Equation (3.28) to give

$$Z_1 = Z_3 = \begin{cases} 120\pi \cdot \cos \theta_i; & \text{for parallel pol.} \\ 120\pi / \cos \theta_i; & \text{for perpendicular pol.} \end{cases}$$

$$Z_2 = \begin{cases} 60\pi \cdot \sqrt{1 - \sin^2 \theta_i / 4}; & \text{for parallel pol.} \\ 60\pi / \sqrt{1 - \sin^2 \theta_i / 4}; & \text{for perpendicular pol.} \end{cases}$$

For the parallel polarization, using (3.36) to yield:

$$Z_{in} = 60\pi\sqrt{1 - \sin^2\theta_i/4}\,\frac{3\cos\theta_i + j\sqrt{1 - \sin^2\theta_i/4}\,\tan\left(\beta_2 d/\sqrt{1 - \sin^2\theta_i/4}\right)}{\sqrt{1 - \sin^2\theta_i/4} + j3\cos\theta_i\tan\left(\beta_2 d/\sqrt{1 - \sin^2\theta_i/4}\right)}$$

This is the load impedance to transmission line 1; thus, the reflection coefficient

$$\Gamma_{//} = \frac{Z_{in} - 120\pi\cdot\cos\theta_i}{Z_{in} + 120\pi\cdot\cos\theta_i}$$

For the perpendicular polarization, we use the same approach to obtain

$$Z_{in} = 60\pi/\sqrt{1 - \sin^2\theta_i/4}\,\frac{3\sqrt{1 - \sin^2\theta_i/4} + j\cos\theta_i\tan\left(\beta_2 d/\sqrt{1 - \sin^2\theta_i/4}\right)}{\cos\theta_i + j3\sqrt{1 - \sin^2\theta_i/4}\,\tan\left(\beta_2 d/\sqrt{1 - \sin^2\theta_i/4}\right)}$$

$$\Gamma_\perp = \frac{Z_{in} - 120\pi/\cos\theta_i}{Z_{in} + 120\pi/\cos\theta_i}$$

a. When the incident angle is set as the variable, the reflection coefficients for both polarizations are plotted in Figure 3.8. The important observations are:
   - The reflection coefficients are about the same at small incident angles, and at 90°;
   - There are two troughs for parallel polarizations but one for perpendicular polarization.
   - The reflection coefficient of perpendicular polarization is always greater than (or the same as) that of parallel polarization.
b. When the frequency is set as a variable, the reflection coefficients for both polarizations are plotted in Figure 3.9. We can see that:
   - The reflection coefficients are periodic functions of the frequency. The period for both polarizations is the same, 350 MHz in this particular case (for the normal incident, the period is 375 MHz), i.e., its wavelength is half of the effective thickness. Thus, *the reflection coefficient is minimized when the thickness of the wall is an integer of half of the effective wavelength.* This important conclusion can be used for antenna radome and housing design.
   - *The reflection reaches the maximum when* $d_e = \lambda/4 + n\lambda/2$, $n = 0, 1, 2\ldots.$
   - The reflection coefficients are very small for low frequencies. This is why low-frequency signals can easily penetrate into buildings.
   - Again, the reflection coefficient of perpendicular polarization is always greater than (or the same as) that of parallel polarization.

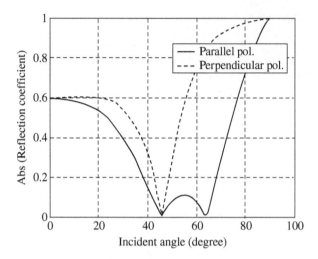

Figure 3.8   Reflection coefficient of a wall as a function of the incident angle

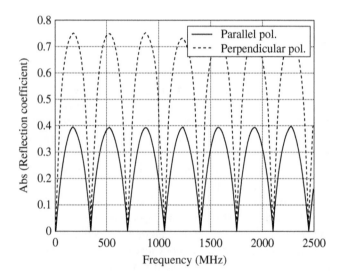

Figure 3.9   The reflection coefficient of a wall as a function of the frequency

In this example, since the conductivity is zero, the waves which are not reflected by the wall will pass through the wall without attenuation and the transmission coefficient can be calculated using the reflection–transmission Equation (3.31).

It is clear to us now that the radiowave reflection and transmission are complicated. The coefficients are functions of frequency, incident angle, polarization, as well as the dielectric properties of the media.

3.3.2 Diffraction and Huygens' Principle

Diffraction is the apparent bending and spreading of waves when they meet an obstacle. It can occur with any type of wave, including sound waves, water waves, and electromagnetic waves. As a simple example of diffraction, if you speak into one end of a cardboard tube, the sound waves emerging from the other end spread out in all directions, rather than propagating in a straight line like a stream of water from a garden hose.

The foundation of the diffraction theory is based on *Huygens' Principle*, which states that *each point on a primary wave front can be considered as a new source of a secondary spherical wave*, as suggested in Figure 3.10. The new source is considered as an equivalent source, which can be expressed mathematically as

$$
\begin{aligned}
J_S &= \hat{n} \times H \\
M_S &= -\hat{n} \times E
\end{aligned}
\tag{3.38}
$$

where J_S and M_S are the equivalent surface electric current and magnetic current on an imaginary surface S, respectively, $\hat{n}$ is the unit vector, normal to the surface S. Thus, the original problem can be replaced by an equivalent source which produces the same electric and magnetic fields outside the (enclosed) surface S. Huygens' principle is therefore closely linked to Love's *equivalence principle* which was developed by replacing an actual radiating source by an equivalent source. This principle is very useful for analyzing aperture-type and slot antennas which will be discussed in Chapter 5.

When the distance from the obstacle to the receiver (Rx) is much larger than the height of the obstacle, which could be a mountain or a building, the relative (to the direct ray) power density S_r can be approximated by [1]

$$
S_r = \frac{\lambda}{4\pi^2 h^2} \left(\frac{r_1 r_2}{r_1 + r_2} \right)
\tag{3.39}
$$

where λ is the wavelength. r_1 and r_2 are the distances to the transmitter and receiver, respectively, and h is the height of the obstacle above the reference line. This relative power density has got no unit. The area around the obstacle can be divided into three regions: (1) the first region has a direct ray from the Tx and a reflected ray from the obstacle, the total field at the receiving point changes significantly since the summation of the two rays can be constructive or destructive depending on their relative phase. (2) The second region is dominated by the direct ray; thus,

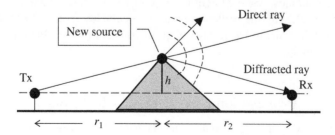

Figure 3.10 Radiowave diffraction over a knife-edge obstacle

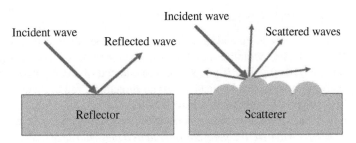

Figure 3.11   Radiowave reflection and scattering

the field at the receiving point is inversely proportional to the distance. (3) The third region has got no direct ray, and the received field is coming from the diffraction.

The diffraction by obstacles is an important mechanism for radio propagation, especially when there is no line-of-sight path available. Radio broadcasting signals can pass over hills/ mountains, and mobile radio signals can reach to the street; these are just some of the examples benefiting from radiowave diffraction. Radars are able to detect various targets, which is also partially due to diffraction of the radar signal from the target.

3.3.3   Scattering

Unlike the other propagation mechanisms where the size of the medium or the obstacle is much larger than the wavelength, *scattering* occurs when the obstacle is comparable or even small than the wavelength. It is the process by which small particles/obstacles in a medium of different dielectric properties diffuse a portion of the incident wave in all directions. In scattering, there are no energy transformation results, only a change in the spatial distribution of the radiation, as shown in Figure 3.11. Back scattering has been employed for radar applications, such as target detection and weather forecast.

Along with absorption, scattering is a major cause of the attenuation of radio propagation by the atmosphere. Scattering varies as a function of the ratio of the particle/obstacle diameter to the wavelength of the wave. When this ratio is less than about one-tenth, *Rayleigh scattering* occurs in which the scattering coefficient varies inversely as the fourth power of the wavelength – this result can be used to explain why the sky is red at sunrise and sunset but blue at midday (the wavelength of red light is long, and the wavelength of blue light is short; thus, most transmitted light is red but most scattered light is blue). At larger values of the ratio of the particle diameter to the wavelength, the scattering varies in a complex fashion described by the *Mie theory*; at a ratio of the order of 10, the laws of *geometric optics* (where the wave can be treated as a ray) begin to apply.

3.4   Radiowave Propagation Characteristics in Media

Radiowaves propagating through a radio channel may undergo reflection, transmission or refraction, diffraction, and scattering. Attenuation or absorption is another important aspect that we have not yet properly discussed. The same radiowave propagating through different media may exhibit very different features. In this section, we are going to briefly examine radiowave propagation characteristics in some common media.

3.4.1 Media Classification and Attenuation

From electromagnetics point of view, materials can be classified as conductive, semi-conductive, and dielectric media. The electromagnetic properties of materials are normally functions of the frequency, so are the propagation characteristics. Recall that, in Chapter 1, we introduced the complex permittivity and defined the loss tangent as the ratio of the imaginary to the real parts of the permittivity, which is Equation (1.36):

$$\tan \delta = \frac{\varepsilon''}{\varepsilon'} = \frac{\sigma}{\omega \varepsilon}$$

It has taken the frequency as well as the normal permittivity and conductivity of the medium into account. The specific classifications are given in [1] as

- Conductor: $\quad \tan \delta = \dfrac{\sigma}{\omega \varepsilon} > 100$

- Semi-conductor: $\quad 0.01 < \tan \delta = \dfrac{\sigma}{\omega \varepsilon} < 100$

- Dielectric: $\quad \tan \delta = \dfrac{\sigma}{\omega \varepsilon} < 0.01$

The classification of some common media as a function of frequency is shown in Figure 3.12, where M is determined by $\tan \delta = 10^M$. It is important to note that:

- The medium classification is indeed frequency-dependent;
- Most materials are in the conducting region at low frequencies (<10 kHz) but in dielectric region at high frequencies (>1 GHz). The ground and water are just two of the examples.
- Figure 3.12 is just an approximation to illustrate material properties over a wide frequency range. The real-world materials are more complicated and may be different from Figure 3.12 in some frequency bands. A typical complex dielectric permittivity spectrum is shown in Figure 3.13. Cole-Cole or Debye (a special case of Cole-Cole) relaxation model is often used to approximate the frequency response of a dielectric material (although the model was originally introduced to describe dielectric relaxation in polymers). The permittivity ε and conductivity σ are not constant but functions of frequency as well.

This classification is useful to evaluate the EM properties of a medium in terms of the loss tangent but not accurate to classify if a medium is lossy or not. Recall that the attenuation constant (it is in fact not a constant) is given by (3.4) as

$$\alpha = \omega \sqrt{\mu \varepsilon} \left[\frac{1}{2} \left(\sqrt{1 + \frac{\sigma^2}{\varepsilon^2 \omega^2}} - 1 \right) \right]^{1/2} = \omega \sqrt{\mu \varepsilon} \left[\frac{1}{2} \left(\sqrt{1 + \tan^2 \delta} - 1 \right) \right]^{1/2} \qquad (3.40)$$

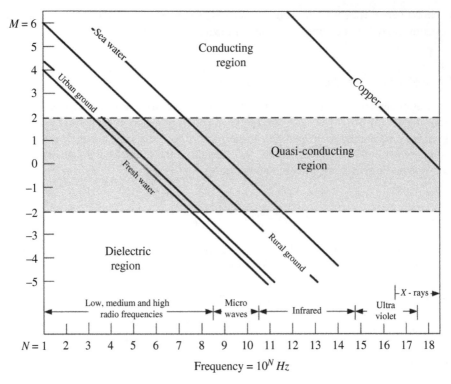

Figure 3.12   Classification of media as a function of frequency ($\tan\delta = 10^M$)

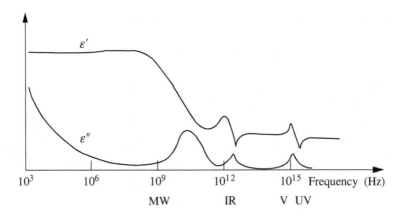

Figure 3.13   A typical complex permittivity spectrum

The loss tangent is just a term in the bracket. The attenuation constant is actually proportional to the frequency if the loss tangent is fixed. Generally speaking, the dominant feature of radio-wave propagation in media is that the attenuation increases with the frequency – this seems to contradict with the material classification. Figure 3.12 is not a good reference for propagation loss characteristics, but a useful plot of the loss tangent against the frequency.

It is now clear that many common media may be considered as conducting materials at low frequencies and dielectric materials at high frequencies by some people, but their attenuation increases with frequency. This is why, generally speaking, low frequencies have been used for longer distance communications and high frequencies are employed for shorter distance communications.

3.4.1.1 Propagation Through Ionosphere

It is important to take into account the propagation characteristics of the ionosphere when long distance and satellite type of communications are considered. The *ionosphere* is the region above the troposphere (where the air is), from about 80 to 400 km above the earth. It is a collection of ions, which are atoms that have some of their electrons stripped off leaving two or more electrically charged objects. The sun's rays cause the ions to form which slowly recombine.

The propagation of radiowaves in the presence of ions is drastically different from that in air, which is why the ionosphere plays an important role in most modes of propagation. The major effects of the ionosphere on radiowaves are:

- Reflection at low frequencies (up to about 30 MHz). Thus, the ionosphere and the earth can form a kind of waveguide and let wave propagate over a very long distance using ground/surface mode.
- Scattering, refraction, and absorption when high-frequency waves (above 100 MHz) pass through it.
- Faraday rotation: the wave polarization plane/line is rotated through the ionosphere. The amount of rotation is dependent on the thickness and charge density of the ionosphere which are functions of time (and where the sun is) and is also dependent on the frequency. The rotation is small at very high frequencies (>10 GHz, that is why linear polarizations may be employed for satellite communications or broadcasting if the frequency is high enough).

3.4.1.2 Propagation in Rain

Rain is an undesired medium of a radio channel. It causes a considerable amount of problems every year for radio communications and radar systems. The major effect of rain on radiowaves is attenuation due to absorption and scattering over a wide range of the spectrum. The attenuation depends on a number of things, and it can be represented by [2]

$$A = aR^b, \quad \text{dB/km} \tag{3.41}$$

where R is the rain rate in mm/h, and a and b are constants that depend on frequency and temperature of the rain, respectively. The temperature dependence is due to the variation of dielectric permittivity of water with temperature.

A light drizzle corresponds to a rain rate of 0.25 mm/h; light rain to 1 mm/h, moderate rain to 4 mm/h, heavy rain to 16 mm/h, and cloud bursts up to many cm/h. Figure 3.14 is an illustration of how the attenuation is linked to the rain rate and frequency [2].

It is clearly shown that the attenuation around 10 GHz is much smaller than 30 and 100 GHz. At a rain rate of 10 mm/h, the attenuation is increased from 0.2 dB/km at 10 GHz to about 2 dB/km at

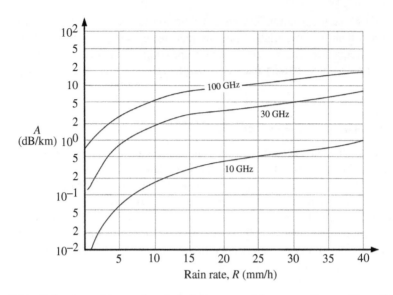

Figure 3.14 Rain attenuation as a function of rain rate and frequency. Source: From Collin [2]

30 GHz (10 times more). If we take the sky noise temperature into account, a low noise and low loss radio 'window' exists between 1 and about 15 GHz – this is why most satellite communication systems operate within this frequency band.

3.4.1.3 Propagation in Snow

Some parts of the world have no snow at all, but some other parts have to face the snow problem most of the time. The attenuation in dry snow is an order of magnitude less than that in rain for the same precipitation rate. But, attenuation by wet snow is comparable to that in rain and may even exceed that of rain at millimeter wavelengths. It is difficult to specify the attenuation in any simple form.

3.4.1.4 Propagation Through Fog

Fog is another unfriendly medium we have to deal with. The attenuation equation is the same as that for the rain but with much smaller attenuation.

3.5 Radiowave Propagation Models

For radio communication system designers, it is important to be able to predict the radiowave propagation *pathloss*, which is defined as the difference between the power transmitted and the power received at the destination. Using the pathloss information, the designer is able to optimize the system and ensure sufficient radio coverage. That is to establish the link budget and figure out where to place the antenna and what the required transmitted power is to cover the desired area. Since this is a very important subject, a comprehensive study has been undertaken by many researchers over the years, some propagation models have been developed for various scenarios. Here, we are going to discuss some basic propagation models which are closely linked to antennas.

3.5.1 Free Space Model

If there is just one ray between the transmitting antenna and receiving antenna (the line-of-sight case), such as satellite/space communications as shown in Figure 3.15, the received power can be obtained using the well-known *Friis' transmission formula*, which will be discussed in Chapter 4, that is

$$P_r = P_t \left(\frac{\lambda}{4\pi r} \right)^2 G_t G_r \tag{3.42}$$

where P_r is the received power, P_t is the transmitted power, G_t and G_r are the gains of the transmitting and receiving antennas, respectively; and r is the distance between the transmitting and receiving antennas. Without considering the antenna performance (let $G_t = G_r = 1$), the pathloss L_P can be found from Equation (3.42) as

$$L_P = 10 \log_{10} \left(\frac{P_t}{P_r} \right) = 20 \log_{10} f + 20 \log_{10} r - 147.6 \, \text{dB} \tag{3.43}$$

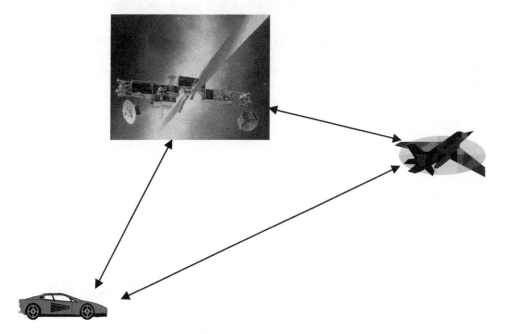

Figure 3.15 Free space communications

where f is the frequency. It means that the pathloss is proportional to the frequency square and the distance square. The larger the distance, the larger the pathloss; the higher the frequency, the larger the pathloss. It is 20 dB/decade.

3.5.2 Two-ray Model/Plane Earth Model

As shown in Figure 3.16, the two-ray model can be applied to many terrestrial communications scenarios, such as radio broadcasting and radio communications in rural environments. The transmitted signals reach the destination via the line-of-sight path and the path reflected by the ground. The result becomes complicated because the signals may be combined constructively and destructively, depending on the reflection coefficient of the ground and the phase difference of these two rays.

The path difference between these two rays is (using the image theory to be discuss in the Chapter 4 and assuming $d \gg h_1 + h_2$)

$$\Delta r = \sqrt{(h_1 + h_2)^2 + d^2} - \sqrt{(h_1 - h_2)^2 + d^2} \approx \frac{2h_1 h_2}{d}$$

When this difference is half of the wavelength, the phase difference is 180° (out of phase), the separation is

$$d_f = \frac{4h_1 h_2}{\lambda} \tag{3.44}$$

and is called the *first Fresnel zone distance*. It is useful to note the following results:

- When the separation of the two antennas is smaller than this distance, $d < d_f$, the pathloss is about 20 dB/decade;
- When the separation is greater than this distance, $d > d_f$, the pathloss is 40 dB/decade because it can be shown [3] that, in this case, the received power

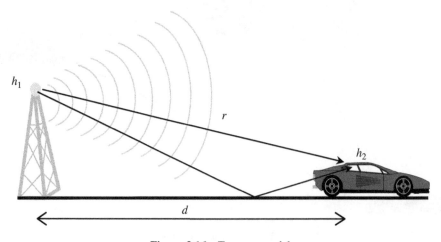

Figure 3.16 Two-ray model

$$P_r = P_t \left(\frac{h_1 h_2}{d^2} \right)^2 \tag{3.45}$$

Hence, the pathloss expressed in dB is

$$L_P = -10 \log_{10} \left(\frac{h_1 h_2}{d^2} \right)^2 = 40 \log_{10}(d) - 20 \log_{10}(h_1 h_2), \text{dB} \tag{3.46}$$

It is apparent that the higher the base-station antenna and the mobile antenna, the smaller the pathloss. Thus, there is no wonder why all mobile operators have wanted to erect their antennas as high as possible!

3.5.3 Multipath Models

For the multipath case (more than two rays), there are no analytical equations which have been obtained to give an accurate prediction of the radio propagation pathloss. But, empirical and statistical representations are available for various scenarios.

Most of the popular outdoor pathloss prediction tools are based on Okumura and Hata's formulation, which was based on a huge amount of measured data for the frequencies between 100 MHz and 3 GHz [3].

For indoor scenarios, many researchers have shown that indoor pathloss actually obeys the distance power law as

$$L_P(d) = L_P(d_0) + 10n \log_{10} \left(\frac{d}{d_0} \right) + X, \text{dB} \tag{3.47}$$

where d_0 is the reference distance (normally it is 1 m), the power index n depends on the frequency, surroundings and building type, and X represents a normal random variable in dB having a standard deviation of σ dB. Typical values for various buildings are provided in Table 3.1 [3]. It is apparent that the value n is normally (but not always) larger than 2, which is the free space case. It also shows that the higher the frequency, the larger the n for the same scenario.

For a multipath environment, in addition to the pathloss, there are many other important characteristics, which include at least the following two:

- Multipath fast fading: the received signal changes significantly (> 30 dB) over a very short distance (few wavelengths), resulted from the complex and vector summation of signals.
- Delay spread: multi-copies of the original signal arrive at the destination at different time through different paths, which may cause dispersion and inter-symbol interference (ISI, the equalizer is a device developed to remove all delayed waves and combat the ISI problem). The RMS (root-mean-square) delay spread T_{RMS} (in seconds) is used to characterize this feature. The delay spread is often employed to define the channel's coherence bandwidth B_C (similar to a filter's bandwidth – a radio channel can be considered as a filter in certain sense). They are linked by the following equation:

Table 3.1 Pathloss exponent and standard deviation measured in different buildings

Building	Frequency (MHz)	n	σ (dB)
Retail stores	914	2.2	8.7
Grocery store	914	1.8	5.2
Office, hard partition	1500	3.0	7.0
Office, soft partition	900	2.4	9.6
Office, soft partition	1900	2.6	14.1
Factory LOS			
Textile/Chemical	1300	2.0	3.0
Textile/Chemical	4000	2.1	7.0
Paper/Cereals	1300	1.8	6.0
Metalworking	1300	1.6	5.8
Suburban home			
Indoor street	900	3.0	7.0
Factory OBS			
Textile/Chemical	4000	2.1	3.7
Metalworking	1300	3.3	6.8

$$B_C = \frac{1}{2\pi T_{RMS}}, (Hz) \qquad (3.48)$$

If the channel bandwidth is greater than the message bandwidth, all the frequency components in the message will arrive at the receiver with little or no distortion and the ISI can be negligible. The channel is referred to as *flat fading*.

If the channel bandwidth is smaller than the message bandwidth, all the frequency components in the message will arrive at the receiver with distortion and ISI. In this case, the channel is referred to as *frequency selective fading*.

Statistically, the power density function (PDF) of the received (short-term) signal envelope follows certain distribution, depending on the specific propagation channel. According to the PDF distribution function, the channel can be classified as the following three channels:

- When there is a line-of-sight ray, it follows the Gaussian distribution and this channel is therefore called the *Gaussian channel*;
- When there is a partial line-of-sight ray (the path is partially blocked by obstacles such as trees), it follows the Rician distribution and this channel is therefore called the *Rician channel*;
- When there is no line-of-sight ray, it follows the Rayleigh distribution and this channel is therefore called the *Rayleigh channel*.

For the same signal to noise ratio (SNR), a communication system can obtain the best bit error rate (BER) from a Gaussian channel and the worst BER from a Rayleigh channel.

Antenna diversity techniques and MIMO antennas, which will be discussed later, are developed specifically for multipath environments. The radio communication system performance can be improved significantly by using these antenna technologies.

3.6 Comparison of Circuit Concepts and Field Concepts

We have now introduced both circuit concepts and field concepts, which may seem to be completely different. However, if we compare these two sets of parameters, some interesting correspondence can be obtained. For example, the product of the voltage and current is power in circuit concepts, whilst the cross product of the electric field and magnetic field is the power density in field concepts; the ratio of the voltage to the current is the impedance, whilst the ratio of the electric field to the magnetic field is the intrinsic impedance. If we take another look at how the electric field and magnetic field are distributed around a two-wire transmission line as shown in Figure 3.17. It is not difficult to conclude that *the fundamental correspondences of these two sets of concepts are*:

- the voltage V corresponding to the electric field E via $V = C_1 \int_{Line1}^{Line2} \boldsymbol{E} \cdot \boldsymbol{ds}$
- the current I corresponding to the magnetic field H via $I = C_2 \oint \boldsymbol{H} \cdot \boldsymbol{ds}$

C_1 and C_2 are just two constants. A list of the correspondence of the circuit and the field concepts is presented in Table 3.2.

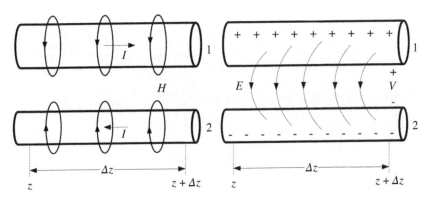

Figure 3.17 The linkage of the circuit concepts and the field concepts

Table 3.2 Correspondence of the circuit concepts and the field concepts

	Circuit	Field
Complex voltage	$V(t, z)$ in V	$E(t, z)$ in V/m
Complex current	$I(t, z)$ in A	$H(t, z)$ in A/m, or $J(t, z)$
Complex power flow	VI^* in W	$E \times H^*$ in W/m^2
Impedance	Z in Ω	η in Ω
Conductance	$1/R$ in S	σ in S/m
Current	V/R in A	$J = \sigma E$ in A/m^2

3.6.1   Skin Depth

Now let us take the skin depth as an example to see the difference between the circuit concepts and the field concepts in applications.

3.6.1.1   Skin Depth – Field Concepts

From the field concept of view, a radiowave traveling into a lossy medium can be described by Equation (3.7). That is

$$E = \hat{x} E_0 e^{-\alpha z + j(\omega t - \beta z)}$$

The *skin depth* is defined as the distance δ through which the amplitude of a traveling plane wave decreases by factor $1/e$ and is therefore

$$\delta = \frac{1}{\alpha} \tag{3.49}$$

For a good conductor ($\tan \delta = \dfrac{\sigma}{\omega \varepsilon} > 100$), the attenuation constant can be approximated as

$$\alpha = \omega \sqrt{\mu \varepsilon} \left[\frac{1}{2} \left(\sqrt{1 + \frac{\sigma^2}{\varepsilon^2 \omega^2}} - 1 \right) \right]^{1/2} \approx \omega \sqrt{\mu \varepsilon} \left[\frac{1}{2} \frac{\sigma}{\varepsilon \omega} \right]^{1/2} = \sqrt{\frac{\omega \mu \sigma}{2}}$$

The skin depth can therefore be expressed as

$$\delta \approx \sqrt{\frac{2}{\omega \mu \sigma}} = \sqrt{\frac{1}{\pi f \mu \sigma}} \tag{3.50}$$

This means that

- the higher the frequency, the smaller the skin depth;
- the larger the permeability, the smaller the skin depth;
- the larger the conductivity, the smaller the skin depth.

For a good conductor, the permittivity has little effects on the skin depth.

It should be highlighted that the wave amplitude is reduced by a factor of $1/e$, or 37%, or 8.686 dB over one skin depth.

3.6.1.2   Skin Depth – Circuit Concepts

From the circuit concept of view, the current density J in an infinitely thick plane conductor decreases exponentially with depth from the surface as follows:

$$J = J_0 e^{-z/\delta}$$

where δ is a constant called the *skin depth*. This is defined as the depth below the surface of the conductor at which the current density decays to $1/e$ (about 37%) of the current density at the surface. Mathematically, it is the same as Equation (3.50). As a result, when calculating the per-unit length resistance of a conducting wire,

$$R = \frac{1}{S\sigma}$$

where S is now the effective area rather than the whole cross section of the wire when the wire radius $r > \delta$. In Figure 3.18, this effective area can be approximated as $S = \pi r^2 - \pi(r-\delta)^2 \approx 2\pi r\delta$ (not $S = \pi r^2$). Thus, the per-unit length resistance is now

$$R \approx \frac{1}{2\pi r\delta\sigma} = \frac{\sqrt{f\mu}}{2r\sqrt{\pi\sigma}} \tag{3.51}$$

It has become apparent that the skin effect causes the effective resistance of the conductor to increase with the frequency of the current. To illustrate the effects of skin depth, the skin depth and resistance of a 3 cm gold rectangular track of a device (as part of a microstrip line or CPW) are given in Table 3.3 (conductivity $\sigma = 4.1 \times 10^7$ S/m). The width and thickness of the track are 16 and 7 μm, respectively. The skin depth is small, and the resistance changes significantly with the frequency.

The effects of skin depth are not just limited to the resistance, it may cause other problems such as overheat in certain areas of the circuit or antenna, which should also be taken into account in system design.

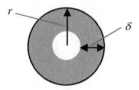

Figure 3.18 A conducting wire of skin depth δ

Table 3.3 Skin depth and resistance of a gold track of dimensions 7 μm × 16 μm × 30,000 μm

Frequency (MHz)	Skin depth (μm)	Resistance (Ω)
1	78.60	6.53
1000	2.49	8.16
2500	1.57	11.72
10,000	0.79	21.72
20,000	0.56	30.07
40,000	0.39	41.91

3.7 Summary

In this chapter, we have obtained the wave solutions and introduced a set of concepts and parameters for field analysis, which include the plane wave, polarization and intrinsic impedance.
Careful examinations on radiowave reflection, transmission, diffraction and scattering have
been undertaken, and the focus has been on the analysis of the radio reflection and transmission,
the two most common propagation mechanisms. Radio propagation models for various scenarios and channel characteristics have been introduced and discussed. The comparison of the field
concepts and circuit concepts has revealed the interesting correspondence between these two
sets of concepts.

References

1. J. D. Kraus and D. A. Fleisch, *Electromagnetics with Application*, 5th edition, McGraw-Hill, 1999.
2. R. E. Collin, *Antennas and Radiowave Propagation*. McGraw-Hill, Inc., 1985.
3. T. S. Rappaport, *Wireless Communications*, 2nd edition, Prentice Hall, 2002.

Problems

Q3.1. Explain the concept of a plane wave with the aid of a diagram.

Q3.2. Compared with other forms of waves (such as sound waves), what is the unique feature
of EM waves?

Q3.3. Radio propagation constant is given by Equation (3.3), prove that the attenuation and
phase constants can be expressed as

$$\alpha = \omega\sqrt{\mu\varepsilon}\left[\frac{1}{2}\left(\sqrt{1 + \frac{\sigma^2}{\varepsilon^2\omega^2}} - 1\right)\right]^{1/2}$$

$$\beta = \omega\sqrt{\mu\varepsilon}\left[\frac{1}{2}\left(\sqrt{1 + \frac{\sigma^2}{\varepsilon^2\omega^2}} + 1\right)\right]^{1/2}$$

respectively.

Q3.4. For a lossy/conducting medium $\tan\delta = \frac{\sigma}{\omega\varepsilon} > 100$, find simplified expressions for the
attenuation and phase constants. If this medium is sea water, find its attenuation and
phase constants, phase velocity and wavelength in the medium at 1 MHz. Make a comparison with their free space counterparts.

Q3.5. If the electric field of a wave can be expressed as

$$E = 2e^{\,j(\omega t + \beta z)}\hat{x}$$

where t is the time and z is the distance along z-axis.

a. What are ω and β? Propagation direction? Polarization?

b. Verify this is a solution of the wave equation;

c. Obtain the wave velocity in free space and magnetic field H;

d. Find the power flow density of the wave;

e. If this wave is reflected by a perfect conductor at $z = 0$, obtain the reflection coefficient and write down the expression of the reflected wave.

Q3.6. Explain the concept of circular polarization, give an example of a right-hand circular polarization and suggest a typical application for such a signal.

Q3.7. What is Brewster's angle? And what is the critical angle?

Q3.8. Explain the concept of skin depth. If a copper box is employed to host a sensitive high data rate digital circuit and the minimum electric field attenuation required at 1 GHz is 100 dB, find the minimum thickness of the box, which is equivalent to how many skin depths?

Q3.9. An important application of $\lambda/4$ matching plate is for eliminating reflections. Prove that a wave is matched through a thick slab of $\varepsilon_r = 4$ dielectric by means of two $\lambda/4$ plates of $\varepsilon_r = 2$ on each side of the slab.

Q3.10. If a plane wave at 1 GHz propagates normally to a round plane having constants: relative permittivity $\varepsilon_r = 9$, conductivity $\sigma = 0$, and relative permeability $\mu_r = 1$, find

a. what the intrinsic impedance of the medium is; and

b. how much energy (in % and dB) is reflected.

If a layer of paint is used to improve the matching, what are its desired characteristics such as ε_r and thickness?

If a layer of paint is used to achieve the maximum reflection, what are its desired characteristics?

Q3.11. A right-hand circularly polarized plane wave is reflected by a good conductor. What is the polarization of the reflected wave? If this wave is reflected by a concrete ground plane, comment on the polarization of the reflected wave again and justify you comments.

Q3.12. Explain the concept of scattering, why is the sky red at sunrise (and blue at noon)?

Q3.13. A mobile radiowave at 1800 MHz of linear polarization is reflected by a 20 cm brick wall with a relative permittivity of 4.5 and conductivity of 0. If the incident angle is $30°$,

a. Calculate the reflection and transmission coefficients if the wave has perpendicular polarization;

b. Calculate the reflection and transmission coefficients if the wave has parallel polarization;

c. If the incident wave has neither perpendicular nor parallel polarization, how can we calculate the reflection coefficient?

Q3.14. Discuss the effects of fog on radio propagation and explain how to obtain the attenuation factor using a mathematical expression. Illustrate how the attenuation changes against frequency with the aid of a diagram.

Q3.15. The indoor radio propagation channel is of interest to many wireless engineers. Path loss and delay spread are two of the most important parameters of a radio channel. Explain the concepts of these two parameters.

Q3.16. Radio propagation models are very important tools for radio system designers and planners. Derive the path loss for the free space model and two-ray model, respectively, and then compare them in terms of the variation against the frequency and distance.

Q3.17. Radiowave absorbing materials are widely used for antenna measurements. A lossy mixture of a high-μ (ferrite) and a high-ε (barium titanate) material can be used effectively for wave absorption with the ratio μ/ε equal to that for free space. Let a 1 GHz plane wave to be incident normally on a solid ferrite-titanate slab of thickness $d = 2$ cm and $\mu_r = \varepsilon_r = 30(1 - j)$. The medium is backed by a flat conducting sheet. How much is the reflected wave attenuated with respect to the incident wave? Expressed it in dB.

4

Antenna Basics

We have introduced circuit concepts and field concepts and studied transmission lines and radiowaves in Chapters 2 and 3. In this chapter, we are going to study antenna theory, see how antennas are linked to radiowaves and transmission lines, and introduce the essential and important parameters of an antenna from both the circuit point of view and field point of view. It is hoped that, through this chapter, you will become familiar with the antenna language and gain a better understanding of antennas. At the end, you will know how it works and what design parameters and considerations are.

4.1 Antennas to Radiowaves

In Chapter 2, we introduced Maxwell's equations which reveal the fundamental relations between the electric field, magnetic field, and the sources. Again, we only discuss a single-frequency source case (an arbitrary case can be considered as the combination of many single-frequency sources), thus Maxwell's equations can be written as

$$
\begin{aligned}
\nabla \times \boldsymbol{E} &= -j\omega\mu\boldsymbol{H} \\
\nabla \times \boldsymbol{H} &= \boldsymbol{J} + j\omega\varepsilon\boldsymbol{E} \\
\nabla \cdot \boldsymbol{E} &= \rho/\varepsilon \\
\nabla \cdot \boldsymbol{H} &= 0
\end{aligned}
\tag{4.1}
$$

Using a similar process of deriving Equation (3.2), we can obtain

$$
\nabla^2 \boldsymbol{E} + \omega^2\mu\varepsilon\boldsymbol{E} = j\omega\mu\boldsymbol{J} + \nabla(\rho/\varepsilon)
\tag{4.2}
$$

Antennas: From Theory to Practice, Second Edition. Yi Huang.
© 2022 John Wiley & Sons Ltd. Published 2022 by John Wiley & Sons Ltd.
Companion website: www.wiley.com/go/huang_antennas2e

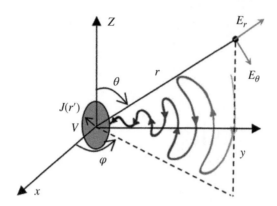

Figure 4.1   Coordinates and radiowaves generated by a time-varying source. Source: From Collin [1]

This is an equation which links the radiated electric field (no magnetic field) directly to the source (current and charge density). To solve this equation, boundary conditions are required. For an open boundary, which means that the field vanishes when the distance from the source V to the field point becomes infinite as shown in Figure 4.1, the solution of Equation (4.2) in a uniformed medium (μ and ε) can be yielded as [1]

$$E(r) = -j\omega\mu \int_V J(r') \frac{e^{-j\beta\cdot|r-r'|}}{4\pi|r-r'|} dv' + \frac{1}{j\omega\varepsilon} \nabla \left(\nabla \bullet \int_V J(r') \frac{e^{-j\beta\cdot|r-r'|}}{4\pi|r-r'|} dv' \right) \qquad (4.3)$$

where r is the distance vector from the origin to the observation point and r' is from the origin to the source point. *This equation gives the radiated electric field from a time-varying current J (the time factor $e^{j\omega t}$ is omitted here) and is the very foundation of the antenna theory – it reveals how the antenna is related to radiowaves.* Only a time-varying current (vibrating charges as shown in Figure 4.1) can generate a radiowave – not DC current or static charges. *The antenna design is all about how to control the current distribution J hence to obtain the desired radiated field E.* The antenna theory could be summarized by this single but complex equation which includes vector partial differentiation and integration. It is normally not possible to yield an analytical expression of the radiated field.

4.1.1   Near Field and Far Field

However, the analytical solution is obtainable for some very simple cases. For example, when the source is an *ideal current element* with length Δl and current value I, the current density vector can be expressed as

$$J = \hat{z}I\Delta l$$

This current is electrically short (i.e. $\Delta l \ll \lambda$) and fictitious, but very useful for antenna analysis. Replacing J with this representation in Equation (4.3), the radiated electric field can be found as

$$E_r = 2\frac{I\Delta l}{4\pi}\eta\beta^2 \cos\theta \left(\frac{1}{\beta^2 r^2} - \frac{j}{\beta^3 r^3}\right)e^{-j\beta r}$$

$$E_\theta = \frac{I\Delta l}{4\pi}\eta\beta^2 \sin\theta \left(\frac{j}{\beta r} + \frac{1}{\beta^2 r^2} - \frac{j}{\beta^3 r^3}\right)e^{-j\beta r} \tag{4.4}$$

$$E_\varphi = 0$$

where η is the intrinsic impedance of the medium. Use this and the first equation in Equation (4.1) to yield the magnetic field:

$$H_r = 0; \quad H_\theta = 0$$

$$H_\varphi = \frac{I\Delta l}{4\pi}\beta^2 \sin\theta \left(\frac{j}{\beta r} + \frac{1}{\beta^2 r^2}\right)e^{-j\beta r} \tag{4.5}$$

It is apparent that, for this simple case, the electric field has E_θ and E_r components, whilst the magnetic field has just H_φ component. The other field components are zero. E_θ, E_r, and H_φ are complex numbers linked to frequency, distance, and angle θ.

When the angle θ is $90°$, the magnetic field magnitude reaches the maximum. The electric field, magnetic field, and their ratio are shown in Figure 4.2 as a function of βr for a given frequency. It is interesting to note that

- The electric field is always greater than the magnetic field;
- When $\beta r > 1$, the ratio of $|E/H| = \eta$ is about 377 Ω in free space.
- When $\beta r < 1$, as the distance increases, the electric field reduces at a much faster rate (60 dB/decade) than the magnetic field (40 dB/decode).

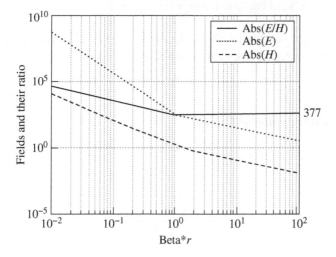

Figure 4.2   The fields E, H, and E/H as a function of βr at a fixed frequency

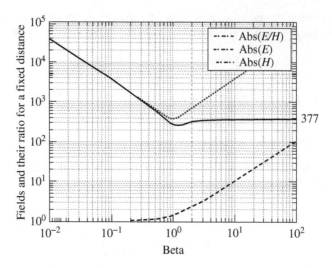

Figure 4.3   The fields E, H, and E/H as a function of β at a fixed distance ($r = 1$)

When the distance is fixed, the electric field, magnetic field, and their ratio are shown in Figure 4.3 as a function of β (which is proportional to the frequency). It is evident that

- $\beta r = 1$ i.e. $r = \dfrac{\lambda}{2\pi}$ is still an important point.
- The electric field first reduces as β (or frequency) increases to the point $\beta r = 1$ and then changes to increase with β (or frequency) after this point.
- The magnetic field is a constant before the point $\beta r = 1$ and then increases with β (or frequency) at the same rate as the electric field when $\beta r > 1$.
- The ratio of E/H reduces (at the same rate as the electric field) first as β increases to about $\beta r = 1$; it then becomes a constant for $\beta r > 1$.

These conclusions are very useful for applications, such as radio frequency identification (RFID) and electromagnetic compatibility (EMC) where the operation may take place near the source/antenna.

4.1.1.1   Far Field (Fraunhofer Region)

When $\beta r \gg 1$, i.e. $r \gg \dfrac{\lambda}{2\pi}$, we have $\dfrac{1}{\beta r} \gg \dfrac{1}{\beta^2 r^2} \gg \dfrac{1}{\beta^3 r^3}$, Equation (4.5) can therefore be simplified as

$$E_\theta = \frac{jI\Delta l}{4\pi r}\eta\beta\sin\theta e^{-j\beta r}$$

$$E_r \approx 0; \quad E_\varphi = 0$$

$$(4.6)$$

and the magnetic field is reduced to

$$H_\varphi = \frac{jI\Delta l}{4\pi r}\beta \sin\theta e^{-j\beta r}$$

$$H_r = 0; \quad H_\theta = 0$$

(4.7)

The fields are now simpler; it is important to note the following points:

- There is now just one electric field component and one magnetic field component;
- Both fields are inversely proportional to the distance r;
- The electric field and magnetic field are in phase and orthogonal to each other. Unlike the near field to be discussed, the power density function (the cross product of these two fields) is not a complex number but a real value which is inversely proportional to the distance square r^2:

$$\boldsymbol{S} = \boldsymbol{E} \times \boldsymbol{H}^* = \hat{r}\left(\frac{I\Delta l}{4\pi r}\beta \sin\theta\right)^2 \eta$$

(4.8)

- The ratio of E/H is η, *the intrinsic impedance*, which is the same as that of the plane wave in Chapter 2;
- The fields are proportional to $\sin\theta$. They are zero at $\theta = 0°$ and $180°$, but maximum at $\theta = 90°$.

Comparing the far field with the plane wave, we can see that they are basically the same except that the far field amplitude is inversely proportional to the distance whilst the amplitude of the plane wave is constant. Thus, the far field can be considered as a *local plane wave*.

It should be pointed out that the far field condition is actually not that straightforward. The condition of $r \gg \lambda/2\pi$ was introduced for electrically small antennas and is just a function of the frequency and not linked to the antenna dimensions. When the antenna size D is electrically large, $D > \lambda$, the common definition of the *far field condition* is

$$r > \frac{2D^2}{\lambda}$$

(4.9)

This is obtained under the condition that the maximum phase difference from any point on the antenna to the receiving point is less than $\pi/8$.

If an antenna cannot be considered as electrically large, the recommended far field condition is

$$r > 3\lambda \gg \lambda/2\pi$$

(4.10)

and some people use 10λ or other conditions which are also aimed to ensure that the field is really far enough to be considered as a local plane wave. For example, a half-wavelength dipole has maximum dimension $D = 0.5\lambda$, $2D^2/\lambda = 0.5\lambda$ is still too small to be considered as in its far field. Thus, this additional condition is necessary.

The far field condition is defined with certain ambiguity. Care must be taken when conducting antenna far field measurements. If the separation between the transmitting and receiving antennas is not large enough, significant errors could be generated.

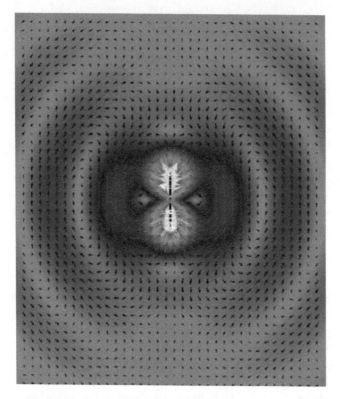

Figure 4.4   The electric field around a current element antenna

4.1.1.2   Near Field

When the far field conditions are not met, the field is considered as the *near field*. The near field of a current element (a dipole in this case) is illustrated in Figure 4.4 where the arrows represent the directions of the electric fields, since both E_θ and E_r have a number of terms linked to the frequency and distance and they are comparable in terms of magnitudes, Equations (4.4) and (4.5) cannot be simplified. E_θ reaches the maximum when $\theta = 90°$, while E_r peaks when $\theta = 0°$ and $180°$ for a fixed distance.

The region for $r < \lambda/2\pi$ is normally called the *reactive near field*. The field changes rapidly with distance. From its power density function $S = E \times H^*$ which is now a complex number using Equations (4.4) and (4.5), we can see that

- It contains both the radiating energy (the real part) and reactive energy (the imaginary part – it does not dissipate energy which is like a capacitor or inductor). The latter is normally dominant in this region.
- It has components in r and φ directions. The former is radiating away from the source, and the latter is reactive around the antenna.

Also, as shown in Figure 4.2, when it is close to the source (a short current element antenna), the electric field strength is much greater than the magnetic field strength. Their ratio is decreased to η as the distance increases.

Table 4.1 Near field and far field conditions

Antenna size D	$D \ll \lambda$	$D \approx \lambda$	$D \gg \lambda$
Reactive near field	$r < \lambda/2\pi$	$r < \lambda/2\pi$	$r < \lambda/2\pi$
Radiating near field	$\lambda/2\pi < r < 3\lambda$	$\lambda/2\pi < r < 3\lambda$ and $2D^2/\lambda$	$\lambda/2\pi < r < 2D^2/\lambda$
Far field	$r > 3\lambda$	$r > 3\lambda$ and $2D^2/\lambda$	$r > 2D^2/\lambda$

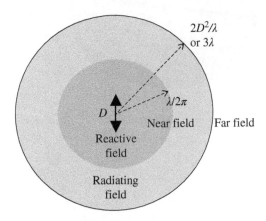

Figure 4.5　Radiated field regions of an antenna of max dimension D

It should be pointed out that, if the antenna is not electrically small, $r < \lambda/2\pi$ to define the reactive near field is not accurate. For simplicity, we can consider that this distance r is the distance from the surface of the antenna, not the center of the antenna.

The region between the reactive near field and the far field is a transition region and is known as the *radiating near field* (or *Fresnel Region*) where the reactive field becomes smaller than the radiating field. A brief summary of the three field regions is given in Table 4.1 and Figure 4.5. For EMC engineers, the near field is of great interest, but for antenna engineers, the far field is much more important than the near field.

4.1.2　Radiation Pattern

From the field point of view, we need to be able to not only distinguish the near field from the far field but also characterize the field, especially the far field, so as to obtain more detailed information on the radiated field characteristics. The most important parameters include such as radiation pattern, beamwidth, directivity, gain, efficiency factor, effective aperture, polarization, and bandwidth. The antenna temperature and radar cross section are key parameters for some applications.

The *radiation pattern* of an antenna is a plot of the radiated field/power as a function of angle at a fixed distance which should be large enough to be considered in its far field. The three-dimensional (3D) radiation pattern of the electrically short current element is plotted in Figure 4.6. 3D pattern is an excellent illustration of the radiated field distribution as a function of angle θ and φ in space. Unfortunately, it is hard and also very time-consuming to measure the

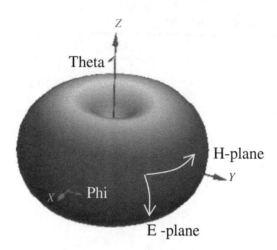

Figure 4.6   The 3D radiation pattern of an electrically short current element

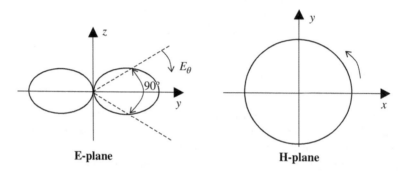

Figure 4.7   The E-plane and H-plane patterns of an electrically short current element

3D pattern of an antenna in practice. Most of the antennas have certain symmetrical feature; thus, in reality, the most important patterns are the radiation patterns in the two main planes: E-plane and H-plane. The E-plane is the lane that the electric field E lies in, while the H-plane is the plane that the magnetic field H is in. For the ideal current element case, the electric field is E_θ and magnetic field is H_φ; thus, the E-plane pattern is the field E_θ measured as a function of θ when the angle φ and distance are fixed, while the H-plane pattern is the field E_θ (not the magnetic field H_φ) measured as a function of φ when the angle θ and distance are fixed. The E-plane (at $\varphi = 0$) and H-plane (at $\theta = \pi/2$) patterns of the short current element are shown in Figure 4.7. Obviously, this antenna has an *omni-directional pattern* in the H-plane; it is a desirable feature for many mobile antennas since the antenna is not sensitive to orientation. Another special case is called the *isotropic antenna*, which has the same radiation power at all angles. This is a hypothetical case and cannot be realized in practice, but we do sometimes use it as a reference for analysis.

It should be pointed out that *the H-plane pattern is actually a measure of the electric field, not magnetic field*. When we talk about radiation pattern, it always means the electric field or

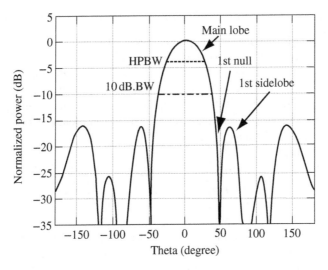

Figure 4.8 A radiation pattern illustrated in a conventional 2D plot

the power (which is proportional to the electric field square) pattern. When the patterns are plotted in the linear scale, the field pattern and power pattern may look very different. However, *when the patterns are plotted in the logarithmic scale (dB plot), both the normalized field and power patterns are the same* since $10 \log(P/P_{max})$ is the same as $20 \log(E/E_{max})$. Thus, in practice, we often plot the patterns in dB scale, which is also easy to see details of the field or power over a large dynamic range, especially easy to see some minor sidelobes. It should be pointed out that if the electric field has more than one component, we normally need to plot each component pattern. The total radiation pattern is useful for gaining the overall view of the radiation.

In addition to the 3D and 2D polar plots, another popular way to plot the radiation pattern is given in Figure 4.8 (also called the rectangular plot or universal plot). In this example, the horizontal axis is the rotation angle θ and the vertical axis is the radiated power in dB which is normalized to the maximum – this is a very common practice since the radiation pattern is about the relative power distribution as a function of angle and the absolute value is not important as long as it is above the noise floor.

The radiation pattern contains a lot of useful information about the radiation characteristics of the antenna, some cannot be quantified (such as the shape of the pattern) and some can be quantified. Some of the most important ones are:

- The *half-power beamwidth* (HPBW) of the main lobe, also called the *3 dB beamwidth*; or just the *beamwidth* (to identify how sharp the beam is);
- The *10 dB beamwidth* or *first null beam width* (FNBW) (another one to capture the main beam shape);
- The *first side lobe level* (expressed in dB, relative to the peak of the main beam);
- The *front to back ratio* (the peak of the main lobe over the peak of the back lobe, another attempt to identify the directivity of the antenna).
- Null positions (sometimes used for anti-interference and positioning).

4.1.3   Directivity, Gain/Realized Gain and Radiation Efficiency

From the field point of view, the most important quantitative information of the antenna is *the directivity*, which is a measure of the concentration of radiated power in a particular direction. It is defined as the ratio of the radiation intensity in a given direction from the antenna to the radiation intensity averaged over all directions. The average radiation intensity is equal to the total radiated power divided by 4π (the total surface area of a unit sphere). If the direction is not specified, the direction of maximum radiation is implied. Mathematically, the directivity (dimensionless) can be written as

$$D = \frac{U(\theta,\varphi)}{U(\theta,\varphi)_{av}} = \frac{4\pi U(\theta,\varphi)}{P_t} = \frac{4\pi U(\theta,\varphi)}{\oiint\limits_{\Omega} U d\Omega} \tag{4.11}$$

where P_t is the total radiated power in W and U is the *radiation intensity* in W/unit solid angle and is linked to the power density S (it has a unit of W/m^2 and was discussed in Chapter 3, such as Equation (3.16)) by distance square, that is

$$U = r^2 S \tag{4.12}$$

Example 4.1   Directivity
The radiated power density of the electrically short current element is given earlier as (Equation 4.8):

$$S = E \times H^* = \hat{r} \left(\frac{I\Delta l}{4\pi r} \beta \sin\theta \right)^2 \eta$$

Determine the directivity of the antenna as a function of the directional angles, find the maximum directivity and express it in decibels.

Solution
For a given frequency, the power density is S, the radiation intensity is therefore

$$U = r^2 \cdot \left(\frac{I\Delta l}{4\pi r} \beta \sin\theta \right)^2 \eta = \left(\frac{I\Delta l}{4\pi} \beta \right)^2 \eta \sin^2\theta = U_0 \sin^2\theta$$

where $U_0 = \left(\frac{I\Delta l}{4\pi} \beta \right)^2 \eta$. The total radiated power is given by

$$P_t = \oiint\limits_{\Omega} U d\Omega = \int_0^{2\pi} \int_0^{\pi} U_0 \sin^2\theta \cdot \sin\theta d\theta d\varphi = U_0 \cdot 2\pi \int_0^{\pi} \sin^3\theta d\theta$$

$$= -U_0 \cdot 2\pi \int_0^{\pi} (1 - \cos^2\theta) d\cos\theta = U_0 \cdot 2\pi \cdot \frac{4}{3} = \frac{8\pi}{3} U_0$$

Thus, the directivity as a function of angles is

$$D = \frac{4\pi U}{P_t} = \frac{4\pi U_0 \sin^2\theta}{8\pi U_0/3} = 1.5 \sin^2\theta$$

The maximum directivity occurs at $\theta = \pi/2$, it is

$$D_0 = 1.5 = 1.76 \text{ dBi}$$

It is convention to use the isotropic antenna as the reference; thus, the directivity is normally expressed in **dBi**, and in this case, the maximum directivity is 1.76 dBi. It means that the antenna radiates 1.76 dB more at the maximum direction than an isotropic antenna when the same amount of power is radiated. It also means that the antenna will radiate less power at some other directions than an isotropic antenna to ensure that the same amount of power is radiated for both antennas. Sometimes, a half-wavelength dipole antenna is used as a reference; in this case, the directivity is expressed in **dBd**, where d is for dipole.

For a directional antenna with one main lobe, the (maximum) directivity is closely linked to the half-power beamwidth in the E- and H-planes. If the beamwidths are known, there is an approximation which can be employed to calculate the directivity:

$$D \approx \frac{4\pi}{\theta_{HP}\varphi_{HP}} \tag{4.13}$$

where θ_{HP} and φ_{HP} are the half-power beamwidths at the two principle orthogonal planes, expressed in radians. If they are given in degrees, this equation can be rewritten as

$$D \approx \frac{41,253°}{\theta_{HP}\varphi_{HP}} \approx \frac{41,000°}{\theta_{HP}\varphi_{HP}} \tag{4.14}$$

The validity of these two equations is based on the assumption that the pattern has only one major lobe. They are not suitable to calculate the directivity of e.g. the short current element, since this kind of antenna is not directional (but omni-directional) in the H-plane and we cannot find the HPBW in this plane.

In practice, the total input power to an antenna can be easily obtained, but the total radiated power by an antenna is actually hard to get. The *gain* of an antenna is introduced to solve this problem. It is defined as the ratio of the radiation intensity in a given direction from the antenna to the total input power accepted by the antenna divided by 4π. If the direction is not specified, the direction of maximum radiation is implied. Mathematically, the gain (dimensionless) can be written as

$$G = \frac{4\pi U}{P_{in}} \tag{4.15}$$

where U is again the *radiation intensity* in W/unit solid angle and P_{in} is *the total input power accepted by the antenna* in Watts. It should be pointed out that the input power to the antennas (also called the supplied power) is different from the input power accepted by the antenna when the feed line is not matched with the antenna. The voltage reflection coefficient Γ at the antenna input and the *match efficiency* $= 1 - |\Gamma|^2$ (Equation (4.41)) will be discussed later in this chapter. Basically, if the feed line is matched with the antenna, $\Gamma = 0$ and matching efficiency is 100%. Another closely linked parameter is called the *realized gain* of an antenna which is the product of the matching efficiency and antenna gain. When an antenna is well matched with the feed line, the realized gain is the same as the antenna gain.

Comparing Equations (4.11) and (4.15), we can see that the gain is linked to the directivity by

$$G = \frac{P_t}{P_{in}} D = \eta_r D \tag{4.16}$$

where η_r is called the *radiation efficiency of antenna* and it is the ratio of the total radiated power to the input power accepted by the antenna:

$$\eta_r = \frac{P_t}{P_{in}} \tag{4.17}$$

This efficiency has taken both the conductor loss and dielectric loss into account, but not the impedance mismatch between the feed line and the antenna. Further discussion, from the circuit point of view, will be presented in Section 4.2 which includes another formula to calculate this factor.

Example 4.2   Gain and total radiated power
If the radiation efficiency of the antenna in Example 4.1 is 50%, the antenna is well matched with the input with a power of 1 W, find:
   a. the antenna gain and
   b. the total radiated power (TRP).

Solution
   a. From Example 4.1, we know the directivity $D = 1.5$. Since the radiation efficiency $\eta_r = 50\% = 0.5$, the antenna gain

$$G = \frac{P_t}{P_{in}} D = \eta_r D \quad = 0.75 \quad \text{or} \quad -1.25 \text{ dBi.}$$

   The gain is smaller than 1 (or 0 dBi)! This is resulted from the small efficiency factor which is very common for electrically small antennas.
   b. Since the input power is 1 W and the antennas is well matched with the power supplier, the total input power accepted by the antenna is therefore 1 W and the total radiated power (TRP):

$$P_t = \eta_r P_{in} = 0.5 \times 1 = 0.5 \ (W) = -3 \text{ dBW} = 27 \text{ dBm}$$

Note, the TRP is a very important parameter in practice, and we will discuss it further in the coming chapters.

4.1.3.1 EIRP

The antenna gain is often incorporated into a parameter called *effective isotropic radiated power*, or *EIRP* which is the amount of power that would have radiated by an isotropic antenna to produce the peak power density observed in the direction of maximum antenna gain, that is

$$EIRP = P_{in}G = P_t D \tag{4.18}$$

where P_{in} is the input power accepted by the antenna. The *EIRP* has a unit of watts but is often stated in dBW (or dB for convenience in practice). Since the directivity of an antenna is always >1, the *EIRP* is therefore greater than the TRP (P_t) of an antenna. The advantage of expressing the power in terms of the *EIRP* is that the path loss between the transmitting antenna and the receiving antenna can be easily obtained by the ratio of the transmitted *EIRP* to the received *EIRP*; this is why in the mobile radio industry the *EIRP* is widely used.

Another useful application of the *EIRP* is on setting radiation limited. For example, according to EU regulations, the power allowed to be transmitted at 2.45 GHz ISM band is the *EIRP* = 100 mW or 20 dBm. If a transmitter of 10 mW is used, the highest antenna gain can only be 10 (it is also 10 dBi), otherwise the EU power limit would be exceeded (cable loss and mismatch are assumed to be very small). Here, we have used the *EIRP* to select the right antenna.

From the input power and the knowledge of the gain of an antenna, it is possible to calculate its radiated power and field strength values at a direction of interest. For example, if the input power accepted by the antenna = 10 dBm, the *EIRP* from an isotropic antenna in all direction would be 10 dBm. However, the *EIRP* of an antenna with maximum gain $G = 7$ dBi would be $10 + 7 = 17$ dBm which is 7 dB more than that from the isotropic antenna at the maximum direction. Please note for both cases, the input power is fixed at 10 dBm, the difference in the *EIRP* is resulted from their antenna gains.

Another closely related term to *EIRP* is the *effective radiated power, ERP*, which is also widely used in the industry; the radiated power is calculated using a half-wavelength dipole rather than an isotropic antenna as reference, thus

$$ERP\,(\text{dBW}) = EIRP\,(\text{dBW}) - 2.15\,\text{dBi} \tag{4.19}$$

Example 4.3   EIRP and ERP
Find the *EIRP* and ERP of the antenna in Example 4.2

Solution
Since the gain of the antenna is $G = 0.75$ and the power into the antenna $P_{in} = 1$ W, we have

$$EIRP = P_{in}G = 1 \times 0.75 = 0.75 = -1.25\,\text{dBW, which is greater than } P_t \text{ in Example 4.2.}$$

$$ERP(\text{dBW}) = EIRP(\text{dBW}) - 2.15\,\text{dBi} = -3.40\,\text{dBW.}$$

4.1.4   Effective Aperture and Aperture Efficiency

A common question about antennas is how the antenna directivity and gain are linked to its physical dimensions. Since the field/current on the antenna aperture is not uniform, thus the concept of antenna effective aperture is introduced to serve this purpose. The *effective aperture*

A_e is less than the physical aperture A_p; the *aperture efficiency* is defined as the ratio of these two, that is

$$\eta_{ap} = \frac{A_e}{A_p} \tag{4.20}$$

It is normally in the range of 50–80%, and the effective aperture can be found as

$$A_e = \frac{\lambda^2}{4\pi} D \tag{4.21}$$

Please note that in some books, the antenna gain is used instead of the directivity in this definition. Thus, the effective aperture in (4.21) can be considered as the maximum effective aperture where the antenna radiation efficiency is 100%. We can use it to calculate the power received by an antenna without knowing its efficiency. If we need to compute the received power at the antenna port, the antenna loss should be taken into account and the antenna gain should be used here as we will show later. In general, the effective aperture of an antenna is proportional to its directivity (and gain), and the directivity can also be expressed in terms of the aperture size and efficiency:

$$D = \frac{4\pi}{\lambda^2} A_e = \frac{4\pi}{\lambda^2} \eta_{ap} A_p \tag{4.22}$$

If the power density S at the receiving antenna is known, we can use the following equation to estimate the received power:

$$P_r = S A_e \tag{4.23}$$

Example 4.4   Effective aperture
The directivity of a pyramidal horn antenna of aperture width a and height b is

$$D = 6.4 \frac{ab}{\lambda^2}$$

Find its aperture efficiency. If the power density around the antenna is $1\,W/m^2$, find the received power of the antenna.

Solution
Its effective aperture is

$$A_e = \frac{\lambda^2}{4\pi} D = \frac{6.4}{4\pi} ab = 0.5093ab$$

Its physical aperture $A_p = ab$, thus the aperture efficiency is

$$\eta_{ap} = \frac{A_e}{A_p} = 50.93\%$$

If the incoming power density is 1 W/m², the received power is

$$P_r = SA_e = 0.5093ab \, (\text{W})$$

This means the larger the aperture, the larger the received power.

4.1.4.1 Effective Height and Antenna Factor

If an antenna is a wire-type antenna (such as a dipole), *the effective height* (h_e) may be used to replace the effective aperture. The effective height is proportional to the square root of effective aperture. It is defined as the ratio of induced open-circuit voltage V on the antenna to the incident electric field E or

$$h_e = \frac{V}{E} \propto \sqrt{A_e} \; (\text{m}) \tag{4.24}$$

Another closely related parameter is the *antenna factor* (A_F) which is defined as the ratio of the incident electric field E to the induced voltage V_0 at the antenna terminal (input port) when it is connected to a load/cable (50 Ω by default). This induced voltage is different from the V in Equation (4.24). Thus

$$A_F = \frac{E}{V_0} \; (1/\text{m}) \tag{4.25}$$

It links the radiated/received field strength to the voltage at the antenna terminal and is a very useful parameter for field strength and voltage conversion in field measurement applications. Thus, it is a widely used device descriptor in the EMC area. However, it is not part of standard antenna terminology. Antenna factor reflects the use of an antenna as a field measuring device or probe. It converts the reading of voltage on equipment (such as a receiver) to the field at the antenna. It is also linked to the gain and effective aperture of antenna. Since the received power at the antenna input port is the same delivered to the load of the antenna:

$$P_{in} = P_r \eta_r = SA_e \eta_r = \sqrt{\frac{\varepsilon}{\mu}} E^2 \cdot \frac{\lambda^2}{4\pi} D\eta_r = \sqrt{\frac{\varepsilon}{\mu}} E^2 \cdot \frac{\lambda^2}{4\pi} G = \frac{V_0^2}{R}$$

Here, the antenna loss has been taken into account by the radiation efficiency (hence, the gain is used). The load impedance R should be 50 Ω for a well-matched system (this assumption is an approximation in reality). Thus, the antenna factor is roughly

$$A_F = \frac{E}{V_0} = \sqrt{\frac{377 \cdot 4\pi}{\lambda^2 G \cdot R}} \approx \frac{9.73}{\lambda \sqrt{G}} \tag{4.25a}$$

It is a function of frequency and antenna gain. In practice, the antenna impedance is hardly 50 Ω; the antenna factor may need to be measured rather than calculated using Equation (4.25a). More information on antenna factor can be found in such as [3, 4].

4.1.5   Other Parameters from the Field Point of View

4.1.5.1   Polarization

Radiowaves are generated by antennas. The antenna polarization is the same as the polarization of its radiating wave. As discussed in Chapter 3, there are basically three polarizations: linear, circular, and elliptical polarizations. Which polarization is generated depends on how the current moves in the antenna. For a linear polarization, the current should travel along one axis; for a circular polarization, two orthogonal currents with a 90° phase off-set should be created on the antenna. The polarization of the current element used in this section is a linearly polarized antenna (along z-axis).

In practice, mixed polarizations may be found in many antennas since an antenna has to meet many requirements. Trade-offs may have to be made, and as a result, a pure linearly polarized or circularly polarized antenna may not be possible or necessary. For example, most of the mobile phone antennas are not purely linearly polarized since they are employed to receive signals with mixed polarizations. However, for a line-of-sight communication system, the polarization has to be matched in order to achieve the maximum efficiency of the whole system. Two orthogonally polarized antennas cannot communicate with each other due to polarization mismatch. Also, for many satellite communication systems (such as the GPS), a circular polarization may be employed in order to combat Faraday rotation in ionosphere. More discussion will be provided in Section 5.4.

4.1.5.2   Bandwidth

Many antenna parameters are functions of frequency. When the frequency is changed, the radiation pattern may also be changed, which may result in the changes of the directivity, gain, HPBW, and other parameters. Also, we may need to consider antenna bandwidth from impedance matching point of view. Thus, it is important to ensure that the right parameters are chosen when the antenna bandwidth is considered.

4.1.5.3   Antenna Temperature

The radiation from different sources is intercepted by antennas and appears at their terminals as an antenna temperature, which is defined as [5]

$$T_A = \frac{\int_0^{2\pi} \int_0^{\pi} T_B(\theta, \varphi) G(\theta, \varphi) \cdot \sin\theta d\theta d\varphi}{\int_0^{2\pi} \int_0^{\pi} G(\theta, \varphi) \cdot \sin\theta d\theta d\varphi} \tag{4.26}$$

where T_B is the source brightness temperature of the radiation source and G is the antenna gain pattern.

When an antenna is used for receiving purpose, the antenna temperature may be required for system consideration and signal to noise ratio (SNR) calculation. In this case, the source is remote (such as the sky or Mars) and the antenna is just a load to the system and may be viewed as a remote temperature measuring device. But this temperature has nothing to do with the physical temperature of the antenna if the antenna is a lossless device, and the temperature

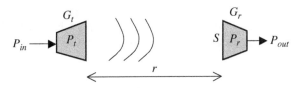

Figure 4.9 Transmitting and receiving antennas for Friis transmission formula

is equal to the source (such as the distant sky) temperature. In reality, the antenna has certain loss. An extreme case is that the antenna is completely lossy; the antenna temperature is therefore equal to its physical temperature. But generally speaking, its physical temperature is partially liked to the antenna temperature. The larger the effective aperture/gain, the larger the antenna temperature.

4.1.5.4 Friis Transmission Formula and Radar Cross Section

Another use of the aperture concept is for the derivation of the Friis transmission formula, which relates the power received to the power transmitted between two antennas separated by a distance r, which is large enough to be considered in the far field of the antenna. As shown in Figure 4.9, from Equations (4.12) and (4.15), the radiated power density from an antenna with gain G_t at distance r can be expressed as

$$S = \frac{U}{r^2} = \frac{DP_t}{4\pi r^2} = \frac{P_{in}G_t}{4\pi r^2} \tag{4.27}$$

This equation clearly shows how the radiated power density is linked to the input power, antenna gain, and distance. From Equations (4.21) and (4.23), we can obtain the received power at the receiving antenna:

$$P_r = SA_e = \frac{P_{in}G_t}{4\pi r^2} \cdot \frac{\lambda^2}{4\pi} D_r$$

To obtain the output power at the receiving antenna, we must take the radiation efficiency of the receiving antenna η_{er} into account, thus

$$P_{out} = P_r \eta_{er} = \frac{P_{in}G_t}{4\pi r^2} \cdot \frac{\lambda^2}{4\pi} D_r \eta_{er} = \frac{P_{in}G_t}{4\pi r^2} \cdot \frac{\lambda^2}{4\pi} G_r$$

That is

$$P_{out} = \left(\frac{\lambda}{4\pi r}\right)^2 G_t G_r P_{in} \tag{4.28}$$

This is the *Friis transmission formula*, and it links the input power of the transmitting antenna to the output power of the receiving antenna which is proportional to the wavelength square and the gains of the two antennas and inversely proportional to the distance square. $(\lambda/4\pi r)^2$ is called the *free space loss factor* – resulting from the spherical spreading of the radiated energy. This formula can also be expressed using the effective apertures of both antennas (A_{et} and A_{er}) as

$$P_{out} = \eta_{et}\eta_{er}\frac{A_{et}A_{er}}{\lambda^2 r^2}P_{in} \qquad (4.28a)$$

Friis transmission formula has been widely used for communications and radar applications. However, it is important to note that this formula is only accurate for free space (i.e., no reflections from the ground or a building, for example) and the polarization of both antennas are the same (i.e. no polarization mismatch).

For radar application, Friis formula can be related to the *radar cross section (RCS)*, σ, which is defined as the ability of a target to reflect the energy back to the radar and it is the ratio of the backscattered power to incident power density, that is

$$\sigma = \frac{\text{scattered power}}{\text{incident power density}} \qquad (4.29)$$

It is a far field parameter and expressed in m^2 or dBsm (decibels square meter). The typical RCS for a jumbo-jet aircraft is about $100\ m^2$, for a fighter aircraft about $4\ m^2$, for an adult man $1\ m^2$, and for a bird $0.01\ m^2$. It is a rather complex function of frequency since the scattered power depends on various things.

The power density in Equation (4.27) can be considered as the incident power density on the target, and the part reflected back to the radar is $\sigma S/4\pi r^2$ and the output power at the radar antenna is

$$P_{out} = \eta_e A_e \cdot \sigma S/4\pi r^2 = \frac{\sigma\lambda^2}{(4\pi)^3 r^4}G_t^2 \cdot P_{in} \qquad (4.30)$$

Thus, the received power is proportional to λ^2, but now inversely proportional to r^4, not r^2. This is a very important conclusion which has been widely used for radar applications.

4.2   Antennas to Transmission Lines

In this section, we are going to an antenna from another angle, the circuit point of view which is completely different from the field point of view.

4.2.1   *Input Impedance and Radiation Resistance*

As we mentioned earlier, the antenna is a transition device linking the transmission line and radiowaves; thus from the circuit point of view, the antenna is just a load to a transmission line. As shown in Figure 4.10, the most important parameter is therefore its impedance.

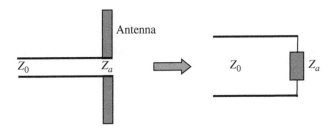

Figure 4.10   Antenna input impedance model

Antenna input impedance (Z_a) is defined as the impedance presented by an antenna at its terminals or the ratio of the voltage to current at its terminals. Mathematically, the input impedance is

$$Z_a = \frac{V_{in}}{I_{in}} = R_a + jX_a \qquad (4.31)$$

where V_{in} and I_{in} are the input voltage and current at the antenna input, respectively. The input impedance is a complex number; the real part consists of two components:

$$R_a = R_r + R_L \qquad (4.32)$$

where R_r is the radiation resistance and R_L is the loss resistance of the antenna.

The *radiation resistance* is the equivalent resistance which would dissipate the same amount of power as the antenna radiates when the current equals the input current at the terminals. If the total radiated power is P_t, then

$$R_r = \frac{P_t}{I_{in}^2} \qquad (4.33)$$

The loss resistance is from conductor loss and dielectric loss if lossy materials are used. If the antenna is directly connected to a source of impedance Z_s, for a matched load, the following condition should be met:

$$Z_s = Z_a^* = R_a - jX_a \qquad (4.34)$$

In reality, the antenna is normally connected to a short transmission line with a standard characteristic impedance of 50 or 75 Ω. Thus, the desired antenna input impedance is 50 or 75 Ω.

4.2.2   Reflection Coefficient, Return Loss, and VSWR

Impedance matching is extremely important. Since the antenna is just a load to a transmission line from the circuit point of view, we know that the reflection coefficient, return loss, and voltage standing wave ratio (VSWR) can be used to judge how well a load is matched with the transmission line, from the transmission line theory in Chapter 2. All these three parameters are interlinked. The specific representations are as follows:

- Reflection coefficient (a complex number, also expressed as S_{11} in practice):

$$\Gamma = \frac{Z_a - Z_0}{Z_a + Z_0} \tag{4.35}$$

- Return loss (expressed in dB):

$$L_{RT} = -20 \log_{10}(|\Gamma|) = -20 \log_{10}\left|\frac{Z_a - Z_0}{Z_a + Z_0}\right| \tag{4.36}$$

- *VSWR* (a ratio between 1 and infinity)

$$VSWR = \frac{1 + |\Gamma|}{1 - |\Gamma|} \tag{4.37}$$

We normally just need to use one of them to judge the impedance matching. The commonly required specification of an antenna is $S_{11} < -10\,\text{dB}$, or $L_{RT} > 10\,\text{dB}$, or $VSWR < 2$. These requirements are actually very close as seen in Table 2.1. For mobile phone antennas, it is also common to specify $S_{11} < -6\,\text{dB}$, or $L_{RT} > 6\,\text{dB}$ or $VSWR < 3$.

If we take a dipole as an example, its impedance can be found and shown in Figure 4.11a. When it is measured again a 50-Ω port, its reflection coefficient and *VSWR* are given in Figure 4.11b. It is interesting to note that:

- The antenna is well matched to $50\,\Omega$ when its length is about half wavelength, that is, its resistance is close to $50\,\Omega$ and its reaction is about 0, while its reflection coefficient is very small ($< -10\,\text{dB}$) and *VSWR* is also very small (< 2).
- When the antenna is resonant (i.e. the reactance is zero), the impedance matching is not necessary very good. For example, when the antenna is about one wavelength, the impedance matching is poor.
- When the antenna is about 1.5 wavelength in length, the impedance match is improved and there is another resonance point.
- For a given antenna, it may work well at a number of frequencies from the impedance matching point of view.

4.2.3 Other Parameters from the Circuit Point of View

4.2.3.1 Radiation Efficiency, Reflection Efficiency, and Total Efficiency

The radiation efficiency was defined by Equation (4.17) and can be calculated using the radiated power and the input power accepted by the antenna. Since they are directly linked to the radiation resistance and loss resistance, thus, the radiation efficiency factor can also be calculated using the following equation

$$\eta_r = \frac{P_t}{P_{in}} = \frac{R_r}{R_r + R_L} \tag{4.38}$$

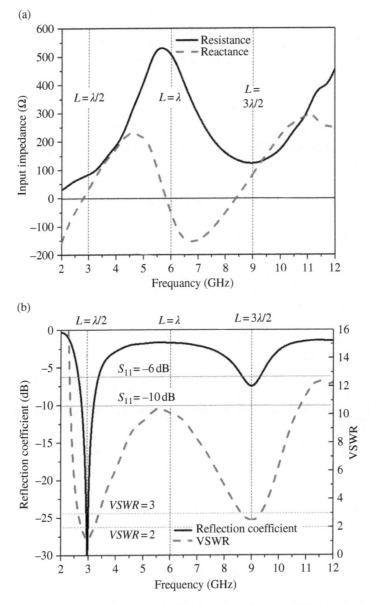

Figure 4.11   A dipole antenna performance: (a) The input impedance and (b) the reflection coefficient and VSWR

From the efficiency point of view, the larger the radiation resistance, the larger the efficiency factor. *But from the system matching point of view, we need the radiation resistance to match with the characteristic impedance of the line and the loss resistance to be zero if possible.* This is very requirement for an antenna from the circuit point of view.

If the input impedance of an antenna is not matched with the feeder impedance, then a part of the signal from the source will be reflected back, there is a reflection (mismatch) loss. It is characterized by the matching efficiency as mentioned earlier which is defined as the ratio of the input power accepted by the antenna to the source supplied power (P_s):

$$\eta_m = \frac{P_{in}}{P_s} = 1 - |\Gamma|^2 \tag{4.39}$$

Thus, the total efficiency of the antenna system (feed and the antenna) is the product of the two efficiencies:

$$\eta_t = \frac{P_t}{P_s} = \eta_m \eta_r \tag{4.40}$$

This has taken the feed-antenna mismatch into account.

Example 4.5   Matching, total efficiency, and radiated power

If the efficiency of the antenna in Example 4.1 is 50%, *VSWR* at the antenna input is 3 and the input/supplied power is 1 W, find:

    a. the matching efficiency;
    b. the total efficiency of the antenna;
    c. the total radiated power.

Solution

    a. *VSWR* $= 3$, we can easily obtain the voltage reflection coefficient using Equation (2.42) as

$$\Gamma = \frac{3-1}{3+1} = 0.5$$

this means a matching efficiency of the antenna is $1 - 0.5^2 = 0.75 = 75\%$.

    b. Since the radiation efficiency is 50%, thus the total efficiency

$$\eta_t = \frac{P_t}{P_s} = \eta_m \eta_r = 0.75*0.5 = 0.375 = 37.5\%$$

    c. Since the input power is 1 W, the total radiated power:

$$P_t = \eta_t P_s = 0.375 \times 1 \text{ (W)} = -4.26 \text{ dB}$$

In Example 4.2, the antenna was well matched with the source and the radiated power was -3 dB; thus, the impedance mismatch in this case has resulted in 1.26 dB loss in radiated power.

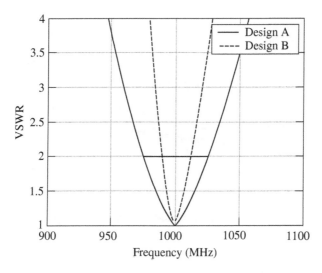

Figure 4.12   VSWR against the frequency for two designs

4.2.3.2   Bandwidth

From the circuit point of view, the antenna bandwidth is a simple but important parameter. It is normally defined as the frequency bandwidth with $L_{RT} > 10$ dB or $VSWR < 2$. Take Figure 4.12 as an example, which is a zoomed version of Figure 2.19, the center frequency of the antenna is 1000 MHz. For the required $VSWR < 2$, the bandwidth for Design A is from 975 to 1026 MHz, whilst the bandwidth for Design B is from 990 to 1012 MHz, and the relative bandwidths for these two designs are 5.1 and 2.2%, respectively. Both are narrow band.

Unlike other devices (e.g. filters and amplifiers), an antenna is a special and complex device in an RF/microwave system. The antenna bandwidth is a very special parameter since we need to consider the parameter from both the circuit point of view ($VSWR$ or L_{RT}) and the field point of view (radiation pattern). It is normally easy to meet one of the requirements but difficult to meet both requirements – this is particularly true for a broadband antenna, which normally means that its $VSWR$ and radiation pattern (e.g. HPBW) have a relative bandwidth of at least 20%. In some applications, a much wider bandwidth is required.

4.3   Summary

In this chapter, we have introduced the basic theory of antennas – the essence about antenna design is how to control the current distribution which determines the radiation pattern and input impedance. A detailed discussion on antenna near field and far field has been conducted. All important parameters of an antenna, which have been grouped from the field point of view and the circuit point of view, have been defined and discussed. A summary of these parameters is given in Figure 4.13. The Friis transmission formula and RCS have also been addressed.

It is worth pointing out that an antenna can be viewed as a two-port network; thus, the field quantities (such as the radiation pattern) are therefore related to the transmission coefficient (S_{21}), while the circuit quantities (such as the input impedance and $VSWR$) are determined by the reflection coefficient (S_{11}) of the network which will be further addressed in later chapters.

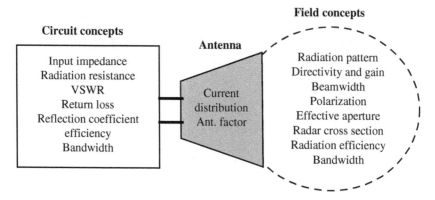

Figure 4.13   A summary of most important antenna parameters

References

1. R. E. Collin, *Antennas and Radiowave Propagation*. McGraw-Hill, Inc., 1985.
2. J. D. Kraus and R. J. Marhefka, *Antennas for All Applications*, 3rd edition, McGraw-Hill, Inc., 2002.
3. A. A. Smith, Jr., 'Standard-site method for determining antenna factors,' *IEEE Trans. Electromagnet. Compat.*, 24(8), 316–322, 1982.
4. R. P. Clayton, *Introduction to Electromagnetic Compatibility*, Wiley Interscience, 1992, pp. 202–206.
5. C. A. Balanis, *Antenna Theory: Analysis and Design*, 2nd edition, John Wiley & Sons, Inc., 1997.

Problems

Q4.1. If you were asked to use one equation to explain the antenna theory, which one would it be? Why?

Q4.2. Explain how a radiowave could be generated.

Q4.3. What are the most important antenna parameters?

Q4.4. An ideal element current source is given as $J = \hat{z} I \Delta l$, derive its radiated electric field using spherical coordinates.

Q4.5. What are the near field and far field of an antenna? What are their major differences?

Q4.6. Find the far field condition of a half-wavelength dipole.

Q4.7. Find the far field condition for a satellite dish antenna of a dimension 60 cm operating at 12 GHz.

Q4.8. Explain the concept of *Antenna Factor*. If the electric field around an antenna is 1 V/m and its antenna factor is 2/m, find the voltage at its matched output.

Q4.9. The radiated power density of an antenna in its far field can be expressed as:

$$ S = E \times H^* = \hat{r}\left(\frac{C}{r}\sin\theta\sin\varphi\right)^2, \quad \text{for} \quad 0 < \varphi < \pi $$

where C is a coefficient determined by the frequency and antenna dimensions. Find the directivity of the antenna as a function of the directional angles, the maximum directivity and express it in decibels.

Q4.10. A mono-static radar is employed to detect a target of a typical RCS of $4\,\text{m}^2$. Its operational frequency is 10 GHz, the transmitted power is 1000 W. If its antenna gain is 30 dBi, find the operational range/distance of this radar.

Q4.11. Compare the usefulness of the reflection coefficient, $VSWR$, and return loss for antenna characterization.

Q4.12. An antenna has an input impedance of $73 + j10\,\Omega$. If it is directly connected to a $50\,\Omega$ transmission line, find the reflection coefficient, $VSWR$, and return loss.

5

Popular Antennas

Chapter 4 has covered the basics of antennas and laid down the foundation for us to gain a better understanding of antennas. In this chapter, we are going to examine and analyze some of the most popular antennas using relevant antenna theories, to see why they have become popular, what their major features and properties (including advantages and disadvantages) are, and how they should be designed.

Since the start of radio communications over 100 years ago, thousands of antennas have been developed. They can be categorized by various criteria:

- in terms of the bandwidth, antennas can be divided into narrowband and broadband antennas;
- in terms of the polarization, they can be classified as linearly polarized and circularly polarized antennas (or even elliptically polarized antennas);
- in terms of the resonance, they can be grouped as resonant and traveling wave antennas;
- in terms of the number of elements, they can be classified as element antennas and array antennas.

In this book, we separate them according to their physical structures into wire-type antennas and aperture-type antennas. This is because different types of antennas exhibit different features and can be analyzed using different methods and techniques. Since antenna arrays can be formed by both types and possess some special features, they will be discussed in Section 5.3.

5.1 Wire-Type Antennas

Wire-type antennas are made of conducting wires and are generally easy to construct, thus the cost is normally low. Examples include dipoles, monopoles, loops, helixes, Yagi–Uda, and log-periodic antennas.

Antennas: From Theory to Practice, Second Edition. Yi Huang.
© 2022 John Wiley & Sons Ltd. Published 2022 by John Wiley & Sons Ltd.
Companion website: www.wiley.com/go/huang_antennas2e

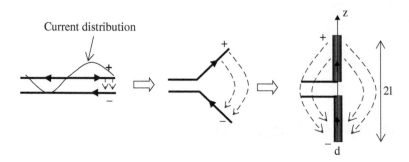

Figure 5.1 Evolution of a dipole of total length $2l$ and diameter d

5.1.1 Dipoles

Dipoles are one of the simplest but most widely used antennas. Hertz used them for his famous experiment. As shown in Figure 5.1, a dipole can be considered as a structure evolved from an open-end two-wire transmission line. A typical structure of a dipole consists of two metal wires which are normally of equal length.

In Chapter 4, an ideal short element of constant current was discussed. In reality, we cannot find such an antenna, since the current is zero at the end of a line. As we now understand that, for the antenna analysis and design, it is all about the current distribution. Once the current distribution is known, other parameters, such as the radiation pattern and input impedance, can be obtained.

5.1.1.1 Current Distribution

For an open-end transmission line, the reflection coefficient at the end ($z = 0$) is 1. From Equation (2.20), the current distribution along such a line can be expressed as:

$$I(z,t) = \text{Re}\left(\frac{1}{Z_0}\left(A_1 e^{j(\omega t - \beta z)} - A_1 e^{j(\omega t + \beta z)}\right)\right) = \frac{2}{Z_0}A_1 \sin(\omega t) \sin(\beta z) \tag{5.1}$$

This is a standing wave with peaks at $z = \lambda/4 + n\lambda/2$ (n are integers). If the dipole is very thin (ideally zero diameter), the current distribution can be approximated by this equation. That is, if the coordinate origin is chosen at the center of the dipole in Figure 5.1, the current can be expressed as:

$$I(z) = = \begin{cases} I_0 \sin(\beta(l-z)), & 0 \le z \le l \\ I_0 \sin(\beta(l+z)), & -l \le z \le 0 \end{cases} \tag{5.2}$$

where I_0 is the *maximum possible current* (different from the current at the feeding point) on the line. If the dipole length is shorter than $\lambda/2$, the maximum current on the dipole is less than I_0. This could be very confusing. The time term $\sin(\omega t)$ is omitted from the representation. Thus,

Table 5.1 Summary of some dipole characteristics

Dipole length $2l$	0.1λ	0.5λ	λ	1.5λ
Current distribution				
Radiation pattern				
Directivity	1.5 or 1.76 dBi	1.64 or 2.15 dBi	2.4 or 3.8 dBi	About 2.3
HPBW	$90°$	$78°$	$47°$	NA
Input impedance	R: very small ($\sim 2\,\Omega$) jX: capacitive	R: $\sim 73\,\Omega$ jX: $\sim 0\,\Omega$	R: very large jX: $\sim 0\,\Omega$ for thin dipole	R: $\sim 100\,\Omega$ jX: $\sim 0\,\Omega$ for thin dipole
Note	jX sensitive to the radius	$R + jX$ not sensitive to the radius	$R + jX$ sensitive to the radius	$R + jX$ sensitive to the radius

the current along the dipole is a time-varying standing wave. From Equation (5.2), we can see that the current is

- zero at the end of the pole;
- $I_0 \sin(\beta l)$ at the feeding point;
- a periodic function with a period equal to one wavelength.

The current distributions for the dipole of length $2l = \lambda/10$, $\lambda/2$, λ, and 1.5λ are shown in Table 5.1. Note that the current distributions on both poles are symmetrical (even function as shown in Equation (5.2)) about the antenna center.

5.1.1.2 Radiation Pattern

Replacing the current density in Equation (4.3) by Equation (5.2) to give

$$E(r) = \left(-j\omega\mu + \frac{\nabla\nabla\bullet}{j\omega\varepsilon} \right) \int_{-l}^{l} \hat{z} I(z) \frac{e^{-j\beta r + j\beta z \cos\theta}}{4\pi r} dz \qquad (5.3)$$

After some mathematical manipulation, the radiated electric field in the far field can be obtained approximately as:

$$E_\theta \approx j\eta \frac{I_0 e^{-j\beta r}}{2\pi r} \left(\frac{\cos(\beta l \cos\theta) - \cos(\beta l)}{\sin\theta} \right) \qquad (5.4)$$

and plotted in Table 5.1. Here, η is the intrinsic impedance. The magnetic field can be obtained as:

$$H_\varphi = \frac{E_\theta}{\eta} \approx j\frac{I_0 e^{-j\beta r}}{2\pi r}\left(\frac{\cos\left(\beta l\cos\theta\right) - \cos\left(\beta l\right)}{\sin\theta}\right) \tag{5.5}$$

Thus, the power density is

$$S = E \times H^* = \hat{r}\frac{\eta I_0^2}{4\pi^2 r^2}\left(\frac{\cos\left(\beta l\cos\theta\right) - \cos\left(\beta l\right)}{\sin\theta}\right)^2 \tag{5.6}$$

Using Equation (4.12), the radiation intensity is

$$U = r^2 S = \frac{\eta I_0^2}{4\pi^2}\left(\frac{\cos\left(\beta l\cos\theta\right) - \cos\left(\beta l\right)}{\sin\theta}\right)^2 \tag{5.7}$$

when βl is small ($< \pi/4$), $\cos(\beta l\cos\theta) \approx 1 - (\beta l\cos\theta)^2/2$, $\cos(\beta l) \approx 1 - (\beta l)^2/2$ and we have

$$\left(\frac{\cos\left(\beta l\cos\theta\right) - \cos\left(\beta l\right)}{\sin\theta}\right) \approx \frac{1}{2}(\beta l)^2\sin\theta$$

Thus, for short dipoles, the radiated field and radiation intensity are

$$E_\theta \approx j\eta\frac{I_0 e^{-j\beta r}}{4\pi r}(\beta l)^2\sin\theta \approx j\eta\frac{I_{in}e^{-j\beta r}}{4\pi r}(\beta l)\sin\theta$$

$$U \approx \frac{\eta I_{in}^2}{16\pi^2}(\beta l)^2\sin^2\theta \tag{5.8}$$

where $I_{in} = I_0\sin\beta l$ is the dipole input current and more useful than I_0 for short dipoles.

The E-plane radiation patterns for the length $2l = \lambda/10, \lambda/2, \lambda$, and 1.5λ are shown in Table 5.1. It is apparent that the current distribution indeed determines the radiation pattern. When the dipole length is less than a wavelength, the currents on both poles have the same polarity and there is only one lobe on both sides and no sidelobes. But when the length is greater than a wavelength, the currents on the dipole become complicated; they travel in two opposite directions which results in the split of the radiation pattern. "+" and "−" in the plots indicate opposite directions of the electric field in space.

5.1.1.3   Directivity and Gain

Using Equation (4.11), we can calculate the directivity by:

$$D = \frac{4\pi U(\theta,\varphi)}{P_t} = \frac{4\pi U(\theta,\varphi)}{\int_0^{2\pi}\int_0^\pi U\sin\theta d\theta d\varphi} = \frac{2U(\theta,\varphi)}{\int_0^\pi U\sin\theta d\theta} \tag{5.9}$$

It is not possible to find a simple expression for the directivity. A numerical solution can be found once the frequency and dipole length are known.

The gain of the antenna can be obtained by Equations (4.15) and (4.16). We may need to know what material is used to make the antenna in order to calculate the gain (and radiation efficiency). An example is given in Example 5.1.

The directivities for the dipole of length $2l = 0.1\lambda$, 0.5λ, λ, and 1.5λ are also shown in Table 5.1. The half-power beamwidths (HPBWs) for these antennas are given in the table as well. The directivity increases with the dipole length and reaches the first maximum (about 3.2) when the length is around 1.25λ (not shown in the table) [1, 2]. The relation between the directivity and beamwidth in Equation (4.13) cannot be applied to this kind of antenna. For some antennas, their HPBWs may be about the same, but the radiation patterns could be actually very different. Thus, it is essential to obtain the 2D or even 3D radiation pattern and not to purely rely on the HPBW to judge the radiation characteristics of an antenna.

5.1.1.4 Radiation Resistance and Input Impedance

Once the current distribution is obtained, we can calculate the radiation resistance using Equation (4.33), that is

$$R_r = \frac{P_t}{I_{in}^2} = \frac{2\pi \int_0^\pi U \sin\theta d\theta}{(I_0 \sin\beta l)^2} \tag{5.10}$$

We can find a complicated but explicit expression of the radiation resistance as:

$$R_r = \frac{\eta}{2\pi} \{ C_E + \ln(\beta l/2) - C_i(\beta l/2) + 0.5 \sin(\beta l/2)[S_i(\beta l) - 2S_i(\beta l/2)] \\ + 0.5 \cos(\beta l/2)[C + \ln(\beta l/4) + C_i(\beta l) - 2C_i(\beta l/2)] \} \tag{5.11}$$

where

$C_E = 0.5772$ is Euler's constant;

$C_i(x) = \int_\infty^x \frac{\cos y}{y} dy$ is cosine integral;

$S_i(x) = \int_0^x \frac{\sin y}{y} dy$ is sine integral.

This complex expression is not really suitable for applications. A lot of efforts have been made to make it simpler and easier to use. One of them can be found in [3]. If the loss resistance is neglected, the input impedance of a dipole less than a wavelength can be approximated as:

$$Z_a \approx f_1(\beta l) - j\left(120\left(\ln\frac{2l}{d} - 1 \right) \cot(\beta l) - f_2(\beta l) \right) \tag{5.12}$$

where

$$f_1(\beta l) = -0.4787 + 7.3246\beta l + 0.3963(\beta l)^2 + 15.6131(\beta l)^3$$
$$f_2(\beta l) = -0.4456 + 17.0082\beta l - 8.6793(\beta l)^2 + 9.6031(\beta l)^3$$

d is the diameter of the dipole.

This formula is valid only when the dipole length is not much longer than a half-wavelength with an error up to 0.5 Ω. In practice, this is the most useful range. If the impedance for a longer dipole is to be calculated, the best approach is to use numerical computation (software packages are widely available on the market), which is addressed in Chapter 6.

For an electrically short dipole, Equation (5.12) can be further simplified as:

$$Z_a \approx 20(\beta l)^2 - j120 \left(\ln \frac{2l}{d} - 1 \right) / (\beta l) \tag{5.13}$$

This expression clearly indicates that:

- the radiation resistance (the real part of Z_a in this case) is proportional to the square of the dipole length and inversely proportional to the wavelength squared.
- the reactance of the input impedance is a function of the radius and length of the dipole. The smaller the radius, the larger the amplitude of the reactance. It is proportional to the wavelength.
- the loss resistance of the antenna has not been taken into account.

The input impedances for the dipole of length $2l = \lambda/10$, $\lambda/2$, λ, and 1.5λ are also shown in Table 5.1. It should be pointed out that the reactance of the input impedance is very sensitive to the radius of the dipole. The only exception is when the length is near half of the wavelength, the very first resonant length. Although $2l = \lambda$ and 1.5λ may also be considered as resonant cases, they are very sensitive to the change of the radius. If the radius is not negligible, they are actually not resonant. A good illustration of the input impedance as a function of the normalized dipole length (and frequency) is given in Figure 5.2, where three dipole antennas of different radii are given. We can see that:

- all dipoles are resonant near half-wavelength (due to the nonzero radius, the resonant length is slightly short than $\lambda/2$), but not at one wavelength.
- the fatter (larger radius) the dipole, the broader the bandwidth.

5.1.1.5 Why the Half-Wavelength Dipole Is the Most Popular Dipole?

Based on the discussions above, we can conclude that there are basically the following four reasons that make the $\lambda/2$ dipole very popular:

- its radiation pattern is omnidirectional in the H-plane which is required by many applications (including mobile communications);
- its directivity (2.15 dBi) is reasonable, larger than the short dipole although smaller than that of the full-wavelength dipole;
- the antenna is longer than a short dipole but much shorter than the full-wavelength dipole; hence, it is a good trade-off between the directivity and size;
- it is the shortest resonant dipole antenna, and its input impedance is about 73 Ω (not sensitive to its radius) which is well matched with a standard transmission line of characteristic impedance 75 Ω or 50 Ω (with $VSWR < 2$). This is probably the most important and unique reason (a

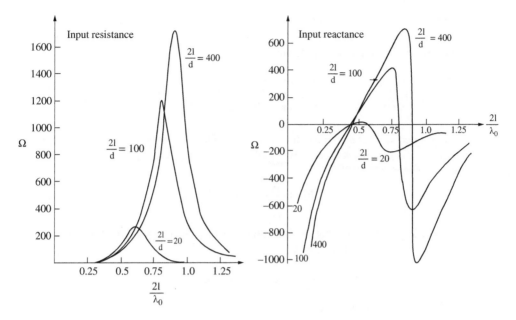

Figure 5.2 Input impedance as a function of the normalized dipole length

full-wavelength dipole could be a resonant antenna as well, but it is longer, and its impedance is not good).

Thus, for overall performance, the half-wavelength dipole antenna is a definite winner.

Example 5.1 Short Dipole
A dipole of the length $2l = 3$ cm and diameter $d = 2$ mm is made of copper wire ($\sigma = 5.7 \times 10^7$ S/m) for mobile communications. If the operational frequency is 1 GHz,

a. obtain its radiation pattern and directivity;
b. calculate its input impedance, radiation resistance, radiation efficiency, and gain;
c. if this antenna is also used as a field probe at 100 MHz for electromagnetic compatibility (EMC) applications, find its radiation efficiency again and express it in dB.

Solution
Since the frequency is 1 GHz, the wavelength $\lambda = 30$ cm, $2l/\lambda = 0.1$, and $\beta l = 0.1\pi$.

a. It is a short dipole, use Equation (5.8) to yield the radiation pattern. The 3D and the normalised E-plane patterns are shown in Figures 4.6 and 5.3, respectively. The H-plane pattern (φ) is omnidirectional. The HPBW in the E-plane is indeed $90°$ (from $45°$ to $135°$ in the figure) as indicated in Table 5.1. This plot is almost exactly the same as that obtained using Equation (5.8) with an error < 0.03 dB.
   Using Equations (5.8) and (5.9), we find directivity being 1.5 or 1.76 dBi.
b. Equation (5.13) can be employed to obtain the input impedance (without considering the loss resistance), the result is

$$Z_{a\_noloss} = 1.97 - j652 \ \Omega$$

(If (5.12) was used, the result would be $Z_a = 2.34 - j634\ \Omega$, which is slightly different).
To calculate the loss resistance, we need to find out the skin depth first. Using Equation (3.50), the skin depth in this case is

$$\delta \approx \sqrt{\frac{1}{\pi f \mu \sigma}} = \sqrt{\frac{1}{\pi \times 10^9 \times 4\pi \times 10^{-7} \times 5.7 \times 10^7}} = 2.1 \times 10^{-6}\ (m)$$

It is much smaller than the radius of the wire. Thus, we need to use Equation (3.51) to compute the loss resistance of the dipole:

$$R_L \approx \frac{2l\sqrt{f\mu}}{d\sqrt{\pi\sigma}} \approx 0.04\ \Omega$$

Since Equation (5.13) has not taken the loss resistance into account, the more accurate input impedance is therefore,

$$Z_a = 1.97 + 0.04 - j652 = 2.01 - j652\ \Omega$$

and the radiation efficiency is

$$\eta_r = \frac{R_r}{R_r + R_L} = \frac{1.97}{1.97 + 0.04} = 98.01\%$$

Using Equation (4.16) to obtain the gain of the antenna: $1.5 \times 0.98 = 1.47$, or 1.67 dBi.
c. The frequency is now 100 MHz and the wavelength $\lambda = 300$ cm, thus $2l/\lambda = 0.01$ and $\beta l = 0.01\pi$. This is an electrically small antenna; we can use the same approach as in (b) to obtain:
The antenna input impedance (without Ohmic loss):
$Z_{a\_noloss} = 0.0197 - j6524.3\ \Omega$
The skin depth: $\delta \approx 6.67 \times 10^{-6} < d/2$
The loss resistance: $R_L \approx 0.0126\ (\Omega)$
Thus, the radiation efficiency is

$$\eta_r = \frac{R_r}{R_r + R_L} = \frac{0.0197}{0.0197 + 0.0126} = 60.99\% \quad \text{or} \quad -2.14\ dB$$

This efficiency is much lower than that at 1 GHz. Normally, electrically small antennas have low efficiencies and high reactances, which is why they are not popular in applications even though they are small in size. Sometimes, we do employ short dipoles for reception where signal-to-noise ratio (SNR) is more important than the radiation efficiency.

In addition to cylindrical dipoles, there are many other forms of dipoles developed for various applications. For example, as shown in Figure 5.4, the *biconical antenna* offers a much wider bandwidth than the conventional cylindrical dipole; a *bow-tie antenna* has a broad bandwidth and low profile, while a *sleeve dipole* can be nicely fed by a coaxial cable and exhibits a wide bandwidth. A $\lambda/2$ *folded dipole* may be viewed as a superposition of two half-wavelength dipoles with an input impedance of around 280 Ω, which is close to the standard characteristic

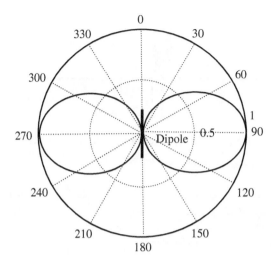

Figure 5.3 E_θ radiation pattern of a short dipole as a function of θ (E-plane)

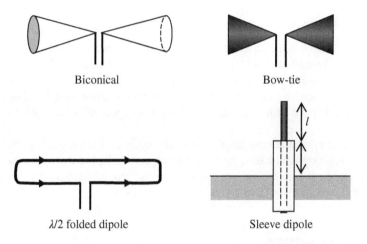

Figure 5.4 Some popular forms of dipole antennas

impedance of 300 Ω for a two-wire transmission line. That is why it was widely used for TV reception in the good old days (in the 1950s and 1960s). More details about these and other forms of dipoles, as well as the effects of parameters such as the feeding gap, may be found in books such as [2–4].

5.1.2 Monopoles and Image Theory

The monopole antenna is half of the dipole antenna as shown in Figure 5.5. There are a lot of similarities between them, but there are also some differences. The best way to investigate the monopole is to utilize image theory.

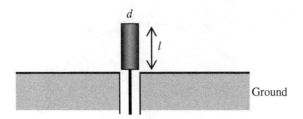

Figure 5.5   A monopole antenna with a coaxial feed line

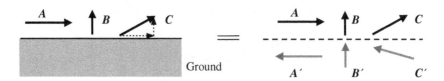

Figure 5.6   The image theory

5.1.2.1   Image Theory

The image theory states that if there is a current *A/B/C* above an infinite perfect conducting ground plane, the ground will act as a mirror to generate its image, *A'/B'/C'* as shown in Figure 5.6. Where *A* and *B* are the two basis currents, any other case, such as *C*, may be viewed as a combination of these two currents. The field at any point above the ground plane is equivalent to the field generated by the current *A/B/C* and its image *A'/B'/C'* without the presence of the ground plane.

The image current is of the same amplitude as the original source and its direction are determined by the boundary conditions, such as the tangential electric field must be zero at the boundary. Using the image theory, we can remove the ground plane and treat the problem as it were in the free space but had a pair of current sources.

5.1.2.2   Monopole Antennas

Applying the image theory to the monopole in Figure 5.5, we can see that it is equivalent to the case of *B–B'* in Figure 5.6, which is a dipole with length $2l$ in free space. The current distribution along the pole is the same as the dipole which has already been discussed earlier, thus the radiation pattern is the same above the ground plane and can be represented by Equation (5.7) with $0 \leq \theta \leq \pi/2$. Since the power is only radiated to the upper half-space and the power to the lower half-space is reflected back to the upper space, this results in an increased directivity. The directivity of a monopole is therefore twice that of its dipole counterpart – this can also be obtained using Equation (5.9) when the range of the integral is changed to $0 \leq \theta \leq \pi/2$.

The input impedance is changed as well. For a dipole, the voltage across the input points of the poles is $V - (-V) = 2V$ while for a monopole, the voltage is between the pole and the ground plane which is V. The input currents for both the dipole and monopole are the same. Thus, the input impedance of the monopole is half of that of its corresponding dipole. A summary of the

Table 5.2 Summary of some monopole characteristics

Monopole length l	$\lambda/20$	$\lambda/4$	$\lambda/2$	$3\lambda/4$
Current distribution				
Radiation pattern				
Directivity	3.0 or 4.76 dBi	3.28 or 5.15 dBi	4.8 or 6.8 dBi	About 4.6
HPBW	$45°$	$39°$	$23.5°$	NA
Input impedance	R: very small ($\sim 1\,\Omega$) jX: capacitive	R: $\sim 37\,\Omega$ jX: $\sim 0\,\Omega$	R: very large jX: $\sim 0\,\Omega$ for thin monopole	R: $\sim 50\,\Omega$ jX: $\sim 0\,\Omega$ for thin monopole
Note	jX sensitive to the radius	$R + jX$ not sensitive to the radius	$R + jX$ sensitive to the radius	$R + jX$ sensitive to the radius

characteristics of some monopoles is given in Table 5.2. Compared with the dipole, the monopole has the following advantages:

- the size is half of the corresponding dipole;
- the directivity doubles that of its corresponding dipole;
- the input impedance is half of its corresponding dipole.

The quarter-wavelength monopole is a resonant antenna, like the half-wavelength dipole, its input impedance is about $37\,\Omega$, which matches well with the $50\,\Omega$ standard transmission line (with $VSWR = 1.35 < 2$). All these good reasons have made the quarter-wavelength monopole one of the most popular antennas. We can find applications almost everywhere, from radio broadcasting towers to mobile phones.

Just like dipoles, there are many derivatives of monopole antennas developed to suit different applications. Some are shown in Figure 5.7. More detailed discussion will be presented in Chapter 8.

5.1.2.3 Effects of Ground Plane

The above analysis and characteristics of monopoles are based on the assumption that there is an infinite perfect conducting ground plane. But in reality, we normally do not have an infinite ground plane, or the ground plane is not a perfect conductor (like the Earth). What are the effects on the monopole?

All parameters of the monopole (radiation pattern, gain, and input impedance are just some examples) may be affected by the ground plane.

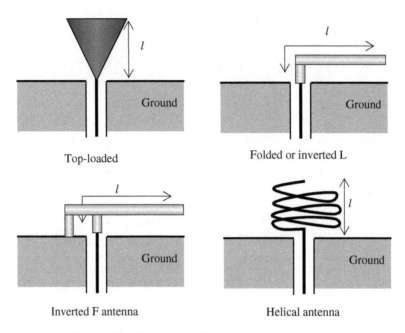

Figure 5.7   Some popular forms of monopole antennas

If the conducting ground plane is of limited size, the radiated power will leak to the lower half of the space, which means that the radiation pattern is changed. There may be side or even back lobes. The edge of the ground plane will diffract the waves which results in many back lobes (the number of back lobes is closely linked to the ground plane size) as shown in Figure 5.8. The maximum angle is changed (tilted toward the sky) and the directivity is reduced. Also, the input impedance may be changed. If the ground plane is not large enough, it can act as a radiator rather than a ground plane. As a rule of thumb, the diameter of the ground plane should be at least one wavelength.

If the ground plane is very large but not made of a good conductor, all the antenna properties are affected, especially the directivity and gain (reduced); the angle to the maximum radiation is also tilted toward the sky, although there are still no lobes in the lower half-space. To improve the reflectivity of the ground, metal meshes are sometimes employed. A good discussion can be found in [2].

5.1.3   Loops and Duality Principle

Loops are another simple and versatile wire-type antenna. They take many different configurations which include circular, square, rectangular, triangular, elliptical, and other shapes.

While the dipole is considered as a configuration evolved from an open-end transmission line, the loop can be viewed as a configuration evolved from a short-end transmission line, as shown in Figure 5.9. Thus, for an electrically small loop, the current is large and the voltage is small – this results in a small input impedance which is very different from a short dipole whose impedance, more precisely the reactance, is very large. We could use the conventional

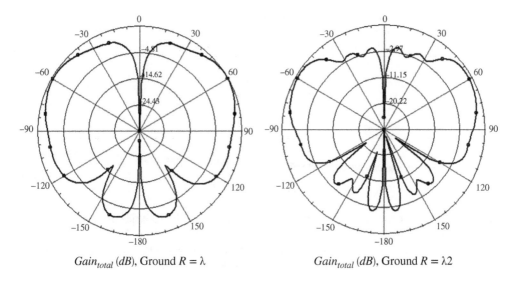

Gain$_{total}$ (dB), Ground $R = \lambda$ Gain$_{total}$ (dB), Ground $R = \lambda 2$

Figure 5.8 Effects of the ground plane on the radiation pattern of a monopole

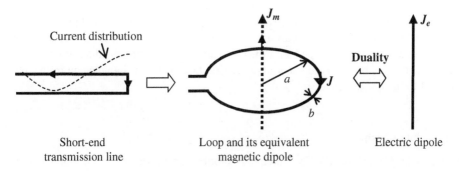

Short-end   Loop and its equivalent   Electric dipole
transmission line   magnetic dipole

Figure 5.9 Transmission line to loop antenna, and its corresponding dipoles

method to obtain loop's properties by finding its current distribution first, which was time-consuming and cumbersome. Since we have already studied dipoles properly, we are going to use another method, the duality principle, to find the characteristics of a loop, which is easier and more straightforward.

5.1.3.1 Duality Principle

Duality means the state of combing two different things which are closely linked. In antennas, the *duality theory* means that it is possible to write the fields of one antenna from the field expressions of the other antenna by interchanging parameters.

From Chapters 1 and 4, we know that the first two Maxwell's equations are

Table 5.3 Duality relationship between System 1 and Systems 2

System 1 with electric current source	System 2 with magnetic current source
J_e	J_m
E_1	H_2
H_1	$-E_2$
ε_1	μ_2
μ_1	ε_1

$$\nabla \times E_1 = -j\omega\mu_1 H_1$$
$$\nabla \times H_1 = J_e + j\omega\varepsilon_1 E_1 \tag{5.14}$$

where E_1 and H_1 are the electric and magnetic fields generated by electric current density J_e in Medium 1 (ε_1 and μ_1).

Now suppose a fictitious magnetic current source with magnetic current density J_m in Medium 2 (ε_2 and μ_2), the Maxwell's equations for this new scenario can be written as:

$$\nabla \times H_2 = j\omega\varepsilon_2 E_2$$
$$\nabla \times E_2 = -J_m - j\omega\mu_2 H_2 \tag{5.15}$$

where E_2 and H_2 are the electric and magnetic fields generated by J_m in Medium 2 (ε_2 and μ_2).

Thus, System 1 with the electric current and System 2 with the magnetic current are duals and their parameters can be exchanged as shown in Table 5.3. Note that the duality links not only the field parameters but also the material properties; there is also a sign change from the magnetic field to the electric field.

5.1.3.2 Small Loops

The implementation of the duality principle is simple. For electrically small loops (circumference $C = 2\pi a < \lambda/10$), the current can be considered as a constant. It was found that a magnetic current $I_m \Delta l$ is equivalent to a small loop of radius a and constant electric current I_0 provided that:

$$I_m \Delta l = j\omega\mu S I_0 \tag{5.16}$$

where $S = \pi a^2$ (area of the loop). This means that their radiated fields are identical. As shown in Figure 5.9, using the duality principle, it is possible to write the fields for the magnetic current from the field expressions of the electric current antenna by interchanging parameters. From Equations (4.6) and (4.7), the far field of a short electric current is given as:

$$E_\theta = \frac{jI\Delta l}{4\pi r}\eta\beta \sin\theta e^{-j\beta r}$$

$$H_\varphi = \frac{jI\Delta l}{4\pi r}\beta \sin\theta e^{-j\beta r} \tag{5.17}$$

$$E_r = E_\varphi = 0; \quad H_r = H_\theta = 0$$

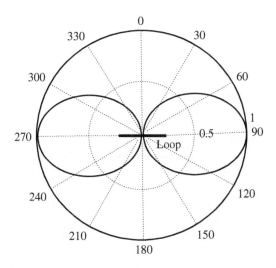

Figure 5.10 E_φ radiation pattern of a small loop as a function of θ (H-plane)

Use Table 5.3 to yield the far fields of a small loop:

$$H_\theta = \frac{jI_m\Delta l}{4\pi r\cdot\eta}\beta\sin\theta e^{-j\beta r}$$

$$E_\varphi = -\frac{jI_m\Delta l}{4\pi r}\beta\sin\theta e^{-j\beta r} \tag{5.18}$$

$$H_r = H_\varphi = 0; \quad E_r = E_\theta = 0$$

These are the far fields of a short loop. The radiation pattern looks the same as a short dipole, as shown in Figure 5.10, except that *the polarization is now E_φ (not E_θ, but still linearly polarized)* which is in the plane of the loop. The maximum radiation occurs at $\theta = \pi/2$, which is also in the loop plane.

The directivity of a small loop is the same as that of a small dipole, which is 1.5 or 1.76 dBi. Using the radiated field expressions in (5.18), the radiation resistance is found as:

$$R_r = \frac{P_t}{I_0^2} = 20\pi^2(\beta a)^4 = 20\pi^2\left(\frac{C}{\lambda}\right)^4 \tag{5.19}$$

where C is the circumference of the small loop. If it is not a circle, C is then the total length of the loop. Comparing Equation (5.19) with the radiation resistance of a dipole, given in Equation (5.13), we can see that the loop radiation resistance is more sensitive to changes of the length and wavelength (power of 4, not power of 2). For a loop and a dipole of the same length, the radiation resistance of the loop antenna is much smaller. For example, if the length is 0.1λ, the radiation resistance of the dipole is 7.89 Ω, while the radiation resistance of the loop is 0.02 Ω which is much smaller.

The resistance is determined by the length normalized to the wavelength, not the absolute length. However, the reactance is different. The reactance of the small loop can be approximated by [2]:

$$X_A = \omega\mu a \left(\ln\left(\frac{8a}{b}\right) - 2 \right) + \frac{a}{b}\sqrt{\frac{\omega\mu}{2\sigma}} \qquad (5.20)$$

where b is the diameter of the wire as shown in Figure 5.9. This equation has taken both the external inductive reactance (the first term) and internal reactance (the second term) of loop conductor into account. For impedance matching, a capacitor in series is required for a small loop.

Equation (5.18) is the far field of an electrically small loop, obtained using the far fields of a short current/dipole element and duality principle. Using the same approach, we can obtain the near fields of an electrically small loop antenna. If we compare the fields of the small loop and dipole antennas, an interesting finding is that, in the near field, the magnetic field could actually be much larger than its electrical field for a small loop antenna (when $\beta r < 1$) which is just the opposite for a short current/dipole antenna. This may partially explain why a loop antenna is called a magnetic antenna while a dipole antenna is called an electrical antenna.

5.1.3.3 Loop Antenna: General Case

If a loop cannot be considered as being small, the current distribution cannot therefore be regarded to be constant. As a result, many properties of the loop are changed and results obtained for a constant current cannot be applied to real antennas. It has been shown, when the circumference of the loop becomes comparable with the wavelength, its maximum radiation shifts to its axis ($\theta = 0$ and π) which is perpendicular to the plane of the loop [5]. This is very different from that of a small loop. There is no simple mathematical expression for the radiated field from such a loop antenna. Computer simulations are really the best choice if a 3D radiation pattern is required. The directivity of a circular loop as a function of the circumference in wavelengths (C/λ) for $a/b = 40$ and $\theta = 0$ is shown in Figure 5.11. The maximum directivity is about 4.5 dBi when the circumference is near 1.4λ. It should be pointed out that the directivity is fairly independent of the radius of the wire (b) for $C/\lambda < 1.4$.

The *one-wavelength loop* (the circumference $C = \lambda$) is commonly referred to as a resonant loop. The current in the loop is approximately $I_0 \cos\varphi$ (maximum at the input) and is approximately equivalent to two $\lambda/2$ dipoles separated by $2a$ ($=\lambda/\pi$) as shown in Figure 5.12. E_θ, E_φ, and total radiation patterns of this antenna are plotted in Figure 5.13. E_θ is zero in the horizontal plane $\theta = \pi/2$ and in the vertical plane $\varphi = 0$, π, while E_φ is small in the vertical plane $\varphi = \pi/2$, $3\pi/2$. The maximum is indeed shifted to the axis, which can be easily explained using the model in Figure 5.12. The directivity of the antenna is around 2.2 or 3.4 dBi (larger than that for a $\lambda/2$ dipole). The input resistance is about 100 Ω which is a reasonable value for matching to a standard 50 or 75 Ω transmission line. However, the input reactance in this case is actually about $-100\,\Omega$ (not sensitive to the change of radius). The resonance occurs when C is slightly larger than λ, as seen in Figure 5.14.

The input impedance of a loop with the circumference $C = 6$ cm (the corresponding frequency is 5 GHz) is shown in Figure 5.14. The input resistance reaches the first maximum

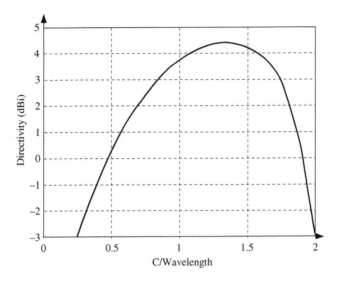

Figure 5.11 Directivity of a loop as a function of the normalised circumference

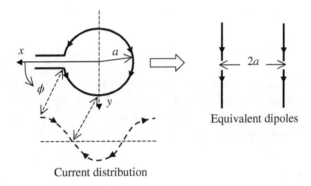

Figure 5.12 Current distribution in a resonant loop and its equivalent pair of dipoles

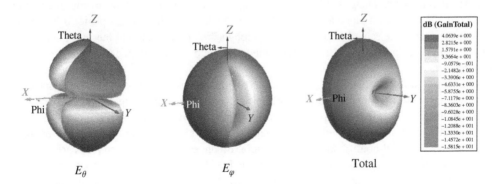

Figure 5.13 The radiation patterns of a loop with $C = \lambda$

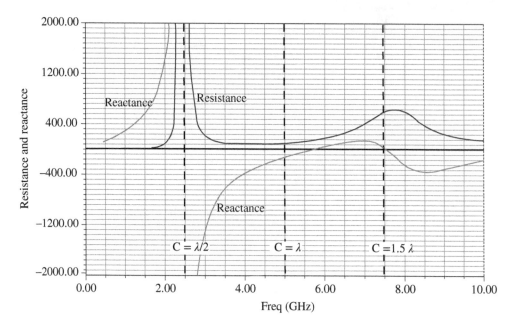

Figure 5.14   The input impedance of a loop with $C = 6$ cm

at about $C/\lambda = 0.5$ (at 2.5 GHz, it is referred to as an antiresonant loop), and the second maximum is at approximately $C/\lambda = 1.5$ (it is not at 9 GHz but near 8 GHz in Figure 5.14). It is interesting to note that the resistance curves for the loop and dipole (see Fig. 5.2) are similar, and the reactive curves are also similar but with a reference position shift for about $\lambda/2$. This is not difficult to understand using the transmission line theory: the open and short points are shifted by $\lambda/4$ on a transmission line which is equivalent to a length of $\lambda/2$ in a loop.

5.1.3.4   Discussion

The loop antenna has some very interesting features in terms of the input impedance and radiation pattern as shown in Figures 5.14 and 5.15. When the loop size is changed from electrically small to about one wavelength, the total radiation is gradually changed from an omnidirectional pattern in the horizontal plane to an omnidirectional pattern in the vertical plane. If its size is further increased, the radiation pattern will no longer be omnidirectional but will have many lobes. Thus, the loop antenna radiation pattern is much more complicated than its dipole counterpart.

In addition to the circular loop, other forms of loop antennas have also been developed for practical applications. Since the analysis of polygonal (such as square, rectangular, triangular, and rhombic) loops is much more complex, they have received much less attention.

Generally speaking, loop antennas have many attractive characteristics. They are of low profile and present a well-controlled radiation pattern. Balanced and unbalanced feeds are possible (coaxial cables can be used to feed the loop or half-loop [2]). Loop antennas are usually employed for linear polarization, although its shape is circular. It has been found that a loop antenna can also radiate circularly polarized waves if a gap is introduced on the loop [6–8].

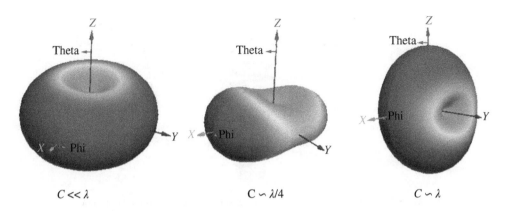

Figure 5.15 Total radiation patterns of loops of different sizes

A one-wavelength loop with a planar reflector has a unidirectional pattern with a relatively high directivity (near 9 dBi) and good input impedance which is of course dependent on the distance to the reflector [3]. For mobile radio communications, the loop is often used in pagers but hardly used in mobile transceivers. This may be due to its high resistance and reactance which make it difficult to match to standard 50 Ω transmission lines over moderate bandwidths. With a ground plane or body, the loop antenna should be an attractive candidate for mobile hand portables. We expect to see more loop and its derivative antennas to be used in mobile wireless communication systems.

5.1.4   Helical Antennas

As indicated earlier, the helical antenna may be viewed as a derivative of the dipole or monopole antenna, but it can also be considered as a derivative of a loop antenna. In this section, we are going to introduce this special antenna which has some very interesting and unique features, as Professor Kraus (who made a significant contribution to this subject) stated [4]: "not only does the helix have a nearly uniform resistive input over a wide bandwidth but it also operates as a super-gain end-fire array over the same bandwidth! Furthermore, it is noncritical with respect to conductor size and turn spacing. It is easy to use in arrays because of almost negligible mutual impedance."

Geometrically, as shown in Figure 5.16, *a helical antenna* consists of a conductor wound into a helical shape. It is a circularly/elliptically polarized antenna. A helix wound like a right-hand (clockwise) screw radiates or receives right-hand circularly polarized waves, whereas a helix wound like a left-hand (anticlockwise) screw radiates or receives left-hand circularly polarized waves. Although *a helix can radiate in many modes*, the axial (end-fire) mode and the normal (broadside) mode are the ones of the most interests. The following notations are used for the analysis:

D = diameter of helix;
s = spacing between turns;
n = number of turns;

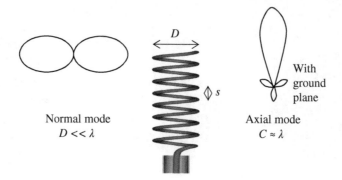

Figure 5.16   Helical antenna and its radiation patterns of two radiation modes

C = circumference of helix = πD;

L = length of one turn = $\sqrt{C^2 + s^2}$.

5.1.4.1   Normal-Mode Helix

The normal (broadside) mode occurs when the diameter of the helix is much smaller than the wavelength ($D \ll \lambda$) and the total length is also smaller than the wavelength. It behaves more like a dipole (or monopole with a ground plane) antenna. The radiation pattern is broadside to the helix axis. In this case, *the helix may be treated as the superposition of n elements, each consists of a small loop of diameter D and a short dipole of length s*. The axes of the loop and dipoles coincide with the axis of the helix. The far field of a small loop is given in Equations (5.18) and (5.16) as:

$$E_\varphi = \frac{\omega \mu S I_{in}}{4\pi r} \beta \sin \theta e^{-j\beta r}$$
$$E_r = E_\theta = 0 \tag{5.21}$$

The far field of a short dipole is shown in Equation (5.8) as:

$$E_\theta \approx j\eta \frac{I_{in}}{4\pi r} (\beta s) \sin \theta e^{-j\beta r} \tag{5.22}$$

The E_θ and E_φ components of dipoles and loops, respectively, are $90°$ out of phase. The combination of them gives a circularly or elliptically polarized wave. The axial ratio is

$$AR = \frac{|E_\theta|}{|E_\varphi|} = \frac{\eta s}{\omega \mu S} = \frac{s}{2\pi f \cdot \sqrt{\varepsilon\mu} \cdot \pi (D/2)^2} = \frac{2\lambda s}{\pi^2 D^2} = \left(\frac{2s}{\lambda}\right) \Big/ \left(\frac{\pi D}{\lambda}\right)^2 \tag{5.23}$$

When the circumference is equal to

$$C = \pi D = \sqrt{2s\lambda} \tag{5.24}$$

the axial ratio becomes unity, and the radiation is circularly polarized. Otherwise, the radiation is elliptically polarized.

From Equations (5.21) and (5.22), we know that the radiation pattern is indeed a typical number eight (8) in the vertical plane as shown in Figure 5.16 and omnidirectional in the horizontal plane. The directivity is again about 1.5 and the HPBW is $90°$, similar to that of a short dipole or small loop.

The input impedance is very sensitive to changes of frequency; the bandwidth is therefore very narrow. In practice, the helix is normally used with a ground plane, and the polarization is predominately vertical and the radiation pattern is similar to that of a monopole. In this case, the radiation resistance approximately is given by $(25.3 \ ns/\lambda)^2$ [9]. These antennas tend to be inefficient radiators (depending on size and materials) and are typically used for mobile communications where reduced size is a critical factor.

5.1.4.2 Axial-Mode Helix

The axial (end-fire) mode occurs when the circumference of the helix is comparable with the wavelength ($C = \pi D \approx \lambda$) and the total length is much greater than the wavelength. This has made the helix *an extremely popular circularly polarised broadband antenna* at the VHF and UHF band frequencies. In this mode of operation, there is only one main lobe and its maximum is along the axis of the helix and, as shown in Figure 5.16 (with a ground plane), there may be some sidelobes. The recommended parameters for an optimum design to achieve circular polarization are

- normalized circumference: $3/4 < C/\lambda < 4/3$;
- spacing: $s \approx \lambda/4$;
- pitch angle: $12° \le \alpha = \tan^{-1}(s/C) \le 15°$;
- number of turns: $n > 3$.

In contrast to the normal-mode helix which has a current almost uniform in phase over the antenna, the phase of the axial-mode helix current shifts continuously along the helix like a traveling wave. Because the circumference is about one wavelength, the currents at opposite points on a turn are about $180°$ out of phase. This cancels out the current direction reversal introduced by the half-turn. Thus, the radiation from opposite points on the helix is nearly in-phase. This is essentially the same as the one-wavelength loop as shown in Figure 5.12. It is therefore not surprising to have the maximum radiation along the axis.

The radiation pattern of the axial-mode helix can be modeled using antenna array theory, discussed later in Section 5.3. Basically, each turn can be considered as an element of the array with a radiation pattern of $\cos \theta$. The normalized total radiation pattern is

$$E = A \cos \theta \frac{\sin \left[(n/2)\Psi \right]}{\sin \left[\Psi/2 \right]} \tag{5.25}$$

where the first term is the normalization factor: $A = 1/n$ for ordinary end-fire radiation and $A = \sin(\pi/(2n))$ for Hansen–Woodyard (HW) end-fire radiation (*the axial-mode helix was found to approximately satisfy Hansen–Woodyard condition* – to be discussed in Section 5.3.2) which is larger than $1/n$ to reflect the increased directivity – it is achieved by increasing the phase change between element sources. The second term in the equation ($\cos \theta$) is the element radiation pattern, and the last term represents the array factor of a uniform array of n elements. $\Psi = \beta(s \cos \theta - L/p)$ and p is the relative phase velocity given by [2, 4]:

$$p = \frac{v}{c} = \begin{cases} \dfrac{L/\lambda}{s/\lambda + 1}; & \text{for ordinary end-fire radiation, } \Psi = -2\pi \\[4mm] \dfrac{L/\lambda}{s/\lambda + \dfrac{2n+1}{2n}}; & \text{for HW end-fire radiation, } \Psi = -2\pi - \pi/n \end{cases} \tag{5.26}$$

It has been found that the HPBW is roughly

$$HPBW \approx \frac{52° \lambda \sqrt{\lambda}}{C\sqrt{ns}} \tag{5.27}$$

It is inversely proportional to C and $\sqrt{ns}$. The beamwidth between first nulls is about

$$FNBW \approx \frac{115° \lambda \sqrt{\lambda}}{C\sqrt{ns}} \tag{5.28}$$

Using the link between the directivity and HPBW in Equation (4.14), we obtain the directivity:

$$D \approx 15 C^2 ns/\lambda^3 \tag{5.29}$$

It is proportional to C^2, n, and s. Thus, the effects of n on directivity are significant when n is small but not significant when n is large (for example, if n is increased from 2 to 4, the directivity is doubled. If n is increased from 20 to 22, the increment is still 2, but the directivity has little change). If we take the minor lobes and the details of the pattern shape into account, a more realistic estimation is

$$D \approx 12 C^2 ns/\lambda^3 \tag{5.30}$$

The axial ratio is found to be

$$AR = (2n + 1)/(2n) \tag{5.31}$$

The input impedance of an axial-mode helix with a ground, which is greater than one wavelength, is almost resistive with values between 100 and 200 Ω. The estimated value within 20% accuracy is expressed by:

$$R = 140 C/\lambda \tag{5.32}$$

A suitable matching circuit is required by the antenna if it is to be connected to a 50 Ω transmission line. One way to bring the input impedance down to the desired value is to use the first 1/4 turn as an impedance transformer. A microstrip is an ideal structure for this purpose, since the ground plane could be the same as that of the antenna. The dielectric substrate height h is linked to the strip width w and feed line characteristic impedance Z_0 by [4]:

$$h = \frac{w}{\frac{377}{\sqrt{\varepsilon_r}Z_0} - 2} \tag{5.33}$$

where ε_r is the relative permittivity of the substrate. This improved-matching modification may result in bandwidth reduction as a trade-off.

Example 5.2   Axial Helix
Design a circularly polarized helix antenna of an end-fire radiation pattern with a directivity of 13 dBi. Find out its input impedance, *HPBW, AR,* and radiation pattern.

Solution
Using the recommended design parameters for an axial-mode helix, we can choose: $s = \lambda/4$ and $C = \lambda$, which gives us the pitch angle:
$12° \le \alpha = \tan^{-1}(s/C) = 14.0362° \le 15°$.
   Since the required directivity is

$$D = 13dBi = 20 \approx 12C^2 ns/\lambda^3$$

Thus, $n = 6.667 \approx 7$ turns. Now we can use the equations above to obtain
   Input impedance: $R = 140C/\lambda = 140 \,\Omega$;
   Half-power beamwidth: $HPBW \approx \dfrac{52°\lambda\sqrt{\lambda}}{C\sqrt{ns}} = 39.3°$;
   Axial ratio: $AR = (2n + 1)/(2n) = 1.0714$;
To calculate the radiation pattern, we first find $L = \sqrt{C^2 + s^2} = 1.03\lambda$; and

$$p = \frac{v}{c} = \begin{cases} \dfrac{L/\lambda}{s/\lambda + 1} = 0.8295; & \text{for normal end-fire radiation, } \Psi = -2\pi \\[4mm] \dfrac{L/\lambda}{s/\lambda + \dfrac{2n + 1}{2n}} = 0.7844; & \text{for HW end-fire radiation, } \Psi = -2\pi - \pi/n \end{cases}$$

Thus, for the normal end-fire radiation:

$$\Psi_1 = \beta(s\cos\theta - L/p) = 2\pi(0.25\cos\theta - 1.03/0.8295) = 1.5708\cos\theta - 7.8019$$

$$E_1 = \frac{1}{7}\cos\theta\,\frac{\sin[3.5\Psi_1]}{\sin[0.5\Psi_1]}$$

(a)

(b)

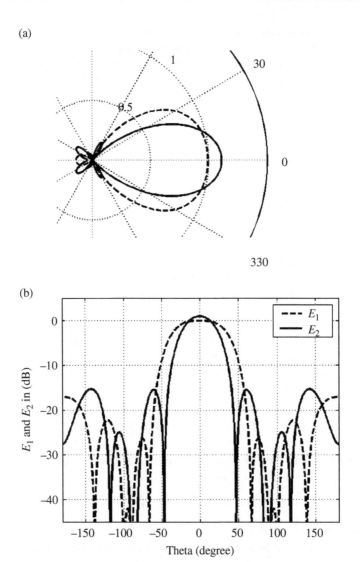

Figure 5.17   Radiated field patterns of two different antennas (a) the polar plot of E_1 (dashed line) and E_2 (solid line) in linear scale; (b) E_1 and E_2 field patterns in dB scale

and for HW end-fire radiation which can be achieved by this helix antenna:

$$\Psi_2 = \beta(s\cos\theta - L/p) = 2\pi(0.25\cos\theta - 1.03/0.7844) = 1.5708\cos\theta - 8.2505$$

$$E_2 = \sin\left(\frac{\pi}{14}\right)\cos\theta\frac{\sin[3.5\Psi_2]}{\sin[0.5\Psi_2]}$$

Both patterns (E_1: dashed line for normal end-fire array and E_2: solid line for the helix) are plotted in Figure 5.17. We can see that HW design can indeed provide an increased directivity

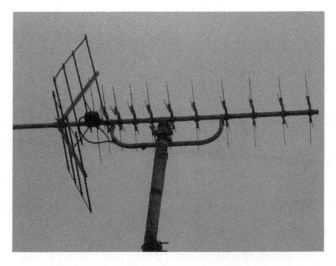

Figure 5.18 A Yagi–Uda TV reception antenna

(by a factor of 1.115 or 0.918 dB in this case). The logarithmic presentation in rectangular plot shows more and clearer details of the radiation than that of the polar plot in linear scale. It is also interesting to note that the increased directivity pattern E_2 has a sharper main beam but higher sidelobe levels (SLLs) than that of ordinary end-fire pattern E_1.

5.1.5 Yagi–Uda Antennas

The Yagi–Uda antenna (also known as a *Yagi*) is another popular type of end-fire antenna widely used in the VHF and UHF bands (30 MHz to 3 GHz) because of its simplicity, low cost, and relatively high gain. The most noticeable application is for home TV reception to be found on the rooftops of houses. A typical one is shown in Figure 5.18.

Yagi and Uda were two Japanese professors who invented and studied this antenna in the 1920s. S. Uda made the first Yagi–Uda antenna and published the results in Japanese in 1926 and 1927, and the design was further developed and published in English by his colleague Professor Yagi a year later [10] (that is why it was first called Yagi antenna but not fair to Uda). Since then, a significant of amount of work has been done theoretically and experimentally. A lot of data and results are available in the public domain.

The main feature of this type of antenna is that it consists of three different elements: the driven element, reflector, and director as shown in Figure 5.19. Some people consider the Yagi–Uda antenna as an array since it has more than one element. However, it has just one active element and feed port; all the other elements (the reflector and directors) are parasitic. Thus, some people consider it as an element antenna rather than an antenna array. The main characteristics and design recommendations of these elements can be summarized as follows:

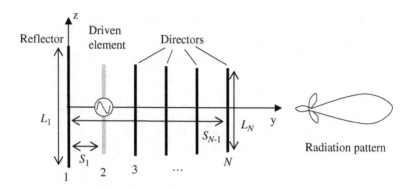

Figure 5.19 Configuration of a Yagi–Uda antenna and its radiation pattern

- *The driven element* (feeder) is the very heart of the antenna. It determines the polarization and center frequency of the antenna. For a dipole, the recommended length is about 0.47λ to ensuring a good input impedance to a $50\,\Omega$ feed line.
- *The reflector* is normally slightly longer than the driven resonant element to force the radiated energy toward the front. It exhibits an inductive reactance. It was found that there is not much improvement by adding more reflectors to the antenna, thus there is only one reflector. The optimum spacing between the reflector and the driven element is between 0.15 and 0.25 wavelengths. It has a large effect on the front-to-back ratio and antenna input impedance.
- *The directors* are usually 10–20% shorter than the resonant driven element and appear to direct the radiation toward the front. They are of capacitive reactance. The director to director spacing is typically 0.25–0.35 wavelengths, with larger spacing for longer array and smaller spacing for shorter arrays. The number of directors determines the maximum achievable directivity and gain.

5.1.5.1 Operational Principle

The special configuration (long reflector and short directors) has made the Yagi–Uda antenna radiate as an end-fire antenna. The simplest three-element Yagi–Uda antenna (just one director) already shows an acceptable end-fire antenna pattern. The radiation toward the back seems to be blocked/reflected by the longer element, but not just by the reflector, the reflector, and director produce push-and-pull effects on the radiation. Induced currents are generated on the parasitic elements and form a traveling wave structure at the desired frequency. The performance is determined by the current distribution in each element and the phase velocity of the traveling wave.

5.1.5.2 Current Distribution

The current distribution on the driven element is determined by its length, frequency, and interaction/coupling with nearby elements (mainly the reflector and first director), while the current distribution in parasitic elements is governed by the boundary condition: the total tangential electric field must be zero on the conducting surface. This results in induced currents and they may be viewed as the second sources of the radiation. Analytical and numerical methods have been employed to obtain the current distribution along each element [2]. The results show that the current distribution on each element is similar to that of a dipole. As expected, the dominant current is on the driven element; the reflector and the first director carry less current, and the currents on other directors are further reduced and they appear to be of similar amplitude which is typically less than 40% of that of the driven element.

5.1.5.3   Radiation Pattern

Once the current is known, the total radiated field can be obtained using antenna equation (4.3), which could be difficult and numerical computations may be required.

For a dipole Yagi–Uda antenna, the radiation from element n is given by Equation (5.4):

$$E(\theta)_n \approx j\eta \frac{I_n e^{-j\beta r}}{2\pi r} \left(\frac{\cos(\beta l_n \cos\theta) - \cos(\beta l_n)}{\sin\theta} \right) \tag{5.34}$$

where I_n is the maximum current and l_n is half the length of the nth dipole. Thus, the total radiation pattern is the field superposition from all the elements and may be expressed as:

$$E(\theta) \approx j\eta \frac{e^{-j\beta r}}{2\pi r} \sum_{n=1}^{N} I_n \left(\frac{\cos(\beta l_n \cos\theta) - \cos(\beta l_n)}{\sin\theta} \right) e^{j\beta S_{n-1} \cos\theta} \tag{5.35}$$

where r is the center of the reflector to the observation point and spacing $S_0 = 0$. It is apparent that each element length and spacing, weighted by its maximum current, affects the total radiation. This approach is the same as the one we are going to use for analyzing antenna arrays later.

One important figure of merit in Yagi–Uda antenna is the *front-to-back ratio* of the pattern. It was found that it was very sensitive to the spacing of the director. It varies from trough to peak and from peak to trough repetitively as a function of the spacing [2].

5.1.5.4   Directivity and the Boom

The directivity can be obtained using Equation (4.11) as we did for dipoles. A simpler estimation of the maximum directivity of Yagi–Uda antenna is proposed by us as:

$$D = 3.28N \tag{5.36}$$

The coefficient 3.28 is resulted from doubling the directivity (1.64) of a half-wave dipole. Since there are N elements, the maximum is obtained when they are combined constructively as 3.28N. The reason for introducing the factor of 2 is due to the fact that the radiation pattern is now unidirectional end fire. The radiation is redirected to just half of the space by the reflector and directors (very little to the other half), which is somewhat similar to the effect of a conducting ground plane. For the simplest three-element Yagi–Uda antenna, $N = 3$, thus $D = 9.84 = 9.93$ dBi $= 7.78$ dBd. When the number is doubled to 6, 3 dB more gain can be obtained ($D = 12.93$ dBi $= 10.78$ dBd). However, if three more directors were added to the antenna, the improvement on the directivity would be just 1.76 dB. Just like the helical antenna, *the effect of the number of elements on the directivity is significant when N is small, but not significant when N is large.*

There is a misperception on how the directivity is linked to the length of the boom. Some people think the directivity is proportional to the length of the boom and not the number of elements. The reality is that, as indicated by Equation (5.36), the number of properly placed elements determines the directivity. Of course, the more the elements, the longer the boom, and hence the larger the directivity. But the point is that the directivity is determined by the number of elements, not the length of the boom.

Another misperception is the directivity or gain of the Yagi–Uda antenna could be infinite when the number of elements goes to infinite as shown in Equation (5.36). The reality is that the maximum directivity or gain of the Yagi–Uda antenna can be achieved is less than 20 dBi due to the weak coupling when the element N is over certain number. Thus, (5.36) is valid when N is up to about 25.

Some characteristics of the antenna may be affected by the boom if it is made of metal, for example the input impedance. The boom should be insulated from the elements, otherwise compensation is required. Since the orientation of the boom is orthogonal to the antenna elements, the effects on the radiation pattern should be small.

5.1.5.5 Input Impedance

The input impedance is very sensitive to the spacing to the nearest two elements and their lengths. *As a rule of thumb, the closer the spacing, the smaller the input impedance.* The coupling reduces the impedance. If a half-wavelength dipole is used as the driven element, its impedance could be reduced from 73 Ω to as little as 12 Ω when the spacing is down to 0.1λ. This could be a problem if the antenna is to be matched with a 50 Ω or even 75 Ω transmission line. A simple solution in practice is to replace a half-wavelength dipole by a folded dipole which has a typical impedance of 280 Ω. Thus, the final impedance of a Yagi–Uda antenna could be reduced to match with a standard 50 or 75 Ω transmission line after taking the coupling into account.

5.1.5.6 Antenna Design

Generally speaking, Yagi–Uda antennas have low impedance and relatively narrow bandwidth (a few percent). Improvement in both can be achieved at the expense of other parameters (such as directivity and sidelobes). The length and diameters of the elements as well as their respective spacing determine the optimum characteristics. Since there are so many variables, the optimum design of a Yagi–Uda antenna is a very complicated task. Extensive and comprehensive experimental and theoretical investigations into Yagi–Uda antennas have been conducted. The most well-known experimental study was conducted by Viezbicke at NBS (now NIST, National Institute of Standards and Technology) [11]. He has produced a lot of data and results. Some are summarized in Table 5.4, along with our estimated maximum directivity. These data are extremely useful for practical antenna designs.

Table 5.4 Optimized elements for Yagi–Uda Antennas (the normalized diameter $d/\lambda = 0.0085$, spacing $S_1 = 0.2\lambda$)

Boom length/λ	0.4	0.8	1.2	2.2	4.2	Note
L_1/λ	0.482	0.482	0.482	0.482	0.475	Reflector
L_2/λ		$\lambda/2$ folded dipole ~0.47				Driven element
L_3/λ	0.442	0.428	0.428	0.432	0.424	Director
L_4/λ		0.423	0.420	0.415	0.424	
L_5/λ		0.428	0.420	0.407	0.420	
L_6/λ			0.428	0.398	0.407	
L_7/λ				0.390	0.403	
L_8/λ				0.390	0.398	
L_9/λ				0.390	0.394	
L_{10}/λ				0.390	0.390	
L_{11}/λ				0.398	0.390	
L_{12}/λ				0.407	0.390	
L_{13}/λ					0.390	
L_{14}/λ					0.390	
L_{15}/λ					0.390	
Spacing/λ	0.20	0.20	0.25	0.20	0.308	Between directors
D in dBd	7.1	9.2	10.2	12.25	14.2	Measured
D in dBi	9.2	11.3	12.3	14.35	16.3	Measured
Estimated D_{max} in dBi	9.93	12.14	12.93	15.95	16.91	Equation (5.36)

It should be pointed out that the element diameter is fixed to 0.0085λ in Table 5.4. If this condition cannot be met in reality, slight changes on the element length should be made and relevant design curves for this purpose were produced in [11] and some were reproduced in [2, 12].

These results were obtained experimentally. Analytical and numerical investigations have shown that further improvements could be achieved. For example, Cheng and Chen [13, 14] used a perturbational technique to vary each spacing and element length and have shown that the directivity of a six-element Yagi–Uda antenna was increased from the initial value (10.92 dBi) to 12.89 dBi which is very close to our estimated maximum directivity 12.93 dBi.

Some optimization methods, such as genetic algorithm and particle swarm, have recently been employed to optimize and miniaturize the design [15–17], good and interesting results have been obtained. It has been shown that optimum gains may be obtained with very small spacings. This is a very attractive feature for modern mobile communication systems, where size is particularly important.

In addition to the dipole, other antenna types (especially loops) have also been used to make Yagi–Uda antennas for various applications.

5.1.6 Log-Periodic Antennas and Frequency-Independent Antennas

A very similar configuration to Yagi–Uda antennas is the *log-periodic antenna* as shown in Figure 5.20. It produces similar end-fire radiation pattern and directivity (typically between

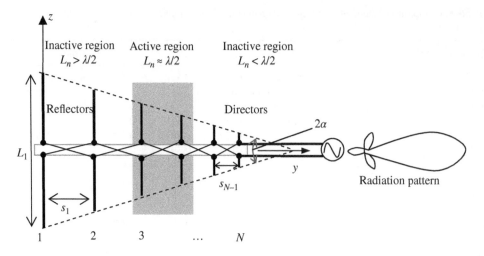

Figure 5.20 The configuration of a log-periodic antenna and its radiation pattern

7 and 15 dBi) and, as the Yagi–Uda, is widely used in the VHF and UHF bands. However, there are two major differences between them:

- *Bandwidth*: the log-periodic antenna has a much wider bandwidth than the Yagi–Uda;
- *Feeder*: each element of the log-periodic antenna is connected to the source and can be seen as a feed of the antenna (i.e., each element is active), while there is only one feed (driven element) and all the others are parasitic elements in the Yagi–Uda antenna.

5.1.6.1 Operational Principle of Log-Periodic Antennas

As seen in Figure 5.20, the log-periodic dipole antenna (LPDA) consists of many dipoles of different lengths. The antenna is divided into the so-called *active region* and *inactive regions*. The role of a specific dipole element is linked to the operating frequency: if its length, L, is around half of the wavelength, it is an active dipole and within the active region; if its length is greater than the half-wavelength, it is in an inactive region and acts as a reflector while if its length is smaller than the half-wavelength, it is also in an inactive region but acts as a director, which is very similar to the Yagi–Uda antenna. The difference is that the driven element shifts with the frequency – this is why this antenna can offer a much wider bandwidth than the Yagi–Uda. A traveling wave can also be formed in the antenna. The highest frequency is basically determined by the shortest dipole length (L_N), while the lowest frequency is determined by the longest dipole length (L_1).

The reason for this antenna being called a *log-periodic antenna* is due to the fact that its input impedance is a periodic function of the logarithm of the frequency. Other parameters that undergo similar variations include the radiation pattern, directivity, and beamwidth.

5.1.6.2 Antenna Design

The geometrical dimensions of the log-periodic antenna follow some pattern and condition, which is another difference from the Yagi–Uda antenna. For analysis, the following notations are used:

L_n = the length of element n, and $n = 1, 2, \ldots, N$;
s_n = the spacing between elements n and $(n + 1)$;
d_n = the diameter of element n;
g_n = the gap between the poles of element n.

They are related to the *scaling factor*:

$$\tau = \frac{L_2}{L_1} = \frac{L_{n+1}}{L_n} = \frac{s_{n+1}}{s_n} = \frac{d_{n+1}}{d_n} = \frac{g_{n+1}}{g_n} < 1; \tag{5.37}$$

and the *spacing factor*:

$$\sigma = \frac{s_1}{2L_1} = \frac{s_n}{2L_n} < 1. \tag{5.38}$$

As shown in Figure 5.20, two straight lines through the dipole ends form an angle 2α which is a characteristic of frequency-independent structure. This angle α is called the *apex angle* of the log-periodic antenna which is a key design parameter and can be found as:

$$\alpha = \tan^{-1}\left(\frac{(L_n - L_{n+1})}{2s_n}\right) = \tan^{-1}\left(\frac{L_n(1-\tau)}{2s_n}\right) = \tan^{-1}\left(\frac{(1-\tau)}{4\sigma}\right) \tag{5.39}$$

These relations hold true for any n. From the operational principle of the antenna and the frequency point of view, Equation (5.37) corresponds to:

$$\tau = \frac{L_{n+1}}{L_n} = \frac{f_n}{f_{n+1}} \tag{5.40}$$

Taking the logarithm of both sides, we have

$$\log f_{n+1} = \log f_n - \log \tau \tag{5.41}$$

This means that the resonant frequency in log-scale is increased by every |log τ|. Thus, the performance of the antenna is periodic in a logarithmic fashion and hence the name log-periodic antenna as mentioned earlier.

This seems to have too many variables. In fact, *there are only three independent variables for log-periodic antenna design*. These three parameters, which can be chosen from such as the directivity, length of the antenna, apex angle, the upper frequency, and the lower frequency, should come with the design specifications. Extensive investigations have been conducted

Table 5.5 Optimum design data for log-periodic antenna

Directivity/dBi	Scaling factor τ	Spacing factor σ	Apex angle α
7	0.782	0.138	$21.55°$
7.5	0.824	0.146	$16.77°$
8	0.865	0.157	$12.13°$
8.5	0.892	0.165	$9.29°$
9	0.918	0.169	$6.91°$
9.5	0.935	0.174	$5.33°$
10	0.943	0.179	$4.55°$
10.5	0.957	0.182	$3.38°$
11	0.964	0.185	$2.79°$

and some optimum designs have been obtained. The most noticeable work was carried out by Carrel [18] and a correction was later made by Butson and Thompson [19]. Computed contours of constant directivity (gain) versus σ and τ were given in these references [2, 3, 12] (but some references have not incorporated the correction of [19]). Here a summary of the optimum design data (with correction) is produced in Table 5.5 which can be used to aid the antenna design. It is apparent that, the higher the directivity, the larger the scaling factor and spacing factor, but the smaller the apex angle.

Another important aspect of the design is the antenna input impedance which can be tuned by changing the diameter d of the element and the feeding gap g between the two poles:

$$g = d \cosh (Z_0/120) \tag{5.42}$$

where Z_0 is the characteristic impedance of the feed line to be connected (the desired impedance). More details can be found from [2, 18].

In practice, the most likely scenario is that the frequency range is given from f_{min} to f_{max}, the following equations may be employed for design:

$$L_1 \geq \frac{\lambda_{max}}{2} = \frac{c}{2f_{min}}; \quad L_N \leq \frac{\lambda_{min}}{2} = \frac{c}{2f_{max}} \tag{5.43}$$

and

$$\frac{f_{min}}{f_{max}} = \frac{L_N}{L_1} = \tau \frac{L_{N-1}}{L_1} = \tau^{N-1} \tag{5.44}$$

Another parameter (such as the directivity or the length of the antenna) is required to produce an optimized design.

Once the geometrical dimensions are obtained, it is desirable to find the radiation pattern. Unfortunately, there is no analytical solution; the best approach is to use some numerical methods or software to complete this task. More discussion on this subject will be presented in Chapter 6.

Example 5.3   Log-Periodic Antenna Design

Design a log-periodic dipole antenna to cover all UHF TV channels, which is from 470 MHz for channel 14 to 890 MHz for channel 83. Each channel has a bandwidth of 6 MHz. The desired directivity is 8 dBi.

Solution

The given three parameters are: f_{min} = 470 MHz, f_{max} = 890 MHz, and D = 8 dBi.

Since the desired directivity is 8 dBi, from Table 5.5, we can see that, for the optimum design, the scaling factor τ = 0.865, the spacing factor σ = 0.157, and the apex angle α = 12.13°. The latter can also be obtained using Equation (5.39).

Because the frequency range is known, using Equation (5.44) yields

$$N = \log\left(\frac{f_{min}}{f_{max}}\right) / \log(\tau) + 1 = \log\left(\frac{470}{890}\right) / \log(0.865) + 1 = 4.40 + 1 = 5.40 \approx 6$$

That means at least six elements are required. On the safe side, we should use seven or even eight elements to sure the desired directivity will be achieved.

If N = 8, we can afford to start from a lower frequency, say 400 MHz, thus

$$L_1 = \frac{c}{2f_{min}} = \frac{300}{800} = 0.375 \text{ (m)},$$

and $L_2 = \tau L_1 = 0.865 * 0.375 = 0.324$ (m), ..., $L_8 = 0.136 < \dfrac{c}{2f_{max}} = \dfrac{300}{2 \times 890} = 0.169$ (m).

The spacing can be obtained using (5.38), that is

$$s_n = 2L_n\sigma = 0.314L_n$$

n = 1–7. (s_1 = 0.118; ..., s_7 = 0.049)

The total length of the antenna is

$$L = \sum_{n=1}^{7} s_n = 0.557 \text{ (m)}$$

In practice, a Yagi–Uda antenna is preferred due to its simplicity (hence cheaper) and higher directivity for the same size. A typical gain of 12 dBi is required and more elements may be used.

In theory, the dipole diameter and feed gap at an element should also be scaled as indicated in Equation (5.37). But in practice, this is sometimes hard to achieve. Because these two parameters do not affect the radiation pattern and directivity, we may just use the same or few different metal poles/tubes to make the antenna. The gap is linked to how each element is fed and Equation (5.42) may be used to tune the impedance. It is not uncommon to use a constant gap, which means a matching section/circuit may be required if the antenna is not well matched with the feed line.

It should be pointed out that for the same number of elements, the Yagi–Uda antenna can produce a higher directivity than its log-periodic counterpart. For example, in Example 5.3, a six-element LPDA can offer 8 dBi gain, but a six-element Yagi–Uda antenna can achieve more than 12 dBi gain. This is due to the fact that the Yagi–Uda is a narrowband antenna and all elements are designed to contribute to the radiation constructively at the desired frequency, while the log-periodic antenna is a broadband antenna and elements within the inactive regions may not contribute to a specific frequency as efficiently and constructively as that in a Yagi–Uda antenna. There is a trade-off between the directivity and bandwidth for fixed N. Question 5.10 will reinforce this concept.

In addition to the LPDA, there are other forms of log-periodic antennas, such as log-periodic loop antennas. The operation and design principle is the same.

5.1.6.3   Frequency-Independent Antennas

If the characteristics (such as the impedance and radiation pattern) of an antenna are not a function of frequency, this antenna is called *frequency-independent antenna*. There are basically three methods to construct such an antenna.

Scaling
If the antenna structure is built with a scaling factor τ, like the log-periodic antenna, its properties will be the same at f and τf. By making the scaling factor close to 1, the properties of the antenna at any frequency will therefore be the same. In practice, even with τ not very close to 1, good frequency-independent characteristics are observed. The LPDA is a good example.

Angle Conditions
If the antenna shape is specified only in terms of angle, the impedance and pattern properties of the antenna will be frequency-independent. Typical examples include various (wire, conical, and planar) spiral antennas as shown in Figure 5.21 [2–4, 12]. Mathematically, the equation for a logarithmic wire spiral can be expressed as:

$$r = r_0 a^{\varphi} \tag{5.45}$$

where

r = radial distance to a point on the spiral;
r_0 = the radius for $\varphi = 0$, determining the upper frequency bound.
φ = angle with respect to x-axis;
a = a constant controlling the flaring rate of the spiral. When $a > 1$, it is right-handed while $0 < a < 1$, it is left-handed.

A second wire spiral which is identical in form to the first one with a rotation of π (now $\varphi \geq \pi$) can be generated by:

$$r = r_0 a^{\varphi - \pi} \tag{5.46}$$

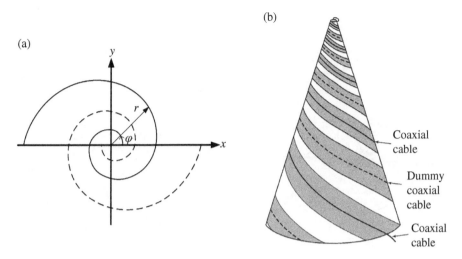

Figure 5.21   Spiral antennas: (a) the wire type, and (b) conical type

These two spirals form a circularly polarized frequency-independent antenna. The impedance, pattern, and polarization remain nearly constant over a wide range of frequencies. The
typical value of the *expansion ratio* $a^{2\pi}$ (the radius increases over one turn of the spiral) is
4. Experimental results suggest that a spiral of one and a half turns is about the optimum. Since
the lower frequency bound is determined by the maximum radius, thus the bandwidth of a typical spiral is $r_0 a^{3\pi} : r_0 \approx 8 : 1$.

The input impedance is determined by the geometrical dimensions, especially around the
antenna feeding. A typical value is between 120 and 180 Ω.

The radiation pattern is bidirectional with two wide beams broadside to the plane of
the spiral, which is approximately $\cos \theta$ when the z-axis is normal to the plane of the
antenna.

Self-complementary

If an antenna structure is identical to its complimentary structure, its input impedance is frequency independence. This will be proved later in this chapter using *Babinet's principle*:
the product of the input impedances of the original and the complementary structures is a constant $= \eta^2/4$ ($\eta \approx 377 \, \Omega$ in free space). The impedances for both structures without truncation
should be identical, thus the input impedance is therefore $\eta/2$ ($\approx 188 \, \Omega$ in free space). Two
examples are shown in Figure 5.22.

In theory, these frequency-independent antennas should be infinitely large to cover the
whole spectrum. In reality, the size has to be finite and therefore they have to be truncated
to cover just the desired frequency range. The truncation has to be carefully done in order
to minimize the effects on antenna performance. This means that the current has to be attenuated and be negligible at the point of truncation – this is why the tail of planar spiral antenna is
of a tapered truncation as shown in Figure 5.22(a).

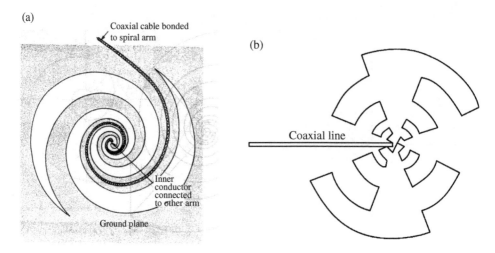

(a)

(b)

Figure 5.22 Two examples of self-complimentary antennas: (a) planar spiral and (b) log-periodic toothed planar antenna

5.2 Aperture-Type Antennas

There is another group of antennas that are not made of metal wires but plates to form certain configurations that radiate/receive EM energy in an efficient and desired manner; this group is called *aperture-type antennas*. They are often used for higher-frequency applications than wire-type antennas. Typical examples include horn antennas and reflector antennas as shown in Figure 5.23, which will be discussed in this section. Although we can still make use of the conventional approach to analyze this type of antenna, i.e. to obtain the current distribution first and then calculate the radiated field and input impedance, a new method which is particularly suitable for analysing aperture-type antennas will be introduced in this section.

5.2.1 Fourier Transform and Radiated Field

It is well known that the Fourier transform is a very useful tool to convert a signal from the time domain, $x(t)$, to the frequency domain, $X(f)$. The mathematical expression is

$$X(f) = \int_{-\infty}^{+\infty} x(t)e^{j2\pi ft}dt \qquad (5.47)$$

Now let us deal with the radiation from an aperture source S_a located in the $z = 0$ plane. As shown in Figure 5.24, it is assumed that the electric field on this aperture is known as $E_a(x, y)$, we need to find out the radiated field in the half-space $z > 0$.

This is a two-dimensional problem. It is demonstrated that we can employ the Fourier transform to obtain the radiation in the far field – it is normally a more efficient and effective method than the conventional current distribution approach which uses the current distribution to characterize the antenna. Thus, the two-dimensional Fourier transform is to be utilized to solve this problem.

Figure 5.23 Some aperture-type antennas (reflector, TEM horn, double-ridged horn, and pyramidal horn antennas)

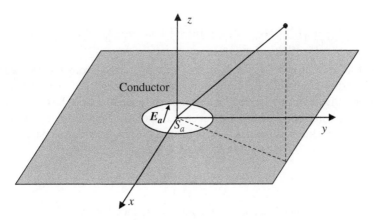

Figure 5.24 Radiation from an aperture source in the $z = 0$ plane

It has been shown that the radiated field from such an aperture source is linked to a plane wave with vector amplitude A propagating in the direction of the propagation vector k by the following expression:

$$E(x,y,z) = \frac{1}{4\pi^2} \int_{-\infty}^{+\infty} \int_{-\infty}^{+\infty} A\left(k_x, k_y\right) e^{-jk\cdot r} dk_x dk_y \tag{5.48}$$

where

$$k \cdot r = \left(k_x \hat{x} + k_y \hat{y} + k_z \hat{z}\right) \cdot (x + y + z) = k_x x + k_y y + k_z z;$$

k_x, k_y, and k_z are propagation constants along x-, y-, and z-directions, respectively; they are linked to the wave number $\beta = \dfrac{2\pi}{\lambda} = \sqrt{k_x^2 + k_y^2 + k_z^2}$, and

$$k_x = \beta \sin\theta \cos\varphi; \quad k_y = \beta \sin\theta \sin\varphi; \quad k_z = \beta \cos\theta \tag{5.49}$$

$$A\left(k_x, k_y\right) = A_t\left(k_x, k_y\right) + A_z\left(k_x, k_y\right)\hat{z};$$

$$A_z\left(k_x, k_y\right) = -\frac{k \cdot A_t\left(k_x, k_y\right)}{k_z} \quad \text{and}$$

$$A_t\left(k_x, k_y\right) = A_x\left(k_x, k_y\right)\hat{x} + A_y\left(k_x, k_y\right)\hat{y} = \int_{S_a}\!\!\!\int E_a(x, y)e^{j\left(k_x x + k_y y\right)}dxdy \tag{5.50}$$

This is the Fourier transform of the aperture field. Thus, Equation (5.48) means that the radiated field E can be obtained using the Fourier transform of the aperture field. The calculation could be complicated, since double integrals are involved.

In the far field, there is no field component along the propagation direction, i.e. TEM waves. Equation (5.48) can be much simplified as [1, 2]:

$$E(r) = j\beta\frac{e^{-j\beta r}}{2\pi r}\left[\left(A_x \cos\varphi + A_y \sin\varphi\right)\hat{\theta} + \left(A_y \cos\varphi - A_x \sin\varphi\right)\cos\theta \cdot \hat{\varphi}\right] \tag{5.51}$$

A coordinate transform from Cartesian coordinates to spherical coordinates has been made here to simplify the expression of the field, which is inversely proportional to the distance, thus the radiated power is inversely proportional to the distance square which is the same as the radiated field from a short wire-type antenna. However, this radiated field has two equally important components in the θ and φ directions – it is not the case for most wire-type antennas.

The magnetic field in the far field (a local plane wave) is given by:

$$H = \hat{r} \times E/\eta \tag{5.52}$$

where η is the intrinsic impedance of the medium as defined before.

Thus, once the source aperture field distribution is known, the radiated fields can be obtained using these equations. Other relevant characteristics (such as the directivity and *HPBW*) can also be calculated using the formulas provided in Chapter 4. A summary of the characteristics of a few rectangular aperture antennas is presented in Figure 5.25, which shows clearly how the aperture field distribution controls the radiation pattern. It is interesting to note that *there is a trade-off between the HPBW and the side-lobe level (SLL): the narrower the HPBW, the higher the SLL*. It is also indicated that *the maximum aperture efficiency occurs for a uniform aperture distribution, while the maximum beam efficiency (the ratio of the solid angle of the main beam to the total beam solid angle) occurs for a highly tapered aperture distribution* – a compromise is normally one of the most important design aspects.

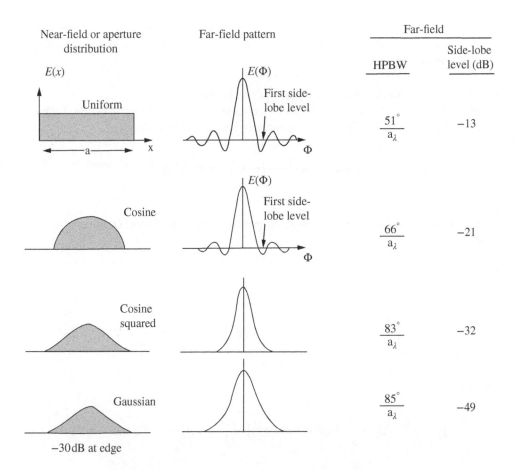

Figure 5.25   Aperture antenna radiation characteristics ($a_\lambda = a/\lambda$)

It can be shown that, for aperture-type antennas, the directivity can also be obtained using the following formula, which is straightforward and powerful:

$$D = \frac{4\pi}{\lambda^2} \frac{\left| \int\limits_{S_a} \int E_a ds \right|^2}{\int\limits_{S_a} \int |E_a|^2 ds} \qquad (5.53)$$

If the aperture distribution is of uniform amplitude, then this equation reduces to

$$D_u = \frac{4\pi}{\lambda^2} A_p \qquad (5.54)$$

where A_p is physical aperture area of the antennas which was used to find the effective aperture size and efficiency in Chapter 4. Using Equation (4.22), we know that, in this case, the aperture efficiency is 100% and the directivity is the largest obtainable from such a physical size. It shows that *the larger the aperture, the larger the directivity – this is a general conclusion for aperture-type antennas.*

Example 5.4   Radiation from an Open-Ended Waveguide

An open waveguide aperture of dimensions a long x and b along y located in the $z = 0$ plane as shown in Figure 2.40. The field in the aperture is TE_{10} mode and given by:

$$\mathbf{E_a} = \hat{y}E_0 \cos\frac{\pi x}{a}, \quad |x| \le \frac{a}{2}, \quad \text{and } |y| \le \frac{b}{2}$$

Find

a. the radiated far field and plot the radiation pattern in both the E- and H-planes;
b. the directivity.

Solution

a. Using Equation (5.50) to yield

$$A_t(k_x, k_y) = \int\int_{S_a} \mathbf{E_a}(x,y)e^{j(k_x x + k_y y)}dxdy = \hat{y}E_0 \int_{-b/2}^{b/2}\int_{-a/2}^{a/2} \cos\frac{\pi x}{a}e^{j(k_x x + k_y y)}dxdy$$

Because

$$\int_{-b/2}^{b/2} e^{j(k_y y)}dy = \frac{1}{jk_y}\left(e^{jk_y b/2} - e^{-jk_y b/2}\right) = \frac{2}{k_y}\sin\left(k_y b/2\right) = b\,\text{sinc}\left(k_y b/2\right)$$

and

$$\int_{-a/2}^{a/2} \cos\frac{\pi x}{a}e^{j(k_x x)}dx = 2\pi a\frac{\cos\left(k_x a/2\right)}{\pi^2 - \left(k_x a\right)^2},$$

We have

$$A_t(k_x, k_y) = \hat{y}2\pi ab E_0\text{sinc}\left(k_y b/2\right)\frac{\cos\left(k_x a/2\right)}{\pi^2 - \left(k_x a\right)^2} \tag{5.55}$$

Note that the special function:

$$\text{sinc}(\alpha) = \frac{\sin\alpha}{\alpha} \tag{5.56}$$

Caution: Matlab (a well-known popular software package) defines $\text{sinc}(\alpha) = \dfrac{\sin\pi\alpha}{\pi\alpha}$ which is different from Equation (5.56) – this should be noted if you use Matlab.

The far field can be obtained using Equation (5.51) as:

$$
\begin{aligned}
\boldsymbol{E}(r) &= j\beta\frac{e^{-j\beta r}}{2\pi r}\left[A_y\sin\varphi\cdot\hat{\boldsymbol{\theta}} + A_y\cos\varphi\cos\theta\cdot\hat{\boldsymbol{\varphi}}\right]\\
&= j\beta\frac{e^{-j\beta r}}{r}abE_0\mathrm{sinc}\left(k_yb/2\right)\frac{\cos\left(k_xa/2\right)}{\pi^2 - \left(k_xa\right)^2}\left[\sin\varphi\cdot\hat{\boldsymbol{\theta}} + \cos\varphi\cos\theta\cdot\hat{\boldsymbol{\varphi}}\right]
\end{aligned}
\tag{5.57}
$$

In the $\varphi = 0$ plane (H-plane), i.e. the xz plane, we have $k_x = \beta\sin\theta$, $k_y = 0$, and

$$
\begin{aligned}
\boldsymbol{E}(r) &= j\beta\frac{e^{-j\beta r}}{r}abE_0\frac{\cos\left(k_xa/2\right)}{\pi^2 - \left(k_xa\right)^2}\left[\cos\theta\cdot\hat{\boldsymbol{\varphi}}\right]\\
&= j\beta abE_0\frac{e^{-j\beta r}}{r}\frac{\cos\left(\beta a\sin\theta/2\right)}{\pi^2 - \left(\beta a\sin\theta\right)^2}\cos\theta\cdot\hat{\boldsymbol{\varphi}}
\end{aligned}
\tag{5.58}
$$

In this plane, the radiated field is actually $E_y(r)$, parallel to the aperture field $\boldsymbol{E}_a$.

While in the $\varphi = \pi/2$ plane (E-plane), i.e. the yz plane, we have $k_x = 0$, $k_y = \beta\sin\theta$, and

$$
\boldsymbol{E}(r) = j\beta\frac{e^{-j\beta r}}{r}abE_0\mathrm{sinc}\left(k_yb/2\right)\frac{1}{\pi^2}\left[\hat{\boldsymbol{\theta}}\right] = j\beta abE_0\frac{e^{-j\beta r}}{\pi^2 r}\mathrm{sinc}\left(\frac{\beta b\sin\theta}{2}\right)\cdot\hat{\boldsymbol{\theta}}
\tag{5.59}
$$

The first SLL for the *sinc* function is -13.2 dB, which is the value for a uniform aperture distribution (along the y-direction).

For a standard waveguide as given in Table 2.5, the typical values at the center frequency for βa and βb are

$$
\beta a \approx \frac{4\pi}{3};\quad\text{and}\ \beta b \approx \frac{2\pi}{3}
$$

Thus, the far-field radiation patterns at the two principal planes can be plotted in Figure 5.26. As expected, it has a unidirectional pattern, and the maximum radiation is normal to the waveguide open end. The pattern in the H-plane is close to $\cos\theta$, while the pattern in the E-plane is very broad. It is also clearly shown from the E-plane pattern that a ground plane is required; otherwise, the field will be radiated to the back of the open waveguide.

b. The directivity can be obtained using Equation (5.53), i.e.

$$
D = \frac{4\pi}{\lambda^2}\frac{\left|\iint\limits_{S_a}\boldsymbol{E}_a ds\right|^2}{\iint\limits_{S_a}|\boldsymbol{E}_a|^2 ds} = \frac{4\pi}{\lambda^2}\frac{\left|\int_{-b/2}^{b/2}\int_{-a/2}^{a/2}E_0\cos\frac{\pi x}{a}dxdy\right|^2}{\int_{-b/2}^{b/2}\int_{-a/2}^{a/2}\left|E_0\cos\frac{\pi x}{a}\right|^2 dxdy} = \frac{4\pi}{\lambda^2}\frac{8}{\pi^2}ab \approx \frac{4\pi}{\lambda^2}(0.81ab)
$$

Again, it is proportional to the aperture physical size (ab) and this result indicates that the aperture efficiency is reduced to 81% compared with uniform excitation. Since the waveguide size is small, the directivity is therefore small. The only way to increase it is to flare out its ends into a horn. Figure 5.27 shows three different horns with increased directivity.

(a)   (b)

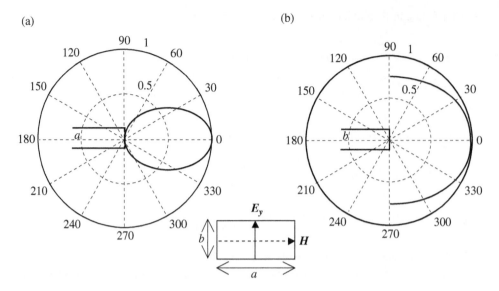

Figure 5.26   Typical radiation patterns of an open waveguide in the H- and E-planes

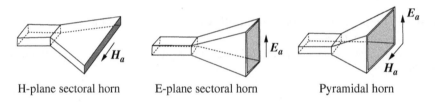

H-plane sectoral horn   E-plane sectoral horn   Pyramidal horn

Figure 5.27   Three horn antennas

5.2.2   Horn Antennas

There is no doubt that horn antennas are the simplest and one of the most widely used micro-wave antennas – the antenna is nicely integrated with the feed line (waveguide) and the performance can be easily controlled. They are mainly used for standard antenna gain and field measurements, feed element for reflector antennas, and microwave communications. The horn can take many different forms: pyramidal horns (shown in Figure 5.27) and conical horns are the most popular horn antennas – the former is most suitable for linear polarization and the latter for circular polarization. Since the basic theory is covered in the previous section, the focus of this section is on design and performance estimation.

5.2.2.1   Pyramidal Horns

The open-ended waveguide has a small directivity and broad beamwidth as demonstrated in Example 5.4, thus it is not suitable for most practical applications. The pyramidal horn is hence evolved from the open waveguide and it is flared in both the E- and H-planes, as shown in

Figure 5.28, which results in narrow beamwidths in both principal planes. The question is how to obtain the optimum design for a specified gain/directivity at a desired operating frequency.

Ideally, the phase of the field across the horn mouth should be constant in order to obtain the desired pattern with minimized sidelobes. This requires a very long horn. However, the horn should be as short as possible for practical convenience. *An optimum design is therefore a compromise in which the difference in the path length along the edge, l_E, and the center of the horn, R_2, is made about 0.25λ.*

All dimensional parameters are shown in Figure 5.28. The directivity to be achieved is D at the operational wavelength λ. The feed waveguide dimensions are of width a and height b. To make such a pyramidal horn, we need to know three of the parameters (A, l_H, R_1, and R_H) in the H-plane and three of the parameters (B, l_E, R_2, and R_E) in the E-plane. The objective of the design is to determine these unknown dimensions.

In the H-plane, the dimensions are linked by:

$$l_H^2 = R_1^2 + (A/2)^2$$
$$R_H = (A-a)\sqrt{(l_H/A)^2 - 0.25} \tag{5.60}$$

and the maximum phase difference in the H-plane is

$$l_H - R_1 \approx \frac{1}{2R_1}\left(\frac{A}{2}\right)^2 = t\lambda \tag{5.61}$$

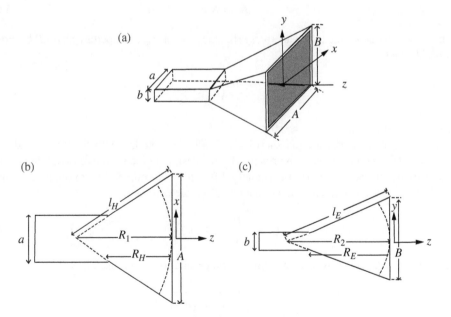

Figure 5.28 Pyramidal horn antennas with dimensional parameters

where t is the maximum phase deviation over the wavelength in the H-plane, i.e. $t = (l_H - R_1)/\lambda$. The larger the value of t, the broader the beamwidth. It was found $t = 3/8 = 0.375$ (but 0.25 for the E-plane) gives the optimum design [2, 3, 12, 20], which means

$$A = \sqrt{3\lambda R_1} \tag{5.62}$$

In the E-plane, the dimensions are linked by:

$$l_E^2 = R_2^2 + (B/2)^2$$
$$R_E = (B - b)\sqrt{(l_E/B)^2 - 0.25} \tag{5.63}$$

and the maximum phase difference in the E-plane is

$$l_E - R_2 \approx \frac{1}{2R_2}\left(\frac{B}{2}\right)^2 = s\lambda \tag{5.64}$$

where s is the maximum phase deviation in the E-plane; the larger the value of s, the broader the beamwidth. It was found $s = 1/4 = 0.25$ gives the optimum design which means

$$B = \sqrt{2\lambda R_2} \tag{5.65}$$

When both the E- and H-planes are put together to form the pyramidal horn, the following condition has to be satisfied in order to make it physically realizable and properly connected to the feed waveguide:

$$R_E = R_H \tag{5.66}$$

The directivity is related to the aperture efficiency factor η_{ap} and aperture size AB by Equation (4.22), that is

$$D = \frac{4\pi}{\lambda^2}\left(\eta_{ap}AB\right) \tag{5.67}$$

For the optimum gain horn, $\eta_{ap} \approx 0.51 = 51\%$.

We have eight independent Equations (5.60a, 5.60b, 5.61, 5.63a, 5.63b, 5.64, 5.66, and 5.67) to solve eight unknown variables in principle for the design problem to be resolved. Since most of them are not linear equations, it is actually difficult (if not impossible) to obtain a solution analytically. After some mathematical manipulation, we can yield

$$(B - b)\frac{B}{s} = (A - a)\frac{A}{t} \tag{5.68}$$

The solution to B (which is positive, thus the other solution is not useful) is

$$B = \frac{b + \sqrt{b^2 + 4sA(A - a)/t}}{2} \tag{5.69}$$

Replacing B in Equation (5.67) to give a desired design equation:

$$A^4 - aA^3 + \frac{tbD\lambda^2}{4\pi s\eta_{ap}}A = \frac{tD^2\lambda^4}{16s\pi^2\eta_{ap}^2} \tag{5.70}$$

This fourth-order equation in A can be solved using numerical methods. Alternatively, it may also be attempted by trial and error. For the optimum design, use a first guess approximation [12, 20]:

$$A = 0.45\lambda\sqrt{D} \tag{5.71}$$

Once the value of A is obtained, we can easily calculate other dimensions of the structure. The HPBW for an optimum horn in the H-plane is [12]

$$HP_H = \varphi_{HP} \approx 78° \frac{\lambda}{A} \tag{5.72}$$

In the E-plane, it is

$$HP_E = \theta_{HP} \approx 54° \frac{\lambda}{B} \tag{5.73}$$

These are approximate values; slightly different approximations are given in [4]. It is also noted that the directivity obtained using these equations and Equation (4.14) is different from that using Equation (5.67) unless $\eta_{ap} = 0.78$ (it should be 0.51 for the optimized horn). It may be due to the fact that Equation (4.14) is an approximation for a single-lobe antenna and this one has sidelobes.

Now we are going to use the following example to summarize the procedure of designing an optimum pyramidal horn.

Example 5.5   Optimum Horn Design
Design a standard gain horn with a directivity of 20 dBi at 10 GHz. WR-90 waveguide will be used to feed the horn.

Solution
The directivity is $D = 20$ dBi $= 100$, wavelength $\lambda = 30$ mm, and the dimensions of the waveguide are $a = 22.86$ mm and $b = 10.16$ mm.

Step 1: *Compute the dimension A from the design Equation (5.70).*
As suggested above, the parameters for the optimum horns are

$$\eta_{ap} = 0.51; \quad s = 0.25; \quad t = 0.375$$

and the design Equation (5.70) becomes

$$A^4 - 22.86A^3 + 214020A = 2.9581 \times 10^8$$

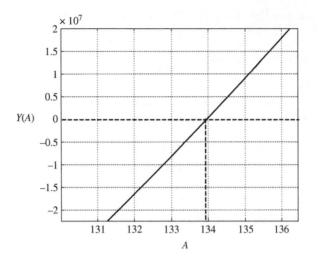

Figure 5.29   Function $Y(A)$ versus A

Using Equation (5.71), we obtain an initial guessing value:

$$A = 0.45\lambda\sqrt{D} = 135 \ (\text{mm})$$

and plot the function:

$$Y(A) = A^4 - 22.86 \ A^3 + 214020 \ A - 2.9581 \times 10^8$$

around $A = 135$ (mm). As seen in Figure 5.29, the actual solution is readily found as

$$A \approx 133.96 \ (\text{mm})$$

Step 2: *Use Equation (5.68) to yield B*
    In this case $B = 104.82$ mm
Step 3: *Find remaining dimensions from:*

> *Equation (5.61) or (5.62) for R_1;*
> *Equation (5.64) or (5.65) for R_2;*
> *Equation (5.60) for l_H and R_H;*
> *Equation (5.63) for l_E and R_E;*

Thus, we can get

> $R_1 = 199.41$ mm;
> $R_2 = 183.14$ mm;
> $l_H = 210.36$ mm and $R_H = 165.38$ mm;
> $l_E = 190.49$ mm and $R_E = 165.38$ mm.

Step 4: *Check if R_H and R_E are the same. If not, it means that the solution of A in Step 1 is not accurate enough.*

From the results in Step 3, we can see that R_H and R_E are identical, thus the design is very good. However, if we used the guessing value $A = 0.45\lambda\sqrt{D} = 135\,(\text{mm})$ as the solution, and it would give $R_H = 168.21$ mm and $R_E = 162.73$ mm. They are obviously different, it means that the design is to be revised.

As the TE_{10} mode is the field pattern for such an antenna, the aperture field distribution is still the same as that of open-ended waveguide (TE_{10} mode as well) but with a different phase term, this is

$$E_a = \hat{y}E_0 \cos\frac{\pi x}{A} e^{-j\beta/2\left(x^2/R_1 + y^2/R_2\right)}, \quad |x| \le \frac{A}{2}, \text{ and } |y| \le \frac{B}{2} \tag{5.74}$$

The radiation pattern is closely related to the phase error parameters s and t. There is no simple expression for the pattern. Universal radiation patterns, as shown in Figure 5.30 where a weighting factor $(1 + \cos\theta)/2$ is not included, have been produced for design purposes and the detailed treatment of this subject can be found in [2]. The radiation pattern of this design is shown in Figure 5.31. The sidelobes in the E-plane are much higher than those in the H-plane – a common feature of this type of antenna, since the aperture field in the E-plane is uniform (not tapered).

The HPBWs in the two principal planes for this antenna are

$$HPBW_H = \varphi_{HP} \approx 78°\frac{\lambda}{A} = 17.47°$$

$$HPBW_E = \theta_{HP} \approx 54°\frac{\lambda}{B} = 15.45°$$

It should be pointed out that:

- the optimum horn is designed for a specific frequency and gain;
- for a horn of fixed length, there is a maximum directivity/gain obtainable which normally corresponds to the optimum design. Under such a condition, the directivity is proportional to the aperture size. Otherwise, this conclusion may not be valid. For example, a horn with a

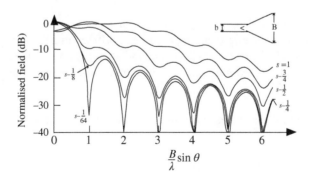

Figure 5.30 E-plane universal patterns for E-plane sectorial and pyramidal horns

(a)   (b)

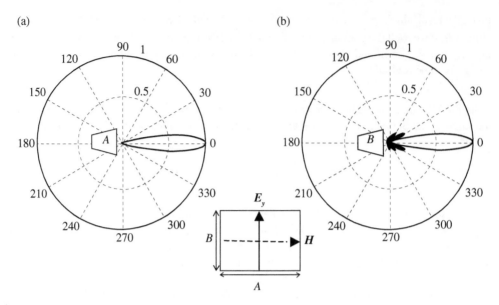

Figure 5.31   Radiation patterns of a pyramidal horn antenna

larger aperture size than the optimum one has larger phase error parameters which result in smaller directivity.

- There are two special cases: in Figure 5.28, if $A = a$, the structure is called the *E-plane sectorial horn antenna;* if $B = b$, the structure is called the *H-plane sectorial horn antenna*. They are both shown in Figure 5.27. They have a very broad beamwidth in one of the principal planes and a very narrow beamwidth in the other plane (the plane opened up). The design procedure is the same as that of the pyramidal horn.

5.2.2.2   Other Horn Antennas

In addition to the pyramidal horn, there are many other horn antennas. The well-known ones include

- *The circular horn*: a simple circularly polarized antenna;
- *The corrugated horn*: it can provide reduced edge diffraction and cross-polarization, and improved aperture efficiency (to 75–80%) and radiation pattern symmetry.
- *The single- or double-ridged horn*: a directional broadband antenna using tapered ridge as shown in Figure 5.23.

Detailed discussion of these antennas can be found in references such as [3].

5.2.3   Reflector and Lens Antennas

The typical gain of a practical horn antenna is up to 20 dBi or so. Higher gains are obtainable, but it means a much larger and heavier horn which is not suitable for most applications. As an

Figure 5.32   Paraboloidal and Cassegrain reflector antennas. Source: Xinhua / Alamy Stock Photo

alternative in practice, reflector and lens antennas can offer much higher gains than horn antennas and are normally relatively easy to design and construct. They are probably the most widely used antennas for high-frequency and high-gain applications in radio astronomy, radar, microwave and millimeter wave communications, and satellite tracking and communications.

Although reflector and lens antennas can take various geometrical configurations, the most popular shape is the paraboloid – because of its excellent ability to produce a pencil beam (high gain) with low sidelobes and good cross-polarization characteristics in the radiation pattern. The largest fully steerable reflector in the world is the 100-m diameter radio telescope of the Max Planck Institute for Radioastronomy at Effelsberg in Germany, whereas the largest reflector antenna is the 500 m radio telescope in China at the time of writing. As displayed in Figure 5.32, the typical feed for such an antenna is a horn antenna which can be placed at the focal point in front of the reflector (front-fed) or at the back (vertex) of the reflector (Cassegrain feed). The latter is known as the *Cassegrain reflector antenna* and was invented to avoid using a long feed line and to minimize the feed blockage problems of a conventional paraboloidal antenna (the sidelobes can be reduced as well). Due to its popularity, the emphasis of this section will be on the paraboloidal reflector antenna. The design principles can be applied to other reflector antenna designs. Many research and design papers have been published in this area and some of the most referenced papers can be found in [21]. A special section on reflector antennas and designs is provided in Section 8.9.

5.2.3.1   Paraboloidal Reflector Characteristics

As Figure 5.32 shows, a front-fed paraboloidal antenna consists of a reflector and a feed antenna. The surface of a paraboloidal reflector, which is illuminated by the feed antenna, is formed by rotating a parabola about its axis. Its surface is therefore a paraboloid of revolution, and rays emanating from the focus of the reflector are transformed into plane waves – this means that it is highly directional. If the reflector was extremely large (infinite) and the feed was a point source at the focal point having radiation only toward the paraboloid, all the radiated EM energy would be directed in one direction (the z-direction in Figure 5.32) with zero beamwidth. Here we have assumed that the radiated waves can be treated as rays (otherwise, the conclusion would not be valid), i.e. the structure is much greater than the wavelength. This unique feature of the paraboloid has made it extremely suitable for high-frequency and high-gain applications.

A good example is the world largest single antenna FAST (Five-hundred-meter Aperture Spherical radio Telescope)[1], as shown in Figure 5.32(c), which was built in Guizhou, China between 2011 and 2017. It has a fixed primary reflector located in a natural sinkhole in the landscape, focusing radio waves on a receiving antenna in a "feed cabin" suspended 140 m above it. The cabin containing the feed antenna suspended on cables above the dish is also moved using a digitally controlled winch by the computer control system to steer the instrument to receive from different directions. The reflector is made of perforated aluminum panels supported by a mesh of steel cables hanging from the rim. It has a novel design, using an active surface made of metal panels that can be tilted by a computer to help change the focus to different areas of the sky. Its working frequency is from 70 MHz to 3.0 GHz, with the upper limit set by the precision with which the primary can approximate a parabola. It could be improved slightly, but the size of the triangular segments limits the shortest wavelength which can be received. Although the reflector diameter is 500 m, only a circle of 300 m diameter is actually used (held in the correct parabolic shape and "illuminated" by the receiver) at any one time. Thus, the whole 500 m aperture could be used to form many 300 m parabolic (not spherical) dish antennas, looking at different directions and different times.

5.2.3.2   Analysis and Design

In practice, it is not possible to make the reflector infinitely large (actually we always try to make it as small as possible) and truncation has to take place. Also, the feed antenna cannot be a point source, which means that the actual performance of the antenna will be different from the ideal one. How to design a good paraboloidal antenna is in fact a difficult task.

Let us assume that the diameter of the reflector is $2a$ and the focal length is F. Thus, any point on this paraboloid must satisfy the following condition:

$$r = \frac{2F}{1 + \cos\theta} = F \sec^2(\theta/2) \quad \theta \le \theta_o \tag{5.75}$$

---

[1] Xinhua/Alamy Stock Photo

where the subtended/angular aperture angle θ_0 (also known as *edge angle*) is determined by the reflector diameter and focal length:

$$\theta_0 = \tan^{-1}\left|\frac{a}{F - a^2/(4F)}\right| \tag{5.76}$$

Here there are two parameters, diameter $2a$ and focal length F, to be decided which does not seem to be too bad. However, another element of the reflector antenna is the feed which is an antenna on its own and it has many variables. Thus, the complete design of the reflector antenna is actually a complex task since there are many parameters which could be changed to meet the specifications, some are independent, and some are interlinked. *The reflector design problem consists primarily of matching the feed antenna pattern to the reflector. The usual goal is to have the feed pattern about 10 dB down in the direction of the rim*, that is the *edge taper* = (the field at the edge)/(the field at the center) $\approx$ 10 dB. Feed antennas with this property can be constructed for the commonly used $F/2a$ value of 0.3–1.0. Higher values lead to better cross-polarization performance but require a narrower feed pattern and hence a physically larger feed antenna.

The antenna analysis can be conducted using the Fourier-transform-based *aperture distribution method* introduced in Section 5.1. Alternatively, the *current distribution method* commonly used for wire-type antenna analysis and the equivalence principle can be employed to obtain the equivalent source current at the aperture. Both methods yield results that are essentially the same for the principle polarized radiation pattern and agree quite well with experimental results as long as the aperture is large enough in terms of the wavelength. Detailed analysis can be found in [21]. As a high-gain antenna, the most important parameter is obviously the directivity/gain. Thus, we are going to see how to estimate this parameter without using expensive computer simulation tools.

Aperture Efficiency and Directivity
Since the feed antenna is closely linked to the reflector, the aperture efficiency should surely reflect this linkage. Let $g(\theta)$ be the power radiation pattern of the feed located at the focus – it is circularly symmetrical (not a function of ϕ). It has been shown that the aperture efficiency is given as [2, 21]:

$$\eta_{ap} = \cot^2\left(\frac{\theta_0}{2}\right)\left|\int_0^{\theta_0} g(\theta) \tan\left(\frac{\theta}{2}\right) d\theta\right|^2 \tag{5.77}$$

As expected, it is determined by both the reflector and the feed. The maximum aperture efficiency is around 82%, which is higher than that of a pyramidal horn.

The aperture efficiency is generally viewed as the product of the

1. *Spillover efficiency*: the fraction of the total power intercepted and collimated by the reflector; it reduces the gain and increases the SLLs;
2. *Taper efficiency*: the uniformity of the amplitude distribution of the feed pattern over the surface of the reflector;
3. *Phase efficiency*: the phase uniformity of the field over the aperture plane; it affects the gain and sidelobes.

4. *Polarization efficiency*: the polarization uniformity of the field over the aperture plane;
5. *Blockage efficiency:* by the feed. It reduces gain and increases SLLs. The support structure can also contribute to the blockage.
6. *Random error efficiency:* over the reflector surface.

Some of these requirements contradict with each other. For example, the larger the reflector angular aperture angle θ_0, the larger the spillover efficiency, but the smaller the taper efficiency. The optimum aperture efficiency is the best trade-offs from all aspects.

Once the aperture efficiency is yielded, the directivity can be readily obtained using Equation (4.22):

$$D = \frac{4\pi}{\lambda^2} \eta_{ap} \left(\pi a^2 \right) \tag{5.78}$$

The analysis here has not taken the feed antenna efficiency into account, which is about 70–80% if it is a horn antenna. Thus, the overall reflector efficiency factor is in the region of 50–70%. It is very common to make a reflector antenna with a gain of over 30 dBi. Some of the world largest antennas have a gain over 70 or even 80 dBi which is almost impossible for a single wire-type antenna to achieve – remember the typical gain for a Yagi–Uda and log-periodic antenna is about 10–15 dBi.

The HPBW can be estimated by rule of thumb:

$$HPBW \approx 70° \frac{\lambda}{2a} \tag{5.79}$$

The beamwidth depends also on the edge illumination. Typically, as the edge attenuation increases, the beamwidth widens and the sidelobes decrease.

Design Considerations and Procedures

In addition to the aperture efficiency and directivity/gain, cross-polarization, phase errors (the maximum fractional reduction in directivity is $\delta^2(1 - \delta^2/4)$, where δ is the phase error), *HPBW* (which can be estimated using Equation (4.14) once the directivity is known), and sidelobes are also to be considered in design. The priority is really down to the specific application.

The design procedure starts with selection of the feed antenna which determines the antenna polarization and the reflector F and $2a$. As mentioned earlier, the usual goal is to let the feed pattern be about 10 dB down in the direction of the reflector rim. The feed and reflector are inter-linked; an iterative process may be required to ensure that the feed antenna pattern is well matched with the reflector. Once the feed pattern and the reflector are known, the radiated field can be calculated, where complex integrations are involved [2, 21]. Normally, computer software is required to accomplish such a task. For example, the software package GRASP, a product of TICRA, is well suited for this application. This package is a set of tools for analyzing reflector antennas and antenna farms. The program is based on PO (Physical Optics)/GTD (Geometrical Theory of Diffraction) calculation. More about software for antenna designs will be given in Chapter 6.

Example 5.6   Edge Taper and Spillover Efficiency

A circular parabolic reflector has $F/2a = 0.5$. The field pattern of the feed antenna is $E(\theta) = \cos \theta$, $\theta < \pi/2$. Find the edge taper, spillover efficiency, and aperture efficiency.

Solution

The edge angle is given by Equation (5.76):

$$\theta_0 = \tan^{-1}\left|\frac{a}{F - a^2/(4F)}\right| = \tan^{-1}\left|\frac{(F/2a)/2}{(F/2a)^2 - 1/16}\right| = \tan^{-1}\left|\frac{0.25}{(0.5)^2 - 1/16}\right| = 53.13°$$

The aperture illumination function is

$$A(\theta) = E(\theta)\frac{1}{r}e^{-j\beta r} = \cos(\theta)\frac{1}{r}e^{-j\beta r}$$

Since r is given by Equation (5.75), thus

$$|A(\theta)| = \frac{\cos(\theta)(1 + \cos\theta)}{2F}$$

The edge taper is therefore:

$$\frac{A(\theta_0)}{A(0)} = \frac{\cos(\theta_0)(1 + \cos(\theta_0))}{(1 + 1)} = 0.4800 = -6.3752 \text{ dB}$$

The spillover efficiency is

$$\eta_s = \frac{P_{intercepted}}{P_{radiated}} = \frac{\int_0^{\theta_0} g(\theta)\sin\theta d\theta}{\int_0^{\pi} g(\theta)\sin\theta d\theta} = \frac{\int_0^{\theta_0} E(\theta)^2 \sin\theta d\theta}{\int_0^{\pi} E(\theta)^2 \sin\theta d\theta} = \frac{\int_0^{\theta_0} \cos^2(\theta)\sin\theta d\theta}{\int_0^{\pi} \cos^2(\theta)\sin\theta d\theta} \approx 0.78$$

Thus, the spillover loss is about -1.08 dB.

The aperture efficiency can be calculated using Equation (5.77). It is about 0.59.

If a uniform aperture illumination is needed, we have:

$$\frac{A(\theta)}{A(0)} = \frac{E(\theta)(1 + \cos(\theta))}{E(0)(1 + 1)} \equiv 1$$

The feed antenna pattern should be

$$E(\theta) = \frac{2}{1 + \cos\theta} \tag{5.80}$$

It means a trough at the center ($\theta = 0$), which is hard to achieve in practice.

Figure 5.33   An example of an offset parabolic reflector radar antenna

5.2.3.3   Offset Parabolic Reflectors

To eliminate some of the deficiencies of symmetrical configurations, the offset parabolic reflector design has been developed for single- and dual-reflector systems, which reduces aperture blockage and offers the advantage of allowing the use of larger $F/(2a)$ ratio while maintaining acceptable structure rigidity. The configuration is also widely employed, most noticeably for domestic satellite receiving antennas which have a typical gain of about 30 dBi. It is also utilized in radar and other applications.

Figure 5.33 is an example of an offset parabolic reflector used for a radar positioning application where the *HPBW* is required to be very narrow in the horizontal plane but wider in the vertical plane. GRASP was employed to predict the performance. The simulated radiation patterns are shown in Figure 5.34 and the main specifications for the H-plane are

• The gain:	28.0 dB (27.9 dB, measured)
• The HPBW:	4.0° (4.4°)
• 10 dB beamwidth is:	6.8° (7.6°)
• The 1st SLL:	−19.5 dB (−26.3 dB)

Measured results are given in the brackets. It is apparent that the simulated results are very close to the measured ones (except the sidelobes in the H-plane). This indicates that our simulation and design tool worked well for this antenna.

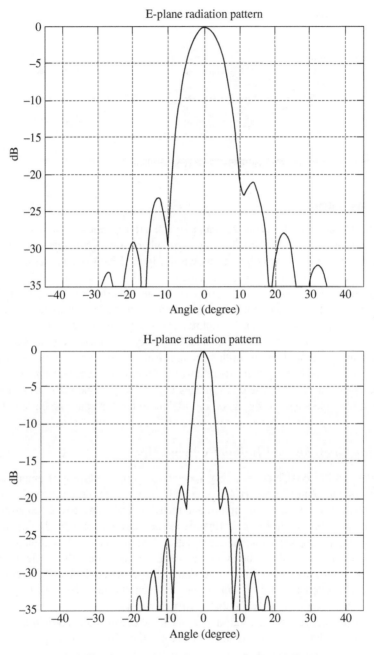

Figure 5.34 Computed radiation patterns in E- and H-planes

Some other reflector antennas, such as flat sheet reflector antennas and corner reflectors, are also used in practice. More information about reflector antennas can be found in Section 8.9.

Lens antennas

(a)   (b)

Source or
primary
antenna

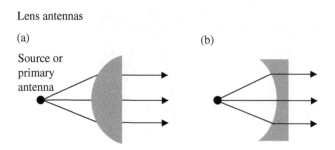

Figure 5.35   Two types of lens antennas

5.2.3.4   Lens Antennas

Just like reflector antennas, lens antennas can also convert a spherical wave into a plane wave to produce high gains and pencil beams. They are only suitable for high-frequency (> 4 GHz) applications. There are basically two types as shown in Figure 5.35: (a) delay lenses, in which the electrical path length is increased by the lens medium (using low-loss dielectrics with a relative permittivity greater than 1, such as Lucite or polystyrene), and (b) fast lenses, in which the electrical path length is decreased by the lens medium (using metallic or artificial dielectrics with a relative permittivity smaller than 1). The source or primary antenna is normally a tapered horn antenna as for a reflector antenna. Lens antennas have very similar characteristics to paraboloidal reflector antennas discussed earlier in this section. But they do offer several advantages over reflector antennas, which make them more attractive for some aerospace applications such as Earth Observation and radars. These advantages include wide-angle scan, very low sidelobes, and low feed horn blockage. A very good discussion on lens antenna can be found in [4].

5.2.4   Slot Antennas and Babinet's Principle

Slot antennas can be considered as a very special group of aperture-type antennas. They are very low profile and can be conformed to basically any configuration, thus they have found many applications, for example, on aircraft and missiles. They may be conveniently energized with a coaxial transmission line or a waveguide as shown in Figure 5.36. The open-end waveguide analyzed in Example 5.4 may be viewed as a slot antenna if a ground plane is used. The aperture distribution method can be employed again to yield the radiated fields. For narrow slots, there is another way to do it: we are going to use the equivalence principle to obtain the radiation characteristic. This is simple and convenient, and the knowledge gained for wire-type antennas can be readily applied to this case.

Recall the equivalence principle introduced in Chapter 3, the radiated field by the slot is the same as the field radiated by its equivalent surface electric current and magnetic current which were given by Equation (3.38), that is,

$$J_S = \hat{n} \times H, \quad M_S = -\hat{n} \times E$$

where E and H are the electric and magnetic fields within the slot, and $\hat{n}$ is the unit vector normal to the slot surface S.

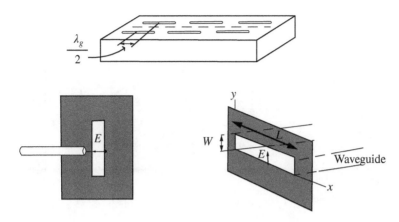

Figure 5.36 Slot antennas fed by coaxial cable and waveguide

Take a half-wavelength slot as an example, since its equivalent electric surface current $J_S = \hat{n} \times H = 0$, the remaining source at the slot is its equivalent magnetic current $M_S = -\hat{n} \times E$ (it would be $2M_S$ if the conducting ground plane were removed using the imaging theory). This is the same kind of equivalent magnetic source as a loop, as discussed in Section 5.1.3. The current distribution along this half-wavelength magnetic source is the same as the current distribution along a half-wavelength dipole. We can therefore use the duality principle introduced in Section 5.1.3 to find the radiated field of the slot from the dipole field.

The normalized electric and magnetic fields of a half-wavelength dipole can be obtained using Equations (5.3) and (5.4) as:

$$E_\theta = \frac{\cos\left[\pi/2 \cos\theta\right]}{\sin\theta}, \quad \text{and} \quad H_\varphi = \frac{\cos\left[\pi/2 \cos\theta\right]}{\sin\theta} \tag{5.81}$$

Using the duality relation in Table 5.3, we obtain the normalized radiated field from a half-wavelength slot antenna as:

$$E_\varphi = \frac{\cos\left[\pi/2 \cos\theta\right]}{\sin\theta}, \quad \text{and} \quad H_\theta = \frac{\cos\left[\pi/2 \cos\theta\right]}{\sin\theta} \tag{5.82}$$

Thus, the only change in the field from dipole to slot is that E_θ is replaced by E_φ, and H_φ is replaced by H_θ. The radiation pattern shape is not changed as seen in Figure 5.37. Of course, if the slot antenna only radiates to half of the space, then the radiation pattern should only appear in half of the space.

Another very important parameter of the antenna is the input impedance. Again, if we utilized another theory, Babinet's principle, this parameter can also be easily obtained using our existing knowledge. *Babinet's principle* states that *the field at any point behind a plane having a screen, if added to the field at the same point when the complementary screen is substituted, is equal to the field at the point when no screen is present.* Obviously, it is meant for optics. If this principle is extended to antennas and radio propagation and the equivalent transmission line model is employed, as we did in Chapter 3 for radiowave transmission and reflection analysis,

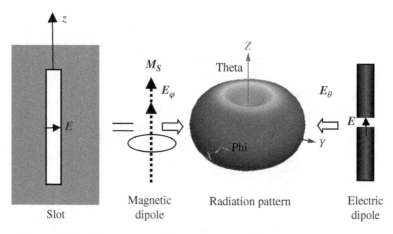

Figure 5.37 Slot antenna, radiation pattern, and its complimentary dipole

we can get the following important relationship for the input impedances of complementary antennas:

$$Z_{slot}Z_{dipole} = \frac{\eta^2}{4} \tag{5.83}$$

It means that the product of the impedances of two complementary antennas is the constant $\eta^2/4$ and this is the interpretation of the Babinet's principle in antennas.

Since the impedance for a half-wavelength dipole is about 73 Ω, the corresponding slot has an impedance of

$$Z_{slot} = \frac{\eta^2}{4Z_{dipole}} = \frac{377}{4 \times 73} \approx 486 \ (\Omega)$$

When we discussed the frequency-independent antennas earlier in this chapter, the self-complementary configurations as shown in Figure 5.22 are two of them. Because their impedances are the same, in free space, this type of antenna has a constant impedance of

$$Z_a = \sqrt{\frac{\eta^2}{4}} = 188.5 \ (\Omega) \tag{5.84}$$

Slot antennas, especially slot waveguide antennas where the antenna and feed line are integrated and exhibit high power-handling capacity, are often used to form an antenna array for applications such as airborne radar. Figure 5.38 is such an example: 26 waveguides and over 300 slot elements are employed to form this antenna array. Each slot is carefully designed to ensure that the field aperture distribution follows a Gaussian distribution as shown in Figure 5.25. It has a very high gain (over 35 dBi) and low sidelobes (below − 25 dB). More will be presented on antenna arrays in Section 5.3.

Figure 5.38   An airborne slot waveguide antenna array

5.2.5   *Microstrip Antennas*

A microstrip antenna, also known as a patch antenna, consists of a metal patch on a substrate on a ground plane, as shown in Figure 5.39. Different feed configurations, including aperture coupled, microstrip line feed, and coaxial feed, are also shown in the figure. The patch can take various forms to meet different design requirements. Typical shapes are rectangular, square, circular, and circular ring. The microstrip antenna is low profile, conformable to planar and nonplanar surfaces, simple and cheap to manufacture using modern printed circuit technology, compatible with monolithic microwave integrated circuit (MMIC) designs, and mechanically robust.

In addition, it is very versatile in terms of resonant frequency, input impedance, radiation pattern, and polarization. All these have made it an extremely popular modern antenna for frequencies above 300 MHz (from UHF band). A huge amount of research papers and many books have been published over the years in this area [22–25]. The major disadvantages of this type of antenna are relatively low efficiency (conducting, dielectric, and especially surface wave losses), low power-handling capability (not suitable for high-power applications), poor polarization purity, and relatively narrow frequency bandwidth. However, significant improvements in, for example, broadening the bandwidth have been made (by stacking and other methods in order to meet the demand from the booming wireless communications industry [25]). Considerable attention has been paid recently on how to make this kind of antenna tunable and reconfigurable by adding loads/switches on or between the patch and ground plane. The future for this antenna seems bright and exciting.

In this section, a rectangular patch antenna is chosen as an example for investigation since it is the most popular printed antenna. We are going to examine the operational principles, major characteristics, and design procedures. As usual, a design example will be given at the end of the section.

(a)

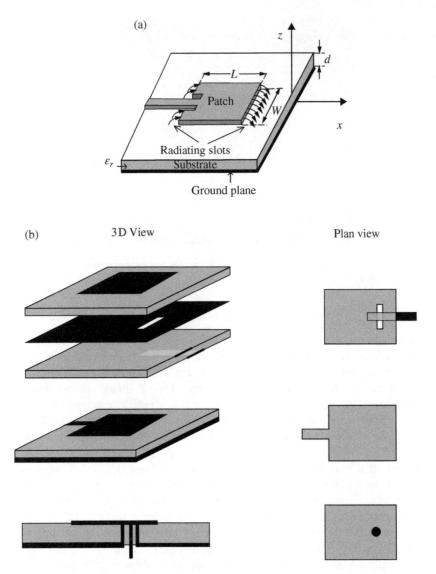

(b)   3D View   Plan view

Figure 5.39 Microstrip antennas and their feeds (a) a microstrip antenna with its coordinates (b) three feeding configurations: coupling feed, microstrip feed, and coax feed

5.2.5.1 Operational Principles

The rectangular antenna dimensions and coordinates are displayed in Figure 5.39(a). Usually, the patch length L is between $\lambda_0/3$ and $\lambda_0/2$ and its width W is smaller than λ_0 (it cannot be too small, otherwise the antenna becomes a microstrip line which is not a radiator), while its thickness t is extremely small. The substrate of thickness d ($\ll \lambda_0$) uses the same kind of materials as listed in Table 2.4. Their relative permittivity is normally between 2 and 24.

To be a resonant antenna, the length L should be around half of the wavelength to act as a half-wavelength dipole. Another explanation is that in this case, the antenna can be considered as a *λ/2 transmission line resonant cavity with two open ends where the fringing fields from the patch to the ground are exposed to the upper half-space (z > 0) and are responsible for the radiation.* This radiation mechanism is the same as the slot line, thus there are two radiating slots on a patch antenna as indicated in Figure 5.39(a). This is why the microstrip antenna can be considered as an aperture-type antenna. Of course, we can still use the current distribution and antenna equation to analyze the operational principle of this antenna. A detailed analysis using the theory of characteristic modes is provided in Section 6.5 which shows clearly what the possible characteristic modes are and how they could be generated (or not generated) to achieve the desired performance.

Since the fringing fields at the ends are separated by $λ/2$, which means that they are $180°$ out of phase but equal in magnitude. Viewed from the top of the antenna, both fields are actually in phase for the x-components which leads to a broadside radiation with a maximum in the z-direction.

5.2.5.2 Analysis and Design

As a resonant cavity, there are many possible modes (as waveguides), thus a patch antenna is multimode and may have many resonant frequencies. The fundamental and dominant mode is TM_{100} (a half-wave change along the x-axis and no changes along the other two axes).

Radiation Pattern and Directivity
The radiation comes from the fringing fields at the two open ends as discussed above, which is equivalent to two slot antennas separated by a distance L. It can be proved that the far-field electric field can be expressed as:

$$E = E_0 \,\text{sinc}\left(\frac{\beta W}{2} \sin\theta \sin\varphi\right) \cos\left(\frac{\beta L}{2}\sin\theta\cos\varphi\right)\left(\hat{\theta}\cos\varphi - \hat{\varphi}\cos\theta\sin\varphi\right) \qquad (5.85)$$

where β is the free space wave number. The first factor is the pattern factor for a uniform line source of width W in the y-direction (similar to Example 5.4) and the second factor is the array factor for the two-element slots separated by L in the x-direction (to be discussed in the next section). For both components, the peak is at $\theta = 0$, which corresponds to the z-direction. It has a broadside unidirectional pattern.

The radiation patterns in the two principal planes are

a. E-plane ($\varphi = 0°$):

$$E = \hat{\theta}E_0 \cos\left(\frac{\beta L}{2}\sin\theta\right) \qquad (5.86)$$

b. H-plane ($\varphi = 90°$):

$$E = -\hat{\varphi}E_0\text{sinc}\left(\frac{\beta W}{2}\sin\theta\right)\cos\theta \qquad (5.87)$$

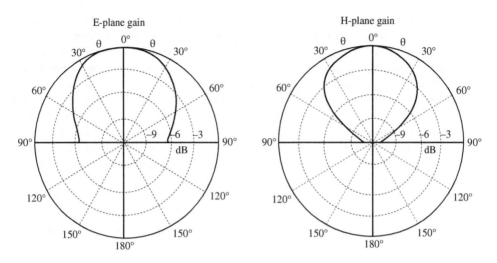

Figure 5.40 Typical radiation patterns of a resonant rectangular patch antenna

These results have neglected the substrate effects and slot width but are good enough as estimates. The typical radiation patterns in the E- and H-planes are shown in Figure 5.40. As discussed in Section 5.1.2, if the ground plane is finite, leaky radiation toward the lower half-space will occur.

With the radiated field, we are able to compute the directivity using Equation (4.11). Asymptotically, the directivity of the microstrip antenna can be expressed as:

$$D = \begin{cases} 6.6 = 8.2 \text{ dBi}, & W \ll \lambda_0 \\ 8W/\lambda_0, & W \gg \lambda_0 \end{cases} \tag{5.88}$$

The larger the width, the larger the directivity. These are approximations as well; the actual directivities are slightly lower than these values.

Input Impedance and Bandwidth
According to Equation (40) of [23], the typical impedance at the edge of a resonant rectangular patch ranges from 100 to 400 Ω, the radiation impedance of a patch at the edge can be approximated as:

$$Z_a \approx 90 \frac{\varepsilon_r^2}{\varepsilon_r - 1} \left(\frac{L}{W} \right)^2 (\Omega) \tag{5.89}$$

Here, the radiation efficiency (a function of the thickness) is assumed to be 100%. Thus, the impedance is determined by three parameters. For a PTFE (Teflon)-based substrate with the relative permittivity of 2.1, to obtain a 50 Ω input impedance, we need $(L/W) = 0.3723$. Since $L = 0.49\lambda$, we therefore obtain $W = 1.316\lambda$.

It was also found that an empirical formula can be used to estimate the impedance fractional bandwidth for $VSWR < 2$ [23]:

$$\frac{\Delta f}{f_0} = \frac{16}{3\sqrt{2}} \frac{\varepsilon_r - 1}{\varepsilon_r^2} \frac{Ld}{\lambda W} \approx 3.77 \frac{\varepsilon_r - 1}{\varepsilon_r^2} \frac{Ld}{\lambda W} \tag{5.90}$$

Thus, the bandwidth is proportional to the thickness of the substrate. It also indicates that the higher the permittivity, the smaller the bandwidth which means there is a trade-off between the size (Ld/W) and bandwidth.

Design Equations and Procedures
Because of the fringing effects, electrically the patch of the antenna looks larger than its physical dimensions; the enlargement on L is given by [2]:

$$\Delta L = 0.412d(\varepsilon_{reff} + 0.3)(W/d + 0.264)/\left[(\varepsilon_{reff} - 0.258)(W/d + 0.8)\right] \tag{5.91}$$

where the effective (relative) permittivity is

$$\varepsilon_{reff} = \frac{\varepsilon_r + 1}{2} + \frac{\varepsilon_r - 1}{2\sqrt{1 + 12d/W}} \tag{5.92}$$

It is related to the ratio of d/W. The larger the d/W, the smaller the effective permittivity.
The effective length of the patch is now

$$L_{eff} = L + 2\Delta L \tag{5.93}$$

The resonant frequency for the TM_{100} mode is

$$f_r = \frac{1}{2L_{eff}\sqrt{\varepsilon_{reff}}\sqrt{\varepsilon_0\mu_0}} = \frac{1}{2(L + 2\Delta L)\sqrt{\varepsilon_{reff}}\sqrt{\varepsilon_0\mu_0}} \tag{5.94}$$

An optimized width for an efficient radiator is

$$W = \frac{1}{2f_r\sqrt{\varepsilon_0\mu_0}}\sqrt{2/(\varepsilon_r + 1)} \tag{5.95}$$

Now the design problem: if the substrate parameters (ε_r and d) and the operational frequency (f_r) are known, how can we obtain the dimensions of the patch antenna (W and L)?

Based on these simplified formulas, we can adopt the following design procedure to design the antenna:

Step 1: Use Equation (5.95) to find the width W;
Step 2: Calculate the effective permittivity ε_{reff} using Equation (5.92);
Step 3: Compute the extension of the length ΔL using Equation (5.91);
Step 4: Determine the length L by solving Equation (5.94) for L, and the solution is

$$L = \frac{1}{2f_r\sqrt{\varepsilon_{reff}}\sqrt{\varepsilon_0\mu_0}} - 2\Delta L \tag{5.96}$$

Example 5.7   Design a Rectangular Patch

RT/Duroid 5880 substrate ($\varepsilon_r = 2.2$ and $d = 1.588$ mm) is to be used to make a resonant rectangular patch antenna of linear polarization;

a. design such an antenna to work at 2.45 GHz for Bluetooth applications;
b. estimate its directivity;
c. if it is to be connected to a 50 Ω microstrip using the same PCB board, design the feed to this antenna;
d. find the fractional bandwidth for $VSWR < 2$.

Solution

a. Follow the design procedure suggested above to obtain:

$$W = \frac{1}{2f_r\sqrt{\varepsilon_0\mu_0}}\sqrt{2/(\varepsilon_r + 1)} = \frac{300}{2 \times 2.45}\sqrt{2/(2.2 + 1)} = 48.40 \text{ (mm)}$$

$$\varepsilon_{reff} = \frac{\varepsilon_r + 1}{2} + \frac{\varepsilon_r - 1}{2\sqrt{1 + 12d/W}} = \frac{2.2 + 1}{2} + \frac{2.2 - 1}{2\sqrt{1 + 12 \times 1.588/48.4}} = 2.108$$

$$\Delta L = 0.412 \times 1.588(2.108 + 0.3)\left(\frac{48.4}{1.588} + 0.264\right)/\left[(2.108 - 0.258)\left(\frac{48.4}{1.588} + 0.8\right)\right]$$

$$= 0.84 \text{ (mm)}$$

and

$$L = \frac{1}{2f_r\sqrt{\varepsilon_{reff}}\sqrt{\varepsilon_0\mu_0}} - 2\Delta L = \frac{300}{2 \times 2.45\sqrt{2.108}} - 2 \times 0.837 = 40.49 \text{ (mm)}$$

Thus, the designed patch should have $L = 40.49$ mm and $W = 48.40$ mm.

b. Since the wavelength at 2.45 GHz is 122.45 mm $> W$, using Equation (5.88) gives the maximum directivity, which is about 6.6 or 8.2 dBi.
c. The input impedance is given by Equation (5.89) and is

$$Z_a = 90\frac{\varepsilon_r^2}{\varepsilon_r - 1}\left(\frac{L}{W}\right)^2 = 90\frac{2.2^2}{2.2 - 1}\left(\frac{40.49}{48.40}\right)^2 = 254.04 \ \Omega$$

which does not match well with a 50 Ω standard microstrip and therefore a quarter-wavelength transformer is used to connect them (refer to Chapter 2). The characteristic impedance of the transition section should be

$$Z_T = \sqrt{50 \times 254} = 112.69 \ \Omega$$

Thus, we can use Equation (2.80) to determine the transition line width W_T, which is:

$$Z_0 = \frac{60}{\sqrt{\varepsilon_r}} \ln \left(\frac{8d}{W_T} + \frac{W_T}{4d} \right) = \frac{60}{\sqrt{2.2}} \ln \left(\frac{8 \times 1.588}{W_T} + \frac{W_T}{4 \times 1.588} \right) = 112.69$$

Thus, $W_T = 0.790$ (mm) which is very thin. To obtain its length, we need to calculate the relative effective permittivity for the line which is given by Equation (2.81 or 5.92):

$$\varepsilon_{re} \approx \frac{\varepsilon_r + 1}{2} + \frac{\varepsilon_r - 1}{2\sqrt{1 + 12d/W_T}} \approx 1.706$$

Thus, the length of the transition should be:

$$\frac{\lambda}{4} = \frac{\lambda_0}{4\sqrt{\varepsilon_{re}}} = \frac{122.45}{4\sqrt{1.706}} = 23.437 \text{ (mm)}$$

The width of the 50 Ω microstrip feed line can be found using Equation (2.84), i.e.

$$Z_0 = \frac{120\pi}{\sqrt{\varepsilon_r}(W_m/d + 1.393 + 0.667 \ln (W_m/d + 1.44))} = 50;$$

The solution is $W_m = 4.367$ mm. The final design is shown in Figure 5.41.

Since the width of the quarter-wave transformer might be too thin to make properly in practice, an alternative design is to employ an inset feed as shown in Figure 5.39(a) and use the following design equation [2, 24, 26, and]:

$$R_{in}(x = x_0) = R_{in}(x = 0) \cos^2 \left(\frac{\pi}{L} x_0 \right) \tag{5.97}$$

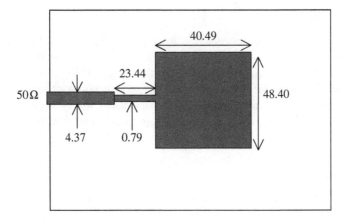

Figure 5.41   A matched resonant patch antenna for 2.45 GHz

Folded dipole	Bow tie	Rectangular slotted	Circular slotted	Circular with "ears"

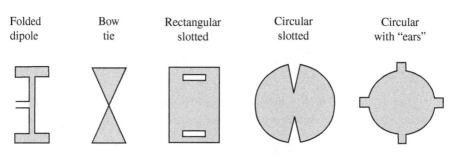

Figure 5.42   A variety of patch antennas

We can find that the recessed distance (the length cutting into the patch) is

$$x_0 = \frac{L}{\pi} \cos^{-1} \sqrt{\frac{50}{112.69}} \approx 0.2796L = 11.32 \,(\text{mm})$$

In this case, no transformer is used and the 50 Ω line can be directly connected to the antenna with an inset of length 11.32 mm, which appears to be a good solution.

d. Using Equation (5.90), we can find the fractional bandwidth for *VSWR* < 2:

$$\frac{\Delta f}{f_0} = 3.77 \frac{\varepsilon_r - 1}{\varepsilon_r^2} \frac{Ld}{\lambda W} = 3.77 \frac{2.2 - 1}{2.2^2} \frac{1.588 \times 40.49}{122.45 \times 48.40} = 0.0101 = 1.01\%$$

The bandwidth is indeed very narrow.

A further in-depth study of this antenna using the theory of characteristic modes is presented in Section 6.5 and it reveals the amazing insight on how this antenna works and demonstrates that it can work well at some other frequencies – which could be controlled by selecting the right feeding points and methods which may result in very different performances.

Some alternative patch antennas are displayed in Figure 5.42. Generally speaking, they can offer broader bandwidth than a rectangular patch.

5.2.5.3   Ground Plane

The ground plane is a part of the antenna. Ideally, the ground plane should be infinite as for a monopole antenna. But in reality, a small ground plane is desirable. As shown in Figure 5.39(a), the radiation of a microstrip antenna is generated by the fringing field between the patch and the ground plane, and the minimum size of the ground plane is therefore related to the thickness of the dielectric substrate. Generally speaking, a $\lambda/4$ extension from the edge of the patch is required for the ground plane, whereas the radius of a monopole ground plane should be at least one wavelength.

5.3   Antenna Arrays

So far, we have studied many different antennas: wire types and aperture types; resonant antennas (such as dipole and patch antennas) and traveling wave antennas (such as Yagi–Uda and periodic antennas). They can all be classified as single-element antennas. Once the frequency is given, everything (the radiation pattern, input impedance, etc.) is fixed. They lack flexibility and the gain is normally very limited. A high-gain antenna means that the aperture size of the antenna has to be very large, which may be a problem in practice. Also, sometimes we need to be able to control the antenna radiation pattern, for example, for tracking or anti-jamming/interference applications. Thus, a single-element antenna is not good enough to meet such a requirement. In this case, an antenna array could be a good solution.

An *antenna array* consists of more than one antenna element and these radiating elements are strategically placed in space to form an array with desired characteristics which are achieved by varying the feed (amplitude and phase) and relative position of each radiating element. The total radiated field is determined by the vector addition of the fields radiated by the individual elements. The total dimensions of the antenna are enlarged without increasing the size of the individual element. The major advantages of an array are

- the flexibility to form a desired radiation pattern;
- the high directivity and gain; and
- the ability to provide an electrically scanned beam (mechanical rotation can be avoided).

The main drawbacks are

- the complexity of the feeding network required; and
- the bandwidth limitation (mainly due to the feeding network)

In this section, we are going to examine the basic operational principle of an antenna array and associated theory, and how to design an array and what the problems and trade-offs are.

5.3.1   Basic Concept

Let us assume that there are N elements in an antenna array as shown in Figure 5.43. The phase and amplitude of each element can be tuned electrically or mechanically using phase shifters

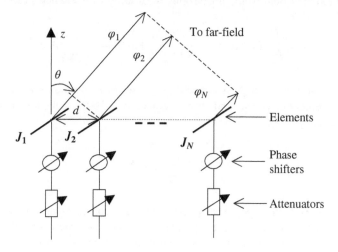

Figure 5.43   A typical antenna array of N elements

and attenuators. Sometimes, these weighting factors are already integrated to the radiating elements and no tuning is required.

The total radiated field can be obtained by summing up the radiated field from each element antenna, i.e.

$$E(r,\theta,\varphi) = \sum_{n=1}^{N} E_n(J_n) \tag{5.98}$$

Each element has many variables, including the element antenna itself and its excitation (amplitude and phase); the overall arrangement of these elements is another variable which can be in many different geometrical forms to achieve the desired radiation characteristics. In practice, all the elements are placed in a specific configuration to form a well-controlled 1D, 2D, or even 3D array (such as linear, circular, rectangular, or elliptical in shape). Arbitrary-shaped arrays are not well studied, and the patterns are not easy to control, thus they are not used.

5.3.2 Isotropic Linear Arrays

Now we are going to use a linear array, the most popular array arrangement, as an example to explore the features of an antenna array. To simplify the analysis, the radiator is assumed to be an isotropic element, but its amplitude and phase are controllable. Since the isotropic element means that its radiation is the same to every direction, from Equation (4.6), we know that the far-field radiation from such isotropic element can be expressed as:

$$E_{ie} = \frac{e^{-j\beta r}}{4\pi r} \tag{5.99}$$

The polarization is not shown here but the field is orthogonal to the propagation direction. Thus, the total radiated field from an isotropic linear array of N elements can be written as:

$$E_{ia} = \sum_{n=1}^{N} E_{ie} A_n e^{j\phi_n} = \frac{e^{-j\beta r}}{4\pi r} \sum_{n=1}^{N} A_n e^{j\phi_n} = E_{ie} \sum_{n=1}^{N} A_n e^{j\phi_n} \tag{5.100}$$

where A_n is the amplitude and ϕ_n is the relative phase (to a common reference as shown in Figure 5.43) of element n. The second term of Equation (5.100) is called the (isotropic) *array factor (AF)*, that is

$$AF = \sum_{n=1}^{N} A_n e^{j\phi_n} \tag{5.101}$$

If we further assume that the spacing between elements is a constant d and all the elements have identical amplitude (say 1) but each succeeding element has a φ_0 progressive phase lead current excitation relative to the preceding one, this array is referred to as *a uniform array*. Using Figure 5.43, the array factor in this case is

$$AF = \sum_{n=1}^{N} A_n e^{j\phi_n} = 1 + e^{j(\beta d \sin\theta + \varphi_0)} + e^{j2(\beta d \sin\theta + \varphi_0)} + \cdots + e^{j(N-1)(\beta d \sin\theta + \varphi_0)}$$

$$= \sum_{n=1}^{N} e^{j(n-1)(\beta d \sin\theta + \varphi_0)} = \sum_{n=1}^{N} e^{j(n-1)\Psi} \tag{5.102}$$

where $\Psi = \beta d \sin\theta + \varphi_0$ is the phase difference (lead) between two adjacent elements and the first element is chosen as the reference antenna. The geometric series can be simplified as:

$$AF = \frac{1 - e^{jN\Psi}}{1 - e^{j\Psi}} = e^{j[(N-1)/2]\Psi} \left[\frac{\sin(N\Psi/2)}{\sin(\Psi/2)} \right] \tag{5.103}$$

The maximum of the above AF is N, and the normalized antenna factor of a uniform array is therefore,

$$AF_n = \frac{1}{N} \left[\frac{\sin(N\Psi/2)}{\sin(\Psi/2)} \right] \tag{5.104}$$

The normalized AF_n in dB as a function of θ in degrees for $\varphi_0 = 0$ is plotted in Figure 5.44 for (a). $N = 20$ and $d = \lambda$, (b). $N = 10$ and $d = \lambda$, and (c). $N = 10$ and $d = \lambda/2$. The maximum at $\theta = 0°$ is the main beam or main lobe. The other maxima are called grating lobes. Generally, grating lobes are undesirable, because most applications such as radars require a single focused beam. The important messages from these plots and Equation (5.104) are

- there are $(N-2)$ sidelobes between the main and first grating lobes (not applicable for small d, to be discussed in Figure 5.47);
- grating lobes appear if $d \geq \lambda$. To avoid any grating lobes, the largest separation between the element should be less than one wavelength;
- the notches between lobes are nulls (in theory);
- the first sidelobe is about -13 dB, the same as that of the uniform aperture antenna.

5.3.2.1 Phased Arrays

From Equation (5.104), we know that the maximum of the radiation occurs at $\psi = 0$ which is

$$\beta d \sin\theta + \varphi_0 = 0 \tag{5.105}$$

or at the angle:

$$\theta = -\sin^{-1}\left(\frac{\lambda\varphi_0}{2\pi d}\right) \tag{5.106}$$

Normally, the spacing d is fixed for an array, and we can control the maximum radiation (or scan the beam) by changing the phase φ_0 and the wavelength (frequency) – this is the principle of *phase/frequency scanned array*. The phase change is accomplished electronically by the use

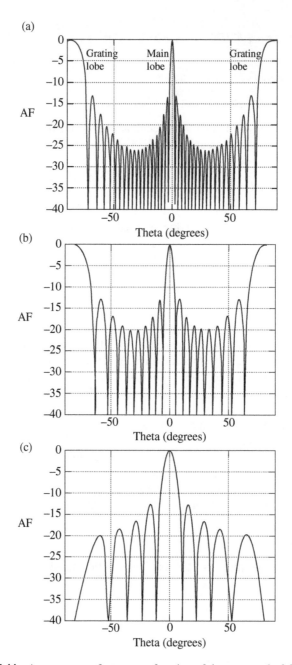

Figure 5.44 Antenna array factors as a function of the scan angle θ for $\varphi_0 = 0$

of ferrite (sensitive to magnetic field) or diode (sensitive to bias voltage) phase shifters, and switched transmission lines are also commonly used. Controlling the beam is one of the most appealing features of arrays. The pattern can be scanned electronically by adjusting the array coefficients (φ_0 or λ). An example of the array pattern (AF) for $N = 10$, $d = \lambda/2$, and $\varphi_0 = 45°$ is

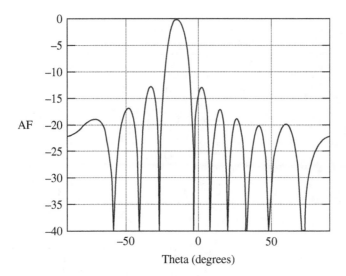

Figure 5.45 Antenna array factors as a function of angle θ for $N = 10$ and $d = \lambda/2$, and $\varphi_0 = 45°$

shown in Figure 5.45, and it is apparent that the maximum is now shifted from $0°$ to another angle ($\theta = -45°/\pi \approx 14.3°$). The angle is controlled by the phase shifter. This approach permits beams to be moved from point to point in space in just microseconds, whereas mechanical scanning of a large antenna could take several seconds or longer.

Phased antenna arrays (1D and 2D) are widely used, especially for military applications. The benefits are obvious, so are the difficulties at the feeding network. Just imagine the wiring for an antenna array with hundreds of elements!

Due to the feasibility and variables within a phased array, it is also known as a *smart antenna*.

5.3.2.2 Broadside and End-fire Arrays

An array is called a *broadside array* if the maximum radiation of the array is directed normal to its axis ($\theta = 0°$ in Figure 5.43); while it is called an *end-fire array* if the maximum radiation is directed along the axis of the array ($\theta = 90°$ in Figure 5.43). For a given frequency, we can control the spacing d and the progressive phase φ_0 to make the array broadside or end fire. The specific conditions are

- Broadside: $\varphi_0 = 0$;
- End fire: $\varphi_0 = \pm 2\pi d/\lambda$;

The major characteristics of the radiation pattern of both arrays are listed in Table 5.6. The polar and rectangular plots of the radiation patterns of the broadside and end-fire arrays for $d = \lambda/2$ and $\lambda/4$ at $N = 10$ are shown in Figures 5.46 and 5.47, respectively.

It should be noted that from these results, for the same geometrical configuration (length and spacing),

Table 5.6 Comparison of the broadside and end-fire antenna arrays

Item	Broadside	End fire
φ_0	0	$\pm 2\pi d/\lambda$
Maxima ($\theta =$)	$\sin^{-1}\left(\pm\dfrac{m\lambda}{d}\right)$, $m = 0, 1, 2, \ldots$	$\sin^{-1}\left(1 - \dfrac{m\lambda}{d}\right)$, $m = 0, 1, 2, \ldots$
HPBW	$\approx 2\left[\dfrac{\pi}{2} - \cos^{-1}\left(\dfrac{1.391\lambda}{\pi Nd}\right)\right]$	$\approx 2\cos^{-1}\left(1 - \dfrac{1.391\lambda}{\pi Nd}\right)$
First null beamwidth (FNBW)	$2\left[\dfrac{\pi}{2} - \cos^{-1}\left(\dfrac{\lambda}{Nd}\right)\right]$	$2\cos^{-1}\left(1 - \dfrac{\lambda}{Nd}\right)$
Directivity	$\approx 2\dfrac{Nd}{\lambda}$, $d < \lambda$	$\approx 4\dfrac{Nd}{\lambda}$, $d < \dfrac{\lambda}{2}\left(1 - \dfrac{1}{2N}\right)$

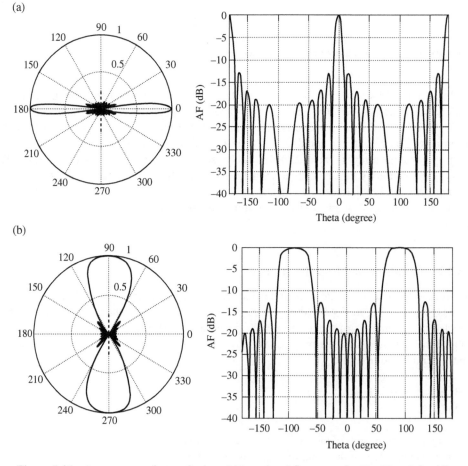

(a)

(b)

Figure 5.46 Antenna array factors for broadside and end-fire arrays for $N = 10$ and $d = \lambda/2$

- the *HPBW* of the end-fire array is always larger than that of the broadside array, so is the first null beamwidth (FNBW);
- the directivity of the end-fire array, as given by Table 5.6, is twice that for the broadside array if the spacing is much smaller than the wavelength.

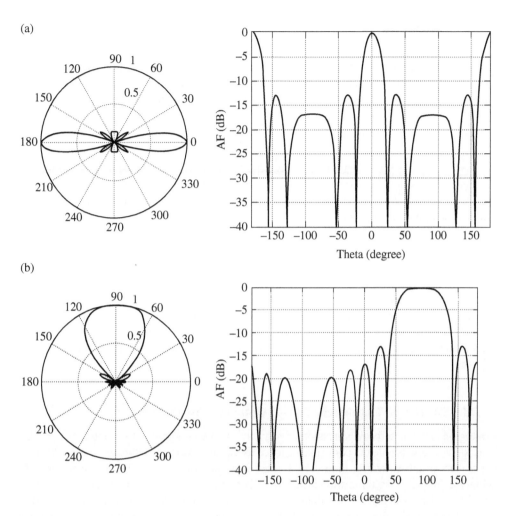

Figure 5.47 Antenna array factors for broadside and end-fire arrays for $N = 10$ and $d = \lambda/4$ (a) broadside $\varphi_0 = 0$ and (b) end-fire $\varphi_0 = -2\pi d/\lambda$

These two conclusions seem to contradict the relation between the directivity and *HPBW* as shown in Equation (4.13). The reality is that broadside arrays can only produce a fan beam (two symmetrical main lobes), while the end-fire arrays can generate a single pencil beam (just one main lobe) when

$$d \le \frac{\lambda}{2}\left(1 - \frac{1}{2N}\right) \tag{5.107}$$

Figure 5.47(b) is just an example to demonstrate this interesting feature, thus Equation (4.13) cannot be applied.

It should also be pointed out that the number of sidelobes for the broadside case is no longer ($N - 2$) when the spacing d becomes small as shown in Figure 5.47(a). Some sidelobes are merged.

5.3.2.3 The Hansen–Woodyard End-fire Array

The end-fire arrays discussed above, which are called the ordinary end-fire arrays, are not optimized to produce the optimum directivity. It is possible to make the main beam narrower and thus increase directivity by changing the inter-element phase shift. Hansen and Woodyard [28] found that the maximum directivity for an end-fire array is obtained when

$$\varphi_0 = \pm (2\pi d/\lambda + 2.92/N) \approx \pm (2\pi d/\lambda + \pi/N) \qquad (5.108)$$

if the array length is much larger than a wavelength. However, it should be pointed out that the condition in Equation (5.108) is not for the maximum possible directivity of any array (which may occur at other directions rather than $\theta = \pm 90°$) but only for end-fire arrays. The directivity can now be approximated as:

$$D \approx 7.28 \frac{Nd}{\lambda} \qquad (5.109)$$

This means an increase of 2.6 dB over the ordinary end-fire array in directivity. This is why the *Hansen–Woodyard Array* is also known as the *increased directivity end-fire array*. When the axial-mode helix was discussed in Section 5.1.4, it was concluded that this type of antenna approximately satisfies the Hansen–Woodyard end-fire condition and increased directivity was obtained.

5.3.2.4 The Dolph–Tschebyscheff (D–T) Optimum Distribution

In addition to the directivity/gain, another very important consideration for antenna arrays is the SLL, since it affects antenna performance (such as the antenna temperature and SNR). As mentioned earlier and shown in Figure 5.48, there is a trade-off between the SLL and gain. To achieve a high gain, a uniform distribution is preferred but it results in a high SLL (about −13 dB), while a well tapered distribution can produce very low SLL or even no sidelobes (such as the binominal distribution) but it exhibits a lower gain. It is shown that the far-field pattern of a linear in-phase ($\varphi_0 = 0$, broadside) array of isotropic point sources can be expressed as a finite Fourier series of N terms. We can use Dolph's procedure [29] to match the Fourier polynomial with a Chebyshev polynomial. This then yields the optimum source amplitude distribution for a specific SLL with all sidelobes of the same level. More details can be found from [4].

5.3.3 Pattern Multiplication Principle

In practice, isotropic sources are to be replaced by practical antennas, which means that the element radiated E_{ie} in Equation (5.100) should be replaced by the radiated field of the antenna element E_e, thus the total radiated field of the antenna array is

$$E_a = E_e \sum_{n=1}^{N} A_n e^{j\varphi_n} = E_e \cdot AF \qquad (5.110)$$

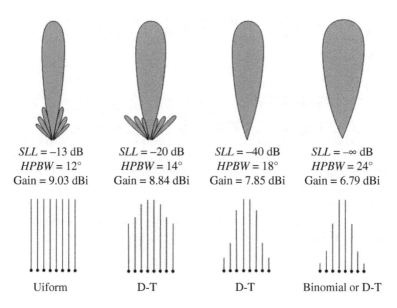

Figure 5.48 The radiation pattern (in half-space), *SLL*, *HPBW*, and gain for four different source distributions of eight in-phase isotropic sources spaced by *λ/2*

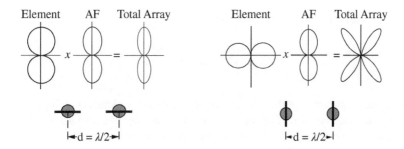

Figure 5.49 Array radiation patterns of two short dipoles separated by *d = λ/2*

This is actually called the *Principle of Pattern Multiplication: the radiation pattern of an array is the product of the pattern of individual element antenna with the (isotropic source) array pattern.*

Let us take a two-element short dipole array as an example. The elements are in-phase and separated by half of a wavelength, and the AF (isotropic array pattern) is therefore a number 8 (in shape) from Equation (5.104). Since the dipole radiation pattern is well known to us now and it is also a number 8 in shape, the total radiation patterns for the horizontally oriented and vertically oriented dipole arrays can therefore be obtained using the pattern multiplication principle, as shown in Figure 5.49.

Using the pattern multiplication principle, we can extend the knowledge and results obtained for isotropic arrays to other antenna arrays. It is now also possible to produce a unidirectional broadside radiation pattern using directional antenna elements.

The product of the directivity of a single element and the directivity of the isotropic array may be used to estimate the directivity of the antenna array, but the accuracy varies with the array geometry [30].

5.3.4   Element Mutual Coupling

When antenna elements are placed in an array, they interact with each other; this interaction between elements due to their close proximity is called the *mutual coupling* which affects the current distribution, hence the input impedance as well as the radiation pattern.

The most significant effects are on the antenna input impedance which is a very important parameter for a single antenna as well as for an antenna array. We can use circuit concepts, which are simpler than field concepts, to analyze the input impedance of element antennas in an array. Each element antenna is considered as a current source. An antenna array of N elements is then treated as an N-port network, and the voltage generated at each element can be expressed as:

$$
\begin{aligned}
V_1 &= Z_{11}I_1 + Z_{12}I_2 + \ldots + Z_{1N}I_N \\
V_2 &= Z_{21}I_1 + Z_{22}I_2 + \ldots + Z_{2N}I_N \\
&\ldots \\
V_N &= Z_{N1}I_1 + Z_{N2}I_2 + \ldots + Z_{NN}I_N
\end{aligned}
\tag{5.111}
$$

where I_n is the current at the nth element and

$$
Z_{nn} = \frac{V_n}{I_n}, \quad n = 1, 2, \ldots, N
\tag{5.112}
$$

is called the *self-impedance* of the nth element when all other elements are open-circuited, or the input impedance when this element is isolated from the other elements (just like it is not in an array). For a half-wavelength dipole, its self-impedance is about 73 Ω. The impedance:

$$
Z_{mn} = \frac{V_m}{I_n}
\tag{5.113}
$$

is called the *mutual impedance* between the elements m and n ($Z_{mn} = Z_{nm}$ by the reciprocity principle to be discussed in Section 5.4.1). Again, using two half-wavelength dipole antennas as an example, the mutual impedance (resistance and reactance) of two parallel side-by-side dipoles is shown in Figure 5.50, while the mutual impedance of two collinear dipoles as a function of the separation between them is plotted in Figure 5.51. Examining these results reveals that:

- the strength of the coupling decreases as spacing increases;
- the parallel configuration couples more than the collinear configuration;
- the far-field pattern of each element predicts coupling strength. The more the radiation, the stronger the coupling.

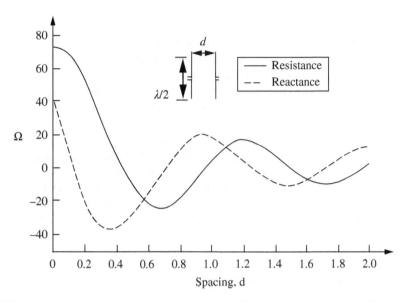

Figure 5.50 Mutual resistance and reactance of two parallel dipoles as a function of the spacing *d* in wavelength

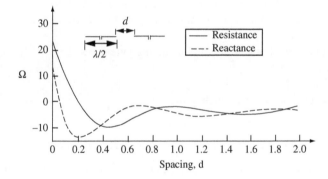

Figure 5.51 Mutual resistance and reactance of two collinear dipoles as a function of the spacing *d* in wavelength

Taking the mutual coupling into account, the input impedance of the *n*th element in the array is

$$Z_n = \frac{V_n}{I_n} = Z_{n1}\frac{I_1}{I_n} + Z_{n2}\frac{I_2}{I_n} + \cdots + Z_{nN}\frac{I_N}{I_n} \tag{5.114}$$

This is the actual input impedance of the element and is also referred to as the *active impedance*.

It should be pointed out that the elements at the array edges have fewer neighboring elements, thus the mutual coupling could vary significantly from edge elements to other elements.

5.4   Discussions

In addition to 1D arrays, 2D antenna arrays have also been developed for practical applications. The slot waveguide array in Figure 5.38 is an example where each element (slot) is carefully designed to ensure the aperture field distribution on the antenna follow the desired pattern (such as a Gaussian distribution) and the impedance (including mutual impedance) is matched with the waveguide. Here, discrete antenna elements are employed to form a continuous aperture field distribution. It is evident that antenna arrays are closely linked to aperture antennas. Some theories developed for aperture antennas may be applied to antenna arrays.

The array feeding network could be very complicated, thus it is a special topic. Generally speaking, just like elements in an electric circuit, the antenna elements in an array may be fed by a parallel circuit, or a series circuit, or the combination of both. But care has to be taken; the feeding network is normally a distributed system at high frequencies. The network may limit the bandwidth of the antenna system and may actually act as a radiator if not properly designed!

5.5   Some Practical Considerations

Up to now, we have covered the most popular antennas: wire-type and aperture-type antennas, and element and array antennas. The reader should have gained a good understanding on their operational principles, major field and circuit characteristics, and the design procedures. However, there are many practical considerations we have not yet studied, and it is not possible to deal with all aspects of antenna design and characterization in this book. In this section, we are going to include some important considerations in practice which include

- the differences between transmitting and receiving antennas;
- antenna feeding and matching;
- polarization; and
- radomes, housings, and supporting structures.

5.5.1   Transmitting and Receiving Antennas: Reciprocity

So far, we have not distinguished between antennas for transmitting and receiving purposes. Are there any differences between them? The short answer is yes and no, depending on what parameters you are talking about. Let us examine these two cases from both the field point of view and the circuit point of view.

Let us assume within a linear and isotropic medium, there exist two sets of sources (J_1, M_1) and (J_2, M_2) which are allowed to radiate in the medium and produce fields (E_1, H_1) and (E_2, H_2), respectively. Using Maxwell's equations, it can be shown that [31]:

$$\int_V (E_1 \cdot J_2 - H_1 \cdot M_2) dv = \int_V (E_2 \cdot J_1 - H_2 \cdot M_1) dv \qquad (5.115)$$

This is called the *Lorentz Reciprocity Theorem*. This can be interpreted as the coupling between a set of fields and a set of sources which produces another set of fields is the same as the

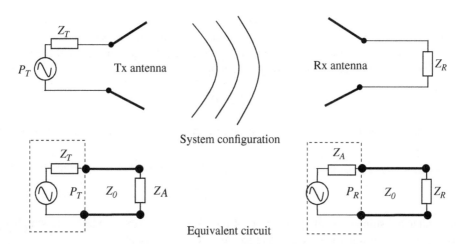

Figure 5.52   A radio transmitting and receiving system and their conventional equivalent circuits

coupling between another set of fields and another set of sources. This can be simply represented as:

$$< 1, 2 > = < 2, 1 > \qquad (5.116)$$

Under this condition, from the circuit point of view, the *mutual coupling impedances between two antennas are the same*:

$$Z_{12} = Z_{21} \qquad (5.117)$$

The self-impedance, i.e., the antenna input impedance, is also the same for the transmitting or receiving mode.

From the field point of view, the same conclusion can be arrived, that is, *the radiated field patterns are the same when the antenna is employed for transmitting or receiving.*

Now let us examine the roles of the antenna at the transmitter and the receiver, respectively. Figure 5.52 shows a typical system configuration for a RF/microwave transmitting and receiving system given in the literature. At the transmitting side, the transmitting antenna is just a load (Z_A) to the feed line with characteristic impedance Z_0. Let us first not take the effects of the transmission line into account as most antenna books do. As discussed in Section 2.3.2 (impedance matching), the antenna impedance and source internal impedance should meet the following condition to achieve the maximum transmission:

$$Z_A^* = Z_T \quad \text{or} \quad R_A - jX_A = R_T + jX_T$$

For simplicity, assume that the imaginary parts of both impedances are zero. If the power produced by the source is P_T, the power radiated by the antenna can be expressed as:

$$P_{rad} = \frac{R_A}{R_T + R_A} P_T \qquad (5.118)$$

At the receiving side, the receiving antenna (which is assumed to be the same as the transmitting antenna) is equivalent to a source with internal impedance Z_A as shown in Figure 5.52. Similarly, to reach the maximum power transfer, the antenna impedance Z_A and the receiver impedance Z_R should meet the following condition:

$$Z_A^* = Z_T \quad \text{or} \quad R_A - jX_A = R_R + jX_R$$

If the power received by the receiving antenna is P_R, the power received by the receiver can be expressed as:

$$P_{rec} = \frac{R_R}{R_A + R_R} P_R \tag{5.119}$$

Here, we have assumed the imaginary parts of both impedances are zero again for simplicity. Since $P_R \leq P_{rad}$, for a perfectly matched receiver, we have $R_A = R_R$, thus

$$P_{rec} = 0.5 P_R \leq 0.5 \frac{R_A}{R_T + R_A} P_T \tag{5.120}$$

This means that the receiver gets half of the power collected by the receiving antenna and the other half is said to be reradiated. If both the transmitting and receiving sides are perfectly matched and there is no propagation path loss, the receiver could only yield 25% of the source power P_T, that is, the maximum power transfer efficiency from the source of the transmitter to the end receiver is 25%, is this correct?

This result can be found in many antenna books, *but it is not correct* – this was actually a classical question. There are different RF/microwave sources, most of them have an AC power to RF/microwave power efficient >50%. For example, a typical consumer microwave oven consumes 1100 W AC power and produces about 700 W of microwave power into the oven with a typical efficiency of 64%, while the other 400 W are dissipated as heat, mostly in the magnetron tube. This obviously does not agree with the theoretical 50% maximum power transfer efficiency. The problem lies in the equivalent circuit of Figure 5.52 where the constant voltage source is not necessarily true (it could be a constant current or power source). For the receiving antenna case, it is a constant power source, not a constant voltage or current source. The received power of the receiving antenna is the product of the power density S and effective aperture of the antenna A_e which is given by Equation (4.23), that is

$$P_R = SA_e$$

A recent study has introduced a new equivalent circuit [32], as shown in Figure 5.53. It has overcome the problem of the conventional equivalent circuit and demonstrated that the maximum power and maximum power transfer efficiency (100% without loss) could be yielded at the same time. In Figure 5.53, the power source consists of both voltage source V_0 and current source I_0, and the internal impedance is the antenna impedance with radiation resistance R_r, loss resistance R_{loss}, and reactance X_A. When the antenna impedance is conjugate-matched with the impedance of the transmission line and receiver, the received power (after deducting the antenna loss power) will be delivered to the receiver $P_{rec} = P_R$, there is no reradiation. When the system is not perfectly matched, any mismatch is subject to the matching efficiency factor

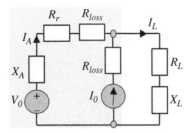

Figure 5.53   A new equivalent circuit for a receiving antenna system

$\eta_m = 1 - |\Gamma|^2$ as discussed in Chapter 4, and the power reflected back to the receiving antenna will go through another round of antenna loss and then be reradiated which is via the antenna radiation resistance in the equivalent circuit. In practice, there are also cable loss and propagation loss, thus, the received power will be much less than the transmitted power, but the power transfer efficiency is no limited to 25% or 50%. More detailed discussion can be found from reference [32].

We can see from this analysis that both transmitting antenna and receiving antenna are equivalent to the same impedance Z_A, although their roles are different: the transmitting antenna is an impedance load, while the receiving antenna is an RF/microwave source at the receiver.

From a system design point of view, the requirements on the transmitting antenna and the receiving antenna are not the same. For transmitting, the paramount importance is to transmit enough power to the desired directions, thus the maximum radiating efficiency and antenna gain are normally the most important parameters. For receiving, the most important requirement is usually a large SNR. Although the efficiency and gain are still important, low sidelobes may be more desirable than other parameters.

It is now clear that antenna properties (such as the radiation pattern and input impedance) are the same when it is used as either a transmitting antenna or a receiving antenna. Thus, there is no need to distinguish it as a transmitting or receiving antenna. However, the design requirements for the transmitting antenna and receiving antenna are usually different in practice. Therefore, the design considerations for a transmitting antenna and for a receiving antenna may be different.

5.5.2   Balun and Impedance Matching

As discussed earlier, an antenna is normally connected to a transmission line and good matching between them is very important. A coaxial cable is often employed to connect an antenna – mainly due to its good performance and low cost. A half-wavelength dipole antenna with impedance of about 73 Ω is widely used in practice. From the impedance matching point of view, this dipole can match well with a 50 or 75 Ω standard coaxial cable. Now the question is: can we connect a coaxial cable directly to a dipole?

As illustrated in Figure 5.54, when a dipole is directly connected to a coaxial cable, there is a problem: a part of the current coming from the outer conductor of the cable may go to the outside of the outer conductor at the end and return to the source rather than flow to the dipole. This undesirable current will make the cable become part of the antenna and radiate or receive

Current distribution

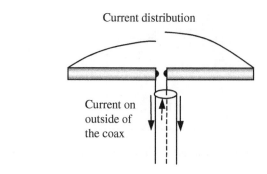

Current on
outside of
the coax

Figure 5.54 A dipole connected to a coax

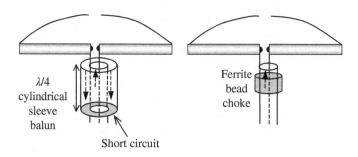

$\lambda/4$
cylindrical
sleeve
balun

Short circuit

Ferrite
bead
choke

Figure 5.55 Two examples of baluns

unwanted signals which could be a very serious problem in some cases. In order to resolve this problem, a balun is required.

The term *balun* is an abbreviation of the two words *bal*ance and *un*balance. It is a device that connects a balanced antenna (a dipole, in this case) to an unbalanced transmission line (a coaxial cable, where the inner conductor is not balanced with the outer conductor). The aim is to eliminate the undesirable current coming back on the outside of the cable. There are a few baluns developed for this important application. Figure 5.55 shows two examples. The sleeve balun is a very compact configuration: a metal tube of $\lambda/4$ is added to the cable to form another transmission line (a coax again) with the outer conductor cable and a short circuit is made at the base which produces an infinite impedance at the open top. The leaky current is reflected back with a phase shift of 180°, which results in the cancelation of the unwanted current on the outside of the cable. This balun is a narrowband device. If the short-circuit end is made as a sliding bar, it can therefore be adjusted for a wide frequency range (but it is still a narrowband device). The second example in Figure 5.55 is a ferrite bead choke placed on the outside of the coaxial cable. It is widely used in the EMC industry and its function is to produce a high impedance, more precisely, high inductance due to the large permeability of the ferrite and act as a filter. With good-quality ferrite beads, a large bandwidth may be obtained (an octave or more). But this device is normally just suitable for the frequencies below 1 GHz which is determined by the ferrite properties. It should be pointed out that a choke is not usually termed a balun – also, at high frequencies it can be lossy, which can reduce the measured antenna efficiency.

There are some other types of baluns; a good discussion can be found in [4]. For example, the tapered balun (see the feed to the TEM horn antenna in Figure 5.23) is broadband, but its length is usually very long (> $\lambda/2$).

When combined with impedance matching techniques, a balun can become an impedance transformer (as well as a balun) [2]. The most well-known examples are the $\lambda/2$ coaxial balun transformer (4 : 1) and ferrite core transformers (n : 1). To connect a folded dipole (shown in Figure 5.4) of impedance about 280 Ω to a coaxial cable of 50 Ω, a $\lambda/2$ coaxial balun transformer could be used to obtain a good impedance match and balanced feed. The general discussion on impedance matching techniques was given in Section 2.3.2.

5.5.3 Antenna Polarization

Radiowave polarization was introduced in Section 3.2.2. There are linearly, circularly, and elliptically polarized waves. Antennas are the device to generate and receive these waves. But how is the antenna polarization linked to the wave polarization?

The antenna polarization is determined by the polarization of its radiating wave. For example, a dipole is a linearly polarized antenna, since its radiating wave has just one component (E_θ). This is relatively easy to understand. The hard question is about circularly polarized waves. How to generate a right-hand or left-hand circularly polarized wave?

5.5.3.1 Circular Polarization

We recall, from Equations (3.22) and (3.23), that a circularly polarized wave must have two orthogonal electric field components. They have the same amplitude but a phase difference of 90°. So far, we have only discussed one circularly polarized antenna: the helical antenna which is also the most popular circularly polarized antenna. It produces two orthogonal components E_θ and E_ϕ (given by Equations (5.21) and (5.22)) which are 90° out of phase.

There are many other antenna designs which can produce circular polarization. Four simple examples are shown in Figure 5.56, where (a) is two orthogonal dipoles which are fed with a 90° phase difference (it can be generated by a phase shifter or $\lambda/4$ physical separation. Which dipole is leading in phase determines the rotation of the circular wave, LCP or RCP; if the radiation in one axial direction is LCP, it is RCP in the opposite axial direction. (b) is a combination of a dipole and a loop which are also of 90° phase difference; (c) is a slotted dipole cylinder and their feeds are also 90° out of phase; and (d) is a circular patched antenna with two symmetrical cuts which ensure that two orthogonal currents are generated with a 90° phase difference. The first three configurations can provide omnidirectional patterns.

(a)   (b)   (c)   (d)

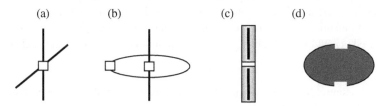

Figure 5.56 Four antennas for circular polarization

Some of the arrangements (such as the helical antenna or design (a) in Figure 5.56) can be placed within a circular waveguide so as to generate a circular TE_{11} mode, and a circular polarized beam can be produced when the waveguide is flared into a conical horn antenna. Some of them can serve as a primary antenna of a reflector antenna so as to produce highly directional and circular radiation.

5.5.3.2 Polarization Match and Mismatch

For a good radio communication system, the polarization must be matched. That is, a linearly polarized wave can be received by a linearly co-polarized (not cross-polarized) antenna, and similarly, a circularly polarized wave should be received by a circularly co-polarized antenna. They must be of the same polarization; otherwise some or even all the received power could be lost due to polarization mismatch. For example, if an incoming wave is of LCP, then the receiving antenna must be of LCP. If it were of RCP, no signals would be produced at the output of the receiving antenna. If it were of linearly polarized, only half of the incoming power could be received and the other half would be lost, i.e. there was a 3-dB loss in power.

If the incoming wave is linearly polarized and the receiving antenna is circularly polarized, is there a problem of polarization mismatch? The short answer is yes, depending on the combination scheme once the signal is received. Generally speaking, the power loss could be up to 3 dB due to polarization mismatch, which is much better than using a linearly polarized antenna with orthogonal polarization! In some applications, the polarisation of the incoming signal is unknown or hard to predict. For example, if a linearly polarized wave is employed for satellite communications, its polarization will be changed when passing through the ionosphere due to Faraday rotation (which is caused by the free electrons in ionosphere, and the polarization rotation is determined by the electrons and the wave frequency) – thus a circular polarization is usually employed. For radio communications in multipath environments (such as indoor), linear polarization is used – because the signal polarization becomes random after multiple reflections and could be said to be statistically uniform. The power loss due to polarization is about 3 dB. If circular polarization is employed, the power loss could be much more significant.

5.5.4   Radomes, Housings, and Supporting Structures

The antenna is a transmitting and receiving device for a radio system and is normally made of metal. It is not an isolated structure but an integrated part of a radio system. Sometimes, a radome is required in order to ensure its efficient and desired operation. The *radome* (radar dome) is defined as the structure which houses an antenna (or the entire radar, including the antenna) and is almost transparent to the radiowaves at the desired frequencies (normally above 1 GHz). It can be constructed in several shapes (spherical, geodesic, planar, etc.) depending on the particular application. Various low-loss construction materials [3] are used such as fiberglass, foams, FTFE, and coated fabric or plastic. The main reasons for using such a radome or housing are

- to protect an antenna from the ravages of the environment: wind, snow, ice, rain, salt, sand, the sun (UV), and even lightning;
- to make the whole structure mechanically sound and viable;

Figure 5.57 Examples of antennas with various radomes/housing. Source: Reproduced by permission of Guidance Ltd, UK

- to conceal the antenna system from spying or the public view; and
- to keep nearby personnel away from being accidentally struck by the fast-moving antenna or affecting the antenna performance.

A few examples of antennas with radomes or housing are shown in Figure 5.57. Radomes are normally much larger than the antenna. Some housing structures are just slightly larger than the antenna and may not be called radomes, such as the mobile phone case, but they play the same rules as radomes. Two of the most important questions asked in practice are

- how these radomes and housing structures are designed;
- what the effects of the radome or housing are on the antenna performance.

5.5.4.1 Design of Radomes and Housings

The antenna radome or housing design is a complex issue. Ideally, it should be incorporated into the antenna design, that is, it should be considered as part of the antenna system. In practice, this is problematic. The radome or housing design may be conducted by another department or company. The basic requirement is that the radome or housing structure should be transparent to the operational frequencies of the antenna, which means that the reflection coefficient and loss must be small and the transmission coefficient must be large (as close to 1 as possible). From Section 3.3.1, we know that this requirement can be achieved when *the thickness of the structure is an integer multiple of half of the effective wavelength in the medium.* This thickness is actually a function of the incident angle as well as the frequency and dielectric properties (ε, μ, σ). The material (such as fiberglass and PTFE) is normally of low loss and low permittivity. Since the radome or housing has to meet both the electrical and mechanical requirements. The structure is therefore often made from a multilayered material, and the sandwich configuration is chosen: hard and reinforced materials are used to form the outer layers and a thick layer of low-permittivity material is placed at the center. This configuration provides the mechanically required robustness and at the same time offers the broad bandwidth and low loss for the RF signals.

5.5.4.2 Effects of Radome and Housing

The effects of the radome and housing on the antenna include at least the following two aspects:

- from the field point of view, the existence of the radome or housing may affect the antenna aperture field distribution, hence the radiation pattern and directivity/gain. A slightly reduced gain (<1 dB) and increased sidelobes are expected in practice, depending on the quality of the radome/housing.
- from the circuit point of view, the load impedance to the antenna feed line may be changed, depending on how close the housing structure and the antenna are. The closer they are, the larger the effect. The effect may be reflected on the input impedance, hence the VSWR. A common observation is that the resonant frequency is shifted downward (lower) when the antenna housing is in place, since the wavelength is decreased inside a medium (by $\sqrt{\varepsilon_r}$).

5.5.4.3 Antenna Supporting Structure

Once an antenna is designed and made, it has to be installed somewhere. TV reception antennas are often erected on a roof, and mobile base station antennas are normally attached to a post or tower as shown in Figure 5.58. Ideally, the antenna sitting or supporting structure should have nothing to do with the antenna performance (or its effects should have been taken into account when designing the antenna). However, the antenna is a radiating device, and the radiation is defined by its radiation pattern. The antenna near field distribution is more complex than that of the far field. Generally speaking, the supporting structure should not be in the radiation region and a dielectric material is preferred over a metal structure, since its interaction with the antenna is much smaller.

As the interaction between the antenna and the housing and supporting structure can only be analyzed case by case, there are no general formulas or equations for the prediction. The accurate estimate of this interaction should be conducted using computer simulation software, which could be slow and expensive. But sometimes, it is worth paying such a cost.

Figure 5.58   Installed antennas and supporting structure

5.6   Summary

In this chapter, we have studied many popular antennas, introduced all relevant theories, and showed how to use these theories for antenna analysis and design. Wire-type antennas, aperture-type antennas, and antenna array have been covered. These antennas have been dealt with from both the field point of view and the circuit point of view – each is unique and special and has found real-world applications. Almost all the antennas introduced have been provided with design procedures or guidelines. Some practical considerations (including balun and housing) have also been presented. In practice, once an antenna is designed, it should be first validated by a computer simulation tool before it is ready for fabrication. This will be addressed in Chapter 6.

References

1. R. E. Collin, *Antennas and Radiowave Propagation*. McGraw-Hill, Inc., 1985.
2. C. A. Balanis, *Antenna Theory: Analysis and Design*, 3rd edition, John Wiley & Sons, Inc., 2005.
3. R. C. Johnson, *Antenna Engineering Handbook*, 3rd edition, McGraw Hill, Inc., 1993.
4. J. D. Kraus and R. J. Marhefka, *Antennas for All Applications*, 3rd edition, McGraw Hill, Inc., 2002.
5. S. Adachi and Y. Mushiake, 'Studies of large circular loop antenna,' *Science Report, Research Inst of Tohoko Univ,* B.9.2, pp. 79–103, 1957.
6. H. Morishita, K. Hirasawa and T. Nagao, 'Circularly polarised wire antenna with a dual rhombic loop,' *IEE Proc.-Microw. Antennas Propag.*, 145(3), 219–224, 1998.
7. R. L. Li, V. Fusco and H. Nakano, 'Circularly polarized open-loop antenna,' *IEEE Trans. Antennas Propag.*, 51(9), 2475–2477, 2003.

8. M. Sumi, K. Hirasawa and S. Shi, 'Two rectangular loops fed in series for broadband circular polarization and impedance matching,' *IEEE Trans. Antennas Propag.*, 52(2), 551–554, 2004.

9. G. Kandonian and W. Sichak, 'Wide frequency range tuned helical antennas and circuits,' *IRE Conv Rec.* Part 2, pp. 42–47, 1953.

10. H. Yagi, 'Beam transmission of ultra-short waves,' *Proc IRE*, 16, 715, 1928.

11. P. P. Viezbicke, 'Yagi Antenna Design,' *NBS Technical Note* 688, December, 1968.

12. W. L. Stutzman and G. A. Thiele, *Antenna Theory and Design*, 3rd edition, John Wiley & Sons, Inc., 2012.

13. C. A. Chen and D. K. Cheng, 'Optimum spacings for Yagi-Uda arrays,' *IEEE Trans. Antenna Propag.*, 21(5), 615–23, 1973.

14. C. A. Chen and D. K. Cheng, 'Optimum element lengths for Yagi-Uda arrays,' *IEEE Trans. Antenna Propag.*, 23(1), 8–15, 1975.

15. S. Baskar, A. Alphones, P. N. Suganthan and J. J. Liang, 'Design of Yagi–Uda antennas using comprehensive learning particle swarm optimization,' *IEE Proc. Microwave Antennas Propag.*, 152(5), 340–346, 2005.

16. S. Lim, and H. Ling, 'Design of a closely spaced, folded Yagi antenna,' *IEEE Antennas Wireless. Propag. Lett.*, 5, 302–305, 2006.

17. Z. Bayraktar, P. L. Werner, and D. H. Werner, 'The design of miniature three-element stochastic Yagi-Uda arrays using particle swarm optimization,' *IEEE Antennas Wireless. Propag. Lett.*, 5, 22–26, 2006.

18. R. L. Carrel, 'Analysis and design of log-periodic dipole antenna,' PhD Thesis, University of Illinois, USA, 1961.

19. P. C. Butson and G. T. Thompson, 'A note on the calculation of the gain of log-periodic dipole antennas,' *IEEE Trans. Antennas Propag.*, 24(1), 105–106, 1976.

20. J. Aurand, 'Pyramidal horns, Part 2: Design of horns for any desired gain and aperture phase error,' *IEEE Antennas Propag. Soc. Newsl.*, 31, 25–27, 1989.

21. S. K. Sharma, S. Sharma, S. Rao and L. Shafai, *Handbook of Reflector Antennas and Feed Systems*, Artech House, 2013.

22. J. R. James and P. Hall, *Handbook of Microstrip Antennas*, IEE, London, 1988.

23. D. R. Jackson and N. G. Alexopoulos, 'Simple approximate formulas for input resistance, bandwidth and efficiency of a resonant rectangular patch,' *IEEE Trans. Antennas Propag.*, 39(3), 407–410, 1991.

24. R. Garg, P. Bhartia and I. Bahl, *Microstrip Antenna Design Handbook*, Artech House Publishers, 2000.

25. Z. N. Chen and M. Y. W. Chia, *Broadband Planar Antennas: Design and Applications.* John Wiley & Sons, Ltd, 2006.

26. K. R. Carver and J. W. Mink: 'Microstrip antenna technology,' *IEEE Trans. Antenna Propag.*, 29(1), 2–24, 1981.

27. R. S. Elliott, *Antenna Theory and Design*, Prentice Hall, Englewood Cliffs, 1981.

28. W. W. Hanson and J. R. Woodyard, 'A new principle in directional antenna design,' *Proc. IRE*, 26, 333–345, 1938.

29. C. L. Dolph, 'A current distribution for broadside arrays which optimizes the relationship between beamwidth and side-lobe level,' *Proc. IRE Waves Electrons*, June, 1946.

30. B. J. Forman, 'Directivity of scannable planar arrays,' *IEEE Trans. Antennas Propag.*, 20, 245–252, 1972.

31. R. F. Harrington, *Time-Harmonic Electromagnetic Fields*, McGraw-Hill, New York, 1961.

32. Y. Huang, A. Alieldin and C. Song, 'Equivalent circuits and analysis of a generalised antenna system,' *IEEE Antenna Propag. Mag.*, April, 2021.

Problems

Q5.1. From an antenna point of view, a two-wire transmission line can be considered as two-wire antennas. Explain why this type of transmission is only suitable for low-frequency, not for high-frequency applications.

Q5.2. Dipole antennas are popular, can you justify why the half-wavelength dipole is the most popular dipole?

Q5.3. Compare the half-wavelength dipole, quarter-wavelength monopole, and half-wavelength monopole in terms of the input impedance, directivity, and radiation pattern.

Q5.4. How can the bandwidth of the half-wavelength dipole be broadened?

Q5.5. Compare a half-wavelength dipole, a half-wavelength slot, and a half-wavelength loop in terms of the input impedance, gain, and radiation pattern.

Q5.6. A short dipole with a length of 3 cm and a diameter of 3 mm is made of copper wire for 433 MHz (an ISM band) applications. Find its input impedance, radiation resistance, and radiation efficiency. If it is to be connected to a 50 Ω coaxial cable, find the reflection coefficient, return loss, and *VSWR*.

Q5.7. Explain the concept of baluns and design a balun for Q5.6 if applicable.

Q5.8. Design a circularly polarized helix antenna of an end-fire radiation pattern with a directivity of 10 dBi. Find out its input impedance, *HPBW, AR*, and radiation pattern.

Q5.9. Explain the design principle of the Yagi–Uda antenna and estimate the maximum directivity of a Yagi–Uda antenna with 10 elements.

Q5.10. Discuss how the directivity is linked to the number of elements, N, scaling factor, and apex angle of a log-periodic dipole antenna (LPDA). If the directivity of a six-element LPDA is 11 dBi, what is the achievable bandwidth?

Q5.11. Design a log-periodic dipole antenna to cover all UHF TV channels, which is from 470 MHz for channel 14 to 890 MHz for channel 83 and each channel has a bandwidth of 6 MHz. The desired directivity is 12 dBi.

Q5.12. Design a gain horn antenna with a directivity of 15 dBi at 10 GHz. WR-90 waveguide is used to feed the horn.

Q5.13. The Sky digital satellite receiving antenna is an offset parabolic reflector antenna with a dish diameter of about 40 cm. If the aperture efficiency factor is about 70%, estimate the directivity of this antenna. The center frequency is 12 GHz.

Q5.14. A circular parabolic reflector has $F/2a = 0.5$. The field pattern of the feed antenna is E $(\theta) = \cos^2\theta$, $\theta < \pi/2$. Find the edge taper, spillover efficiency, and aperture efficiency.

Q5.15. RT/Duroid 6010 substrate ($\varepsilon_r = 10.2$ and $d = 1.58$ mm) is to be used to make a resonant rectangular patch antenna of linear polarization;
  a. design such an antenna to work at 2.45 GHz for Bluetooth applications;
  b. estimate its directivity;
  c. if it is to be connected to a 50 Ω microstrip using the same PCB board, design the feed to this antenna;
  d. find the fractional bandwidth for *VSWR* < 2.

Q5.16. Two half-wavelength dipoles are in parallel and separated by $\lambda/4$. Find the input impedance for each antenna and then obtain and plot the radiation pattern for the following cases:
  a. they are fed in phase;
  b. they are fed 90° out of phase;
  c. they are fed 180° out of phase.

Q5.17. Four half-wavelength dipoles are employed to form a linear array. Find the maximum directivity obtainable for this array and plot its radiation pattern.

Q5.18. Four $\lambda/4$ monopoles with a large ground plane are used to form a linear array. The spacing between elements is 1λ, and all monopoles are to be fed in-phase from a 1-GHz transmitter via a 50 Ω microstrip line.

a. Design a feed system for this array (mutual coupling is ignored);

b. Find its radiation pattern. Is this a broadside or end-fire array?

c. Find its directivity.

Q5.19. Explain the principle of a phase-scanned array.

Q5.20. What is the Hansen–Woodyard end-fire array?

Q5.21. Design a 1.5-GHz circularly polarized antenna that has an omnidirectional radiation pattern in the H-plane.

Q5.22. The bandwidth is an important consideration for any RF devices. If an antenna with an impedance of 300 Ω is matched to a 50 Ω feed line with a quarter-wavelength transformer at the design frequency of 1 GHz. Find the bandwidth for $VSWR < 2$ if the antenna impedance is a constant and then comment on your results if the antenna impedance is not a constant (the real case).

Q5.23. Double $\lambda/4$ transformers can improve the bandwidth of the matching network. Redo Q5.22 using a double $\lambda/4$ transformer.

6

Computer-Aided Antenna Design and Analysis

In the previous chapter, we introduced and analyzed some of the most popular antennas and discussed what the important considerations are from the design point of view. You should now be able to design some basic standard antennas (such as dipoles and horns) without resorting to a computer. However, antenna design is a very challenging subject because there are usually many variables involved, even for a given type of antenna. Due to the significant advance of computer hardware and software over the past three decades, it has now become a standard practice to employ simulation software to aid antenna designs – just as in any other industry. The objective of this chapter is therefore to introduce the basic theory (such as the method of moments) for computational electromagnetics, provide a brief overview of antenna simulation tools on the market, and then demonstrate how to use some industrial standard software to analyze and design antennas. Furthermore, the theory of characteristic modes is introduced to aid antenna design (a patch antenna is used as an example). It is shown that the TCM is an excellent tool to reveal the insight on how an antenna works and also to aid the optimum design and selection of the feed for an antenna.

6.1 Introduction

As the antenna equation (4.3) indicates, the prediction of antenna performance is a very complex issue and the analytical approach is only suitable for antennas of simple geometry. In the past, engineers designed antennas by experience and a certain amount of "black magic." A good understanding of antennas and impedance matching theory and techniques, and the ability to accomplish a good match over the desired frequency range were required. After the prospective antenna was designed and built, there always remained the question if the antenna had met the specifications. Thus, an antenna measurement using a test range had to be conducted in order to verify that all design criteria were met. Next, if all went well on the testing, the prospective antenna would be transferred to the production department. Finally, the production antenna would be tested to verify that the design could be duplicated. However, if the measured results were not good enough, some tuning and modifications had to be made and further tests were required. This process normally included minding, tinkering, cut-and-try, and gluing and screwing parts together until the antenna met the desired specifications. Needless to say, this was a very expensive and time-consuming procedure.

Those days are now gone forever. We live in a competitive market where time and performance are critical. Guesswork and long design cycles are out of the question. New state-of-the-art antennas are required to meet all the design criteria for sophisticated applications that may involve large quantities at competitive cost. Various commercial antenna design software packages have been developed with very good accuracy and multi-functions and are widely used in the industry. The ability to use computer-aided design (CAD) tools has become an essential requirement for a good antenna engineer.

Now let us look back to see how the antenna design software has been evolved. In the mid 1960s, mainframe computers were finally fast enough to conduct antenna modeling. However, it still required an astute antenna design engineer with a good analytic mathematical background to write the equations and computer programs to model antennas. In the 1970s, the Numerical Electromagnetic Code (NEC) program was developed using FORTRAN (a computer computational program) by the US government [1]. This program was accurate and suitable for wire-type antennas but difficult to use and required a large mainframe computer. Furthermore, its use was restricted by the government. This all changed in the mid 1980s when personal computers (PCs) became widely available. Soon, there were several antenna modeling programs available. The primary one was MININEC, a smaller program based on NEC that runs fast on a PC [2, 3]. Also, a PC version of NEC became available. As PCs increased in speed, so did the antenna modeling programs. It was not unusual to wait for hours or more for the analysis of an antenna. Nowadays, PCs with powerful processors can model the same antenna in just a few seconds. In the 1990s, a large number of EM simulation tools and antenna design software packages, based on various methods, were developed and appeared on the market. They are suitable for not only wire-type antennas but also aperture-type antennas. Many of these packages have been updated for every one or two years; new functions and improved algorithms have been included. For example, parametric optimization is now available. These programs will perform the optimization for you. All you need to do is to input a reasonable design along with a properly weighted trade-off or figure-of-merit (FoM). The FoM weighs the relative importance of each antenna parameter such as the gain, pattern, and impedance over a set of input frequencies, and an optimized design will then be produced by the computer. These powerful modeling programs allow the antenna designer to use keystrokes on a PC keyboard instead of getting their hands dirty from hours of cutting and testing the actual antenna. Furthermore, the performance goals can be tested before any actual antenna is constructed. You can now cut-and-try an antenna design for maximum gain, front-to-back ratio, impedance, or a combination of the three by changing element lengths, diameters and spacings using only a keyboard. This significantly shortens the design to completion of production antennas.

The above may give the impression that antenna design is no longer a "black art." While this may be true, it still requires a lot of knowledge and understanding of the interrelated parameters in the antenna design, FoM, as well as the software itself. In fact, as the antenna designs of today get more complicated, so does the work of the antenna designer. The antenna modeling and optimization programs available today do not completely compensate for all variables. The output data file must be properly interpreted by the user. Sometimes, the tolerances of the element dimensions may be difficult to realize. Element correction factors may still be required to compensate for the mechanical mounting of the antenna elements. Sometimes, the simulation results may not be accurate enough due to practical reasons (such as limited computer memory or poor convergence). This is where the antenna designer comes in to play. One must be able to realize what designs can and cannot be built as well as any compensation that is required. After all, computers are just computers and their outputs are just numbers; they will never replace innovative engineers to analyze and design antennas. Modeling and design software does not design an antenna for you. They only predict how the inputted antenna design should perform and leave it up to the ingenuity of the antenna designer to "tweak" or optimize the mechanical dimensions until the desired antenna performance is attained.

In summary, to become a modern professional antenna designer, one has to have a comprehensive understanding of antennas, a good grasp of mechanical engineering principles, and excellent skills of using computer modeling and optimization software to aid the design.

6.2   Computational Electromagnetics for Antennas

Computational electromagnetics (CEM) is a unique subject of interest to all electromagnetics engineers and researchers. It has a very wide range of applications in RF engineering, EMC (electromagnetic compatibility), radar, wireless communications, electrical and electronics engineering, and extending to areas such as biomedical engineering – antennas are just one of the areas. Many methods have been developed over the years; some were first introduced for other applications. Generally speaking, CEM can be divided into numerical methods and high-frequency methods as shown in Figure 6.1. *High-frequency methods are suitable for structures much larger than the wavelength, whereas numerical methods are more suitable for smaller structures.* There is no specific boundary between these two groups. As a rule of thumb, 20λ is usually used as the upper limit for numerical methods, which is really determined by the computation power (memory size and computational time).

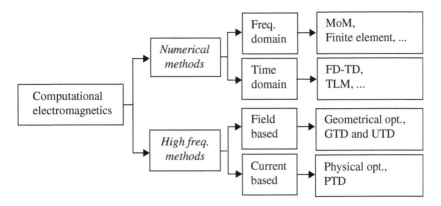

Figure 6.1   Classification of computational electromagnetic methods

Numerical methods can be subdivided into frequency-domain methods (such as the method of moments and the finite element method) and time-domain methods (e.g. the finite-difference time-domain method and the transmission line modeling method), whereas the high-frequency methods can be subdivided into field-based methods (the geometrical optics method) and current-based methods (the physical optics method). Although cost-effective electromagnetic software is readily available and it is now not necessary to write your own software from scratch, a good understanding of the principles on which the software is based is necessary in order to set the relevant parameters properly and avoid the misuse and misinterpretation of the results.

In this section, we are going to first use the method of moments as an example to see how a numerical method can be employed to model and analyze an antenna, and then introduce some other popular methods used in CEM. A general comparison of these methods will be made in Section 6.2.5.

6.2.1 Method of Moments (MoM)

The *method of moments* (MoM, or moment method) was first introduced in mathematics. The basic idea is to transform an integral or differential equation into a set of simultaneous linear algebraic equations (or matrix equation) which may then be solved by numerical techniques [4–6]. It was first applied to electromagnetic problems in the 1960s. Roger Harrington was the person who made the most significant contribution to this area [5, 6]. His book, *Field Computation by Moment Methods* [6], is the first book to explore the computation of electromagnetic fields by the method of moments – the most popular method to date for the numerical solution of electromagnetic field problems. It presents a unified approach to MoM by employing the concepts of linear spaces and functional analysis. Written especially for those who have a minimal amount of experience in electromagnetic theory, theoretical and mathematical concepts are illustrated by examples that prepare readers with the skills they need to apply the MoM to new, engineering-related problems. In this subsection, we are going to introduce the concept of MoM and apply it to model and analyze a dipole antenna.

6.2.1.1 Introduction to MoM

Let us deal with a general case: a linear equation

$$L(F) = g \tag{6.1}$$

where L is a known linear operator, g is a known excitation function, and F is the unknown function to be determined. In physics, L means the system transfer function and g represents the source. The objective is to determine F once L and g are specified.

First, the function F is expanded using a series of known *basis functions (or expansion functions)* $f_1, f_2, f_3, \ldots$ in the domain of operator L, we have

$$F = \sum_n I_n f_n \tag{6.2}$$

where I_n are unknown complex coefficients to be determined and $n = 1, 2, ..., N$. N should be infinite in theory but is a limited number in practice.

Replacing F in Equation (6.1) by Equation (6.2) and using the linearity of L to give

$$\sum_n I_n L(f_n) = g \tag{6.3}$$

The problem becomes how to determine these unknown coefficients I_n.

Second, a set of *weighting functions (or testing functions)*: $W_1, W_2, W_3, ...$ in the domain of L is chosen and then the inner product is formed:

$$\sum_n I_n \langle W_m, L(f_n)\rangle = \langle W_m, g\rangle \tag{6.4}$$

where $m = 1, 2, ..., M$. Again, M should be infinite in theory, but a finite number in practice and a typical, but not unique, *inner product* is defined as

$$\langle x(z), y(z)\rangle = \langle y(z), x(z)\rangle = \int_L x(z)y(z)\ dz \tag{6.5}$$

Third, the inner product is performed on Equation (6.4) for $m = 1$ to M to give the matrix equation:

$$\begin{bmatrix} \langle W_1, L(f_1)\rangle & \langle W_1, L(f_2)\rangle & \cdots & \langle W_1, L(f_N)\rangle \\ \langle W_2, L(f_1)\rangle & \langle W_2, L(f_2)\rangle & & \langle W_2, L(f_N)\rangle \\ \cdots & & & \\ \langle W_M, L(f_1)\rangle & & \cdots & \langle W_M, L(f_N)\rangle \end{bmatrix} \begin{bmatrix} I_1 \\ I_2 \\ . \\ I_N \end{bmatrix} = \begin{bmatrix} \langle W_1, g\rangle \\ \langle W_2, g\rangle \\ . \\ \langle W_M, g\rangle \end{bmatrix} \tag{6.6}$$

or in more compact form:

$$[Z_{mn}][I_n] = [V_m] \tag{6.7}$$

where

$$Z_{mn} = \langle W_m, L(f_n)\rangle \tag{6.8}$$

and

$$V_m = \langle W_m, g\rangle \tag{6.9}$$

They are now readily obtained, and the unknown coefficients can be yielded as

$$[I_n] = [Z_{mn}]^{-1}[V_m] \tag{6.10}$$

The unknown function F can therefore be obtained approximately using Equations (6.2) and (6.10).

The selection of the basis functions and weighting functions is a key to obtain accurate solutions efficiently and successfully. In general, the basis functions should have the ability to accurately represent and resemble the anticipated unknown function. There are many possible basis/weighting sets, but only a limited number is used in practice. The elements of these functions must be linearly independent, so that the N equations in Equation (6.6) will also be linearly independent and give a unique solution. In addition, they should be chosen to minimize the computations required to evaluate the inner product. If both the basis functions and weighting functions are the same, then this special procedure is known as *Galerkin's method*. We are going to use the example below to show how to implement the MoM in practice.

Example 6.1   MoM
Solve the following differential equation using the MoM:

$$\frac{d^2 F(z)}{dz^2} = 1 + 2z^2 \tag{6.11}$$

for $0 \leq z \leq 1$ with the boundary conditions: $F(0) = F(1) = 0$.

Solution
This is a simple boundary problem. The transfer function is

$$L = \frac{d^2}{dz^2},$$

and the source is

$$g(z) = 1 + 2z^2.$$

It is not difficult to obtain the exact solution of this differential equation as

$$F(z) = -\frac{2}{3}z + \frac{1}{2}z^2 + \frac{1}{6}z^4 \tag{6.12}$$

There are many basis and weighting functions which may be suitable for this problem. The selection of these functions is crucial. Here, we are going to use Galerkin's method to solve this problem.

According to the format of the source $g(z)$, it is natural to choose the basis function as $f_n(z) = z^n$. However, you will find in the end that the boundary condition $F(1) = 0$ cannot be met. Taking the boundary conditions into account, we can choose the basis functions and weighting functions as

$$f_n(z) = W_n(z) = z - z^{n+1} \tag{6.13}$$

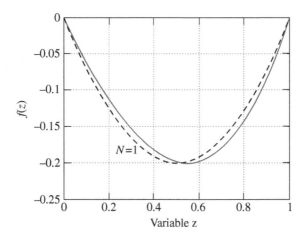

Figure 6.2 Comparison of MoM result for $N = 1$ (dashed line) with the exact solution (solid line)

and the inner product is defined as

$$\langle f_m(z), f_n(z) \rangle = \int_0^1 f_m(z) f_n(z) \ dz. \tag{6.14}$$

Using Equations (6.8) and (6.9), we can obtain

$$Z_{mn} = \langle W_m, L(f_n) \rangle = \int_0^1 (z - z^{m+1}) \frac{d^2}{dz^2} (z - z^{n+1}) dz = -\frac{mn}{m+n+1}$$

$$V_m = \langle W_m, g \rangle = \int_0^1 (z - z^{m+1})(1 + 2z^2) \ dz = \frac{m^2 + 3m}{(m+2)(m+4)} \tag{6.15}$$

For $M = N = 1$, we have $Z_{11} = -\frac{1}{3}$ and $V_1 = \frac{4}{15}$. Using Equation (6.10) to give

$$I_1 = -\frac{4}{5}, \text{ and the solution : } F_{N=1} = \sum_n I_n f_n = -\frac{4}{5}(z - z^2)$$

As shown in Figure 6.2, the result obtained by the MoM (dashed line) is close to the exact solution (solid line). To increase the accuracy, we can increase N.

For $M = N = 2$, we have

$$\begin{bmatrix} -\dfrac{1}{3} & -\dfrac{1}{2} \\ -\dfrac{1}{2} & -\dfrac{4}{5} \end{bmatrix} \begin{bmatrix} I_1 \\ I_2 \end{bmatrix} = \begin{bmatrix} \dfrac{4}{15} \\ \dfrac{5}{12} \end{bmatrix}$$

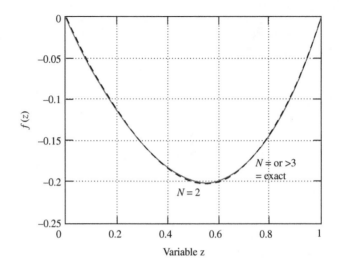

Figure 6.3 Comparison of MoM result for $N = 2$ (dashed line) with the exact solution (solid line)

Thus, we can easily use a computer program (such as Matlab [7]) to solve the matrix equation to obtain

$$\begin{bmatrix} I_1 \\ I_2 \end{bmatrix} = \begin{bmatrix} -\dfrac{3}{10} \\ -\dfrac{1}{3} \end{bmatrix}$$

and

$$F_{N=2} = \sum_n I_n f_n = -\frac{19}{30}z + \frac{3}{10}z^2 + \frac{1}{3}z^3.$$

This is very close to the exact solution as shown in Figure 6.3. If we increase M and N further to say 3, then we can yield

$$\begin{bmatrix} I_1 \\ I_2 \\ I_3 \end{bmatrix} = \begin{bmatrix} -\dfrac{1}{2} \\ 0 \\ -\dfrac{1}{6} \end{bmatrix}$$

thus

$$F_{N=3} = -\frac{2}{3}z + \frac{1}{2}z^2 + \frac{1}{6}z^4.$$

This is now the same as the exact solution given by Equation (6.12). It is not difficult to verify that, when $N > 3$, the solutions are unchanged, the same as Equation (6.12). From Figures 6.2 and 6.3, we can see that the accuracy of the numerical solutions increases with N and this means that results are converged. It is very important to have a convergent solution. If the result does not converge, it normally indicates that the basis and weighting functions have not been properly chosen and are not suitable for the problem.

This example has successfully demonstrated the usefulness of the MoM in solving linear equations. Through this exercise, we have seen that a differential equation has been converted to a matrix equation which is easy to solve and the results are very accurate.

Next, we are going to see how to use the MoM to analyze and model a dipole antenna.

6.2.1.2   Analysis of a Dipole Antenna Using MoM

The MoM has been widely used to solve wire-type antenna problems. Here, we are going to take a dipole of length $2l$ and diameter $2a$ as an example, shown in Figure 6.4, to illustrate the application of MoM in antennas.

The total electric field at any point in the space can be expressed as the sum of the incident field E^i and scattered/radiated field E^s:

$$E^t(r) = E^i(r) + E^s(r) \tag{6.16}$$

Since the dipole is made of a perfect conductor (good conductor in reality), the total tangential electric field must be zero on the surface of the antenna, that is,

$$E^i_z(r) + E^s_z(r) = 0; \text{ on the antenna} \tag{6.17}$$

For a receiving antenna or scatterer, the incident field is usually a plane wave. For a transmitting antenna, the incident field can be seen as the excitation of the antenna and there are two source models: delta gap source and magnetic frill generator [8, 9]. The most used source model is the delta gap model which is also chosen for our investigation. In Figure 6.4, a voltage source

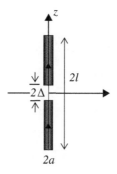

Figure 6.4   A dipole of length $2l$, diameter $2a$, and gap 2Δ

of $2V_0$ (from $+V_0$ to $-V_0$) is applied across the feeding gap 2Δ; the incident field can therefore be expressed as

$$E^i_z(r) = \begin{cases} V_0/\Delta; & \text{for } |z| < \Delta \\ 0; & \text{for } \Delta < |z| < l \end{cases} ; \text{on the antenna} \tag{6.18}$$

The general radiated field can be represented by the antenna equation (4.3) as

$$E^s(r) = \left(-j\omega\mu + \frac{\nabla\nabla\bullet}{j\omega\varepsilon} \right) \int_V J(r') \frac{e^{-j\beta\cdot|r-r'|}}{4\pi|r-r'|} dv' \tag{6.19}$$

On the surface of the antenna, the current density can be replaced by a line-source current I (z), we have

$$\int_{-l}^{l} I(z')K(z,z') \, dz' = -E^i_z(z) \tag{6.20}$$

where the kernel is given by

$$K(z,z') = \frac{1}{j4\pi\omega\varepsilon} \left(\frac{\partial^2}{\partial z^2} + \beta^2 \right) \frac{e^{-j\beta r}}{r} \tag{6.21}$$

Equation (6.20) is known as *Pocklington's integral equation* and has been used extensively for dipole antennas. The problem has now become how to determine the current $I(z)$ by solving this integral equation.

Using the MoM, we need to choose a set of basis functions. The operator in this case is an integral with a complex kernel. Although we can select the entire domain basis functions as in Example 6.1, we prefer to use sub-domain functions which may make the inner product computation much simpler and faster.

The antenna can be divided equally into N segments as shown in Figure 6.5. The unit pulse function is defined as

$$P_n(z) = \begin{cases} 1; & \text{for} -l + \frac{2l}{N}(n-1) < z \leq -l + \frac{2l}{N}n \\ 0; & \text{otherwise} \end{cases} \tag{6.22}$$

and is chosen for the basis functions; thus, the current on the dipole is approximated as

$$I(z) \approx \sum_{n=1}^{N} I_n P_n(z) \tag{6.23}$$

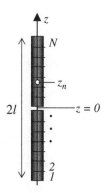

Figure 6.5   A dipole is equally divided into N segments

This expansion in terms of pulse functions is a staircase approximation. The current is constant on each segment. The larger the N, the more accurate the solution should be.

We can use Galerkin's method again, which can produce a reasonable solution. A better weighting function for this case is the Dirac delta function $\delta(z)$, which is

$$W_m(z) = \delta(z - z_m) \tag{6.24}$$

where z_m is the z – coordinate of the center of the segment m, mathematically it is

$$z_m = -l + \frac{2l}{N}(m - 0.5); \quad \text{and} \quad m = 1, 2, ..., N \tag{6.25}$$

A very useful feature of the Dirac delta function is

$$\int f(z)\delta(z - z_0) \ dz = f(z_0) \tag{6.26}$$

This approach is called the *pulse-expansion and point-matching* MoM. The integral equation is enforced at N points along the antenna axis. The inner product can be defined as

$$\langle \, f_m(z), f_n(z) \rangle = \int_{-l}^{l} f_m(z)f_n(z) \ dz \tag{6.27}$$

Thus, the element

$$
\begin{aligned}
Z_{mn} = \langle W_m, L(\, f_n) \rangle &= \int_{-l}^{l} \left[\delta(z - z_m) \int_{-l}^{l} P_n(z')K(z, z') \ dz' \right] dz \\
&= \int_{-l + 2l(n-1)/N}^{-l + 2l(n)/N} K(z_m, z') \ dz'
\end{aligned}
\tag{6.28}
$$

and

$$V_m = \langle W_m, g \rangle = -\int_{-l}^{l} \delta(z - z_m) E^i_z(z) \ dz = -E^i_z(z_m) \tag{6.29}$$

We can let N be an odd number and choose the center segment as the feed to simplify the computation.

It should be pointed out that if the matching point and the source point are at the same place, the distance r will be zero and the kernel will be infinite – this is a problem. A reasonable way to avoid this singularity is to let the current source locate at the center of the wire and the matching point on the antenna surface, that is

$$R_m = r(z_m, z') = \sqrt{a^2 + (z_m - z')^2} \tag{6.30}$$

and Equation (6.28) can be written in a more computational-friendly form as [10, 11]

$$Z_{mn} = \int_{-l + 2l(n-1)/N}^{-l + 2l(n)/N} \frac{e^{-j\beta R_m}}{j4\pi\omega\varepsilon R_m{}^5} \left[(1 + j\beta R_m)(2R_m{}^2 - 3a^2) + (\beta a R_m)^2 \right] \ dz' \tag{6.31}$$

When the segment is small enough, this expression can be further simplified as

$$Z_{mn} \approx \frac{e^{-j\beta R_{mn}}}{j4\pi\omega\varepsilon R_{mn}{}^5} \left[(1 + j\beta R_{mn})(2R_{mn}{}^2 - 3a^2) + (\beta a R_{mn})^2 \right] \frac{2l}{N} \tag{6.32}$$

where

$$R_{mn} = r(z_m, z_n) = \sqrt{a^2 + (z_m - z_n)^2} \tag{6.33}$$

Now the Pocklington's integral Equation (6.20) is transformed into the following matrix equation:

$$[Z_{mn}][I_n] = [V_m] \tag{6.34}$$

All elements of Z_{mn} and V_m can be computed easily, and unknown coefficients I_n can now be obtained without difficulty. Hence, the current distribution along the antenna can be found using (6.23) and is illustrated by Figure 6.6.

Other important antenna parameters can also be obtained. For example, the input impedance is given by

$$Z_a = \frac{2V_0}{I_{N/2}} \tag{6.35}$$

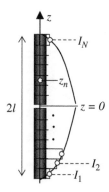

Figure 6.6   Current distribution along a dipole

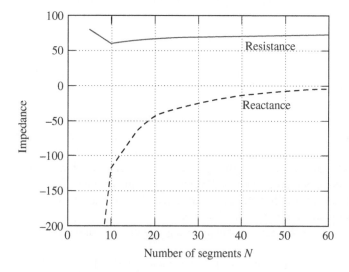

Figure 6.7   Input impedance convergence of the point-matching approach

and the total radiated field (hence the radiation pattern) can be calculated using

$$E^t(r) = E^i(r) + E^s(r) = E^i(r) + \int_{-l}^{l} \sum_{n=1}^{N} I_n P_n(z')K(r, z')dz' \tag{6.36}$$

For a half-wavelength dipole, the convergence of the calculated impedance as a function of the number of segments N is shown in Figure 6.7. When N is small, the impedance is not stable. As $N > 20$, the resistance approaches 80 ohms and the reactance is close to 0 ohms, which are comparable with measured results.

Using this method, we can produce a lot of results for different antennas, without actually making any antennas. From Figure 6.8, which is also shown in Table 5.1, we can clearly see how the current distribution varies with the antenna length in wavelengths and how the current distribution determines the radiation pattern.

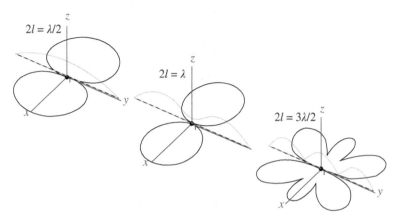

Figure 6.8 Current distributions (along *Y*-axis) and radiation patterns (in *XoY* plane) of three dipoles with length $\lambda/2$, λ, and $3\lambda/2$, respectively

6.2.1.3 Discussion and Conclusions

In this subsection, we have first introduced the MoM and demonstrated how to use it to solve a linear equation and then applied the method to a dipole antenna. The main antenna parameters have been obtained. It is important to note:

- the first aim is to obtain the current distribution when solving an antenna problem numerically and then to find other antenna parameters using the obtained current.
- the selection of the basis and weighting functions is crucial, which may affect the complexity and time of computation as well as the accuracy of the results;
- if the result converges, then the more segments, the more accurate (but with an increased memory requirement and computation time!);
- if the result does not converge, it usually means that the basis and weighting functions are not suitable;
- using the point-matching technique, the segment length should be smaller than $\lambda/10$; it is usually about $\lambda/20$ which is also generally required by other simulation methods;
- there is a trade-off between the accuracy and computation requirements (memory and time) – this is a conclusion that is applicable to all numerical methods.

Although we have only applied the MoM to a dipole antenna, the method is general and can be easily applied to other wire-type antennas. It can even be used to solve aperture-type antenna problems. Another beauty of it is that it can be combined with high-frequency methods to deal with electrically large problems. Some very good commercial software packages are already available on the market and will be discussed later in this chapter.

6.2.2 Finite Element Method (FEM)

The *Finite Element Method (FEM)* was originally introduced for solving complex elasticity, structural analysis problems in civil engineering and aeronautical engineering. Its development can be traced back to the work by Alexander Hrennikoff [12]. The method was often based on an energy principle, e.g., the virtual work principle or the minimum total potential energy principle, which provides a general, intuitive, and physical basis that has a great appeal to structural engineers. A rigorous mathematical foundation was provided by Strang and Fix in their book *An Analysis of the Finite Element Method* published in 1973 [13], and the method has since been generalized into a branch of applied mathematics for numerical modeling of physical systems in a wide variety of engineering disciplines, e.g., electromagnetics and fluid dynamics. It has been used for finding approximate solutions of partial differential equations as well as integral equations and is particularly suitable for problems involving irregular boundaries and non-homogenous material properties.

The FEM may be implemented in the following four steps:

1. Discretization of the solution region into elements (usually triangular in shape).
2. Generation of equations for the fields or potentials at each element.
3. Integration or assembly of all elements.
4. Solution of the resulting system of equations.

Just like the MoM, the problem is also converted into a matrix equation of current elements at the end. The FEM has been employed by a large number of commercial EM simulation packages and becomes one of the most popular and established numerical techniques in engineering. An example of how the region of a loop antenna is discretized into FEM elements is shown in Figure 6.9. The obtained current distribution on the loop is also shown in the figure. In Section 6.4, we are going to use FEM-based software to simulate and design antennas. For more details about this technique, refer to references [14, 15].

6.2.3   Finite-Difference Time-Domain (FDTD) Method

The *Finite-Difference Time-Domain* (FDTD) method belongs to the general class of grid-based differential time-domain numerical modeling methods. It was introduced by Yee in 1966 [16]. Since then, a significant development has been made on improving, implementing, and

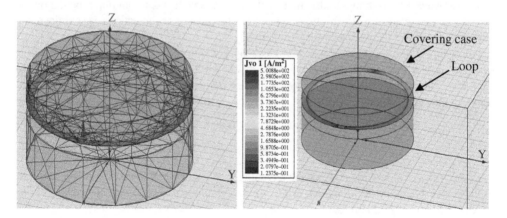

Figure 6.9   FEM simulation of a loop antenna: discretization and current distribution. In this case, the loop is placed in a device (e.g. watch) for wireless communications; the effects of the covering case on the loop can be estimated

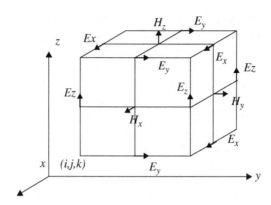

Figure 6.10   FDTD (Yee) cell

spreading this method [17–19]. The boundary conditions were one of the major problems of this method.

The problem domain is discretized into many cells (usually square or rectangular), known as Yee cells/lattices. A typical one is shown in Figure 6.10. The time-dependent partial-differential Maxwell's equations given in Equation (1.29) are discretized using central-difference approximations to the space and time partial derivatives. The resulting finite-difference equations are solved in a leapfrog manner: the electric field vector components in a volume of space are solved at a given instant in time; then, the magnetic field vector components in the same spatial volume are solved at the next instant in time. This process is repeated over and over again, until the desired transient or steady-state electromagnetic field behavior is fully evolved. This scheme has proven to be very robust and remains at the core of many current FDTD software constructs. Furthermore, Yee proposed a leapfrog scheme for marching in time wherein the E-field and H-field updates are staggered so that E-field updates are conducted midway during each time step between successive H-field updates, and conversely. On the plus side, this explicit time-stepping scheme avoids the need to solve simultaneous equations and furthermore yields dissipation-free numerical wave propagation. On the minus side, this scheme mandates an upper bound on the time step to ensure numerical stability. As a result, certain classes of simulations can require many thousands (or more) of time steps for completion.

The accuracy of the computed results is basically determined by the cell size and time steps, which is simpler and easier than making the right choices on basis and weighting functions in the MoM. The cell size is determined by the highest frequency of interest. The biggest dimensions of the cell should normally be smaller than $\lambda_{min}/20$, and the time step size

$$\Delta t \leq \frac{1}{c\sqrt{\dfrac{1}{(\Delta x)^2} + \dfrac{1}{(\Delta y)^2} + \dfrac{1}{(\Delta z)^2}}}.$$ Thus, this is limited by the computational power as for

any other numerical method. One of the advantages of this method is that material properties (ε, μ, σ) can be easily accommodated into the computational scheme, in just the way they appear in Maxwell's equations.

Since it is a time-domain method, the method is extremely suitable for solving wideband problems, but narrow-band problems may take a long time to converge. Today, the FDTD has become the most popular time-domain method in CM. A number of well-known computer simulation tools were developed based on this technique.

6.2.4   Transmission Line Modeling (TLM) Method

The *Transmission Line Modeling* (TLM, also called *transmission line matrix*) method is another well-known time-domain modeling method which was originally introduced by Prof P. B. Jones at the University of Nottingham, UK [20]. The main features of this method are its simplicity and the use of the transmission line model which means that lumped elements are used. The method has been evolved and developed over the years [21–23]. The simulation domain is divided into many nodes; a typical one is shown in Figure 6.11. All fields (*E* and *H*) and material properties are represented by transmission line elements (such as *V* and *I*). The simulation is also implemented in a time-iterative fashion like FDTD. In fact, this method is very similar to the FDTD although originating from different ideas. An interesting comparison between them was made in [24]. Again, there were some commercial simulation packages based on this method; the most notable one was MicroStripes produced by Flomerics which is now acquired by a Siemens company (called Mentor) and has changed the focus to thermal and mechanical simulations. But, there are still some groups working in this area, and open-source codes can be downloaded from some relevant websites [25].

6.2.5   Comparison of Numerical Methods

There are many other numerical methods (e.g. boundary element methods [26]) which are also used for electromagnetic simulations but are less popular (it does not mean less accurate) than

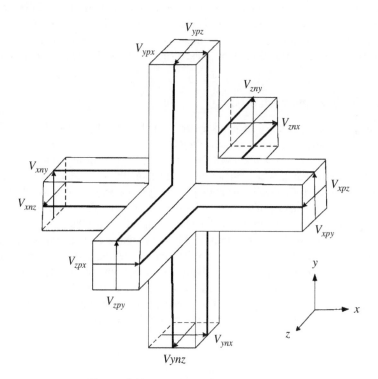

Figure 6.11   A TLM node for simulation

Table 6.1 Comparison of the time- and frequency-domain methods

	Frequency-domain methods		Time-domain methods	
	MoM	FEM	FDTD	TLM
Advantages	Fast at single frequency; Easily combined with other methods to deal with large problems.		Broadband results in one simulation; Good for pulse-type problems.	
Disadvantages	Difficult to deal with pulse-type problems;		Not suitable for electrically large systems	
Note	Most suitable for wire-type antennas	Be careful with very thin wire	Be careful with the boundary conditions	Be careful with thin wires

the ones we have discussed. All these numerical methods need to divide the structure of interest into many cells/elements; some approximations are then used to convert the problem into an easily solvable one (such as a matrix equation). A common problem of these approaches is the huge demand on computer memory and computation time. The accuracy of most commercial packages is quite comparable; the computation efficiency and storage requirements are varied, depending on specific problems. A comparison of frequency- and time-domain methods is given in Table 6.1. Every method and every software package have its advantages and disadvantages, but none is perfect. When these methods are applied to antenna analysis, care has to be taken. For example, the maximum dimension of each cell/element should be smaller than $\lambda/20$ to ensure the accuracy of the results. Some methods (such as the MoM) can easily deal with very thin wire problems, whilst other methods find it difficult to handle such problems. When choosing a method or software, all these aspects should be taken into account.

6.2.6 High-Frequency Methods

Numerical methods are generally of good accuracy and flexibility for various configurations. When applied to electrically large structures, they all have problems such as computer memory, runtime, or convergence – this is true even for very well-written commercial software based on these methods. The modeling is at present too computationally demanding to be useful for such problems. However, high-frequency methods [27] are particularly suitable for electrically large structures where the wave nature of the radiowave need not be considered. Methods used in optics, a well-developed subject, can be employed for electromagnetic and antenna analysis. These high-frequency methods are usually divided into field-based methods and current-based methods.

Geometrical optics (GO) or *ray optics* is a field-based method and describes light propagation in terms of "rays" which are perpendicular to the wave fronts of the actual optic waves [27, 28]. GO provides rules for propagating these rays through an optical system which indicates how the actual wave front will propagate. The basic idea is that the field at distance R away from the source can be obtained using the field in the same tube of rays at distance L away from the source, i.e.

$$|E(R)| = |E(L)| \frac{L}{R} \tag{6.37}$$

This agrees with the conclusion we have obtained for antenna far field where the field is inversely proportional to the distance. A lot of problems may then be solved by ray-tracing method.

It should be pointed out that GO is a significant simplification of optics and fails to account for many important optical effects such as diffraction and polarization. Sometimes, it is inadequate to completely describe the behavior of the EM field, and it is necessary to include the diffracted field – this theory is known as *geometrical theory of diffraction (GTD)*, and the theory for wedge diffraction is called *uniform theory of diffraction (UTD)*. The combination of GO/GTD/UTD has been applied to large antenna analysis and simulation for many years [29, 30]. The hybrid methods using GO/GTD/UTD with numerical methods (e.g. the MoM) have successfully developed to deal with complex large problems such as antennas mounted on the surface of an electrically large structure [31–33].

Physical optics (PO) or wave optics is a current-based method, building on Huygen's principle (which was discussed in earlier chapters), and models the propagation of complex wavefronts through propagation channels, including both the amplitude and the phase of the wave. This technique can account for diffraction, interference, and polarization effects, as well as aberrations and other complex effects; thus, it is more general than GO. Equivalent current sources in the illuminated regions are obtained using the equivalence principle introduced in Chapter 3. For a perfect conductor, the equivalent surface current is

$$J_e = \begin{cases} \hat{n} \times H; & \text{in the illuminated region} \\ 0; & \text{in the shadowed region} \end{cases} \tag{6.38}$$

Once the current is known, the radiated field can be found using an equation such as Equation (4.3). Approximations are still generally used when applied numerically on a computer.

Just like GO, PO has also inherited limitations on dealing with diffraction. The *physical theory of diffraction (PTD)* is developed as an extension of PO that refines the PO surface field approximation just as GTD or UTD refines the GO surface field approximation, more information can be found from [27] as an example.

The typical applications of the high-frequency methods include the calculation of the scattering and diffraction, radar cross section (RCS), effects of finite antenna ground plane, electrically large antennas (such as reflector antennas), and the interaction between antennas and mounting structures.

It is apparent that high-frequency methods are very useful for the analysis of electrically large structures which may not be the antennas themselves but the mounting structures. These methods offer the following advantages:

- there is no limitation on the maximum dimension of the structure for the simulation, nor is there any runtime overhead in increasing the frequency for the same dimensions of the structure;
- it is not memory or runtime intensive;
- it can compute the interaction of an antenna with a structure which takes over from numerical methods when these become unusable.

6.3   Computer Simulation Software

A large number of electromagnetic modeling software packages have been developed and are available on the market. Some are more successful technically and commercially than the others. Some have become industrial standard design and analysis software. It is evolving all the time; thus, it is not possible to give an exhaustive list of the software available. In this book, we can only deal with some of the popular ones. Since these packages have been developed by many people for many years, they are now well-written and normally have a very user-friendly interface for a designer to input the design and the simulated results can be well presented in 2D or even 3D graphs. There is no point for an individual researcher/engineer to develop another general simulation package and reinvent the wheel. However, there is always a demand for specialized software which may perform better than the general-purpose software for specific applications.

Although all the commercial software packages are now user-friendly and relatively easy to operate, you do need to have adequate knowledge of setting up the simulation and interpreting the results. Most of all, you need to have your smart design ready before simulation – the software does not do the design for you but only helps you to validate and improve your design!

In this section, we are going to introduce a few commercial software packages with different complexity and features for antenna designs. Most of the software suppliers nowadays offer free evaluation packages which have limited functionality or a limited period of time for the license, but they are good learning tools for the reader to use and evaluate some simple antenna designs. You are encouraged to explore various packages before any purchase is made. This book does not offer any shopping advice.

6.3.1   Simple Simulation Tools

As mentioned earlier, the MoM-based NEC program was one of the earliest computer programs developed for wire-type antenna modeling and design. Similar window-based commercial packages are readily available. EZNEC [34] and 4NEC2 [35] are just two well-known examples. Demonstration/evaluation versions, which are basic with a very limited number of segments and restricted functionality, are available for free download. The full versions offer excellent capability (over 1000 segments) at an affordable price (EZNEC v. 6.0 at $99 in 2018) or completely free (4NEC2). They are great tools for beginners, amateurs, and even professionals.

If you have access to MATLAB (www.matlab.com), you could use MATLAB Antenna Toolbox™ which is more advanced than EZNEC and 4NEC2 programs and provides functions and applications for the design, analysis, and visualization of antenna elements and arrays. You can design standalone antennas and build arrays of antennas using either predefined elements with parameterized geometry (there is an antenna library) or arbitrary planar elements. One unique attraction is that it can be easily linked to other MATLAB toolboxes for such as global optimization and integration with other devices.

Antenna Toolbox uses the MoM to compute port properties (e.g. the impedance), surface properties (e.g. the current and charge distribution), and field properties (such as the near-field and far-field radiation pattern). You can visualize antenna geometry and analyze results in 2D and 3D. You can also integrate antennas and arrays into wireless systems and use impedance analysis to design matching networks. It provides radiation patterns for simulating beam forming and beam steering algorithms. You can generate Gerber files from your design for manufacturing printed circuit board (PCB) antennas. It has the following key features:

- rapid design and visualization of antennas using predefined or custom elements;
- design of arbitrary PCB structures, and Gerber file generation for antenna manufacturing;
- design of linear, rectangular, conformal, and custom antenna arrays;
- large array analysis using the infinite array or embedded element pattern approach;
- port analysis of impedance, return loss, and S-parameters of antennas and antenna arrays;
- radiation field analysis of the pattern, E and H fields, and beamwidth of antennas, antenna arrays, and custom data;
- surface analysis of antenna and antenna array current, charge, and meshing.

It should be pointed out that the Antenna Toolbox is mainly suitable for wire and PCB antennas. Also, MATLAB is an expensive software package, and one has to pay extra money to access this toolbox. Thus, it may be not a good option for some people.

Overall, EZNEC is easily accessible for all: it is a simple and easy-to-use program for modeling and analyzing many different types of antennas by using many conducting-wire segments. It plots azimuth and elevation patterns; provides the antenna gain, feed-point impedance, SWR, and current distribution; finds and reports beamwidth, 3-dB pattern points, front-to-back ratio, take-off angle, sidelobe characteristics; and more. All information, including patterns, can be displayed on screen or printed. And, it is easy to use with a menu structure, spreadsheet-like entry, and many shortcut features. You describe the antenna (and other nearby structures if desired) as a group of straight conducting wires, choosing the orientation, length, and diameter. Add sources at the feedpoints and, if desired, transmission lines, a realistic ground, and loads to simulate loading coils, traps, or similar components. Antenna descriptions and pattern plots are easily saved and recalled for future analysis. Multiple patterns can be superimposed on a single graph for comparison. You can see the antenna pattern and currents on the same color 3D display, and you can also rotate the antenna display and zoom in for details.

In Section 6.4, we are going to use EZNEC demonstration version as our design tool, to introduce the major features of this kind of software and to see how it can be used to aid the design and analysis of wire-type antennas. For more complicated antennas, a more advanced and full-wave simulation tool is required.

6.3.2   Advanced Simulation Tools

There are quite a few advanced EM simulation tools which have been widely accepted by the antenna and RF/microwave engineering industry. They can be used to simulate both wire-type and aperture-type antennas with complex structures and integrate them with other electronic devices (e.g. filters) and circuits. The typical cost for an industrial license is over $30k/year, while an educational license is about $3k/year (in 2020). Here, we are going to introduce four of the most popular EM software (in alphabetic order) which could be considered as industrial standard simulation tools for antenna designs.

6.3.2.1   CST Studio Suite

CST (computer simulation technology) Studio Suite® is a high-performance 3D EM analysis software package for designing, analyzing, and optimizing EM components and systems. Electromagnetic field solvers for applications across the EM spectrum are contained within a single user interface in CST Studio Suite. The solver types available include general-purpose Time-Domain and Frequency-Domain Solvers for both high-frequency and low-frequency problems, along with a full-wave Integral Equation Solver, Eigenmode and Asymptotic Solvers, a self-consistent Particle-in-Cell (PIC) Solver, statics and multi-physics solvers, and many more specialized solvers as well as hybrid solver combinations. These solvers offer an accurate, versatile approach for tackling various applications. They can be coupled to perform hybrid simulations, giving engineers and researchers the flexibility to analyze whole systems made up of multiple components in an efficient and straightforward way.

Common subjects of EM analysis include the performance and efficiency of installed antennas and filters, electromagnetic compatibility and interference (EMC/EMI), exposure of the human body to fields, electro-mechanical effects in motors and generators, and thermal effects in high-power devices. CST Studio Suite is used by many technology and engineering companies around the world. Simulation allows the use of virtual prototyping, device performance can be optimized, potential compliance issues can be identified and mitigated early in the design process, the number of physical prototypes required can be reduced, and the risk of test failures and recalls minimized.

For antenna designs, the main solvers are:

- Transient Solver – general purpose;
- Frequency-Domain Solver – general purpose;
- Integral Equation Solver – electrically large structures, and RCS;
- Asymptotic Solver – installed performance, and RCS;
- Eigenmode Solver – resonant cavities;
- Multilayer Solver – planar structures;
- Hybrid Task – hybrid simulation with multiple solvers.

These are a good indication of how powerful this simulation tool is. Furthermore, CST Studio Suite has a special software package called **Antenna Magus** which is a software tool to help accelerate the antenna design and modeling process. It increases efficiency by helping the engineer and researcher to make a more informed choice of antenna element, providing a good starting design. Validated antenna models can be exported to CST Studio Suite from a huge antenna database of over 330 antennas; thus, one can get to the customization phase of an antenna design quickly and reliably.

Another good feature of CST is that it offers both CAD-based and voxel-based human body models suitable for EM simulation which is very convenient for the study of body effects on antenna performance or SAR (specific absorption ratio).

6.3.2.2   HFSS

Ansys HFSS (High Frequency Structure Simulator) is another 3D EM simulation software tool for designing and simulating high-frequency electronic products such as antennas, antenna arrays, RF/microwave components, high-speed interconnects, filters, connectors, IC packages, and printed circuit boards. It is based on finite element, integral equation, asymptotic, and advanced hybrid methods to solve a wide range of RF/microwave and high-speed digital electromagnetic problems. It has been widely used to design high-frequency, high-speed electronics found in communications systems, radar systems, advanced driver assistance systems (ADAS), satellites, internet-of-things (IoT) products, and other high-speed RF and digital devices.

HFSS employs versatile solvers (similar to CST) and an intuitive GUI (graphical user interface) to give excellent performance plus deep insight into 3D EM problems. Through integration with Ansys thermal, structural, and fluid dynamics tools, HFSS provides a powerful and complete multi-physics analysis of electronic products, ensuring their thermal and structural reliability. HFSS is synonymous with excellent accuracy and reliability for tackling 3D EM challenges by virtue of its automatic adaptive meshing technique and sophisticated solvers, which can be accelerated through high-performance computing (HPC) technology.

The Ansys HFSS simulation suite consists of a comprehensive set of solvers to address diverse electromagnetic problems ranging in detail and scale from passive IC components to extremely large-scale EM analyses such as automotive radar scenes for ADAS systems. Its reliable automatic adaptive mesh refinement let the designer focus on the design instead of spending time determining and creating the best mesh. It helps engineers to extract circuit parameters (such as the impedance), visualize 3D electromagnetic fields (near- and far-fields), and generate Full-Wave SPICE™ models to evaluate signal quality effectively, including transmission path losses, reflection loss due to impedance mismatches, parasitic coupling, and radiation. HFSS can be dynamically linked to other Ansys software to create a powerful electromagnetic-based design flow and is one of the industrial standard antenna design software packages. The major features include:

- suitable for almost all structures and configurations (wire- or non-wire-types);
- all antenna results are given, and some animated results may also be shown;
- easy to optimize the design (using its parametric function or Optimetrics software);
- good accuracy.

In Section 6.4, we are going to use this software as our design tool, to introduce the major features of this kind of software and see how it can be employed to aid the design and analysis of antennas.

6.3.2.3   FEKO

Altair FEKO is a comprehensive CM software package used widely in the telecommunications, automobile, aerospace, and defense industries. FEKO offers several frequency and time-domain EM solvers under a single license (similar to CST and HFSS). Hybridization of these methods enables the efficient analysis of a broad spectrum of EM problems, including antennas, microstrip circuits, RF/microwave components and biomedical systems, the placement of antennas on electrically large structures, the calculation of scattering as well as the investigation of EMC.

FEKO also offers tools that are tailored to solve challenging EM interactions, including dedicated solvers for characteristic mode analysis (which will be discussed in Section 6.5) and bi-directional cables coupling. Special formulations are also included for efficient simulation of integrated windscreen antennas and antenna arrays.

Overall, FEKO has got the following features:

- Time- and frequency-domain full-wave solvers: MoM, FDTD, FEM, and MLFMM (The Multilevel Fast Multipole Method);
- asymptotic methods: PO, LE-PO (large element physical optics), RL-GO (Ray launching geometrical optics), UTD;
- true hybridization of methods to solve complex and large multi-scale problems;
- unique characteristic mode analysis (CMA) solver to calculate modal currents, eigenvalues, modal significance, and characteristic angles;
- dedicated solvers for wave propagation and radio network coverage analysis;
- fully parallelized and optimized to exploit multi-CPU distributed memory resources;
- GPU-based solver acceleration;
- optimized out-of-core solver to deliver solutions when RAM limits are reached.

As a result, FEKO seems to be able to deal larger systems than most other software on the market. Its CMA solver produces accurate results and is popular in the antenna community. Thus, this software has been used for all CMA examples provided in this book.

6.3.2.4 TICRA

TICRA is a leading provider of cutting-edge reflector antenna modeling software for spacecraft operators and manufacturers, space agencies, Earth stations, defense organizations, and research institutions. Based in Copenhagen, Denmark, and with agents around the world, TICRA's products are trusted worldwide as the reference standard for reflector antenna design and development.

They offer a few special software packages (CHAMP, POS, SATSOFT, DIATOOL, and SNIFT – visit their website https://www.ticra.com/for details) with the main design software called **GRASP** which is for reflector systems, enabling fast and accurate analysis and design of the most advanced reflector antenna systems. Multiple antennas may be defined within the same project, and the general command structure enables the user to define which of those will be considered during a given analysis. This opens for the possibility of making advanced scattering analysis of clusters of antennas. GRASP offers an advanced PO algorithm as the baseline analysis method, supplemented by optional GTD and MoM solvers for advanced applications. The intuitive wizard in GRASP allows for easy setup of single reflectors, Gregorian and Cassegrain systems as well as axially displaced dual reflectors. The wizard generates a good starting point for continued and more elaborate investigation of antenna designs. Further modifications may include changing the surface profile, the rim or edges, the surface material as well as adding other objects to simulate the antenna environment, investigating near fields and more.

The GRASP Student Edition is a reduced-functionality version of the professional GRASP. It gives you the chance to become familiar with the software, and it can help you analyze many realistic antennas. With the GRASP Student Edition, you can:

- define single and dual reflector systems with parabolic, hyperbolic, ellipsoidal, and flat surface shapes;
- define reflector rims as circles, ellipses, or rectangles;

- illuminate the reflectors with a number of analytically given feed patterns, including Gaussian beams;
- perform PO and/or GO-UGTD analysis of single and dual reflectors;
- display pattern cuts and two-dimensional grids.

The Student Edition can be downloaded from:
https://www.ticra.com/software/grasp/grasp-student-edition/

There are many other commercial EM simulation software packages in this very competitive market. There have been some mergers and takeovers. Here, we have listed some other well-known EM simulation tools in alphabet order for your information:

1. **IMST** (https://www.imst.de/) provides design and engineering services in the areas of antennas, wireless communication systems, and RF components. IMST's software tool **EMPIRE** is a very powerful and efficient 3D EM modeling tool for complex designs like radar antennas and millimeter wave frontends.
2. **Mician** (https://www.mician.com/) is recognized as a leading developer of EM software tools for horn antennas and horn antennas with reflectors, particularly single parabolic antenna dual Cassegrain and dual Gregory antennas, displaced axis antennas as well as feed networks and waveguide components.
3. **Remcom** (https://www.remcom.com/) is the provider of the well-known **XFDTD** software. Applications include antenna design and placement, 5G/MIMO, mobile device, biomedical, EMI/EMC, microwave, wireless propagation, military, automotive radar, and more.
4. **SPEAG** (https://speag.swiss/) has a range of antenna-related simulation and measurement tools. EM simulator, such as **SEMcad**, offers various EM phantom models for evaluating the effect of electromagnetic wave in human body and addresses public concerns on the health and safety issues due to today's widespread use of EM waves radiated from various antennas.
5. **Sonnet** (https://www.sonnetsoftware.com/) provides MoM-based EM simulation tools (Sonnet Suites and Sonnet Lite), especially popular for planar and multi-layer circuits and devices. Very easy to use.
6. **WIPL-D** (https://wipl-d.com/) with its flagship software products **WIPL-D** Pro and WIPL-D ProCAD enables users to perform fast and accurate high-frequency simulations of antennas, antenna positioning, microwave circuits, scatterers etc.

6.4   Examples of Computer-Aided Design

In this section, we are going to use two commercial software packages with different complexity and features to design a few antennas. The operation of these packages is similar to other similar software packages. It is hoped that, through these design examples, the reader will gain an in-depth appreciation of these two packages, obtain a better understanding on antenna theory, and be able to use some industrial standard software to aid antenna design.

6.4.1   Wire-type Antenna Design and Analysis

As mentioned earlier, in this subsection, we are going to use EZNEC demonstration version as our design tool, to introduce the major features of this kind of software, and to see how it can be used to aid the design and analysis of wire-type antennas.

EZNEC is a MoM-based powerful but very easy-to-use program for modeling and analyzing various (especially wire-type) antennas in their intended operating environment. It was developed by Roy Lewallen (W7EL) and has been evolved over the years [34]. This window-based program is very user-friendly, and there is a pull-down menu to provide help whenever needed. There are also some examples and a test drive exercise provided with the software package. It takes just a few hours for a new user to become familiar with the program.

6.4.1.1   Design Examples

Example 6.2   14 MHz dipole for Ham radio transceiver
4 MHz (20 m band) is one of the amateur radio (ham radio) bands widely used around the world. Assume that you are going to make a dipole as a transceiver antenna which will be placed in your garden. Use EZNEC to design the antenna and analyze the effects of the ground plane on the antenna input impedance, gain and radiation pattern. Horizontal polarization is assumed.

Solution
14 MHz has a wavelength of 21.4 m. To make a resonant and efficient dipole, the dipole length L should be close to $\lambda/2$, more precisely, $L \approx 0.48\lambda \approx 10$m which is affected by the ground.

Now let's use EZNEC Demo V.50 (it is free!) to aid the analysis and design. On opening the software, we see the user interface as shown in Figure 6.12 which is basically divided into two areas: the one on the left is for data input and output and the other one (center) is to define the antenna to be simulated. The simulation should follow these steps: define the antenna ⇒ view the input antenna ⇒ run the simulation ⇒ check the results.

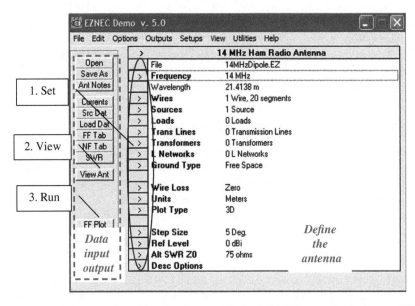

Figure 6.12   EZNEC user interface (Src: Source; FF: far field; NF: near field). Source: EZNEC

The main options to set up the simulation are:

1. *Frequency*: it means the center one, 14 MHz in our case;
2. *Units*: there are five options: m, mm, ft, in, and wavelength; we choose "m"
3. *Wires* that form the antenna: It means entering the end coordinators of each straight wire, diameter, and the number of segments. You can even input the coating material if applicable, as seen in Figure 6.13. Each segment should be smaller than $\lambda/20$. A wire of 10 m is divided into 20 segments as the first attempt for our design. Clicking on "*View antenna*", it gives Figure 6.14 which shows the antenna (3 m above the ground floor $z = 0$) and the coordinates.
4. *Sources*: the source location, amplitude, and phase. Our selection is shown in Figure 6.15. For a center-fed case, it would be better to choose an odd number of segments (such as 19), so the actual position is the specified one.
5. *Loads*: the load impedance at a specific location if applicable;
6. *Transmission line*: location, length, characteristic impedance, and loss if used;

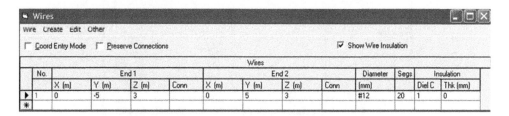

No.	End 1				End 2				Diameter	Segs	Insulation	
	X (m)	Y (m)	Z (m)	Conn	X (m)	Y (m)	Z (m)	Conn	(mm)		Diel C	Thk (mm)
1	0	-5	3		0	5	3		#12	20	1	0

Figure 6.13   Wires input interface. Source: EZNEC

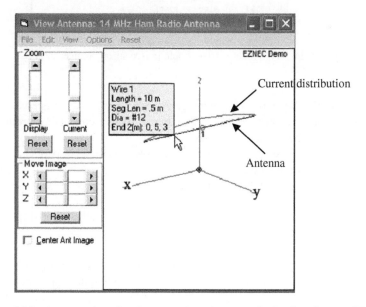

Figure 6.14   Antenna view showing controls and current distribution. Source: EZNEC

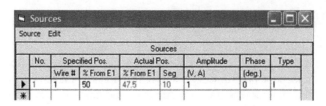

Figure 6.15   Source input interface. Source: EZNEC

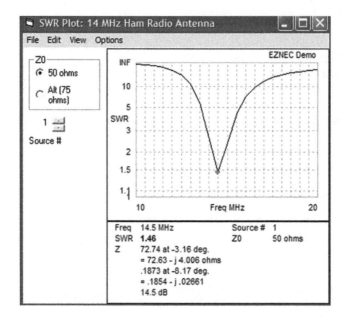

Figure 6.16   VSWR for a 10 m dipole in free space. Source: EZNEC

7. *Transformer and L Networks* for matching: the port locations and impedances;
8. *Ground type*: there are three options: free space, perfect conducting ground, and real ground which may be defined by the user;
9. *Wire loss*: you can choose a predefined material (such as copper or perfect conductor with zero loss) or a user-defined one;
10. *Plot type*: 2D (azimuth or elevation) or 3D.

To analyze the effects of the ground plane, we are going to choose three different grounds: free space, perfect conductor, and real ground.

a. **Free space:** First, assuming that the antenna is in free space, we need to select the ground type as "*free space*". Click on "*SWR*" to define the sweep frequency and then click on "*FF plot*" to start the simulation. The results for 10–20 MHz are given in Figure 6.16. It is shown that this 10 m dipole resonates around 14.5 MHz with an input impedance of $72.63 - j\,4.006$

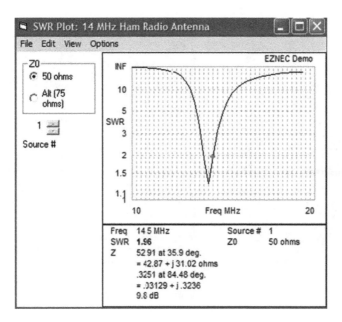

Figure 6.17   VSWR for a 10 m dipole 3 m above a perfect ground. Source: EZNEC

ohms. When it is connected to a 50 ohms transmission line, the VSWR is about 1.46. These numerical results (as well as the radiation pattern in Figure 6.18(a)) are in excellent agreement with our theoretical ones given in Chapter 5.

b. **Perfect ground:** now, we assume that the antenna is 3 m above a perfect ground plane and run the simulation again to obtain Figure 6.17. It shows that the resonant frequency is now shifted down to 14.25 MHz and the input impedance at 14.5 MHz is changed to 42.87 + j31.02 ohms. The comparison of radiation patterns for both cases is presented in Figure 6.18. The maximum gain at $90°$ is increased from 2.11 dBi for free space to 8.58 dBi for this case. Thus, the effects of the ground plane are significant – it acts as a mirror to create another antenna (image) forming a two-element dipole array. This array has increased the gain by a factor of (not 2 but) about 4 (6.4 dB)! This is because the radiation pattern is changed from omni-directional to unidirectional in the *H*-plane (elevation) which is different from a dipole to a monopole case (where the gain is doubled and the radiation pattern is still omni-directional in the *H*-plane when a ground plane is employed). It may not look much like the dipole patterns you may have seen (the textbook pictures usually show a dipole in free space), but it is the kind of pattern in reality.

c. **Real ground:** if we choose the "real ground plane" in the menu which has a conductivity of 0.005 S/m and relative permittivity of 13, the simulation results show that the resonant frequency is around 14.35 MHz and the input impedance at 14.5 MHz is now 61.74 + j14.65 ohms. The radiation pattern is similar to Figure 6.18(b), and the gain is down to 5.38 dBi, while the HPBW is 110°, which is the largest among these three cases. Thus, we can slightly increase the length to 10.2 m (by trial-and-error) to make it resonate at 14 MHz with a VSWR = 1.25.

Since the maximum direction of radiation is at $\theta = 0$, but in practice the wave does not arrive at that angle but more likely at around $45°$, we can therefore change the height of the antenna to

(a)

(b)

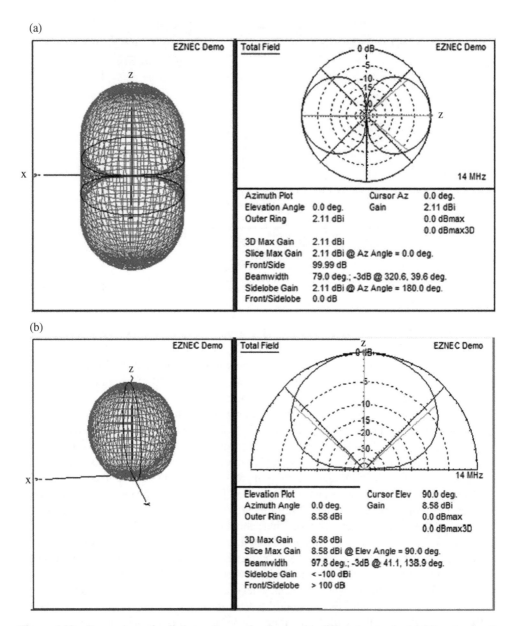

Figure 6.18 Comparison of radiation patterns of a dipole with different grounds. (a) 3D and azimuth plot for a free space case (omni-directional in elevation). (b) 3D and elevation plot for a perfect ground case. Source: EZNEC

tune the radiation pattern. This tuning process can be guided by the antenna array theory introduced in Chapter 5 since the ground acts as a mirror and this becomes a two-element antenna array. As shown in Figure 6.19, when the antenna is above the ground plane by about half of the wavelength ($z = 9$ m in the figure), the 3 dB radiation pattern covers the elevation angle roughly

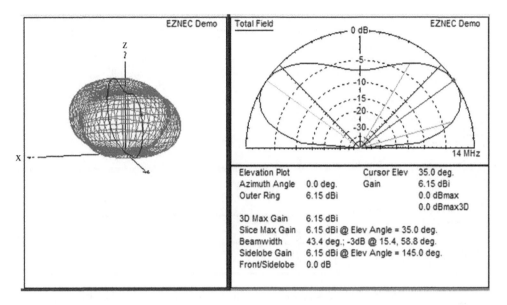

Figure 6.19 Radiation pattern for a dipole of 10.2 m placed 9 m above a real ground plane. Source: EZNEC

from 15° to 60° which matches well with our requirement. The maximum gain is 6.15 dBi (at 35°) which is larger than that of a quarter wave monopole (5.15 dBi) – here, it can be considered as an array!

Example 6.3   Monopole array
Two quarter-wave monopoles are separated by a quarter-wavelength. They are placed above a perfect ground plane and fed with the same amplitude and variable phase. Analyze how the phase difference in the feed affects the array performance and compare with theoretical results.

Solution
For convenience, we choose the frequency to be 300 MHz (so the wavelength is 1 m) and assume the array is made of perfect conducting wire with a diameter of 2 mm. Each quarter-wave monopole is divided into eight segments ($< \lambda/20$). There are now two sources placed at the ends of the monopoles. A perfect ground is chosen as suggested. The EZNEC input interface and the antenna plot with current distribution are shown in Figure 6.20.

Run the program for the initial phase difference $\varphi_0 = 0$, $\pi/2$, π, and $3\pi/2$ to obtain the radiation patterns and the input impedances (click on "*SWR*"). 3D and elevation patterns at $\varphi = 0$ are shown in Figure 6.21, and the input impedances are given in Table 6.2. It took just few minutes on a laptop computer to yield all these results, but it could take a much longer time to obtain the theoretical results, especially when the mutual coupling is taken into account.

It is apparent that the feed phase difference between the two monopoles affects both the radiation pattern and input impedance significantly. When the phase difference is 0° (the same as

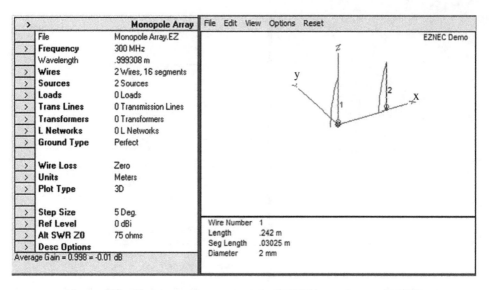

Figure 6.20   Two-monopole array and the EZNEC input. Source: EZNEC

360°) or 180°, the radiation pattern is symmetrical about ZY plane and the two impedances of the monopoles are the same but not 37 ohms, the self-impedance of a $\lambda/4$ monopole. In other cases, the pattern is not symmetrical and the two impedances are not the same – these results might be a surprise to some people, but they are real in practice. This is another proof of how useful and important the simulation tool can be!

It is normally hard to find an analytical solution to the input impedance of an antenna array, but it could be relatively easy to obtain its radiation pattern. We are now going to derive the radiation pattern using the array theory learnt in Section 5.3.

From Equation (5.104), the array factor for this two-element array can be expressed as

$$AF_n = \frac{1}{2}\left[\frac{\sin(2\Psi/2)}{\sin(\Psi/2)}\right] = \cos(\Psi/2) \tag{6.39}$$

where $\Psi = \beta d \sin(\pi/2 - \varphi) + \varphi_0 = \pi \sin(\pi/2 - \varphi)/2 + \varphi_0$ at the azimuth plane, while $\Psi = \pi \sin(\theta)/2 + \varphi_0$ at the elevation plane. Since the quarter-wave monopole antenna pattern is known (using Equation (5.4) and let $0 < \theta < 90°$) and also shown in Table 5.2, we can therefore employ the principle of pattern multiplication to obtain the monopole array pattern:

$$E_\theta \propto \left(\frac{\cos(0.5\pi\cos\theta)}{\sin\theta}\right)AF_n \tag{6.40}$$

The antenna array patterns at the azimuth and elevation planes for the initial phase difference $\varphi_0 = 0°$, $90°$, $180°$, and $270°$ are plotted in Figure 6.22, which are very close to the numerical results presented in Figure 6.21. The only noticeable difference is that, at $\varphi_0 = 90°$, the electric field is zero in theory but very small (-30 dBi) from EZNEC for $\varphi = 0°$ and $\theta = 90°$; and at $\varphi_0 = 270°$, we see the same difference for $\varphi = 180°$ and $\theta = -90°$. These differences can be explained since in the theoretical model, the currents on the monopoles are assumed to be the same, but they are not in reality when they are fed with phase differences (thus, the impedances are different as shown in Table 6.2). The numerical model has taken the difference into account and may therefore provide more accurate results than the theoretical one with approximations/simplifications.

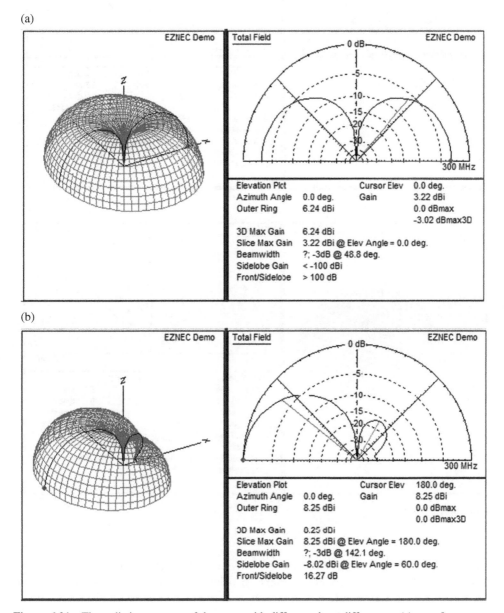

Figure 6.21 The radiation patterns of the array with different phase differences. (a) $\varphi_0 = 0$. (b) $\varphi_0 = 90°$. (c) $\varphi_0 = 180°$. (d) $\varphi_0 = 270°$. Source: EZNEC

(c)

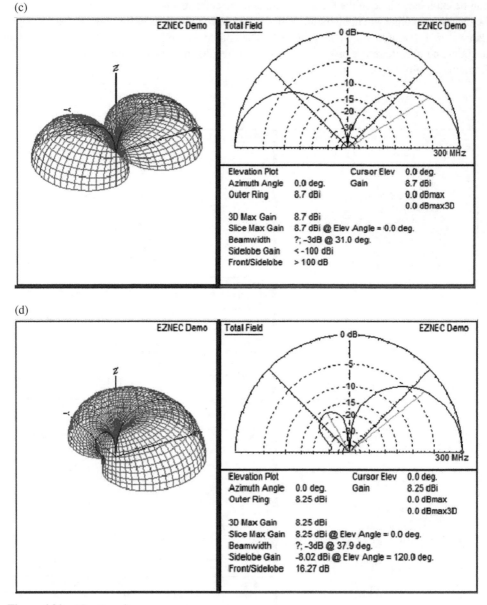

(d)

Figure 6.21   *(Continued)*

This type of software is not just limited to wire-type antennas and can also be used for much more complicated simulations. For example, a helicopter simulation model is shown in Figure 6.23 [34], where the surface is replaced by wire meshes and each wire should be smaller than $\lambda/20$. The computation could become a problem if the frequency of interest is too high. In that case, alternative approach/software is required.

Table 6.2 Input impedance for the two monopoles at 300 MHz

Phase difference φ_0	Monopole 1 (ohms)	Monopole 2 (ohms)
0°	62.74 + j 6.188	62.74 + j 6.188
90°	59.4 + j 46.21	22.76 + j 2.846
180°	19.38 + j 42.87	19.38 + j 42.87
270°	22.76 + j 2.846	59.4 + j 46.21
360°	62.74 + j 6.188	62.74 + j 6.188

(a)   (b)

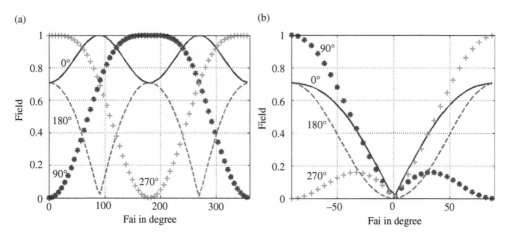

Figure 6.22 Array patterns in the two principle planes for different φ_0. (a) Azimuth plane. (b) Elevation plane

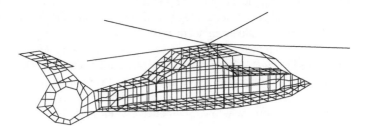

Figure 6.23 A wire model of a helicopter

6.4.2 General Antenna Design and Analysis

Some relatively simple software packages (such as EZNEC) are cheap and very useful for wire-type antenna analysis and design, but they are not very suitable to deal with antennas which have complicated configurations and use various construction materials. More sophisticated software packages (such as CST Microwave Studio, FEKO, XFDTD and HFSS, more details can be found from their websites) are very powerful but not cheap.

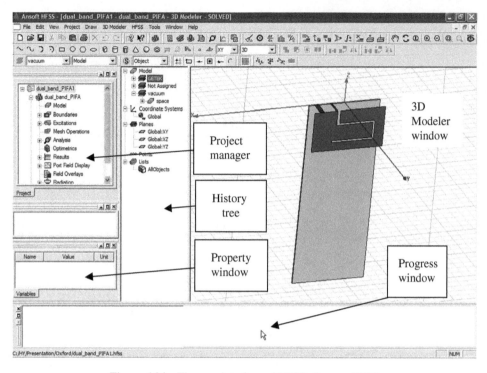

Figure 6.24   The user interface of HFSS. Source: HFSS

In this section, we are going to use HFSS [36] as an example to design a dual-band patch antenna for cellular mobile radio communications. As mentioned earlier, HFSS utilizes a 3D full-wave Finite Element Method (FEM) to compute the circuit and field behaviors of high-frequency and high-speed components. With HFSS, engineers can extract circuit parameters (such as the impedance), visualize 3D electromagnetic fields (near- and far-fields), and generate Full-Wave SPICE™ models to evaluate signal quality effectively, including transmission path losses, reflection loss due to impedance mismatches, parasitic coupling and radiation. HFSS can be dynamically linked to other software to create a powerful electromagnetic-based design flow and is one of the industrial standard antenna design software packages.

Figure 6.24 shows the user interface which is basically divided into five main regions: project manager, history tree, property window, 3D modeler window, and progress window.

6.4.2.1   Design Examples

Now, we are going to use the following example to illustrate how to use this software to aid our antenna design.

Example 6.4   Dual-band GSM antenna
Planar Inverted F Antennas (PIFAs) have been widely reported and are popular in mobile phones. The main reasons for this are that (a) they are of low profile; (b) their radiation patterns are near omni-directional; (c) they are installed above the phone circuitry, "re-using" the space

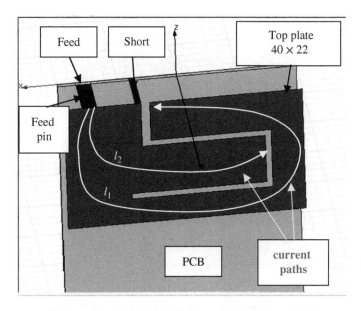

Figure 6.25   A dual-band PIFA antenna. Source: HFSS

within the phone to some degree; (d) they exhibit a low specific absorption rate (and less loss to the head). The objective of this exercise is to design a dual-band PIFA for a GSM mobile handset. It should cover 880–960 MHz (E-GSM 900) and 1710–1880 MHz (DCS 1800 or GSM 1800). The dimensions should be small enough for a standard mobile handset.

Solution
PIFAs are evolved from the inverted F antenna as shown in Figure 5.7. For an inverted F antenna, the total length of the antenna should be close to a quarter-wavelength. In our case, the two center frequencies are: 920 MHz ($\lambda_1 = 326$ mm) and 1755 MHz ($\lambda_2 = 171$ mm), we therefore need to create two resonant paths on the antenna: one is about $l_1 = 81.5$ mm and the other is approximately $l_2 = 42.7$ mm. After careful consideration, we decide to use a PIFA antenna of dimensions 40 mm × 22 mm with a slot to create two current paths which are close to l_1 and l_2, respectively, as shown in Figure 6.25. The detailed initial design is given in Figure 6.26. The PCB size is 40 mm × 100 mm (please note, for many smart phones, the typical size is increased to 70 mm × 150 mm and the space for the antenna is less than 70 mm × 10 mm).

It is almost impossible to use EZNEC to simulate such a structure. However, it is relatively straightforward to employ software such as HFSS to model it. The step-by-step procedures of using HFSS to simulate the antenna are:

1. Launch HFSS: you should see a schematic user interface as Figure 6.24.
2. Set tool options and open a new project if required.
3. Create a 3D antenna model: this is the main part of entering your designed antenna. You need to input the coordinates of every corner. The detailed dimensions are given in Figure 6.26.

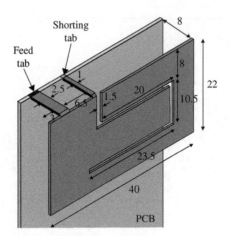

Figure 6.26 A dual-band PIFA (dimensions in mm, slot width constant at 1 mm, apart from where indicated to be 1.5 mm, thickness of antenna plate 0.5 mm)

4. Assign boundaries to all surfaces: there are many options, such as radiation, perfect E, and perfect H.
5. Assign Excitations: there are also many types, 50-ohm lumped port is selected for this antenna analysis.
6. Set up an analysis: you need to select the solution frequency, frequency sweep if required, also the accuracy requirement and number of passes. For our simulation, the frequency is set from 800 MHz to 2000 MHz to cover the desired frequency range.
7. Validate the model: to check if there are any errors in the inputted model.
8. Start the simulation: the real-time progress report is shown in the progress window.
9. View solution data: you should view the convergence first to see if the results are converged, then to create reports on various antenna parameters. If the results are not converged, more iteration is required. The requirement on the computational power (memory and time) could be excessive. This is a major and common problem for such a simulation.

For our design, the convergence plot is shown in Figure 6.27 and the results are converged after eight iterations. The simulation used 250 MB memory and 320 MB disk space. The return loss is illustrated by Figure 6.28. As expected, there are two resonant bands: one is around 910 MHz, and the other one is near 1750 MHz. Figure 6.29 shows the radiation pattern at 1755 MHz, which is close to omni-directional and has a gain/directivity of 4.358 dBi. To further assist our analysis, we can plot other important information, such as the current distribution at 1755 MHz on the plate as shown in Figure 6.30 along with the finite element meshes which were generated adaptively by the software. We can also obtain the animated current/field on the antenna. It is seen that the current is concentrated along the edge of the slot and the pattern is a standing wave at 1755 MHz (resonant). This corresponds to the shorter path identified in Figure 6.25.

For comparison, an antenna with the designed dimensions has been made as shown in Figure 6.31. Due to the special driving point of this antenna, a microstrip line of 50 ohms is created as its feed line which is connected to a SMA connector. A quarter-wave choke is

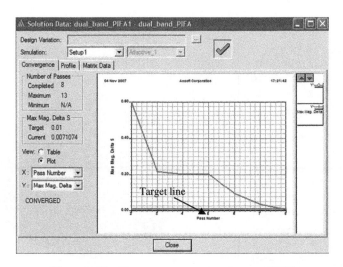

Figure 6.27   The convergence plot. Source: HFSS

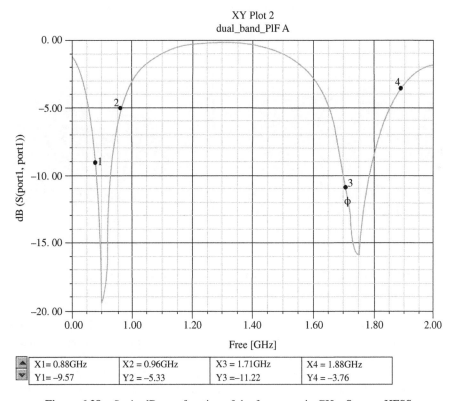

X1= 0.88GHz	X2 = 0.96GHz	X3 = 1.71GHz	X4 = 1.88GHz
Y1= −9.57	Y2 = −5.33	Y3 =−11.22	Y4 = −3.76

Figure 6.28 S_{11} in dB as a function of the frequency in GHz. Source: HFSS

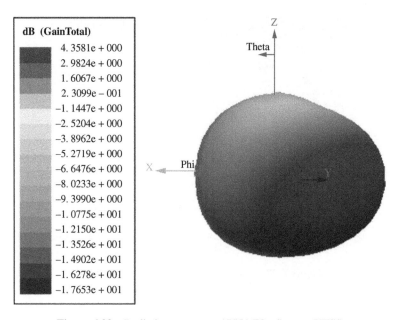

Figure 6.29   Radiation pattern at 1755 MHz. Source: HFSS

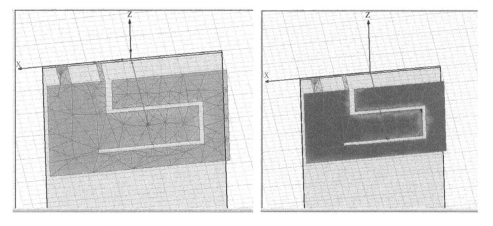

Figure 6.30   Meshes and current distribution on the antenna. Source: HFSS

employed to minimize the effects of the measurement cable. The transmission line that runs from the central connector to the antenna driving point is de-embedded from the measured results. More detailed information about this antenna can be found from [37], and general discussions on how to make accurate antenna measurements are presented in the next chapter.

The simulated and measured impedances of this configuration over the 800–1040 MHz and 1610–2090 MHz bands are shown in Figure 6.32(a) and (b), respectively. Two measured results are shown which are based on the same configuration and measured using the same techniques to give an indication of the manufacturing tolerances involved.

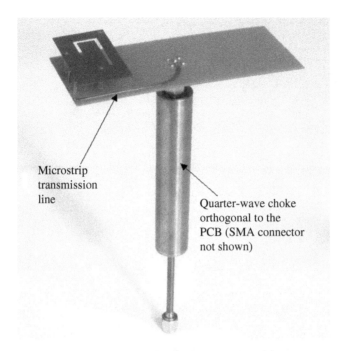

Figure 6.31   The dual-band PIFA antenna with feed line and RF choke

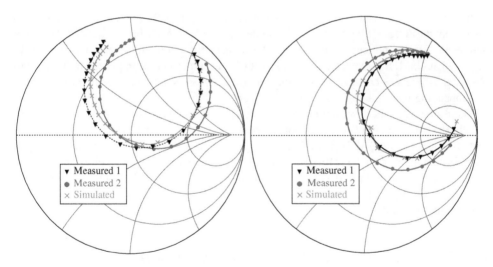

Figure 6.32   Simulated and measured impedances on Smith Chart. (a) Impedance over 800–1040 MHz. (b) Impedance over 1610–2090 MHz

The simulated and measured reflection coefficients (S_{11}) in dB are presented in Figure 6.33. For the resonant frequency, simulations and measurements tie up very well in the 800 MHz band. In the 1800 MHz band, differences in the resonant frequency of approximately 15 MHz are observed. At both frequency bands, the resonant frequency simulation accuracy

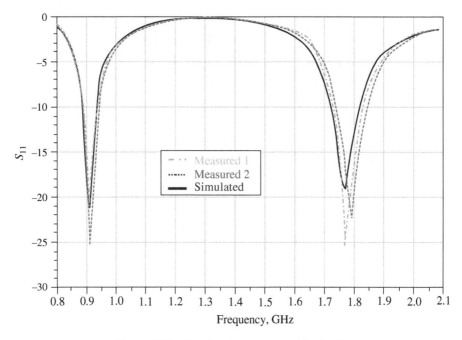

Figure 6.33   Simulated and measured S_{11} in dB

is thought to be of the same order as the manufacturing accuracy. Some phase rotation is apparent at both the 800 MHz band and, to a lesser degree, the 1800 MHz band. The 1800 MHz measurements have a noticeably higher resistance than the simulations, although this is not observed at GSM. It is expected that manufacturing accuracy can be improved in future work. Residual differences between simulations and measurements are then expected to depend on de-embedding inaccuracies and simulation errors.

There are many other useful functions which may be employed to aid our design. For example, the design for Example 6.4 does not completely meet the industrial return-loss requirement (> 6 dB) over the 1800 MHz band. Fine tuning is required: this can be achieved by using the optimization function of the software. We can set a parameter (such as the width of the antenna) as the variable to be optimized and run the computer software to find the optimum value for this parameter. This kind of advanced features can be found from the software suppliers and is not covered by this book.

6.5   Theory of Characteristic Modes for Antenna Design

The above EM simulation software is an excellent tool to aid antenna design. They can estimate the performance of a designed antenna accurately and even optimize the antenna performance, but they cannot provide original ideas and perform systematic design, and they also cannot reveal the physical insight on how an antenna works, although the field and current distributions on the antenna can be obtained to help us to analyze the antenna. From antenna theory, such as the antenna equation in Chapter 4, we know that a time-varying current can radiate an

EM wave which does not tell us how an antenna should be design. Thus, antenna design has mostly been based on previous knowledge and experience; try-and-error has played an important role in the design process which is obviously not very scientific and systematic. A better antenna design process is required.

In this section, we are going to introduce the Theory of Characteristic Modes (TCM) and then demonstrate how it can be used to help antenna designs.

The TCM was first introduced in 1960s [38] and then reformulated, developed, and applied to electromagnetic applications successfully by Harrington and Mautz in 1970s [39–44]. However, it did not receive significant attention in antenna engineering until about 2000s when the TCM was employed to aid antenna design and analysis [45–51]. It has been demonstrated its unique and useful features and become an increasingly powerful tool for antenna design and analysis – It can provide an excellent physical insight and a better understanding of an antenna design; hence, it can help antenna researchers and engineers to optimize the design with confidence. The use of TCM for antenna design is often called *characteristic mode analysis* (CMA).

As detailed in references [38–42], the main idea of the TCM is that any conducting object (such as an antenna) has many (actually an infinite number of) characteristic/eigen modes, all the characteristic modes have orthogonal far field patterns and orthogonal current distributions on the object but not all these modes may be generated for a given design. The total induced currents on the object under a given excitation can be decomposed into these orthogonal characteristic modes weighted by their corresponding complex modal weighting coefficients. Normally, the total current on an object can be approximated by a few significant modes with large weighting coefficients in the band of interest. Therefore, suppressing/exciting a selected mode in general means the reduction/generation of contribution of the mode to the total current by minimizing/maximizing its modal weighting coefficient – this is very similar to the case of a conducting cavity where there are many possible modes for a given cavity but only some cavity modes may be generated which depends on the operational frequency, the excitation and the cavity dimensions. It may be considered as the generalization of the closed cavity modes in free space. In practice, how to implement this idea is not always easy.

In this section, we are going to introduce the TCM and apply it to a number of antenna designs. It will demonstrate that characteristic modes are real current modes that can be computed numerically for conducting bodies of arbitrary shape (Note: the TCM can also be applied to non-conducting bodies). Since characteristic modes form a set of orthogonal functions, they can be used to expand the total current on the surface of the body. The resonance frequency of modes, as well as their radiating behavior, can be determined from the information provided by the eigenvalues associated with the characteristic modes. Moreover, by studying the current distribution of modes, an optimum feeding arrangement can be found in order to obtain the desired radiating behavior and optimized antenna performance.

6.5.1   Mathematical Formulation of Characteristic Modes

Now consider the problem of a conducting body defined by surface S, in an impressed electric field E^i. Since the boundary condition is that the total tangential electric field must be zero on surface S, we have

$$\left[L(J) + E^i \right]_{\tan} = 0 \tag{6.41}$$

where $L(J)$ is the electric field at a point in space produced by the current J on the surface S and can be calculated using the antenna equation. A modal solution for the current J on the conducting body can be obtained using the eigencurrents (also known as characteristic currents) as both the expansion and testing functions in the method of moments as discussed in Section 6.2.1. Following this procedure, we assume J to be a linear superposition of the mode currents:

$$J = \sum_n a_n J_n$$

$$J = \sum_n a_n J_n \tag{6.42}$$

where a_n are weighting coefficients to be determined and J_n are eigencurrents (or eigenfunctions) on the conducting body and orthogonal to each other. These weighting coefficients indicate the significance of the corresponding eigencurrents in the total current on the conducting body. 0 means that the corresponding eigencurrent is not generated on the body. Substituting J in Equation (6.41) with (6.42) to obtain

$$\left[\sum_n a_n L(J_n) + E^i \right]_{\tan} = 0 \tag{6.43}$$

We use the method of moment procedure and take the inner product of (6.43) with each complex conjugate of J_m which results in a set of equations:

$$\left[\sum_n a_n \langle J_m{}^*, L(J_n) \rangle + \langle J_m{}^*, E^i \rangle \right]_{\tan} = 0 \tag{6.44}$$

where $*$ means the complex conjugate, while $< >$ is the inner product. Since the operator L in (6.41) has the dimension of impedance, it can be written as

$$[L(J)]_{\tan} = Z(J) = R(J) + jX(J) \tag{6.45}$$

The real and imaginary parts (R and X) of the impedance operator are linked to the eigenvalue λ_n as [39]

$$X(J_n) = \lambda_n R(J_n) \text{ or } \lambda_n = X(J_n)/R(J_n) \tag{6.46}$$

Both X and R are real symmetric operators; hence, eigenvalues and eigencurrents must be real. The eigenvalue λ_n is a function of frequency. Obviously, when λ_n is zero, it means resonance and the imaginary part is zero. When λ_n is negative or positive, it means capacitive (such as a small dipole) and inductive (such as a small loop), respectively. The impedance operator, eigencurrents, and eigenvalues satisfy the following orthogonality conditions:

$$\langle J_m{}^*, R(J_n) \rangle = \delta_{mn}$$

$$\langle \boldsymbol{J}_m{}^*, X(\boldsymbol{J}_n) \rangle = \lambda_n \delta_{mn} \tag{6.47}$$

where δ_{mn} is the Kronecker delta ($\delta_{mn} = 0$ when $m \neq n$; $\delta_{mn} = 0$ when $m = n$). Since the currents are real and the complex conjugate of a current is the same as the current. The orthogonality conditions do not simply mean that the eigencurrents are orthogonal physically, both physically orthogonal (e.g. one goes to x-direction and the other goes to y-direction) and physically non-orthogonal currents can meet the orthogonal conditions of (6.47) as we will see later in design examples.

It is interesting to note that the complex power density produced by current $\boldsymbol{J}_n$ can be obtained as

$$P = \langle \boldsymbol{J}_m{}^*, Z(\boldsymbol{J}_n) \rangle = (1 + j\lambda_n)\delta_{mn} \tag{6.48}$$

which is a complex function with radiated/loss power (the real part) and storage power (the imaginary part). Applying the impedance operator definition and properties in Equation (6.44) to yield

$$(1 + j\lambda_n)a_n = -\langle \boldsymbol{J}_m{}^*, \boldsymbol{E}^i{}_{\tan} \rangle$$

That is

$$a_n = \frac{-\langle \boldsymbol{J}_m{}^*, \boldsymbol{E}^i{}_{\tan} \rangle}{(1 + j\lambda_n)} \tag{6.49}$$

Using (6.42) and (6.49), one can represent the surface current using the eigenvalues and eigen-currents as

$$\boldsymbol{J} = \sum_n \frac{-\langle \boldsymbol{J}_m{}^*, \boldsymbol{E}^i{}_{\tan} \rangle}{(1 + j\lambda_n)} \boldsymbol{J}_n \tag{6.50}$$

6.5.2 Physical Interpretation of Characteristic Modes

Equation (6.50) shows how the total current on a conducting body is formed by the summation of the characteristic currents (eigencurrents). The term of the inner product in this expression is called the modal-excitation coefficient since it is determined by coupling between the excitation and related eigencurrent. This value indicates if a particular mode is excited by the incident field (or the antenna feed). The eigenvalue given by (6.46) is of utmost importance because its magnitude gives information about how well the associated mode radiates.

From Equation (6.48), we can see that the reactive power is proportional to the magnitude of the eigenvalue. When a mode is at resonance, its associated eigenvalue is zero (and the modal coefficient in (6.50) is maximum), the normalized power density is 1 and has no imaginary part, that is, all energy is radiated and no reactive (electric or magnetic) energy in this case. In addition, the sign of the eigenvalue determines whether the mode contributes to storing magnetic energy ($\lambda_n > 0$, inductive) or electric energy ($\lambda_n < 0$, capacitive) which is useful to know in some cases.

For antenna design, we can use the TCM to find the eigenvalue variation with frequency, as it provides information about the resonance and the radiating properties of the current modes.

However, sometime, the variation of the eigenvalue is not significant to be observed as we will see in the coming example, thus, in practice, other alternative representations of the eigenvalues are preferred.

One is called the *modal significance* (MS) which is defined as

$$MS_n = \left| \frac{1}{1 + j\lambda_n} \right| \tag{6.51}$$

It represents the normalized amplitude of current mode n in the range of [0, 1]. When a mode resonates, its MS = 1; when a mode is not generated, its MS = 0. It is important to note that MS only depends on the shape and size of the conducting object and it does not account for the external excitation. The significance of each mode can be identified by the maximum value of one in the modal-significance curves (vs frequency). This means that the larger its maximum value is, the more effective the associated mode contributes to radiation. The radiating bandwidth of a mode can then be established according to the width of its MS curve near the maximum point. The drawback of this definition is that it does not reveal the capacitive or inductive nature of a specific mode.

Another even-more-intuitive representation of the eigenvalues is based on the use of characteristic angles. The *characteristic angle* (CA), α_n, is defined as

$$\alpha_n = 180° - tan^{-1}(\lambda_n) \tag{6.52}$$

This angle is between 90° and 270°. From a physical point of view, the CA models the phase difference between a characteristic current J_n and the associated characteristic field. When $\lambda_n = 0$, the mode resonates and its CA is 180°. Therefore, when the CA is close to 180°, the mode is a good radiator. When the CA is near 90° or 270°, the mode mainly stores energy. When the angle is between 90° and 180°, this is an inductive case, while the angle is between 180° and 270°, this is a capacitive case. Thus, the radiating bandwidth of a mode can be obtained from the slope at 180° of the curve described by its CA.

6.5.3   *Examples of Using TCM for Antenna Designs*

There have been many published examples of using the TCM to aid antenna design. Here, we are going to use a microstrip patch antenna as an example to demonstrate the usefulness of the TCM. A microstrip antenna was designed in Example 5.7 where the size of the conducting radiator/patch is 40.49 mm × 48.40 mm to work around 2.45 GHz for Bluetooth applications. We will start with this conducting radiator in free space to identify its characteristic modes and then add a substrate and a ground plane to make it a PCB patch antenna.

The computer simulation software CST is selected for this study. The calculated eigenvalues, MS, and characteristic angles of this conducting radiator in free space are obtained and given in Figure 6.34 as functions of frequency. It can be seen that:

1. There are four resonant modes below 8 GHz: at 3.1, 4.2, 5.1, and 7.1 GHz (their corresponding current distributions are provided in Figure 6.35), but none of these modes resonates at 2.45 GHz because the substrate and ground plane are not included when compared with Example 5.7. Another two curves are also shown in Figure 6.34, but their resonant frequencies are above 8 GHz (they are higher modes).

2. It is relatively hard to tell the exact resonant frequencies from the eigenvalue and MS curves because the bandwidths are large for these four modes (i.e. the value around resonant frequency does not change much over a large frequency band). However, it is easier to identify the resonant frequencies from the curves of CA in Figure 6.34(c) (Note that eigenvalue and MS curves are good enough for many cases in practice). The first four modes are of capacitive energy, while the two higher modes are of inductive energy (this means their eigencurrents form loops) when the frequency is below its resonant frequency – this feature is observed in Figure 6.34(a) and (c), but not in Figure 6.34(b).

3. The lowest mode (mode 1) resonates around 3.1 GHz, and its current distribution is along the length (the longer sides of the radiator) as shown in Figure 6.35(a). The eigenwavelength of mode 1 is about 96.77 mm; thus, the length of the radiator (48.40 mm) is half of the lowest (fundamental) eigenmode wavelength. This is equivalent to the half-wavelength dipole case, but there is no feed at the middle.

4. Mode 2 resonates around 4.2 GHz and the current goes along all four sides of the radiator as displayed in Figure 6.35(b). The currents at opposite sides are with the same amplitude but opposite directions (thus, the radiation pattern will have split beams). This is not a desired mode which has been widely used in practice.

5. Mode 3 current is illustrated in Figure 6.35(c) which is similar to the current of mode 1 but rotated by 90°. The current path is shorter; hence, the eigenfrequency is higher than that of mode 1.

6. The eigencurrent of mode 4 is presented in Figure 6.35(d) and can be considered as the second harmonic of mode 1. The current distribution becomes one wavelength along the length of the rectangular radiator: the current direction is changed after a half wavelength. Note that the resonant frequency in this case is about 7.1 GHz, not $2 \times 3.1 = 6.2$ GHz. This is because the width of the radiator (40.49 mm) is comparable with the wavelength (42.25 mm) which has made an impact on the resonant frequency. If this width were much smaller (say 5 mm), the eigenfrequency would be about 6.2 GHz which has been validated by simulation.

7. It can be verified that all these eigencurrents satisfy the orthogonality conditions shown in (6.47) by taking the inner product of any two of the eigencurrents. We can see that some eigencurrents flow in the same direction (e.g. currents of modes 1 and 4), but their phase and values may be different so as to satisfy the orthogonality conditions.

From these results, we can clearly see which mode may play a significant role in radiation at a given frequency. For example, at 3 GHz, the dominant mode is mode 1, while at 5 GHz, the dominant mode is mode 3. However, if the excited current is not vertical at 5 GHz, mode 3 will not be generated and the dominate mode may be mode 1 since its current is horizontal and its MS is very close to 1 (its CA is very close to 180°).

Next, a dielectric substrate and a conducting ground plane are added to next to the rectangular conducting radiator (48.40 mm × 40.49 mm) forming a PCB patch antenna as shown in Figure 6.36. The ground plane size is 120 mm × 100 mm with a substrate of a thickness 1.588 mm and relative permittivity 2.2. The antenna feeding is not included in this stage. We are going to use the knowledge gained from above CMA to aid the antenna design. We will soon find out if this antenna is suitable for 2.45 GHz as demonstrated in Example 5.4, and if it is good for other frequencies as well.

When the TCM is employed to analyze this PCB antenna without a feed, a new set of results can be obtained. The simulated CA as a function of frequency is shown in Figure 6.37 where

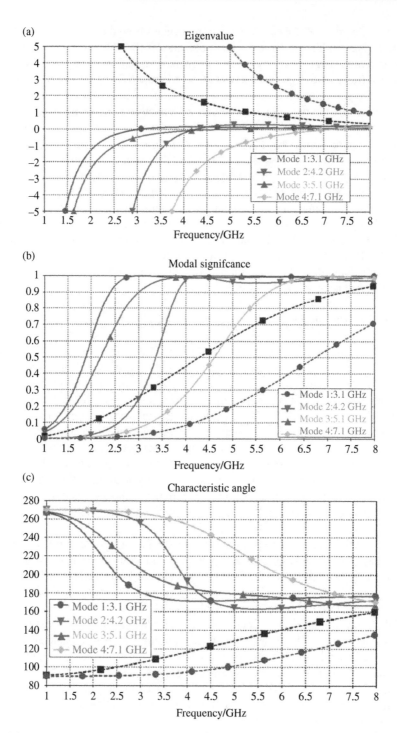

Figure 6.34 Simulated results of a rectangular conducting radiator in free space: (a) Eigenvalue; (b) Modal significance; (c) Characteristic angle

(a)   (b)   (c)   (d)

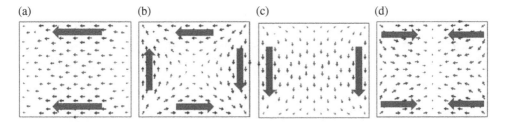

Figure 6.35   The first four resonant modes of a rectangular conducting radiator. (a) Mode 1. (b) Mode 2. (c) Mode 3. (d) Mode 4

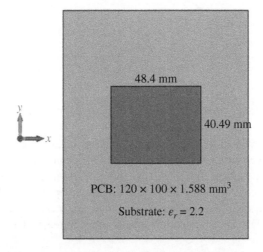

48.4 mm

40.49 mm

PCB: $120 \times 100 \times 1.588$ mm^3

Substrate: $\varepsilon_r = 2.2$

Figure 6.36   A PCB patch antenna with the radiator dimensions 48.40 mm × 40.49 mm

only the first six modes are presented (there are many higher modes). As expected, there are more eigenmodes below 8 GHz than the previous case (the radiator in free space case) since the size of the object is larger. The response is complicated since it is from the whole structure: the ground plane, the substrate, and the patch radiator. Take mode 1 as an example (Note: this mode 1 is for the PCB patch antenna, not mode 1 for the conducting radiator discussed earlier although the current distribution for both modes is the same as that in Figure 6.35(a)), the first (lowest) eigenfrequency is around 1.2 GHz (250 mm in wavelength) which is about twice the length of the ground plane. The CA curve changes significantly around 2.3 GHz which is due to the effect of the patch radiator. The abrupt changes in the CA around 2.0, 2.3, 3.1, and 3.9 GHz for the first few modes in this plot are closely associated with the resonant modes of the antenna as we are going to see next. To become a complete antenna, a feed is required and where to place the feed is a critical decision since it can determine which modes will be generated and how the antenna will perform. Once a feed is added to the structure, the characteristic modes for the antenna should not be changed much. A common approach in practice is to examine the reflection coefficient S_{11} and link the troughs (resonant frequencies) to the characteristic modes.

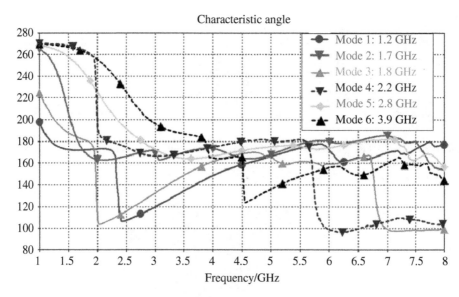

Figure 6.37   The characteristic angle as a function of frequency for the PCB patch antenna without a feed

There are two type of feeding sources: a voltage source and a current source. A voltage source can bridge the gap between two conductors of different potentials. It can be used most effectively at where the voltage difference is the largest. A current source is to inject current (through coupling) to a conductor and can be applied most effectively at where the current is the largest. Thus, a voltage source is employed for a dipole, and a current source is used for a conducting rod or loop. Further examination on the eigencurrents of the first four modes in Figure 6.35 reveals that:

1. At the middle point of the length of the patch, the currents are the strongest for modes 1 and 2, but the smallest for modes 3 and 4.
2. At the middle point of the width of the patch, the currents are the strongest for modes 2 and 3, but the smallest for modes 1 and 4.
3. There are currents near the corners for all four modes. Thus, if a feed is placed near a corner of the patch antenna, all four modes could be generated. Thus, a corner is selected for our feeding point.

Figure 6.38(a) shows the reflection coefficient when the patch antenna is fed at a corner using a voltage source, while Figure 6.38(b) gives the reflection coefficient when the antenna is fed at the same corner through a microstrip line (also a voltage source, but the result is about the same if it is fed by a current source). In order to improve the impedance matching at the desired 2.4–2.45 GHz, a quart-wave transformer as in Example 5.7 has been used. In both cases, the first four resonant frequencies are almost the same (some slight shift is due to the use of impedance matching section) – around 2.0, 2.4, 3.2, and 4.0 GHz. The current distributions on the antenna at these four frequencies are plotted and checked against the eigenmodes illustrated in Figure 6.35, and it is found that they correspond nicely to modes 1, 3, 2, and 4, respectively. The 3D radiation patterns at these frequencies are presented in Figure 6.39. It is

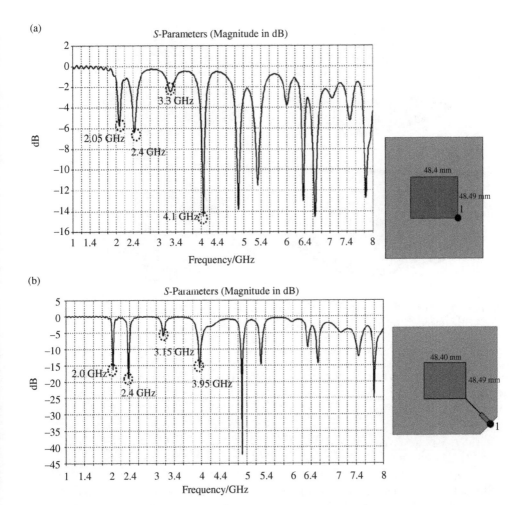

Figure 6.38 The reflection coefficient of the patch antenna with a feed from a corner. (a) The patch antenna excited by a voltage source. (b) The patch antenna excited by a voltage source with an impedance matching section

interesting to note that the resonant frequencies have been changed significantly with the presence of the substrate and ground plane (PCB). In particular, the resonant frequency of mode 3 is now lower than that of mode 2. This could be explained using the cavity mode theory for a patch antenna. The resonant frequency of TM_{mn} (or TM_{mn0} viewed as a cavity) mode for such a rectangular patch antenna can be obtained from [8]:

$$f_{mn} = \frac{c}{2\sqrt{\varepsilon_{reff}}} \sqrt{\left(\frac{m}{L + 2\Delta L}\right)^2 + \left(\frac{n}{W + 2\Delta W}\right)^2} \qquad (6.53)$$

where L and W are the length and width of the rectangular patch, respectively; ε_{reff} is the relative permittivity given by (5.92); ΔL and ΔW are the enlargement on L and W due to the fringing

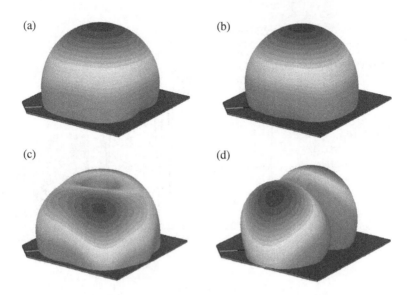

Figure 6.39 3D radiation patterns of the patch antenna at different frequencies. (a) 2.0 GHz (mode 1). (b) 2.4 GHz (mode 3). (c) 3.2 GHz (mode 2). (d) 4.0 GHz (mode 4)

effects as discussed in Section 5.2.5, respectively. The fundamental mode is TM_{10}, and its resonant frequency is the lowest and found to be:

$$f_{10} = \frac{c}{2\sqrt{\varepsilon_{reff}}} \sqrt{\left(\frac{1}{L + 2\Delta L}\right)^2} = \frac{300,000,000}{2\sqrt{2.11}} \times \frac{1000}{48.04 + 0.9} = 2.09 \text{ GHz}$$

Similarly, the resonant frequencies for TM_{01}, TM_{11}, and TM_{20} modes are found to be 2.49, 3.25, and 4.18 GHz, respectively. These resonant frequencies are in good agreement with the four resonant frequencies identified in Figure 6.38 and also the abrupt change frequencies of CA in Figure 6.37. Thus, this antenna can work well at a number of frequencies in addition to the desired 2.4–2.45 GHz – this was not obvious before the analysis of the characteristic modes.

To study this patch antenna further using the TCM, a current source (slot feed) is placed at the middle point of the length and modes 1 and 2 should be generated. The simulated reflection coefficient S_{11} is plotted in Figure 6.40(a), as expected mode 1 (2.0 GHz) and mode 2 (3.2 GHz) are indeed generated but modes 3 and 4 are not generated. Similarly, when a current source is placed in the middle point of the width of the patch antenna, mode 2 (3.2 GHz) and mode 3 (2.4 GHz) are generated as demonstrated by S_{22} in this figure. But, mode 4 (4.0 GHz) is indeed not generated as expected.

Now, a voltage source is employed to excite desired eigenmodes. When the source is placed at the middle point of one side of the patch, the current will flow symmetrically to the left and the right with the same amplitude but opposite directions. Thus, when the voltage source is put at the middle point of the width of the patch, modes 1 and 4 can be generated as shown by S_{11} in Figure 6.40(b), but not modes 2 and 3. When the voltage source is located at the middle point of

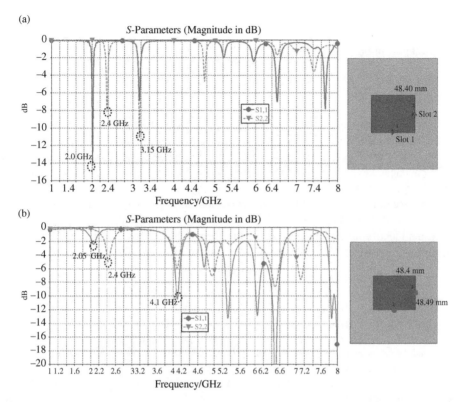

Figure 6.40 The reflection coefficients of the patch antenna with different feeds. (a) Current feeding sources at two different sides. (b) Voltage feeding sources at two different sides

the length of the patch, modes 1 and 2 cannot be generated but modes 3 and 4 are produced as shown by S_{22} in Figure 6.40(b).

It should be pointed out that at some resonant frequencies, the feeder may already be well matched with the antenna (e.g. at 2.0 GHz for S_{11} in Figure 6.40(a)), but at some other resonant frequencies such as 2.4 GHz for S_{22} in Figure 6.40(b), they are not well matched, an impedance matching network is therefore needed as demonstrated in Figure 6.38 and Example 5.7. If no resonant modes are generated at the desired frequency (such as 1.5 GHz), even a matching network is introduced, the antenna may not perform well (narrow band and low efficiency). Thus, a well-designed antenna should have an eigenmode at the desired frequency, and the mode is excited by its feeding source. This is a golden rule of thumb for antenna design using the TCM.

Through these three examples, we have now seen how to use the TCM to aid a rectangular patch antenna design, including how to identify the eigenmodes and use the appropriate excitation. The focus has been on the first four eigenmodes. It has been clearly demonstrated that:

1. These four modes can be fully generated by selecting the feeding point around a corner of the antenna (using either a voltage or a current source).
2. If a current source is chosen, modes 1 and 2 can be generated when it is placed at the middle point of the length; modes 3 and 2 can be produced when it is located at the middle point of the width. In both cases, mode 4 cannot be excited.

3. If a voltage source is chosen, modes 1 and 4 can be generated when it is located at the middle point of the width; modes 3 and 4 can be produced when it is positioned at the middle point of the length. In both cases, mode 2 cannot be excited.

6.6   Summary

In this chapter, we have introduced the CM techniques for modern antenna designs. The focus has been placed on the MoM which gave us insight into how a complex field equation could be solved numerically and how to obtain the desired antenna parameters. A number of commercial software packages have been introduced. Two of them have been employed as examples to illustrate how they could be utilized for antenna design and analysis.

Another very important part of this chapter is about the TCM which is an extremely useful and powerful tool on finding the eigen resonant frequencies and currents of a given antenna structure and identifying the right excitation source and feeding position. Examples have been provided to demonstrate the usefulness of this powerful tool for antenna designs.

References

1. G. J. Burke and A. J. Poggio, "Numerical electromagnetics code (NEC) – method of moments", Technical Document 11, Naval Ocean Systems Centre, San Diego, USA, January 1981.
2. J. Julian, J. M. Logan and J. W. Rockway, "MININEC: a mini-numerical electromagnetics code", Technical Document 516, Naval Ocean Systems Centre, San Diego, USA, 1982.
3. J. Rockway, J. Logan, D. Tan and S. Li, *The MININEC System: Microcomputer Analysis of Wire Antennas*, Artech House, 1988.
4. B. Friedman, *Principles and Techniques for Applied Mathematics*, John Wiley & Sons, New York, 1956.
5. R. F. Harrington, "Matrix methods for field problems", *Proc. IEEE*, 55(2), 136–149, 1967.
6. R. F. Harrington, *Field Computation by Moment Methods*, Macmillan, New York, 1968.
7. MATLAB: http://www.matlab.com
8. C. A. Balanis, *Antenna Theory: Analysis and Design*, 2nd edition, John Wiley & Sons Inc., 1997.
9. W. L. Stutzman and G. A. Thiele, *Antenna Theory and Design*, 2nd edition, John Wiley and Sons, 1998.
10. J. H. Richmond, "Digital computer solutions for rigrous eqations for scattering problems", *Proc. IEEE*, 53, 796–804, 1965.
11. G. A. Thiele, "Wire antennas" in *Computer Techniques for Electromagnetics* by R. Mittra (Ed.), Pergamon Press Ltd, New York, 1973.
12. A. Hrennikoff, "Solution of problems of elasticity by the frame-work method", *ASME J. Appl. Mech.*, 8, A619–A715, 1941.
13. G. Strang and G. J. Fix, *An Analysis of the Finite Element Method*, Prentice-Hall, Englewood Cliffs, 1973.
14. J. L. Volakis, A. Chatterjee and L. C. Kempel, *Finite Element Method Electromagnetics: Antennas, Microwave Circuits, and Scattering Applications*, John Wiley & Sons, 1998, p. 368.
15. O. C. Zienkiewicz and R. L. Taylor, *Finite Element Method (5th Edition) Volume 1 – The Basis*, Elsevier, 2000
16. K. Yee, "Numerical solution of initial boundary value problems involving Maxwell's equations in isotropic media," *IEEE Trans. Antennas Propagat.*, 14, 302–307, 1966.
17. G. Mur, "Absorbing boundary conditions for the finite-difference approximation of the time-domain electromagnetic field equations," *IEEE Trans. Electromagn. Compatibility*, 23, 377–382, 1981.
18. A. Taflove, *Computational Electrodynamics: The Finite-Difference Time-Domain Method*, Artech House, Norwood, MA, 1995.

19. A. Taflove and S. C. Hagness, *Computational Electrodynamics: The Finite-Difference Time-Domain Method*, 3rd edition, Artech House, 2005.

20. P. B. Johns and R. L. Beule, "Numerical solution of 2-dimensional scattering problems using a transmission-line method," *Proc. IEE*, 118(9), 1203–1208, 1971.

21. W. J. R. Hoefer, "The transmission line matrix (TLM) method," in *Numerical Techniques for Microwave and Millimeter Wave Passive Structures* by T. Itoh, (Ed.), 1990.

22. P. B. Johns, "A symmetrical condensed node for the TLM-method," *IEEE Trans. Microwave Theory Tech.*, 35(4), 370–311, 1987.

23. C. Christopoulos, *The Transmission-Line Modelling Method in Electromagnetics*, Morgan and Claypool Publishers, 2006.

24. M. Krumpholz, C. Huber and P. Russer, "A field theoretical comparison of FDTD and TLM," *IEEE Trans. Microwave Theory Tech.*, 43(8), 1935–1950, 1995.

25. TLM. https://en.wikipedia.org/wiki/Transmission-line_matrix_method

26. P. K. Banerjee, *The Boundary Element Methods in Engineering*, McGraw-Hill College, 1994.

27. P. K. Pathak, "Techniques for high-frequency problems," Chapter 5 in *Antenna Handbook* by Y. T. Lo and S. W. Lee, Van Nostrand Reinhold Company, New York, 1988.

28. Eugene Hecht, *Optics*, 4th edition, Pearson Education, 2001.

29. A. R. Lopez, "The geometrical theory of diffraction applied to antenna and impedance calculation", *IEEE Trans. Antennas Propagat.*, 14, 40–45, 1966.

30. G. L. James, *Geometrical Theory of Diffraction for Electromagnetic Waves*, 1st edition in 1976 and 3rd edition, Peter Peregrinus Ltd (IEE now IET), 1986.

31. TICRA: GRASP, http://www.ticra.com/

32. W. L. Stutzman and G. A. Thiele, *Antenna Theory and Design*, 2nd edition, John Wiley & Sons, 1998.

33. G. A. Thiele, "Hybrid methods in antenna analysis", *IEEE Proc.*, 80(1), 66–78, 1992.

34. EZNEC: http://www.eznec.com/

35. 4NEC2: http://www.qsl.net/4nec2/

36. HFSS: https://www.ansys.com/products/electronics/ansys-hfss

37. K. Boyle, M. Udink, A. de Graauw and L. P. Ligthart, "A dual-fed, self-diplexing PIFA and RF front-end", *IEEE Trans. Antennas Propag.*, 55, 373–382, 2007.

38. R. J. Garbacz, "Modal expansions for resonance scattering phenomena," *Proc. IEEE*, 53(8), 856–864, 1965.

39. R. F. Harrington and J. R. Mautz, "Theory of characteristic modes for conducting bodies," *IEEE Trans. Antennas Propag.*, 19(5), 622–628, 1971.

40. R. F. Harrington and J. R. Mautz, "Computation of characteristic modes for conducting bodies," *IEEE Trans. Antennas Propag.*, 19 (5), 629–639, 1971.

41. R. F. Harrington and J. R. Mautz, "Control of radar scattering by reactive loading," *IEEE Trans. Antennas Propag.*, 20 (4), 446–454, 1972.

42. J. R. Mautz and R. F. Harrington, "Modal analysis of loaded N-port scatters," *IEEE Trans. Antennas Propag.*, 21 (2), 188–199, 1973.

43. R. F. Harrington and J. R. Mautz, "Optimization of radar cross section of N-port loaded scatterers," *IEEE Trans. Antennas Propag.*, 22 (5), 697–701, 1974.

44. R. F. Harrington and J. R. Mautz, "Pattern synthesis for loaded N-port scatterers," *IEEE Trans. Antennas Propag.*, 22 (2), 184–190, 1974.

45. R. J. Garbacz and D. M. Pozar, "Antenna shape synthesis using characteristic modes," *IEEE Trans. Antennas Propag.*, 30(3), 340–350, 1982.

46. M. C. Fabres, E. A. Daviu, A. V. Nogueiram and M. F. Bataller, "The theory of characteristic modes revisited: a contribution to the design of antennas for modern applications," *IEEE Antennas Propag. Mag.*, 49(5), 52–68, 2007.

47. M. C. Fabres, Systematic design of antennas using the theory of characteristic modes, PhD Dissertation, Universidad Politecnia de Valencia, 2007.

48. Y. Chen and C. F. Wang, "Synthesis of reactively controlled antenna arrays using characteristic modes and DE algorithm," *IEEE Antennas Wireless Propag. Lett.*, 11, 385–388, 2012.

49. E. Safin and D. Manteuffel, "Reconstruction of the characteristic modes on an antenna based on the radiated far field," *IEEE Trans. Antennas Propag.*, 61(6), 2964–2971, 2013.

50. E. Safin and D. Manteuffel, "Manipulation of characteristic wave modes by impedance loading," *IEEE Trans. Antenna Propag.*, 63(4), 1756–1764, 2015.

51. Y. Chen and C. F. Wang, *Characteristics Modes Theory and Applications in Antenna Engineering*, Wiley, Hoboken, NJ, USA, 2015.

Problems

Q6.1. There are many antenna design software packages available free of charge (for evaluation version). Using the information provided in this chapter as a guide to download at least one software package and get familiar with the software which will be used for the following exercises.

Q6.2. Using your newly downloaded software, to redo Examples 6.2–6.4, and identify if there are any differences in the results.

Q6.3. Using the new software to design a loop antenna which operates at 2.45 GHz and has an omni-directional radiation pattern.
  a. If the loop is electrically small, find the radiation pattern and input impedance, then compare with the theoretical results.
  b. If the loop is about one wavelength and the desired VSWR is smaller than 2 for a 50-ohm feed line, obtain a good design and justify whether a matching circuit is required.
  c. If the loop is made of a conducting wire, discuss the effects of the diameter of the wire on the loop input impedance and gain.

Q6.4. With the aid of software, design a circularly polarized helix antenna of an end-fire radiation pattern with a directivity of 13 dBi and find out its input impedance, *HPBW, AR,* and radiation pattern. Then compare these results with that of Example 5.2.

Q6.5. Design a patch antenna specified by Example 5.7 with the aid of computer software and then compare your results with that of Example 5.7.

Q6.6. The initial design of a 6-element Yagi-Uda antenna is given in Table 5.4. With the aid of antenna software, find its input impedance, radiation pattern, and gain.

Q6.7. With the aid of software, design a circularly polarized antenna for GPS application. The frequency should cover 1559–1610 MHz (L1) band, VSWR < 2 for a 50-ohm feed line, and the radiation pattern should be unidirectional (towards the sky).

Q6.8. Explain the theory of characteristic modes. For a wire dipole antenna, find the first two characteristic modes without using a software simulation package.

Q6.9. Find the first two characteristic modes of a wire loop antenna using the TCM, and then validate your answer by using a software simulation tool.

Q6.10. Find the first two characteristic modes of the top radiator of the PIFA antenna in Figure 6.26 with the aid of a software simulation tool, and then plot the current distributions of these two modes.

7

Antenna Materials, Fabrication, and Measurements

Once an antenna is designed, it should be made and tested. The construction of an antenna could be a complex and critical process since the antenna must meet the electrical and mechanical specifications as well as some other requirements (such as costs). In this chapter, we are going to see what conventional and new materials and fabrication processes are used to make antennas, what antenna measurements should be conducted and how. Good understanding of these topics is an essential part of becoming a good antenna engineer and researcher.

7.1 Materials for Antennas

From a construction point of view, antennas are normally manufactured using conducting materials (e.g. dipoles, loops and horns), low-loss dielectric materials (e.g. dielectric resonant antennas), or a combination of both (e.g. patch antennas). The selection of the right material and a robust construction are important elements of making a successful antenna. The antenna evolution has been heavily influenced by the materials industry.

Traditionally, antennas are relatively large and made of metals. As operational frequencies have increased, many antennas are now relatively small and may be manufactured in a very large quantities, especially for consumer electronic products such as smart phones and laptops. New materials and fabrication processes have therefore been introduced to meet evolving requirements. Key innovations in such as composite materials and novel selective metallization processes have influenced antenna designs. These innovations allow cost-effective realization of three-dimensional (3D) antennas that have good electrical performance, are mechanically robust, and can withstand harsh environmental conditions. Thus, when an antenna is designed, in addition to electrical properties (such as the conductivity, permittivity, permeability, and loss tangent), we have also to take other things into account, which include at least the following aspects:

Antennas: From Theory to Practice, Second Edition. Yi Huang.
© 2022 John Wiley & Sons Ltd. Published 2022 by John Wiley & Sons Ltd.
Companion website: www.wiley.com/go/huang_antennas2e

- **Mechanical considerations**: the material should be strong and robust enough to keep its desired shape under desired working conditions.
- **Environmental considerations**: the material should be resilient to environmental changes and maintain its desired performance – for example, it should not oxidize or erode in order to maintain its good conductivity. The effects of temperature, humidity, rain, and snow should also be taken into account. Sometimes, a layer of paint is applied to meet these requirements. The paint may affect the antenna performance, so a trade-off is necessary.
- **Cost consideration**: as for any other product, the cost of antenna manufacture should be kept to a minimum.
- **Weight consideration**: it should normally be as light as possible for most applications.

7.1.1 Conducting Materials

Conducting materials are the most widely used materials to build antennas. In principle, all conductive materials can be used; however, the antenna efficiency is closely linked to the conductivity of the material: the higher the conductivity, the higher the efficiency. Thus, in practice, we only choose materials of very good conductivity. As a result, copper, brass (an alloy of copper and zinc), bronze (an alloy of copper and tin), and aluminum are widely used to make antennas. The conductivities of some common metals used in antenna construction are given in Table 1.3 in Chapter 1.

To construct prototype antennas (especially those of special shapes) at home or in a lab, metallic tapes or even aluminum foils (from a supermarket) could be used since they are conductive, very flexible, and cheap. However, it is necessary to ensure that the metal thickness is much greater than the skin depth of the material at the desired frequency. The discussion on skin depth was provided in Section 3.6.1.

One of the most overlooked antenna construction considerations is the *galvanic corrosion* that usually occurs when two dissimilar metals are brought into physical contact (mated) during the assembly process and exposed to the weather. There can be serious corrosion at their respective contact points. There is a direct relationship between various types of dissimilar metals when they are mated. Some dissimilar metals (such as copper and brass) when mated cause very little corrosion. There are other metals, however, that react most harshly when matched. Zinc and brass, for example, will cause corrosion with the zinc metal quickly breaking down. There are several ways of reducing and preventing this form of corrosion. One of them is to electrically insulate the two metals from each other. This can be done using plastic or another insulator to separate the metals. Use of absorbent washers that may retain fluid is often counterproductive. Sometimes they have to be in electrical contact, the best solution is to use the same metal throughout the construction. If this is not possible, the next course of action would be to assemble materials that have a close relationship on a galvanic metals table as shown in Table 7.1.

Another related issue is PIM – *passive intermodulation*, which means the generation of interfering signals caused by nonlinearities in the passive components of a radio system (such as antennas and connectors). The interaction of these components generally produces nonlinear effects, especially at the place where two different metals come together. Junctions of dissimilar materials are a prime cause. Even nearby metal objects such as guy wires and anchors, roof flashings, and pipes of a mobile base station can also cause PIM. The result is a diode-like nonlinearity that makes an excellent mixer. As nonlinearity increases, so does the amplitude of the PIM signals. Two signals mix together (amplitude modulation) to produce sum and difference signals, which could be within the same band and cause interference – this is a major antenna concern for mobile base stations and high-power radio systems (Figure 7.1).

Table 7.1 Galvanic metals table

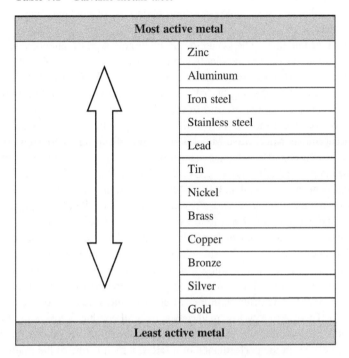

Most active metal	
	Zinc
	Aluminum
	Iron steel
	Stainless steel
	Lead
	Tin
	Nickel
	Brass
	Copper
	Bronze
	Silver
	Gold
Least active metal	

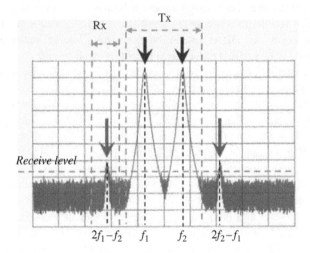

Figure 7.1 Generation of passive intermodulation (PIM) signals

7.1.2 Dielectric Materials

Dielectric materials are used to form the desired antenna shape to protect a metal antenna (a conductor may be inside the dielectric or the dielectric may be covered by a conductor) or to act as a dielectric antenna. The dielectric properties of the materials that are commonly used in antenna design are given in Tables 1.2 and 2.4 (PCB substrates). However, new materials are emerging all the time and some important materials for antennas may have not been included in these tables.

Dielectric resonator antennas (DRAs) are a relatively new type of antennas that have some attractive properties and become more widely used due to the availability of low-loss and high-permittivity materials [1–3]. The typical relative permittivity is from 10 to 100. These antennas have been mainly realized by making use of ceramic materials characterized by high permittivity and high Q-factor (between 20 and 2000). They have also been made using plastic materials (e.g. polyvinyl chloride [PVC] with relative permittivity $\varepsilon_r = 3$ around 1 GHz), Rogers TMMs (Thermoset Microwave Materials, a range of ceramic thermoset polymer composites with $\varepsilon_r = 3$–13), RO3010 laminates (ceramic-filled PTFE composites with $\varepsilon_r = 10.2$), RT-Duriod/Alumina ($\varepsilon_r = 12$), $Sr(Zr_xTi_{1-x})O_3$ ($\varepsilon_r = 25$), and $Ca(Ti_{1-x}Zr_x)O_3$ at $x = 1$ ($\varepsilon_r = 27$). Their loss tangent is normally less than 0.01. Because of the discontinuity between the dielectric and air, DRAs act as resonators whose resonant mode is excited by a specially designed feeding method. The coupling of power to the structure can be achieved using coupling slots, coaxial probes, microstrip, or coplanar transmission lines. Dielectric resonators of any shape could be used for antenna designs. The shape and dielectric properties of the dielectric along with the frequency and excitation determine which resonant mode may be generated, hence the radiation pattern and input impedance. The widely used geometries are rectangles, cylinders, rings, and hemispheres. DRAs offer a number of good features, including small size and reasonable bandwidth. The use of low-loss materials and small conductor losses permit high radiation efficiencies. Unlike conductive antennas, the influence of nearby objects (such as human hands) has limited effect on the performance of a DRA, which is one of the major reasons for DRAs becoming popular in some portable devices.

It is well known that there is normally a trade-off between antenna size and its bandwidth. The smaller the size (which means a larger permittivity), the narrower the bandwidth. This result is applicable to not only DRAs but also other antennas, such as patch antennas – a high permittivity substrate is required if the antenna has to be made small.

When dielectric materials are used for coating and protection purposes, from the antenna design point of view, the main requirement on materials is low loss (in addition to other requirements), which affects the antenna efficiency and input impedance. Another noticeable effect of the dielectric material on the antenna is that the resonant frequency is shifted downward, which will be demonstrated in Section 7.5. Foam dielectrics are widely used as sealers for microwave components, spacers in reflector antennas, and low-reflection supports. Specially loaded foams are often used for low-power applications while ceramics are preferred for high-power applications. Reinforced plastics are often used as structural materials for antennas, feeds, and mounting structures.

PCBs are commonly used for making planar antennas [4]. There is a trend to integrate antennas with the RF front end and circuits. Thus, the development of PCBs has an impact on antenna manufacturing. Liquid crystal polymer (LCP) material provides an alternative to traditional polyimide film for use as a substrate in flexible circuit construction. This development offers improvements in manufacturing and processing of LCP flexible circuits and antennas. LCP film is directly metallized using a vacuum sputtering process. Unlike laminated substrates, the sputtered LCP substrate yields a good metal-to-LCP adhesion, avoids trace line undercut problems, eliminates remaining copper issues, and provides effective circuit patterning resolution. Either subtractive or additive processing can be used, which is good for antenna applications. It has good thermoplastic properties, a low dielectric constant (about 3.0) and loss tangent (less than 0.001) remaining constants over the frequency range of 1 kHz to 45 GHz, with a negligible moisture effect. The properties of multilayer (3D) vertical integration capability, good electrical and mechanical properties, and near-hermetic nature make this substrate a practical choice for the fabrication of antennas [4, 5].

7.1.3   Composites

A composite material is a combination of two or more materials with different physical and chemical properties. They are combined to create a material that is specialized to realize a certain feature, for instance, to become stronger, lighter, or resistant to electricity. They can also be made to improve some parameters such as strength and stiffness. Composite materials are used for constructing antennas (as a whole or part of an antenna). For example, polycarbonate/acrylonitrile butadiene styrene (PC/ABS) alloys are widely used for making mobile phone antennas. They combine the excellent heat resistance of polycarbonate with the improved processibility of ABS.

A typical thermoplastic composite begins with high-performance engineered polymer to which fillers are added to enhance characteristics. For some applications, the polymer is likely to be a high-temperature moldable thermoplastic material, such as grades of polyphenylene sulfide. Composite materials are strong and can be tailored to provide impact resistance, tensile strength, flexural strength, and other desirable properties. However, the choice of composites is affected by operating temperature and fluid resistance requirements, so a good understanding of the expected temperature extremes and environment of the antenna is necessary when designing composite parts.

Advanced carbon-fiber composite (CFC) materials are used in such as the aerospace industry as a replacement for metal because of their higher strength, lower cost, and lighter weight. There are various kinds of CFCs: reinforced continuous carbon fibers, short carbon fibers, carbon black, and carbon nanotube (CNT) [6]. CFCs address the need for lightweight, cost-effective, mass-producible electrically conductive parts (Carbon fiber reinforced polymer has conductivity around 106 S/m). Conductive composites typically offer a 30–40% weight savings over aluminum. For antenna applications, uses of CFCs range from ground planes to enclosures. Due to the favorable mechanical and electrical properties, CNTs have been of interest in nanoelectronics and nanoantenna applications. The density of CNT composites is around 1.4 g/cm, half that of aluminum and more than five times lower than that of copper. Showing high thermal conductivity of about 10 times that of copper, CNTs are attractive for high heat-transfer applications. CNT composites can be made using single-wall nanotubes or multiwall carbon nanotubes. However, CNT dipoles show extremely low efficiency in 2–3 orders of magnitude lower than their corresponding copper dipoles for microwave applications, due to their high resistance per unit length. Therefore, CNT arrays and composites are proposed to improve the efficiency. The electrical conductivity of CNT composites depends on the properties and loading of the CNTs, the aspect ratio of the CNTs, and the characteristics of the conductive network throughout the matrix. It was shown in [6] that the effective conductivity of CNT can be made around 10,000 S/m @ 30 GHz, thus the efficiency of an antenna made of CNT can be comparable with that of a copper antenna. In mobile and military applications where the antenna may be used in harsh environments, replacing metals with a more suitable material increases the system reliability.

Glass fiber composites are moldable and offer an economical solution for producing radomes and antenna substrates. Typical radomes are formed using E-glass reinforcement for economical designs or quartz fiber reinforcement when low loss is critically important. Glass fiber composites offer thinner, lighter parts than non-fiber-reinforced designs. Glass fibers also increase the dielectric constant of most composites, enabling antenna size reduction when these composites are used as substrate materials. Composite materials can also be engineered to provide desired dielectric constants through the addition of various filling materials, such as hollow glass microspheres, conductive particles, or foaming agents. For both carbon fiber and glass fiber composites, firer length is an important design parameter. Longer fibers offer more strength but reduced ability to manufacture small features. Moldable long-fiber composites allow significant thickness reductions while maintaining equivalent strength of short fibers. Continuous-fiber reinforcements are attractive for further weight reduction on designs with large, smooth features.

This is a very active and evolving area. New composite materials of special electromagnetic properties may change the way how we design and make antennas. For example, a new composite material with its permittivity inversely proportional to the frequency square has recently been developed at Liverpool University; this means that the wavelength inside the material is a constant (not a function of frequency), thus it can significantly reduce the size but broaden the bandwidth of an antenna and many other RF/microwave devices. It would be interesting to see how this is going to develop in the years to come.

7.1.4   Metamaterials and Metasurfaces

The prefix *meta* is a Greek word and means "beyond, about," thus a metamaterial indicates that the characteristics of the material are beyond what we see in nature. Metamaterials are artificially crafted composite materials that derive their properties from internal microstructure, rather than chemical composition found in natural materials. The concept might have been introduced long time ago, but the realization and applications have been just about 20 years [7–15]. The interest has been growing significantly over the last decade or so since this new type of materials has a wide range of potential applications from low microwave to optical frequencies.

The core concept of metamaterials is to produce materials by using artificially designed and fabricated structural units made of metallic or dielectric material to achieve the desired properties and functionalities. These sub-wavelength structural units – the constituent artificial "atoms" and "molecules" of the metamaterial – can be tailored in shape and size, the lattice constant and interatomic interaction can be artificially tuned, they can be designed and placed at desired locations. By engineering the arrangement of these small unit cells into a desired architecture or geometry, one can tune the permittivity or permeability of the metamaterial to positive, near-zero, or negative values. Thus, metamaterials can be endowed with properties and functionalities unattainable in natural materials [7–9].

Figure 7.2 shows that materials can be divided into four categories according to their permittivity ε and permeability μ:

1. Conventional double-positive (DPS) materials: both of their permittivity ε and permeability μ are positive, thus, they are called DPS materials. Electromagnetic (EM) waves at the boundary between air and the medium/material are partially reflected and refracted into the medium as shown in Figure 7.2 and discussed in Chapter 3.

2. Epsilon negative (ENG) materials: they display a negative permittivity ε but positive permeability μ. Many plasmas exhibit this characteristic, for example, noble metals such as gold or silver are ENG in the infrared and visible spectrums, but not at RF/microwave frequencies. It is found that some special wire structures and some other structures such as complementary split ring resonators (CSRRs) can demonstrate this characteristic and can be considered as new artificial materials; thus, they are metamaterials. In this case, the refractive index becomes imaginary and EM waves cannot propagate in the material but reflected at the boundary between air and the medium. Any propagation into the medium would be evanescent (that is vanishing in the medium).

3. Mu-negative (MNG) materials: they have a positive ε but negative μ. Gyrotropic or gyromagnetic materials have this characteristic. Again, in this case, the refractive index becomes imaginary and EM waves cannot propagate in the material and will be reflected at the boundary between air and the material. It is found that, at RF/microwave frequencies, split ring resonators (SRRs) have shown such an interesting characteristic, hence they can be considered as metamaterials that have been used for such as filtering and shielding applications.

4. Double-negative (DNG) materials: both the permittivity ε and permeability μ are negative in a given frequency band, which is now commonly known as DNG metamaterials. EM waves at the boundary between air and the medium are partially reflected and partially refracted into the medium at different directions compared with DPS materials as shown in Figure 7.2. In this case, the refraction index n becomes negative, EM waves travel in the *backward* direction in the medium, which is new and not observed in nature materials. Both the reflection and refraction are anomalous, and Snell's law doesn't work in this case. It is found that the combination of thin wires and SRRs as shown in Figure 7.3 can produce this special feature; thus, DNG metamaterials are termed as "left-handed materials (LHM)" due to the left-handedness of electric, magnetic field and wave vector.

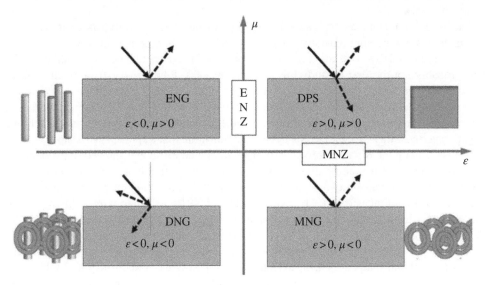

Figure 7.2   Material classification

Figure 7.3 Metamaterial formed by SRRs and wire strips. Source: Jeffrey.D.Wilson@nasa.gov (Glenn research contact) – NASA Glenn Research

Another property not normally found in nature that can be achieved with metamaterials is near-zero refractive index (NZI), which could be Epsilon near-zero (ENZ) or Mu near-zero (MNZ) materials as indicated in Figure 7.2. In this type of materials, either the permittivity or permeability is designed to have its real part close to zero. Materials with unique properties such as these have a wide range of potential applications in electromagnetics at frequencies ranging from the low microwaves to optical, including shielding, low-reflection materials, novel substrates, antennas, electronic switches, "perfect lenses," absorber, and resonators, to name only a few.

The history of metamaterials is actually very interesting. The electrodynamics of hypothetical materials having simultaneously negative permittivity and permeability was first theoretically predicted by Russian scientist Victor Veselago around 1967 [10] although he was probably not the first to study this subject. In 2000, John Pendry at Imperial College London was the first to identify a practical way to make an LHM in which the conventional right-hand rule is not followed. His idea was that metallic wires aligned along the direction of a wave could provide negative permittivity. Another challenge at the time was how to achieve negative permeability ($\mu < 0$). He managed to demonstrate that a split ring (C shape) with its axis placed along the direction of wave propagation could do so and showed that a periodic array of wires and rings could give rise to a negative refractive index. He also proposed a related negative-permeability design, the Swiss roll. A left-handed material was first implemented in a two-dimensional (2D) periodic array of SRRs and long wire strips by David Smith and his team at Duke University around 2000 [12]. The logical approach was to excite the SRRs and wire strips in order to force the structure to behave like magnetic and electric dipoles, respectively. A method was provided in 2002 to realize negative-index metamaterials using artificial lumped-element loaded transmission lines in microstrip technology. At microwave frequencies, the first, imperfect invisibility cloak was realized in 2006. By 2007, experiments that involved negative refractive index had been conducted by many groups. Since then, there have been a large number of theoretical and experimental studies on metamaterials. The effective electromagnetic parameters were also retrieved experimentally and numerically from the transmission and reflection data. Rather than SRRs and wire strips, the left-handed feature has been realized with periodic loading of conventional microstrip transmission lines with series capacitors and shunt inductors. Many microwave circuits have been implemented by using this strategy such as compact broadband couplers, broadband phase shifters, compact wideband filters, compact resonator antennas, left-handed leaky wave antennas, which have a very unique property of backfire-to-endfire frequency scanning capability with broadside radiation, which is not possible for conventional leaky wave antennas [7–9].

It is now known that metamaterials are typically engineered by arranging a set of small sub-wavelength metallic or dielectric scatterers or apertures (not limited to SRR) in a regular or irregular array throughout a region of space, thus obtaining some desired bulk electromagnetic characteristics. The metamaterial itself in general is a composite material. Each unit cell (SRR or other designs) has an individual tailored response to the electromagnetic field. This is similar to how light actually interacts with everyday materials; materials such as glass or lenses are made of atoms, an averaging or macroscopic effect is produced. The cell is designed to mimic the EM response of atoms, only on a much larger scale. In addition, as part of periodic composite structure, these are designed to have a stronger EM coupling than is found in nature. The larger scale allows for more control over the EM response, while each unit is smaller than the radiated electromagnetic wave.

Metamaterials are much more active than ferromagnetic materials (which possess a permanent magnetic moment in the absence of an external field, such as iron and nickel) found in nature due to their special characteristics. The pronounced EM response in such lightweight materials demonstrates an advantage over heavier, naturally occurring materials. Each unit cell can be designed to have its own EM response. The response can be enhanced or lessened as desired. In addition, the overall effect reduces power requirements. There are many applications and potential applications of metamaterials. A high-profile application is the invisibility cloak technology to obscure an object where the material must be able to precisely control the path of light. Although metamaterials already have revolutionized optics and RF/microwave engineering, their performance has been limited by their inability to function over broad bandwidths. Designing a metamaterial that works across an ultra-wideband remains a considerable challenge.

7.1.4.1 Metasurfaces

The fascinating functionalities of metamaterials typically require multiple stacks of material layers, which not only leads to extensive losses but also brings a lot of challenges in fabrication. Many metamaterials consist of complex metallic wires and other structures that require sophisticated fabrication technology and are difficult to assemble. Three-dimensional metamaterials can be extended by arranging electrically small scatterers or holes into a 2D pattern at a surface or interface. This surface version of a metamaterial has been given the name *metasurface*. For many applications, metasurfaces can be used in place of metamaterials. They have the advantage of taking up less physical space than full 3D metamaterial structures do; consequently, they offer the possibility of less-lossy structures as well. Thus, metasurfaces have attracted a greater attention than 3D metamaterials and advanced very rapidly over the past 10 years or so.

Generally speaking, metasurfaces can be divided into two groups: Metasurfaces that have a "cermet" topology with an array of isolated (non-touching) scatterers are called *metafilms*. Metasurfaces with a "fishnet" structure that are characterized by periodically spaced apertures in an otherwise relatively impenetrable surface are called *metascreens*. Other kinds of metasurfaces lie somewhere between these two extremes. For example, a grating of parallel conducting wires behaves like a metafilm in the direction perpendicular to the axes of wires, but like a metascreen in the direction along these axes.

The construction of metasurfaces may not be strictly 2D, there could be a specially designed thickness of the metasurface material as the third dimension. However, this thickness is normally much smaller than the wavelength of operation. The unit cell could be metallic or dielectric, and the whole structure could be formed by the same or different cells arranged in a periodic or nonperiodic way – depending on the application and desired performance.

Over the years, there have been many special period surface structures developed, such as the frequency-selective surface (FSS), high-impedance surface (HIS), and EM bandgap (EBG) structures. A popular question is: are they metasurfaces?

These special structures are named from their specific functions or macroscopic performances while metasurfaces are defined by the bulk material properties: realized effective permittivity and permeability (e.g. DNG and NZI). Thus, they are named from different levels: one is from material level while the other from function level. That is why metasurfaces usually overlap many research areas of conventional 2D periodic structures. For example, frequency-selective performances can be implemented using metasurfaces or non-metamaterials, depending on the specific design. Generally speaking, the conventional FSS is not a metamaterial, but HIS and EBG are normally considered as metamaterials. A well-known mushroom-type HIS is shown in Figure 7.4 where the hexagonal metallic patches are connected to the ground through vias and each unit cell acts an LC resonator [13]. Unlike a conducting ground plane where the reflection coefficient of an EM wave is −1 (the phase is 180°), the reflection coefficient on this HIS at the resonant frequency is +1 (the phase is 0), just like the case where a transmission line is connected to an open circuit whose impedance is very high (+∞ in this case), that is why it is called HIS. This structure can also be considered as an artificial magnetic conductor (AMC) because the tangential magnetic field at the surface is zero, rather than the electric field, as seen on an electric conductor. We can make use of this special feature by placing a wire dipole or monopole antenna in parallel with the surface closely to produce a directional radiation pattern as shown in Figure 7.4, which is not possible for a normal conducting ground plane. The obvious benefits of this arrangement are low profile and high gain. In addition to providing an unusual reflection phase, such an HIS can also be used to manipulate surface waves. It possesses a band gap between the first band, which supports transverse magnetic waves, and the second band, which supports transverse electric waves. The TM band does not reach the TE band edge but stops slightly below it, which results in a band gap between TE and TM surface waves. In other words, no surface waves are supported within the band gap, and this interesting property can be used in applications such as decoupling of nearby antennas. For the HIS antenna in Figure 7.3, another major advantage is that the surface wave on the ground is suppressed; thus, there is very little energy leakage to the other side of the ground plane and the back-lobe of the antenna radiation pattern is therefore much reduced.

Many applications have been explored and investigated using various metasurfaces. Reference [8] showed a wide range of potential applications in electromagnetics (ranging from low microwave to optical frequencies), which include controllable "smart" surfaces, miniaturized

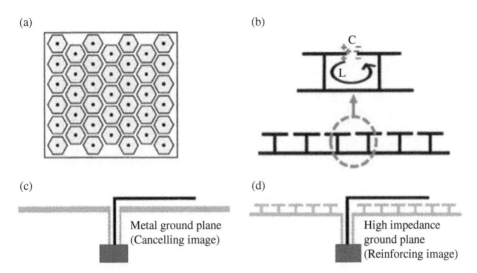

Figure 7.4 A high impedance surface and its application for a low-profile wire antenna: (a) top view of the HIS; (b) side view of the HIS with a zoomed view of a unit cell; (c) a horizontal antenna on a conducting ground plane; (d) a horizontal antenna on the HIS/AMC

cavity resonators, novel wave-guiding structures, angular-independent surfaces, absorbers, biomedical devices, terahertz switches, and fluid-tunable frequency-agile materials. The review paper [9] was focused on physics and applications of metasurface materials at optical frequencies. Good summaries of metasurfaces for antenna applications can be found from such as [14, 15]. From these research papers and findings, we can see clearly that developments of metasurfaces have been closely linked to antennas and resulted in at least the following three results:

1. Antenna-inspired metasurfaces: antenna theory has been used to guide the development of metasurfaces to realize certain functions such as EM wave anomalous reflection, anomalous refraction, polarization conversion, and absorption.
2. Metasurface-assisted antennas: metasurfaces have been exploited to increase the directivity of the antenna, reduce radar cross section (RCS), suppress the interference between radiating elements, or empower the antennas with certain configurability.
3. Metasurface antennas: the integration of metasurfaces and antennas is called metasurface antenna, or *metantenna* for short. A metantenna typically consists of an array of meta-atoms/cells distributed over an electrically large structure, each element is of sub-wavelength in size and separated from its neighboring cells much less than a wavelength. It usually consists of one or more feeds and one metasurface, where the metasurface is an indispensable component acing as the antenna aperture and the feed is usually placed at one end or in the center of the metasurface. In such an antenna configuration, the metasurface is used as the antenna aperture, which radiates (receives) EM waves from (into) the feed. Since metasurfaces usually possess different functions in different bands and for different polarized waves, in addition to EM wave radiation/reception, other functions can be integrated into the antenna as well, such as frequency scanning, low RCS, isolation enhancement, bandwidth expansion, and many more.

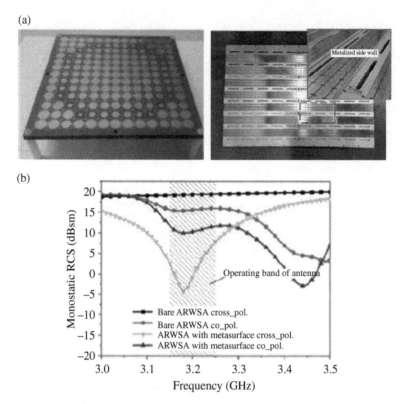

Figure 7.5 Metasurfaces for antenna designs: (a) transmissive metasurface for gain enhancement; (b) meta-atoms used around slot antenna array to reduce RCS. Source: From Wang et al. [15]

Two selected metasurface antennas are given in Figure 7.5. In Figure 7.5(a), the elements comprising square ring and circle patches are designed with different transmission phase, and they form a transmissive phase gradient metasurface. The spherical-like wave fronts radiated from the source antenna can be transformed by the metasurface (as a superstrate of the antenna) into planar ones. As a result, the beam is focused with a much higher gain. For RCS reduction, which is required for radar applications, meta-atoms/cells are used to surround a slot-waveguide antenna array as shown in Figure 7.5(b), the reduction on RCS is more than 6 dB for co-polarization and 15 dB for cross-polarization.

In this section, we have shown that metamaterials and metasurfaces, as new artificial materials, have already been applied to a wide range of real-world applications. They can be well integrated with antennas to achieve what was not possible before. Although it is quite unlikely that they will become popular for all applications, they do provide opportunities for people to use these materials to design new antennas for some special applications. Furthermore, other materials, such as new liquid materials (either conductive or dielectric) and graphene, have also been used for making antennas. A special section on liquid antennas is provided in Chapter 8. New materials are set to grow and will make a significant impact on antennas and related industry in the future.

7.2   Antenna Fabrication

For successful development of an antenna system, the fabrication plays a significant role. In many cases, design criteria face fabrication limits, and fabrication schemes are geared toward practical designs, the two facets go hand in hand.

In the past, antenna fabrication was mainly a mechanical manufacturing process. However, as the operational frequency increases and various "funny" and "strange" designs are produced, antenna fabrication becomes more complex and challenging. Luckily, this change is accompanied by the advancement of manufacturing technology and material engineering. For example, recently some new materials and processes have been developed. Silver inks and deposited coppers are used for making small and lightweight antennas for applications such as RFID [16]. Electrically conductive adhesives can be used to replace conventional soldering. New composite and artificial materials have been developed and provide some unique features as discussed in Section 7.1.

The techniques currently used for antenna fabrication can be separated into two categories: *subtractive methods* and *additive methods*. Typically, subtractive methods include the patterning of material through methods that require removing materials from a host, while additive methods include a direct deposition of material onto a host. The general process flows of these two categorical methods are outlined below:

- Subtractive Methods:
  - Substrate and material preparation
  - Bulk material deposition
  - Masked patterning
  - Sacrificial etching

- Additive Methods:
  - Substrate and material preparation
  - Selective material deposition

The fabrication technology and process for large and small antennas are different. For example, a large satellite dish antenna can be made of fiberglass (light and cheap) or aluminum (more durable and expensive). For fiberglass-based dishes, we may use the following process:

- To make fiberglass suitable for dish manufacture, a sheet molding compound mixture that includes reflective metallic material and ultraviolet scattering compositions is mixed with resin, calcium carbonate, and a catalyst cure. This mixture forms a paste that is poured onto a sheet of polyethylene film that has fiberglass added in chopped form. The result is a sheet layered with the compound paste, fiberglass, and the polyethylene film.
- This sheet is then pressed at certain temperature (typical 30 °C) to mature. To shape the sheet into the desired parabolic shape, it is pressed at high pressure. The dish is then trimmed, cooled, and painted. After the paint has dried, the dish is packed for shipment in sturdy boxes.

For metallic dishes, the common metal of choice is aluminum. This type of dish can be assembled in sections called petals, or all at once. An aluminum plate is perforated with a punching die, creating tiny holes. The size of these holes is contingent on the manufacturer's preference. Larger holes mean greater loss of the signal, so fairly small holes are selected.

Another factor in the selection of hole size is the power of the broadcasting satellite. Newer, more powerful satellites require a hole size that is approximately half that required for older, less powerful satellites. The newly perforated aluminum plate is then heated, stretched over a mold, cooled, and trimmed. A paint powder coating for protection is then applied using an electrostatic charge, in which the paint is given an opposite electrical charge from the plate. The dish or petal is then heated to melt the powder and seal the paint on. The petals are usually sealed together with ribs in the factory.

Mesh petals are made from aluminum that is extruded – forced into a die of the proper shape. They are usually joined together on site by sliding them into aluminum ribs that attach to the hub and then securing them with metal pins. This process is general and can be applied to many aperture antennas.

Sometimes, 3D selective metallization is required for antenna fabrication. The typical method of metallizing specific shapes on 3D surfaces is selective plating. This process requires labor-intensive application of physical masks to the surface of the part followed by a multistep plating process. Because of the high labor content, selectively metallized parts are usually relatively expensive. Alternative processes include laser direct structuring (LDS) and two-shot molded interconnect devices (MID). Both allow cost-effective 3D metallization; however, both are constrained by the range of available substrate materials. In addition, injection molds are required for both these processes, increasing nonrecurring expenses and lead times. To overcome these substrate limitations, a new process is developed, which starts with the application of a sprayable conductive coating to the surface of the part. Next, this coating is cured by radiative or thermal processes. Finally, the coating is ablated to the desired pattern using a computer numerical control (CNC) laser. This process results in 3D conformal shapes with metallization resolutions as fine as 100 microns and allows molded or machined parts to be used as substrates. This 3D selective metallization process can be applied to a wide range of substrates – including plastics, chemically resistant composites, glass, ceramic, and metals – with acceptable adhesion, a temperature range from −65 to +200 °C, and corrosion resistance. The metallization is also durable and withstands shocks, vibration, fluids, and salt spray to the levels required for most aerospace applications. This process enables rapid development and manufacture of robust 3D antennas for harsh environments.

For small antenna fabrication, a wide range of fabrication techniques and processes are developed, such as PCB, MEMS, LTCC, LCP, LDS, and 3D Printing. Here we are going to provide a brief summary of these technologies.

7.2.1   PCB-Based Fabrication

PCB stands for *Printed Circuit Board* and consists of a flat sheet of insulating material (called substrate) and a layer of copper foil laminated to the substrate. Chemical etching divides the copper into separate conducting lines called tracks, pads for connections, vias to pass connections between layers of copper, and features such as solid conductive areas for electromagnetic shielding or other purposes. The surface of a PCB may have a coating that protects the copper from corrosion, the coating is called solder resist or solder mask. A PCB can have single, double, or multiple copper layers. A two-layer board has copper on both sides; the most used PCB has FR4 substrate with a relative permittivity around 4.4 and loss tangent around 0.02 at 100 MHz.

PCB-based fabrication is the most well known and most widely used for planar structures. The fabrication process could have over 10 steps. For PCB antenna fabrication, the main steps are given below:

- *PCB CAM*: The fabrication data generated by computer-aided design is read into the CAM (Computer-Aided Manufacturing) software.
- *Copper patterning*: This step is to replicate the pattern in the fabricator's CAM system on a protective mask on the copper foil PCB layers. Subsequent etching removes the unwanted copper unprotected by the mask. (Alternatively, a conductive ink can be ink-jetted on a blank [nonconductive] board. This technique is also used in the manufacture of hybrid circuits).
- *Subtractive, additive, and semi-additive processes*: Subtractive methods remove copper from an entirely copper-coated board to leave only the desired copper pattern. In additive methods, the pattern is electroplated onto a bare substrate using a complex process. The advantage of the additive method is that less material is needed, and less waste is produced. In the full additive process, the bare laminate is covered with a photosensitive film, which is imaged (exposed to light through a mask and then developed, which removes the unexposed film). The exposed areas are sensitized in a chemical bath, usually containing palladium and similar to that used for through hole plating, which makes the exposed area capable of bonding metal ions. The laminate is then plated with copper in the sensitized areas. When the mask is stripped, the PCB is finished. Semi-additive is the most common process: The un-patterned board has a thin layer of copper already on it. A reverse mask is then applied. Additional copper is then plated onto the board in the unmasked areas; copper may be plated to any desired weight. Tin–lead or other surface platings are then applied. The mask is stripped away, and a brief etching step removes the now-exposed bare original copper laminate from the board, isolating the individual traces. The additive process is commonly used for multilayer boards as it facilitates the plating-through of the holes to produce conductive vias in the circuit board.
- *Chemical etching*: This is usually done with ammonium persulfate or ferric chloride. For plated-through holes, additional steps of electroless deposition are done after the holes are drilled, then copper is electroplated to build up the thickness, the boards are screened, and plated with tin/lead. The tin/lead becomes the resist leaving the bare copper to be etched away. The etchant removes copper on all surfaces not protected by the resist.
- *Drilling*: Holes through a PCB are typically drilled with drill bits made of solid coated tungsten carbide. Coated tungsten carbide is used because board materials are abrasive. High-speed steel bits would dull quickly, tearing the copper and ruining the board. Drilling is done by computer-controlled drilling machines, using a drill file that describes the location and size of each drilled hole. Holes may be made conductive, by electroplating or inserting hollow metal eyelets, to connect board layers. Some conductive holes are intended for the insertion of through-hole-component leads. Others used to connect board layers are called vias.

Additional steps may be required for commercial production, such as plating and coating, solder resist application, silkscreen printing, bare-board test, and assembly. PCB milling or laser etching may be used to remove unwanted parts for prototyping. For laboratory prototyping, the process is normally kept to the minimum.

7.2.2 MEMS

MEMS stands for *microelectromechanical systems* and is a general term for highly miniaturized systems that are created through the integration of many small components, with or without moving parts. MEMS devices emerged in the late 1960s following with the development of lithography-based microfabrication techniques for highly integrated computer circuit components. One of the most outstanding features is their ultrahigh precision and very fine resolution capabilities for many different kinds of materials. On the other hand, these microfabrication technologies usually require a clean room environment and are very expensive, especially if the feature size of the structures is very small. Therefore, these technologies have been used mainly on highly integrated circuit core dies including central processing units (CPUs), memory units, sensors as well as the well-known MEMS switches and actuators. However, in recent years the desire for system miniaturization is getting stronger, especially in the area of mobile electronic devices. System on chip (SoC) and system in package (SiP) are becoming more and more important, the demand for high-frequency antennas makes MEMS fabrication technologies a good choice for antenna fabrication.

MEMS are made up of components typically between 1 and 100 µm in size, and MEMS devices generally range in size from 20 µm to a mm, although components arranged in arrays can be more than 1000 mm^2. Thus, MEMS antennas would be less than 1 mm for frequencies above 100 GHz. Microstructures are fabricated with a combination of additive processes, such as chemical vapor deposition (CVD), spin coating, and electroless plating, in conjunction with subtractive processes, such as wet etching and dry etching, depending on the material and thickness of the structure. With a proper sequential process, it is possible to create 3D structures applicable for antennas through these well-developed standard micromachining techniques. These methods have the capability to be seamlessly integrated into other CMOS chips and circuitry in the form of on-silicon antennas and on-chip antennas. Generally, micromachining can be categorized into surface micromachining and bulk micromachining with regard to the thickness of the structure.

7.2.3 LTCC

LTCC stands for *low-temperature co-fired ceramics*. This low temperature is compared with the temperature used for traditional ceramic technology, which requires high processing temperatures (1600 °C) for what we now refer to as high-temperature co-fired ceramics (HTCC). These temperatures made it impossible to use low-cost and highly conductive metals (Ag, Cu, etc.) that would melt in such conditions. However, LTCC technology is generally fired below 1000 °C due to a special composition of the material. This permits the co-firing with highly conductive materials (silver, copper, and gold) and offers many advantages because of the high performance of the ceramics, offering a wide range of available permittivities and low dielectric losses. It also features the ability to embed passive elements, such as resistors, capacitors, and inductors into the ceramic package minimizing the size of the completed module.

LTCC technology is a multilayer approach [17]. This allows for the low-cost and compact integration of active and passive components as well as the package-level antenna integration as shown in Figure 7.6. These multilayer capabilities and high-performance materials also offer the advantage of making highly efficient multilayer antenna designs possible. The starting material is normally composite green tapes, consisting of ceramic particles mixed with polymer

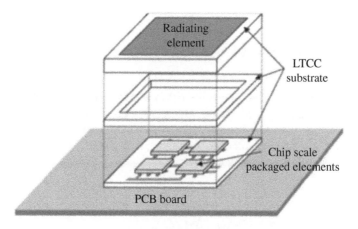

Figure 7.6　LTCC-integrated antenna

binders. The tapes are flexible and can be machined, for example, using cutting, milling, punching, and embossing. Metal structures can be added to the layers, commonly using via filling and screen printing. Individual tapes are then bonded together in a lamination procedure before the devices are fired in a kiln, where the polymer part of the tape is combusted and the ceramic particles sinter together, forming a hard and dense ceramic component. The suitable frequency range is very wide, from 1 to about 100 GHz.

7.2.4　LCP

LCP stands for *liquid crystal polymer* (it is not liquid at room temperature!), which has a high mechanical strength, extremely good chemical resistance, inherent flame retardancy, and good durability in rugged ambient conditions. Although thermotropic LCPs possess a variety of properties that make them attractive candidates for electronic substrates, for a long time, standard LCP processing techniques resulted in films with anisotropic in-plane properties. Currently many electronic designs, especially flexible and conformal ones, including antenna structures, are realized using LCP as a host substrate. It has excellent EM properties at microwave frequencies (low relative permittivity: 2.9 and low loss tangent: 0.002 at 20 GHz), low moisture absorption (<0.004 % by weight), and excellent dimensional stability (< 0.1%) [17].

To fabricate antennas on LCP, standard two-sided PCBs are typically used.

- *Imaging and Etching*: LCP is suitable for traditional photolithography processes due to its inherent chemical/moisture resistance that prevents any undercutting at the polymer–copper interface. In this process, a photoresist is first deposited onto the copper clad laminate, followed by exposure (imaging) and stripping to define the desired antenna pattern. Then, the exposed copper is chemically etched out to leave the desired antenna topology.
- *Drilling*: For large holes (e.g. greater than 8 mils in diameter), mechanical drilling is normally used in standard industry practice. For small holes (e.g. less than 8 mils in diameter), lasers are typically used. For LCP materials, the best vias are made with shorter, more frequent pulses along with a higher number of pulses than that required for some other standard materials.

- *Desmear*: After drilling, a cleaning step is required and typically achieved by a wet chemical permanganate exposure for most RF substrates. However, this method does not work well with LCP due to its extremely high chemical resistance. Instead, plasma treatment is very effective at cleaning holes.

LCP has been one of the most preferable conformal RF/microwave substrates due to the numerous desirable features listed above, which has made it an ideal substrate for flexible antennas, RFID, filters, and many other RF/microwave devices, particularly at high frequencies (e.g. mm waves). Compared with other popular fabrication processes, LCP is not cheap, which could a major drawback for commercial products.

7.2.5 LDS

LDS stands for *laser direct structuring*, which enables electronic assemblies to be made in flexible geometric shapes. Smart phones, hearing aids, and smart watches are becoming smaller and more powerful partially thanks to this process. For almost 20 years, LDS has made it possible to apply electronic conductor paths directly onto plastic parts during series production. The automated manufacturing processes also make this process more economically attractive. This is a stable and reliable process that allows 3D-MID (molded interconnect devices) assemblies to be produced. When using 3D-MID, electronic components can be fitted directly onto a 3D base body, without circuit boards or connecting cables. The base body is manufactured using an injection molding process, whereby the thermoplastic material has a nonconductive, inorganic additive.

The additives in the material are "activated" by direct laser structuring so that the plastic material can accommodate the electrical conductor paths. The laser beam writes the areas intended for the conductor paths and creates a micro-rough structure. The released metal particles form the nuclei for the subsequent chemical metallization. In this way, the electrical conductor paths are applied to the areas marked by the laser. The other areas of the 3D base body remain unchanged. The plastic component can then be assembled in standard surface mount device processes similar to a conventional PCB. It is also suitable for soldering in a reflow oven.

The LDS technology has been proven in numerous everyday applications. For example, it can be found in compact sensors such as pressure sensors. Mobile phones also contain MID based on LDS technology. Millions of phones use these 3D MID as space-saving integrated antennas as shown in Figure 7.7. Other applications can be found, for instance, in medical, air-conditioning, and safety technology.

7.2.6 Printing

Most of the above methods require the use of patterning masks, electronic plating, planar material growth, chemical etchants, and various other procedures to both deposit and remove materials from host substrates. Additive material printing offers an alternative approach to the widespread methods of subtractive electronic fabrication. Mass material deposition and subsequent removal are replaced with select, direct patterning, where the only materials used are materials directly constructing the desired electronic patterns and structures. Through the

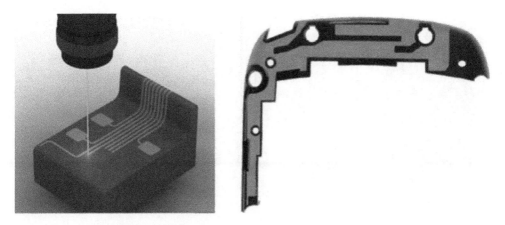

Figure 7.7   LDS technology and an LDS mobile phone antenna

utilization of conductive and insulating ink/filament materials, direct printing methods such as ink-jet and 3D printings are capable of fabricating flexible and rigid multilayer antenna structures in a purely additive process. The entire stack-up of a multilayer antenna system, including conductive metallization and thick dielectric substrates, can be achieved with ink-jet printing, removing the need for multilaminate processing, trimming, and bonding. 3D printing further extends the versatility of printed systems into an unlimited variety of applications, where flexible plastics and conductors can be used to realize wearable, reconfigurable wireless systems with a feasibly lowered cost and improved functionality

Additive manufacturing can reduce the time and material costs in a design cycle and enable on-demand printing of customized parts. Using conductive nanoparticle-based and thick dielectric polymer-based inks, ink-jet printing has realized a variety of antenna structures. Some of the greatest advantages of ink-jet printing are its fully additive nature and host material flexibility. Conductive and dielectric ink materials can be patterned.

New multi-material 3D printers that can print both metal and dielectric materials enable the additive manufacturing of antennas and RF components. Developments in software are critical to leveraging this capability; good tools allow more effort to go toward creation than implementation. The plastic used for these prints is normally polylactic acid (PLA), which is popular for prototyping in part because it has less toxic fume than other plastics. Its relative permittivity is around 2.8 at 1 GHz. Standard silver ink for 3D printer (a room-temperature curing silver conductive ink) has DC conductivity around 3.45 MS/m (for reference, pure silver has a conductivity of 61 MS/m). When making PCB devices and antennas, it is common to consider the trace and space tolerances; that is, the accuracy to which one can maintain a desired trace width and the gap between traces. For a 3D printer to deposit the ink with a 0.25-mm nozzle, the typical accuracy is good above 100 μm. The ability to print highly conductive materials with substrates is a desired feature that is currently rapidly evolving. It is widely accepted that useful RF and antenna structures can be 3D-printed, which has the potential to revolutionize the design, supply, and sustainment phases of an acquisition program. This technology has now been utilized for rapid prototyping and to support customization in the field.

In addition to the above fabrication methods, there are some other techniques used for antenna fabrication. For example, *flexible printed circuit* (FPC) process has been used to

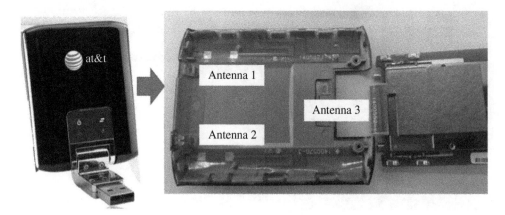

Figure 7.8   A USB mobile data dongle with three antennas fabricated using FPC technology

manufacture small flexible 2D/flat antennas and connections/cables. They are made with a photolithographic technology. An alternative way of making flexible foil circuits or cables is laminating very thin copper strips in between two layers of PET (polyethylene terephthalate). These PET layers, typically 0.05 mm thick, are coated with an adhesive, which is thermosetting, and will be activated during the lamination process. FPCs have several advantages in applications, such as tightly assembled electronic packages, flexible electrical connections during its normal use. However, there are also some disadvantages such as the increased cost and more difficult assembly process. Nevertheless, the flexibility and 2D feature have attracted a lot of applications over the past 30 years. Large credit is due to the efforts of Japanese electronics packaging engineers who have found countless new ways to employ flexible circuit technology. A more recent variation on flexible circuit technology is one called "flexible electronics," which commonly involves the integration of both active and passive functions in the processing. One example is given in Figure 7.8 where three dual-band antennas made by FPC technology are placed inside a small USB mobile data dongle, which is not possible using standard PCB fabrication process.

   In this section we have introduced different antenna fabrication processes. They all have advantages and disadvantages. Trade-offs must be made when choose the fabrication process. Lithographic-based methods are a widespread standard throughout all forms of electronic fabrication and offer the lowest minimum feature sizes; however, high equipment costs, long processing times, and a lack of large-area scalability are factors that must be considered. Additive techniques such as ink-jet and 3D printings offer relatively quick prototyping, lower cost, and seamless integration with large-scale fabrication; however, an increase in minimum feature size must be considered, where printing techniques offer minimum features nearly 10–100 times larger than lithographic techniques. With the current push toward the development of low-cost, ubiquitous wireless systems throughout emerging technologies, the need for improvement within the realm of antenna fabrication is greatly stressed. This improvement should be realized with methods that reduce material waste, increase productive throughput, and exhibit an ease of reconfigurability, while highlighting current trends in miniaturization and functional diversity. The increase of the operational frequency to mm-wave and THz and the demand for antenna on chip and antenna in package will undoubtedly have huge impact on the trend of antenna fabrication in the future.

7.3 Antenna Measurement Basics

Once an antenna is designed and constructed, it is essential to validate the design by proper measurements, which is an essential element of the antenna development process. The most important performance figures of merits are from the impedance and radiation pattern measurements. The input impedance may be specified as a value at a particular frequency and/or as a maximum VSWR or return loss over a range of frequencies (often referred to 50 ohms). The measurement is relatively straightforward. However, the radiation measurement is much more complicated and time-consuming. Typical measurements are the radiation pattern, efficiency, and gain measurements, which should be performed in an antenna test range (to be discussed in next section). The radiation patterns of some antennas (such as those of radar, microwave links, some cellular base stations, and satellites) are highly directional and often have tightly specified envelopes. Not only will the peak gain be specified, but challenging requirements might also be placed on parameters such as polarization purity (axial ratio), sidelobe levels, and efficiency. The testing and evaluation of some of these parameters could be difficult. Details of impedance measurements are given in Section 7.5. The radiation measurements are provided in Section 7.6. Some other measurement issues are discussed in Section 7.7.

7.3.1 Scattering Parameters

In previous chapters, we defined and used the reflection and transmission coefficients for both the circuit and field problem analyses. These concepts can be generalized for network analysis.

A two-port network problem can be illustrated by Figure 7.9, where a_1 and a_2 are the input whilst b_1 and b_2 are the output at Port 1 and Port 2, respectively. This network is characterized by *scattering parameters* or *S-parameters*:

$$[\mathbf{S}] = \begin{bmatrix} S_{11} & S_{12} \\ S_{21} & S_{22} \end{bmatrix} \tag{7.1}$$

which links the input to the output by

$$\begin{bmatrix} b_1 \\ b_2 \end{bmatrix} = \begin{bmatrix} S_{11} & S_{12} \\ S_{21} & S_{22} \end{bmatrix} \begin{bmatrix} a_1 \\ a_2 \end{bmatrix} \tag{7.2}$$

Figure 7.9 A two-port network

Thus, we have:

$$S_{11} = \text{Port 1 reflection coefficient} = b_1/a_1;$$
$$S_{12} = \text{Port 2 to Port 1 transmission coefficient/gain} = b_1/a_2;$$
$$S_{21} = \text{Port 1 to Port 2 transmission coefficient/gain} = b_2/a_1;$$
$$S_{22} = \text{Port 2 reflection coefficient} = b_2/a_2.$$

(7.3)

S-parameters are actually reflection and transmission coefficients for a network of N-port. In this case, $N = 2$. These parameters were originally introduced in optics where optical waves were scattered by objects. The concepts were later extended to radiowaves and RF engineering, but the term of S-parameters has remained unchanged.

It should be pointed out that if a network is passive and it contains only isotropic and loss-free materials that influence the transmitted signal, the network will obey

- the *reciprocity principle*, which means $S_{21} = S_{12}$ or more generally $S_{mn} = S_{nm}$, and
- the *law of power conservation*

$$[S]^H[S] = [I]$$

(7.4)

where $[S]^H$ is the *complex conjugate transpose* of the S-parameters matrix $[S]$ and $[I]$ is the identity/unit matrix. For a two-port loss free network, we have

$$|S_{11}|^2 + |S_{21}|^2 = 1$$
$$|S_{22}|^2 + |S_{12}|^2 = 1$$

(7.5)

which is equivalent to

$$|a_1|^2 + |a_2|^2 = |b_1|^2 + |b_2|^2$$

(7.6)

The input power is the same as the output power, as expected.

Almost all antennas, attenuators, cables, splitters, and combiners are reciprocal (but maybe not loss-free) networks. The networks that include anisotropic materials in the transmission medium such as those containing ferrite components will be nonreciprocal. Although it does not necessarily contain ferrites, an amplifier is also an example of a nonreciprocal network.

Now the question is how S-parameters are linked to antennas. From Figure 7.10, we can clearly see that a transmitting–receiving antenna system in the space can be considered as a

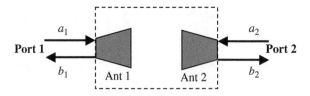

Figure 7.10 The equivalent two-port network of a transmitting–receiving antenna system

two-port network. The transmission and reflection can be characterized using S-parameters. S_{11} and S_{22} are the reflection coefficients of Antenna 1 and Antenna 2, respectively. They indicate how well the antenna feed line is matched with the antenna. S_{21} and S_{12} are the transmission coefficients from one antenna to another. They are determined by the characteristics of both antennas (such as radiation patterns, matching) and the separation between them. It should be pointed out that this equivalent network is not a confined two-port network but a lossy open network since the power radiated from one antenna is not all received by another antenna, thus the conditions in Equations (7.5) and (7.6) do not hold for such an antenna network. However, the reciprocity principle can still be applied to this case.

7.3.2   Network Analyzers

For RF/microwave engineering, the *oscilloscope* is often employed to view signals in the time domain while the *spectrum analyzer* is used to examine signals in the frequency domain. For antenna measurements, the most useful and important piece of equipment is the *network analyzer*, which is basically a combination of a transmitter and a receiver. Normally it has two ports, and the signal can be generated or received from either port. The main parameters that it measures are the S-parameters, i.e. it measures the reflection and transmission characteristics of a network. There are two types of network analyzers:

- *Scalar Network Analyzer* (SNA): measures the amplitude of the parameters of a network, such as VSWR, return loss, gain, and insertion loss.
- *Vector Network Analyzer* (VNA): measures both the amplitude and phase of the parameters of a network.

The VNA is much more powerful than the SNA; in addition to the parameters that can be measured by the SNA, it can also measure some very important parameters such as the complex impedance, which is essential for antenna measurements. Thus, we need a VNA not SNA for antenna measurements.

7.3.2.1   The Configuration of a VNA

A picture of a typical VNA is shown in Figure 7.11. It is seen that the equipment comes with two standard coaxial connectors. The measured results can be saved to a disk or a computer via a USB port at the back. The architecture of a VNA is given in Figure 7.12. It clearly shows that the source signal can be transmitted from either Port 1 or Port 2 to the device under test (DUT), which is controlled by a switch. A part of the signal from the source is provided directly to the reference R, which will be used to compare with the received signal A (from Port 1) or B (from Port 2) by the CPU.

VNAs can perform most of the necessary measurements without the need for any manual tuning. As importantly, they also contain integrated computers and graphical displays, which allow an unprecedented level of data manipulation and display – for example, making it possible to plot calibrated, phase-adjusted impedances directly onto an on-screen Smith Chart. Indeed, network analyzers now have most of the features of typical personal computers: high levels of processing power, high-resolution color screens, familiar operating systems, network

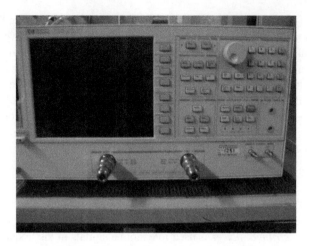

Figure 7.11   A picture of a typical VNA

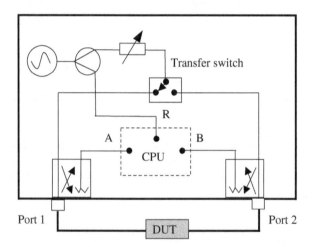

Figure 7.12   Typical configuration of a VNA

connections, etc. Furthermore, most network analyzers support the standardized file formats that are used by circuit simulators, allowing measurements to be performed and then imported into simulations of the system in which the antennas operate.

7.3.2.2   What Can a VNA be Used to Measure?

The VNA is frequency-domain equipment; it can obtain the signal in the time domain using the Fourier transforms. For example, it can be used as a time-domain reflectometer (TDR) to identify discontinuities of an antenna, a transmission line, or a circuit. For antenna measurements, the typical parameters that can be measured by a VNA include:

Transmission measurements:

- Gain (dB)
- Insertion loss (dB)
- Insertion phase (degrees)
- Transmission coefficients (S_{12}, S_{21})
- Distance/length (m)
- Electrical delay (s)
- Deviation from linear phase (degrees)
- Group delay (s)

Reflection measurements:

- Return loss (dB)
- Reflection coefficients (S_{11}, S_{22})
- Reflection coefficients vs distance (Fourier Transform)
- Impedance ($R + j\,X$), which can be displayed on the Smith Chart
- VSWR

Note: these parameters can be displayed on the VNA screen. However, when the data is exported to a computer or saved to a disk, we normally get the complex S_{11} at each sampling frequency.

It should also be pointed out that, although the VNA can be viewed as the combination of a transmitter and a receiver, the received signal is not displayed as its absolute but a relative value, which is different from a conventional receiver. Thus, the VNA cannot normally be used as a receiver or spectrum analyzer.

7.3.2.3 Calibration and Measurement Errors

Using a VNA to perform antenna measurements, a careful calibration must be conducted. The reasons are as follows:

- As a radiator, the antenna should not be placed too close to the VNA (to avoid coupling and interference), that is, not directly connected to the VNA. Thus, a cable and connectors have to be used.
- The cable and connectors introduce attenuation and a phase shift.
- The reading on the VNA is at the default reference plane, but what we want to measure is the reading at the input port of the antenna.

We therefore need to remove the effects of the cable and connectors and shift the measurement reference plane right to the end of the cable – this process is called calibration. The standard calibration needs three terminations for one-port calibration, i.e. short, open, and load/matched. The commonly used calibration for a two-port network is short-open-load-thru (SOLT) calibration. Other calibration methods, such as thru-reflection-line (TRL) calibration [18], short-open-load-reciprocal (SOLR) element, and line-reflection-match (LRM, the best calibration accuracy up to 110 GHz), are also used in practice.

There are various errors that may be introduced to the measurement. The well-known ones include system errors, random errors, and drift errors (it is very important to warm up the equipment before making any measurements). Because of the radiating nature of the antenna, antenna measurements have got many other possible sources of errors. The system errors can be removed by calibration. Some errors (such as random errors) cannot be eliminated. Just like simulation results, measured results are also approximated.

7.4   Antenna Measurement Facilities

Normally we can measure RF/microwave devices (such as filters and amplifiers) in a standard laboratory, but this is not true for antenna measurements, since antennas radiate EM waves in the measurement and the measured results could be affected by the reflected waves in the environment. For example, if a radiation pattern was measured in a standard laboratory, the result could be wrong since the received signal would be the combination of the waves from all possible paths. Thus, where to conduct antenna measurements is a very important issue. An ideal measurement environment should be reflection- and interference-free, and no unwanted signals will be produced to contaminate the measurement. In this section, we are going to introduce the main measurement facilities/environments (also called test ranges) used for antenna measurements.

7.4.1   Open-Area Test Site

An open-area test site (OATS) is, as the name suggests, an outdoor site where no reflectors, other than the ground, are present over a relatively wide area. The most attractive advantage of an OATS is the low cost (and readily available in some companies). A typical site is illustrated in Figure 7.13.

It can be seen from Figure 7.13 that the path difference between the direct and reflected waves is

$$\Delta = \left(\sqrt{l^2 + (h_T + h_R)^2} - \sqrt{l^2 + (h_T - h_R)^2} \right) \tag{7.7}$$

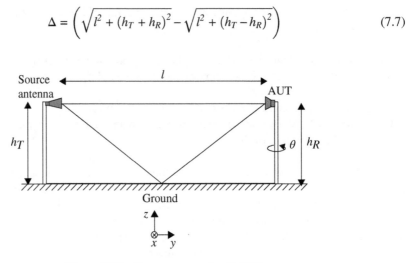

Figure 7.13   Open-area test site (OATS)

where

l: distance between the transmit and receive antennas
h_T: height of the transmit antenna above ground
h_R: height of the receive antenna above ground

Using the binomial expansion and the assumption that $l \gg (h_T + h_R)$ gives

$$l\left[1 + \left(\frac{h_T \pm h_R}{l}\right)^2\right]^{0.5} \approx 1 + \frac{1}{2}\left(\frac{h_T \pm h_R}{l}\right)^2 \tag{7.8}$$

Substituting in Equation (7.7) gives

$$\Delta \approx \frac{2h_T h_R}{l} \tag{7.9}$$

If we assume that $l \gg 2h_T h_R$ and $(h_T + h_R)$, we can neglect the distance-dependent amplitude differences between the two waves and assume that both the direct and reflected components travel over the distance l. We must, however, account for amplitude and phase differences due to the reflection coefficient of the ground. Since the phase term of the wave is $e^{j\beta \cdot r}$, hence the signal at the receiver is given by

$$E = E_D\left[1 + \Gamma \exp\left\{ j\frac{4\pi h_T h_R}{\lambda l}\right\}\right] \tag{7.10}$$

where E_D is the field strength that would be measured at the receiver in free space. Hence, using the Friis transmission formula in Section 3.5, the received power can be written

$$P_R = P_T\left(\frac{\lambda}{4\pi l}\right)^2 G_T G_R\left|1 + \Gamma \exp\left\{ j\frac{4\pi h_T h_R}{\lambda l}\right\}\right|^2 \tag{7.11}$$

Obviously, if the reflection coefficient of the ground is not zero, the received power at a fixed distance l is not uniform. For example, if it is a metal ground plane, the horizontal and vertical reflection coefficients approach -1. This gives

$$P_R = 2P_T\left(\frac{\lambda}{4\pi l}\right)^2 G_T G_R\left\{1 - \cos\left(\frac{4\pi h_T h_R}{\lambda l}\right)\right\} \tag{7.12}$$

which can be simplified to

$$\frac{P_R}{P_T} = 4\left(\frac{\lambda}{4\pi l}\right)^2 G_T G_R \sin^2\left(\frac{2\pi h_T h_R}{\lambda l}\right) \tag{7.13}$$

When $\lambda l \gg 2\pi h_T h_R$, this equation can be further simplified to Equation (3.45) – the two-ray model result. However, generally speaking, the received power is a periodic function of $h_T h_R$.

With the presence of the ground reflection, the power at the antenna under test (AUT) is sensitive to the heights of transmit and receive antennas, their separation, and the wavelength. The field is not suitable for radiation pattern measurements. Thus it is very important to minimize the reflection from the ground plane by raising the antenna height or placing RF-absorbing materials (RAM – more details to follow in the next section) between the two antennas to allow more accurate measurements.

There are other issues to be considered when using an OATS. For example, the site must be located in a place where the ambient level of RF interference is low. This is increasingly difficult due to the density of wireless communications systems and locating a site in a remote area (in terms of radio coverage) causes logistical problems associated with the transport of devices and personnel. On the other hand, the radiated power and frequencies are also regulated (a license may be required to operate) to avoid interference with nearby communication systems. Furthermore, since an OATS is outdoor, testing time is limited by daylight hours and weather conditions. In some countries there is very little daylight during cold winters; often such seasonal variations are inconsistent with business requirements. To counter the problems associated with testing time, measures may be taken to extend the number and duration of operational days. However, weather shields, lighting (and the associated cables) and heating systems are potential sources of reflections and interference. Thus, an OATS is not ideal for most antenna measurements.

7.4.2   Anechoic Chamber

Anechoic chambers were developed so that antenna engineers could enjoy the convenience and productivity of being able to perform measurements indoors. The main obstacle to this is that "indoors" implies "within walls" and the walls reflect antenna signals and, hence, distort radiation pattern measurements. The term "anechoic" simply means without echoes: measures are taken – usually using RAM – to prevent reflections from the walls, ceiling, and floor. In addition, the walls are usually metalized to prevent external interference from entering the measurement room. A typical anechoic chamber is illustrated in Figure 7.14.

The chamber consists of a metallic room with a source antenna at one end that is used to excite an AUT at the other end. The AUT is usually placed in the "quiet zone" (QZ) (no reflection and uniform field area) and mounted on a positioner (or turntable) that is rotated in the azimuth plane (the x-y plane in Figure 7.14) to obtain a 2D radiation pattern. Some chambers

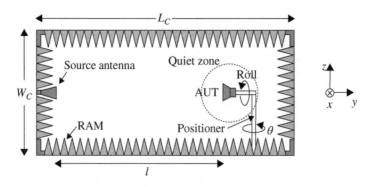

Figure 7.14   Typical anechoic chamber

may be equipped with positioners that are also capable of movement, or "tilt," in the elevation plane while others may have so-called "roll over azimuth" positioners in which the AUT rotates orthogonally to the azimuth rotation, allowing measurement of the full 3D radiation pattern. The source antenna may be rotated about its axis for polarization measurements. The chamber is usually arranged to transmit from the source antenna and receive on the AUT, although this may be reversed if it is more convenient (the reciprocity principle!). The turntable positioner, plotter, receiver, and transmit source are all under computer control.

Some chambers are rectangular in shape; some chambers are tapered from the source to the AUT to prevent the formation of standing waves and to make the wavefront incident on the AUT more planar, increasing the effective chamber length.

Radiation pattern measurement accuracy is limited by the finite reflectivity of the chamber walls, the positioner, and the cables that are used to feed the AUT. Multiple reflections "fill-in" nulls in the radiation pattern and create false sidelobes. The shape of the room is designed to minimize reflections from the walls over a wide range of incident angles and frequencies. The lowest frequency of operation is determined by the length of the absorber – typically the absorber needs to be approximately one wavelength long at the lower end of the operational range. At 300 MHz – the lower end of the UHF band – this corresponds to 1 m, which is too large for installations in normal commercial buildings. The upper end of the frequency range is determined by the composition of the absorbers and their surfaces. Most RAM is made of carbon-loaded polyurethane foam (for the resistivity and reduced weight). Pyramids and wedges are widely used shapes. Pyramids work best at normal incident. A tapered impedance transition from the free space to the back of the absorber is to ensure broadband-absorbing performance.

The dynamic range of an antenna chamber is limited by the source transmit power and receiver sensitivity. This is not a serious limitation provided that the signal is detected in a narrow bandwidth and a wide dynamic range low-noise amplifier is used at the test antenna. Often a network analyzer can be used, perhaps with an additional transmit amplifier.

At low frequencies (< 500 MHz, say), an OATS would normally be used. This is because any chamber would need to be very long to operate in the far field and, also, a large volume of RAM would be needed, since the RAM depth should be a couple of meters or more for low frequencies. However, some very large chambers exist that are capable of operation at lower frequencies and near-field measurements can also be used to reduce the measurable frequency of indoor systems.

Anechoic chambers can be used to measure antenna radiation either in the near or far fields. Near-field measurements are dealt with in the following two sections.

7.4.2.1   An Example of an Anechoic Chamber

A picture of an anechoic chamber at the University of Liverpool is shown in Figure 7.15. The chamber has dimensions of 6.0 m × 3.5 m × 3 m and most of the RAM has a length of 300 mm for frequencies from 1 to 40 GHz with a return loss > 40 dB. The source antenna is a broadband double-ridged horn antenna, and the QZ is about 0.5 m × 0.5 m × 0.5 m for an antenna separation of 3 m, which is suitable for the measurement of most mobile radio antennas. The system setup is given in Figure 7.16 where a VNA is used as the transmitter and receiver. A computer-controlled turntable/positioner (which has two step motors for azimuth and elevation controls respectively) is connected to a controller via an optical link (to minimize measurement errors).

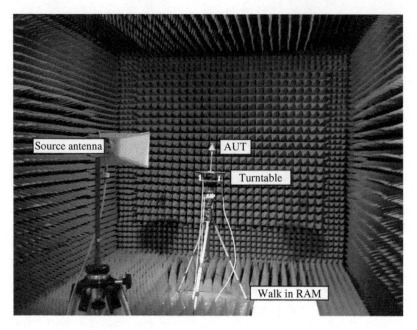

Figure 7.15 An example of an anechoic chamber

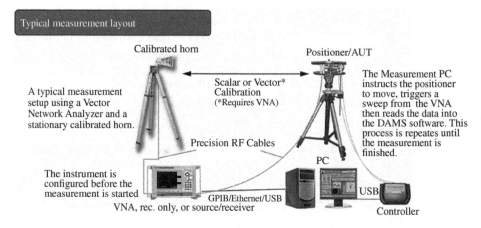

Figure 7.16 Automatic antenna measurement system setup (Diamond Engineering)

This automatic measurement system and its software were developed by Diamond Engineering [19] at a very reasonable price, which is particularly suitable for small antenna measurements.

7.4.3   Compact Antenna Test Range (CATR)/Plane Wave Generator (PWG)

For some electrically large antennas, its far-field distance is too large for a conventional anechoic chamber or even an OATS, a possible solution is to utilize a compact antenna test range (CATR) to generate a plane-wave like uniformed AUT area. CATR is also called a plane wave generator (PWG) in the new IEEE Standard 149. The idea of CATR or PWG is simple: just to replace the source antenna by a large offset parabolic reflector antenna (as discussed in Chapter 5) in a conventional anechoic chamber or OATS! Thus, a CATR/PWG can be considered as an optical system with a source antenna and one or more reflectors, which are much larger than the wavelength of the measurement. It is the most common method to perform measurements of antennas based on plane wave generation. The reflector collimates the spherical wave radiated by the source antenna into an approximately plane wave over a finite volume in the test area (i.e. QZ). The most common PWGs are array or reflector-based. Common reflector shapes are parabolic or cylindrical. The size of the test area is typically half the size of the reflector, and the distance between the reflector and test area is often three times or more of the size of the reflector. The compact range enables the testing of medium to electrically large antennas at a significantly shorter distance than would be required in a traditional far-field test range.

The plane wave generated by the CATR/PWG is approximately uniform over the test area with some taper toward the edges of the QZ, caused by the amplitude taper of the feed. Edge effects from the reflector are often compensated by adding serrations or edge shaping to direct diffractions away from the QZ. Being an optical system, the bandwidth of the CATR/PWG is mainly limited by the finite bandwidth of the feed and the efficiency of the edge treatment [20]. While CATRs/PWGs have been designed to operate at frequencies from as low as 500 MHz to the millimeter and sub-THz range, more details can be found from [21, 22]. An example of a CATR/PWG system, indicating the reflector, feed, and QZ, is shown in Figure 7.17.

The array-based CATR/PWG consists of a suitable arrangement of radiating elements with optimized excitation. The main advantage of the array-based PWG over the reflector-based solution is the reduced physical size for equivalent testing capabilities. At lower frequency, the size advantage is particularly evident as it affects system dimensions and cost. At millimeter waves, such size advantage is less important. However, the array-based CATR/PWG does not require a precisely aligned feed as the conventional CATR/PWG and its performance is not affected by the direct leakage from the feed into the QZ. These features give an added freedom in the implementation of the test system. Figure 7.18 shows the implementation of a millimeter wave array-based CATR/PWG for device tests with and without human. In this system, the DUT

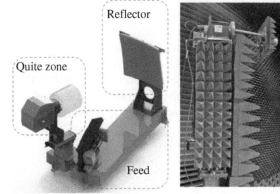
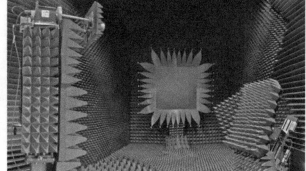

Figure 7.17   An example of CATR/PWG with feed, reflector, and QZ

Figure 7.18 An example of millimeter wave plane wave generator for device with and without human testing

is rotated only in azimuth while the CATR/PWG rotates in elevation. This enables to test devices by themselves, mounted on phantoms emulating a user scenario or even life person testing.

7.4.4   Near-Field Systems

Antenna radiation performance can also be evaluated using a near-field system where the radiated field is scanned on a surface close to the AUT. The measurement distance is generally much smaller than the far-field condition as mentioned in Chapter 4. The far-field radiation pattern is obtained mathematically with the so-called *near-field to far-field transformation* techniques. Sketches of the main scanning geometries, planar, cylindrical, and spherical, are shown in Figure 7.19. The main advantages of the near-field systems are that the short measurement distance can improve the dynamic range (smaller free-space losses) and reduce the size of the anechoic chamber with a consequent cost reduction. If the AUT is an antenna array, it can identify a faulty element easily from the near-field scanning result. There are basically three types of near-field measurement systems: planar, cylindrical, and spherical systems.

7.4.4.1   Planar and Cylindrical Near-Field Systems

Planar and cylindrical near-field scanners are particularly useful for measuring medium and high directive antennas such as apertures, arrays, and reflectors. The former, shown schematically in

(a)

(b)

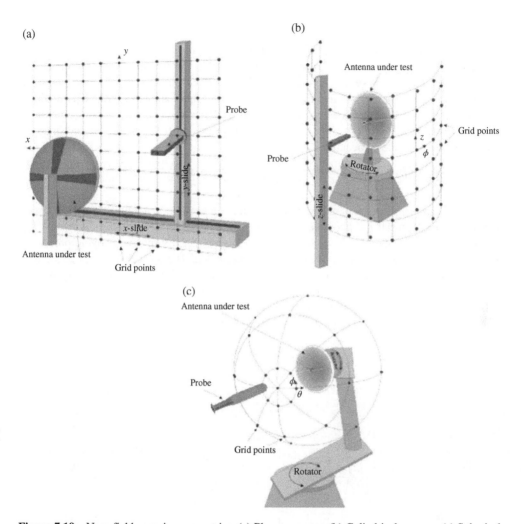

Figure 7.19 Near-field scanning geometries. (a) Planar scanner. (b) Cylindrical scanner. (c) Spherical scanner. (Courtesy of [23])

Figure 7.19(a), is best at measuring antennas with unidirectional radiation such as planar arrays (with many elements in the x and y dimensions) and parabolic antennas. The latter, shown schematically in Figure 7.19(b), is best for measuring arrays with more elements in the y than the x dimension and with omnidirectional radiation in the azimuth plane.

A planar system such as the one shown in Figure 7.19(a) operates by moving a measurement antenna – or probe – across a plane close to and extending beyond the aperture of the antenna. The spacing between this plane and the AUT is large enough to prevent mutual interaction with the probe, but small enough to allow a compact system. In a cylindrical system such as the one shown in Figure 7.19(b), the probe is moved on a vertical slide while the AUT is rotated with respect to its vertical axis (azimuths rotation).

7.4.4.2 Spherical Near-Field Systems

Spherical near-field systems have the advantage of measuring the entire field around the antenna: this allows the total radiated power (TRP) from the antenna to be measured and radiated efficiency and power can be calculated. This makes such systems particularly suitable for the measurement of handset antennas, where efficiency and radiated power are particularly important (since handset antennas operate in a multipath environment, receiving signals from all directions, the gain in any one of these directions is relatively unimportant, since an averaging effect generally occurs).

Spherical near-field measurement systems generally fall into two categories: single-probe and multi-probe systems. The former is in general less complicated electrically than the latter but is slower due to the needed mechanical movement.

The actual implementation of single-probe systems can be done in a multitude of ways, with varying degrees of mechanical complexity, ranging from a fairly simple roll-over-azimuth setup to complex double-gantry arm systems as seen in Figure 7.20. Systems employing probes on telescopic and/or robotic arms have also been implemented. The advantages of such systems are their flexibility and the ability to accommodate the planar and cylindrical scanning geometries.

Multi-probe systems have been developed (at a competitive price) largely due to the computational power of modern computers, which allows each probe to be compensated digitally every time a measurement is made (for all angles and frequencies). A typical commercial

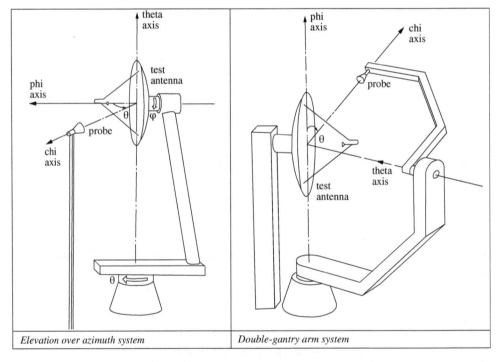

Elevation over azimuth system   *Double-gantry arm system*

Figure 7.20 Different mechanical implementation of a spherical near-field antenna measurements system. Roll-over-azimuth and double-gantry arm systems. (Reproduced by permission of the IET [24])

Figure 7.21   A multi-probe-based spherical near-field measurement MVG system

system is shown in Figure 7.21 [25]. The chamber is large enough for measurements with the complete system present: the phone and the user. The scanning method is fast enough to allow users to stay still for the duration of the measurement (the measurement of the full sphere at an angular resolution of approximately 5° with 20 frequency points is possible in approximately three minutes).

The system consists of a circular array of dual polarized wideband measurement probes (visible as crosses in Figure 7.21), for elevation measurement and a rotating turntable for azimuth measurement. The diameter of the circle of probes is approximately 3.2 m. This, combined with the near-field measurement technique, allows measurement of large radiating structures. The system shown operates from 450 MHz to 6 GHz without the need to replace any parts (as is the case for conventional near-field chambers, where the probes have a limited frequency range). More information about this kind of systems can be found from the final section of this chapter, Section 7.9.3.

7.4.5   Reverberation Chamber

Reverberation chambers, which are electrically large metallic chambers with stirrers inside, have been used for electromagnetic compatibility (EMC) measurements since the 1970s (for radiated emission and immunity tests) [26–28]. From around year 2000, they have also been applied to the measurement of antennas and mobile phones. Unlike an anechoic chamber, no RAM is required for a reverberation chamber, and its boundary condition is perfect reflection, which is the opposite of the anechoic chamber. Thus, the reverberation chamber is very different from other measurement facilities discussed above and not suitable for measuring some antenna parameters (such as the radiation pattern, although it is possible, but complicated [28]). It is not a standard antenna measurement facility. Nevertheless, it is a unique facility and complementary to the anechoic chamber as we will see later. The main advantages of this facility are given below:

- It can be made relatively small and cheap (no absorber).
- A large field strength can be generated with a limited power (good for radiated immunity test).
- The received field/power is sensitive to radiated power (good for radiated emission test and TRP measurement).
- It is a multipath environment and suitable for some special applications.

The operational principle of reverberation chambers for antenna measurements is illustrated in Figure 7.22. The chamber itself is essentially a metal box with a source antenna to generate EM field, which is stirred by the stirrer/stirrers to generate an averaged uniform field in the equipment under test (EUT) area. An AUT is placed in an approximately central location, which is the EUT area. Mode stirrers (large metallic plates or "paddles" that are controlled by motors) are used to randomize the field received by the AUT: i.e. it is as if waves arrive with equal probability from all angles. Both polarizations are also of equal probability, and the field is uniform on average away in the EUT area. This is a reasonable approximation of the multipath field that is typically experienced by a mobile phone. Hence, it can be used to measure antenna diversity gain and characterize multiple-input and multiple-output (MIMO) antennas. The reciprocity theory applies, that is, the AUT can also be the source of radiation. Since it is a shielded environment, a reverberation chamber can be used for antenna efficiency and TRP measurement, which will be discussed later in this chapter. Some commercial products are available on the market [29].

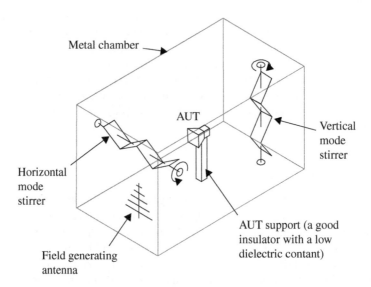

Figure 7.22   Typical reverberation chamber

7.5   Impedance, S_{11}, and VSWR Measurements

There are many important antenna parameters that can be grouped from the circuit point of view and the field point of view as discussed in Chapter 4. From the circuit point of view, impedance and related parameters are of interest. For how to conduct antenna measurement properly, one can refer to a recently revised IEEE standard on test procedures for accurate antenna measurements for details [30]. Here we are going to address some important and practical aspects of antenna measurements.

Up until approximately the late 1980s, most impedance measurements were performed with manually tuned impedance bridges. Such bridges are still used today – for example, operating impedance bridges are used for measurements of AM broadcast antennas due to their ability to be connected "in line" while high powers are transmitted by the antenna. However, for laboratory use, the VNA has become the industry standard equipment for high-frequency (distributed element) impedance measurements, and it is very different from the multimeter for DC and low-frequency (lumped element) impedance measurements.

From the circuit theory provided in Chapter 2, we know that the load impedance and the transmission line characteristic impedance determine the reflection coefficient, VSWR, and return loss. Since the line impedance for the VNA is normally set as 50 ohms, the antenna impedance, reflection coefficient S_{11} (defined by Equation 7.3), and VSWR measurements can be obtained using a VNA, which actually just measures the S_{11} (a complex number) at the port connected to the AUT and then to calculate other desired parameters (e.g. impedance, VSWR, or return loss) using the inside CPU. The standard measurement procedures are as follows:

- Select a suitable cable (low loss and phase stable) for the measurement and ensure that it is properly connected to the VNA (otherwise a significant error could be generated) – this is a major source of measurement error.
- Select the measurement frequency range and suitable number of measurement points over the frequency range.
- Perform the one-port calibration and ensure that the cable is not moved (or errors could be generated) during the calibration (and measurements).
- Conduct the measurements in an environment with little or no reflection (such as an OATS or an anechoic chamber as discussed earlier).
- Record the measured results.

These procedures are straightforward, but care must be taken, and sometimes, additional devices or measures are required – depending on the AUT. For example, if the AUT is a dipole antenna, which is directly connected to the cable, a balun or ferrite bead/choke is required as discussed in Chapter 5, or currents may come back from the outside of the outer conductor of the cable back to the VNA and result in errors. A monopole antenna with a small ground plane may also suffer from this problem as shown in Figure 7.23, where a monopole UWB antenna with a small ground plane was measured with and without a ferrite choke around the cable next to the antenna connector. At lower frequencies, the cable effect ("ringing") was quite obvious. It should also be pointed out that ferrite chokes are good at lower frequencies and may introduce loss, thus not suitable for antenna efficiency measurement. Another well-known case is the impedance measurement of an electrically small antenna, which could be tricky since the

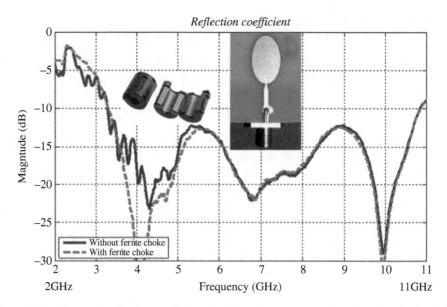

Figure 7.23 Measured reflection coefficient S_{11} of a UWB antenna with and without ferrite choke

magnitude of its measured reflection coefficient could be very close to 1, which causes singularity problem in calculating the impedance.

It should be pointed out that many people consider S_{11} in dB as the return loss, which is wrong. As defined by Equation (2.40), the return loss should be $-(S_{11}$ in dB); thus, there is a sign difference. If S_{11} is -10 dB, the return loss is $+10$ dB.

7.5.1 Can I Measure These Parameters in My Office?

In practice, many people perform this one-port measurement in a standard lab or office – this is fine provided that the antenna is not very directional, and the signal reflected could be considered small. The most sensitive frequency band is around the resonance. For a very directional antenna, one may point the antenna toward a RAM to reduce the strong reflection.

Let us take the dual-band PIFA antenna designed in Example 6.4 as an example. The comparison of the simulated and measured results was provided in Chapter 6. Here the measured S_{11} (in dB) in an office (dashed line) and in an anechoic chamber (solid line) is shown in Figure 7.24, respectively. It is apparent that the difference between them is too small to tell since the directivity of this antenna is small. The only noticeable difference is indeed around the two resonant frequencies (the two troughs).

7.5.2 Effects of a Small Section of a Transmission Line or a Connector

In order to make measurements, it is very often that we have to use a connector and probably a small section of a transmission line that cannot be included in the calibration. As a result, they

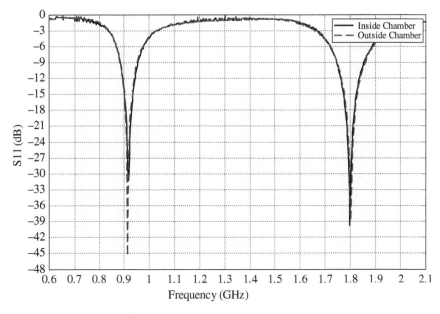

Figure 7.24 Results comparison of a dual-band antenna measured in an office and in an anechoic chamber

may introduce errors to the measured results. Their effects are more significant on the impedance than on the VSWR and return loss. Because the impedance is a complex number and sensitive to the length change of the line while the VSWR and return loss are scalar and not sensitive to the length change.

7.5.3   Effects of Packages on Antennas

The effects of a ground plane and the installation on antennas were discussed early, and it was concluded that both the impedance and radiation pattern might be affected. Here we use the dual-band PIFA antenna again as an example to illustrate how the package, as shown in Figure 7.25, may affect the antenna resonance. The package case is made of a plastic material with low loss over the frequency range of interest. The measured results of S_{11} in dB for the antenna with (solid line) and without (dashed line) the package case are shown in Figure 7.26, which is from 0.7 to 1.2 GHz for the GSM 900 band. It is apparent that the resonant frequency is shifted down from 0.965 to about 0.905 GHz, which is quite considerable. Thus, its package and installation must be taken into account when the antenna is designed.

7.6   Radiation Pattern Measurements

As one of the most important parameters, the radiation pattern has to be measured once an antenna is developed, which is a much more difficult and time-consuming task than the

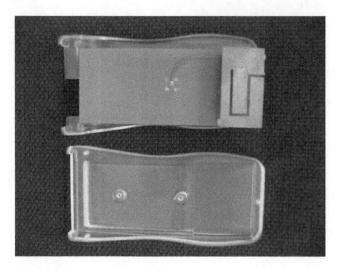

Figure 7.25 A dual-band mobile antenna with its package case

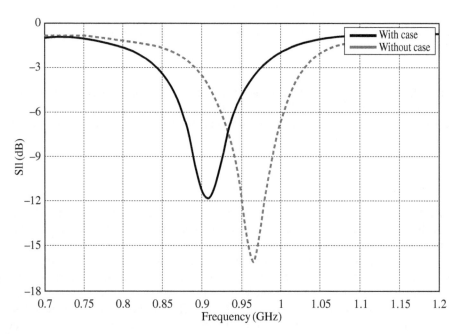

Figure 7.26 The measure S_{11} (in dB) with (solid line) and without (dashed line) the case over 0.7–1.2 GHz

impedance measurement. There are a few methods of measuring antenna radiation patterns. They can be classified as the near-field and far-field measurement methods or the frequency domain and time domain methods [30–32]. Which method is chosen depends primary on the antenna size and location. For example, medium wave broadcast antennas are installed at steel towers. It is possible to measure in the hemisphere above the ground, but a helicopter or hot-air balloon would be required to do so: normally this would be prohibitively expensive. Instead, measurements are normally performed on the ground along "radials" that emanate at the antenna phase center and extend outward (with chosen angle increments) for a few tens of kilometers. Measurements are made at various distances along these radials (at points where access to land is allowed) and radiation patterns are calculated from these measurements, where possible making allowances for the conductivity of the ground. This is something of an extreme example, but it illustrates that the method of antenna measurement can be very specific to a particular antenna type. In this section we will only deal with some common pattern measurement methods, focusing on some of the most widespread applications.

7.6.1   Far-field Condition

Antenna pattern measurements are usually performed in the far field: defined, somewhat nominally, as the separation required between source and AUT such that the phase variation of the wavefront across the aperture of the test antenna is less than $\pi/8$ radians (22.5°), as illustrated in Figure 7.27. In addition to the AUT, a source antenna is required to generate the far field.

Referring to Figure 7.27, it is clear that the circular wavefront from the phase center of the source antenna gives rise to a phase variation across the aperture of the test antenna, of width D, given by

$$\Delta\varphi = \left(\frac{2\pi}{\lambda}\sqrt{l^2 + \frac{D^2}{4}}\,l\right) = \left(\frac{2\pi l}{\lambda}\sqrt{1 + \frac{D^2}{4l^2}}\right) \qquad (7.14)$$

Assuming that $l \gg D/2$, this becomes

$$\Delta\varphi \approx \frac{\pi D^2}{4\lambda l} \qquad (7.15)$$

Defining the far field as the separation, l, required for a phase variation of less than $\pi/8$ across the aperture gives

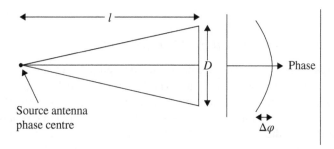

Figure 7.27   Phase variation across an antenna aperture

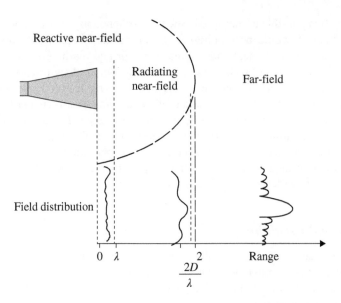

Figure 7.28   Fields from a radiating antenna. (Image provided by Nearfield Systems, Inc)

$$l \geq \frac{2D^2}{\lambda} \tag{7.16}$$

It is very important to note:

- As discussed in Section 4.1.1, Equation (7.16) is for electrically large antennas. For smaller antennas, $l > 3\lambda$ is also required.
- When the antenna is rotated, the distance from one end to the source varies from $l + D/2$ to $l - D/2$. The $1/r$ field dependence due to this rotation can give rise to pattern errors. If a ± 0.5 dB amplitude error is acceptable, then $l > 10D$ is required.
- The radiated field is sensitive to the distance as shown in Figure 7.28. The near-field pattern could be very different from the far-field pattern. The radiation pattern in the far field is stable as long as the distance is large enough to be considered in the far field.

7.6.2   Far-field Radiation Pattern Measurements

For far-field radiation pattern measurements, we can employ an OATS or an anechoic chamber. A typical measurement system setup is shown in Figure 7.16 where a VNA is used as the transmitter and receiver. Since the power from the VNA is normally small, up to 3 dBm, to ensure a good dynamic range (to see peaks and troughs), the separation between the source antenna and the AUT should not be too large, but must be large enough to be considered in the far field of both antennas. If the power from the VNA is not large enough, a power amplifier can be used or to replace the VNA with a transmitter and receiver. Once the preparation is ready, we can follow the following standard procedure to measure the radiation pattern:

- Select a suitable source antenna and place the AUT on the turntable inside the chamber or OATS with the correct polarization, height, and separation distance. It is important to ensure that the AUT is rotated around its phase center, which may not be its feeding port.
- Set the measurement frequency and suitable angle resolution for the rotation. There is a trade-off between the measurement time and angle resolution. A fine resolution is preferred for accurate results, but it may take a longer time.
- Perform the two-port calibration to remove the effects of cables. Unlike the one-port calibration, the cable may be moved for measurement after the calibration. However, the radiation pattern measurement is actually about the relative power distribution at a fixed distance as a function of angle, the measurement can be done without calibration – the drawback is that the measured S_{21} has then included the cable loss and may affect the measurement dynamic range.
- Conduct the measurements and record the measured results.
- Process the measured results and plot the pattern in 2D or 3D. It is common that the radiation pattern is normalized to the peak. The power pattern and field pattern are the same when dB scale is used.

The radiation pattern is 3D. However, in practice, only 2D patterns in the E-plane and H-plane are measured (most of the time) using the far-field measurement system, because it takes a long time to obtain a 3D pattern, which may not be a necessary requirement. 3D radiation patterns are normally obtained using computer simulation tools or near-field measurement systems.

7.6.3   Near-field Radiation Pattern Measurements

For near-field antenna measurements as shown in Figure 7.19, the critical thing is to measure both the field amplitude and phase accurately, which means that a high-spec phase-stable cable is required. The field probe should be specially designed to minimize its effects on the EM field while maximizing this bandwidth. Unlike the far-field measurement system, the radiation pattern obtained in a near-field system is through a near-field to far-field transform as discussed in Section 5.2 using numerical data processing and the fast Fourier transform. The measurement is highly automatic since it is almost impossible to do it manually. The measurement procedure is similar to the far-field system. The main difference is on the selection of the probing distance and measurement area, and the data processing is much more complicated.

A major advantage of using a near-field system is that 3D radiation pattern can be easily obtained. The drawback is that it may take longer time than the far-field system to obtain 2D plots. In terms of the measurement accuracy, it is comparable with the far-field system [30].

7.7   Gain Measurements

The most important figure of merit to describe the antenna radiation performance is the gain, which is directly proportional to the directivity and radiation efficiency. There are several methods of measuring the absolute gain of an antenna, which are discussed below.

7.7.1   Gain Comparison Measurements

This most commonly used method requires three antennas: the source antenna to produce the radiation, a reference antenna of known gain, and the AUT. The signal obtained from the AUT is compared with that from an antenna of known gain (such as a standard-gain horn for higher frequencies and a dipole for lower frequencies) with constant power transmitted from the source antenna. The gain of the antenna under test, G_{AUT} is given by

$$G_{AUT} = \frac{P_{AUT}}{P_{SG}} G_{SG} \tag{7.17}$$

where P_{AUT} and P_{SG} are the powers accepted by the AUT and the standard gain (SG) antenna, respectively (allowing for mismatches) and where G_{SG} is the known gain of the standard gain horn. The power accepted by the antenna may be different from the power received by the receiver since we need to take the impedance matching and loss of the cable into account. When a VNA is used, Equation (7.17) can be rewritten as

$$G_{AUT} = \frac{\left(1 - |S11_{SG}|^2\right)|S21_{AUT}|}{\left(1 - |S11_{AUT}|^2\right)|S21_{SG}|} G_{SG} \tag{7.18}$$

where S_{11} and S_{21} are the reflection and transmission coefficients of the SG antenna and AUT. The advantage of this method is that it does not rely on the knowledge of the path loss between the source and test antennas in the chamber. There is no need to know the source antenna gain either. Accuracy is determined by the positioning of the AUT and standard gain antenna and the calibration of the standard gain antenna. It is difficult and expensive to measure the absolute gain of standard gain antennas to within a few tenths of a dB. In 3D chambers, the directivity of the standard gain horn can be measured (via radiation pattern integration) and assumed to be equal to the gain for horns with a metal only construction (the losses are very low, so the horn can be assumed to be 100% efficient).

This gain measurement technique is also known as gain-transfer or gain-substitution method, and it can be applied both to direct far-field and to near-field measurements [30]. In the latter, two full 3D acquisitions, one for AUT and one for the standard gain antenna, are needed in order to obtain the far-field pattern of both devices and be able to compare them. This method is accurate provided that the known gain is accurate, and the main source of errors could be from the cable, turntable, and misalignment. When trying to obtain the peak gain, it is important to ensure that the peak of the radiation is directed to the source antenna, which is relatively easy for the near-field measurement system since a 3D radiation pattern is available from the measurement.

7.7.2   Two-antenna Measurement

It is possible to use two antennas (a source antenna with known gain and the AUT) to obtain the antenna gain. In this case, it relies on knowledge of the chamber path loss, L, given by (see Section 3.5.1)

$$L = \left[\frac{\lambda}{4\pi l}\right]^2 \tag{7.19}$$

where λ is the wavelength, and l is the separation between the source and test antennas. If the gains of the source and test antennas are denoted by G_S and G_{AUT} respectively, the power received by the latter is given by

$$P_{AUT} = LG_SP_SG_{AUT} \tag{7.20}$$

where P_S is the source antenna power. This equation is the same as Equation (3.42), the Friis transmission formula. If the gain of the source antenna is not known, then identical test antennas may be used at both the source and receive end of the chamber to establish their gain. Alternatively, the return signal from a single test antenna facing a large conducting sheet may be used to measure its gain (the effective chamber length being twice the separation of the antenna and sheet). In this case, it is important that the inherent mismatch of the antenna is taken into account to calculate the antenna gain from the measurement results. This method is only suitable for direct far-field measurement, not the near-field system. The main source of errors may come from the measurement environment. The accuracy of the chamber path loss given by (7.19) is critical. One may use two different separations (in the far field) to check the accuracy of the measurement results.

7.7.3 Three-antenna Measurement

If three antennas (one of them is the AUT) of unknown gains G_1, G_2, and G_3 are used as source and test antennas in all three possible combinations, denoting the source power by P_S, the received powers are

$$\begin{aligned}
P_{12} &= P_SLG_1G_2 \\
P_{23} &= P_SLG_2G_3 \\
P_{13} &= P_SLG_1G_3
\end{aligned} \tag{7.21}$$

Thus, there are three equations in three unknowns so that all three gains may be established. These equations may be easily solved, giving

$$\begin{aligned}
G_1 &= \frac{P_{12}P_{13}}{P_{23}P_SL} \\
G_2 &= \frac{P_{12}P_{23}}{P_{13}P_SL} \\
G_3 &= \frac{P_{23}P_{13}}{P_{12}P_SL}
\end{aligned} \tag{7.22}$$

As for the two-antenna measurement, this method does not rely on the use of a standard gain antenna but does require knowledge of the path loss between the source and test ends of the chamber. We can use the same method to control the source of errors and check the measurement accuracy. More information on this and related methods can be found from such as [30].

7.8 Efficiency Measurements

The antenna *radiation efficiency* is defined as the ratio of the gain and directivity of an antenna, or the ratio of the total power radiated by the antenna (P_t) to the total power accepted by the antenna (P_{in}) at its input terminal, and can be also determined by the radiation resistance and loss resistance as shown in Equation (4.38) in Chapter 4, i.e.

$$\eta_r = \frac{P_t}{P_{in}} = \frac{R_r}{R_r + R_L} \tag{7.23}$$

Thus, the efficiency can be obtained by measuring either the directivity and gain or the radiation resistance and loss resistance.

The antenna gain measurements were discussed in Section 7.7. The directivity is often computed based on the integration of the radiation pattern or based on the half-power beam width as discussed in Section 4.1. This directivity/gain method could be difficult to realize with precision in practice for some cases since the full 3D radiation pattern is required or the estimated directivity is not accurate enough.

Wheeler cap method, which was originally proposed by Wheeler [33], is a popular alternative method for antenna efficiency measurement. It measures the radiation and loss resistances of the antenna in two steps. The first step is to measure the real part of the antenna impedance (included radiation resistance R_r and loss resistance R_L) in an anechoic chamber or OATS. The second step is to measure the loss resistance by placing the AUT inside a small nonresonant metal cap to prevent its radiation as shown in Figure 7.29. In this case, the cap is equivalent to a load to the antenna (a short circuit across the radiation resistance in the ideal cap case; an open circuit in free space case). Thus, the real part of the measured input impedance should be the loss resistance of the antenna. Once the loss resistance and the real part of the input impedance are known, we can use Equation (7.23) to calculate the efficiency.

Wheeler cap method is really attractive for antenna efficiency measurements due to its convenience, accuracy, and low cost. However, the originally recommended radius of the cap should be around $\lambda/2\pi$, and the AUT size is limited approximately to $\lambda/4$ [34]. That is, the method is suitable for electrically small antennas and the cap size should be small enough to avoid resonance. For larger antennas, a larger cap has to be used, which will result in resonance – the cap is therefore no longer equivalent to a short circuit to the antenna but a load

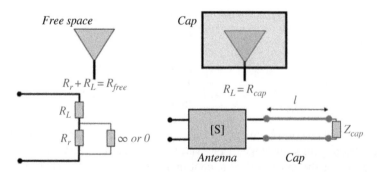

Figure 7.29   Equivalent circuit of Wheeler cap method for antenna efficiency measurement

with loss. Over the years, many people have tried to remove the effect of the cap resonance on the measurement, modified Wheeler cap methods (such as sliding wall cavity method and UWB Wheeler cap method) have been developed, and some good results are obtained [35, 36].

More recently, the reverberation chamber (RC) has been employed for antenna efficiency measurement and significant progress has been made. Like the antenna gain measurement, there are also methods using a reference antenna with or without known efficiency [30, 36]. The main advantage of the RC method is that the AUT can be very large (physically or electrically), even on-body or installed antennas can be accurately measured [37]. However, there are possibly two drawbacks on using the RC method: one is the long measurement time, and the other is that its frequency must be greater than the lowest usable frequency (LUF); thus, it is not suitable for low frequencies, which is complementary to Wheeler cap method. Of course, the cost of an RC is much higher than that of a Wheeler cap. This has led to the development of the *source stirred cap/cavity/chamber (SSC) method* [38]: the AUT is first placed inside an electrically large metal cap/chamber, its impedance is then measured when the antenna is stirred (i.e. moved inside the cap to different locations and orientations) as shown in Figure 7.30. Finally, the efficiency can be calculated using (7.23) at each position/orientation and frequency. All the results are plotted in the same figure. Most calculated efficiencies are not correct since they are affected by the resonance and the loss of the cavity is included. However, the peak values have not (little) affected cavity resonances. Thus, these peaks are the correct antenna efficiencies over the frequency range of interest. This method offers the advantages of the conventional Wheeler cap method in terms of the cost, accuracy, speed, and convenience and is also suitable for ultra-wide frequency band (for both electrically small and large antennas), which is not offered by the conventional Wheeler cap method.

Since antenna efficiency is such an important parameter but not easy to get it accurately while the demand from industry is high (e.g. how to obtain mobile antenna efficiency), it has attracted a

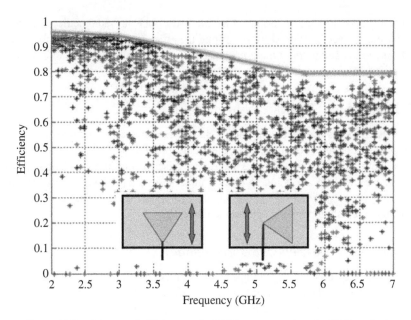

Figure 7.30   Measured efficiency of a UWB antenna using the SSC method

Table 7.2 Comparison of measurement methods for antenna efficiency measurements

Measurement method	Accuracy	Comments
Pattern integration	< 20%	Suitable for wideband antennas. 3D radiation measurement required, time-consuming.
Directivity/gain	< 20%	Suitable for a wide range of antennas, in particular directional antennas.
Wheeler cap	< 3%	Suitable for electrically small antennas, cost-effective, quick, and little data processing.
Sliding wall cavity	< 5%	Suitable for many small antennas, cost-effective. A special cavity required.
UWB Wheeler cap	< 10%	Suitable for wideband and small antennas, cost-effective, quick, and simple data processing.
Source stirred cap/ chamber	< 5%	Suitable for a wide range of antennas, wideband, cost-effective, quick, and little data processing.
RC with a reference antenna	< 10%	Suitable for both large and small antennas. A reference antenna with known radiation efficiency required. Time-consuming, and the frequency >LUF.
RC without a reference antenna	< 10%	Suitable for both large and small antennas. No reference antenna required. A lengthy chamber calibration is required and the frequency >LUF.
Radiometric	< 15%	Suitable for many antennas. Special equipment required.
Q-factor	< 10%	Suitable for electrically small antennas, not as simple as Wheeler cap method.
Calorimetric	< 5%	Suitable for small antennas and systems (e.g. a mobile phone). A special heat-flow comparator required.

lot of attention and investment. Some other methods, such as radiometric and calorimetric methods, have also been developed for measuring the efficiency. They all have their advantages and disadvantages. A summary of these methods is given in Table 7.2. More information can be found from [30, 36].

7.9   Miscellaneous Topics

Antenna measurements are a very important and active subject. Sometimes it is difficult to make accurate measurements since there are many possible sources of errors and various limitations. New antenna developments have also raised some new challenges in measurements. In this section we are going to discuss three special topics: impedance de-embedding techniques, MIMO over-the-air (OTA) testing, and probe array in near-field systems for antenna measurements. They should be of interest to some readers.

7.9.1   Impedance De-embedding Techniques

Sometimes direct measurement of the impedance of an antenna could be a problem. Now let us have another look at the dual-band PIFA antenna in Example 6.4, as shown in Figure 6.25. The feed pin is right at the edge of the PCB board. It is not possible to solder a coaxial connector at this point. For measurement purpose we have to introduce a short microstrip line as shown in Figures 6.31 and 7.31. This feed line has been carefully chosen (not to disturb the antenna current

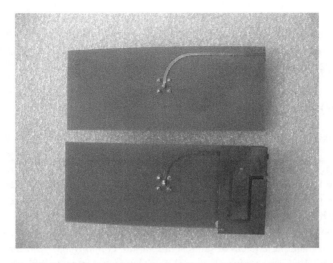

Figure 7.31 A microstrip feed line and a PIFA antenna

distribution) to minimize its effects on the antenna performance. Although the introduction of this additional line may not affect the VSWR and return loss if there is no mismatch and loss, it will certainly change the reading of the impedance. A method to eliminate the effects of the line and obtain the impedance at a desired reference point is to use the impedance de-embedding technique.

The basic idea of impedance de-embedding is to make an identical feed line with an open-circuit load and then measure its reflection coefficient, which will be used to calculate the antenna impedance at the desired reference point.

The reflection coefficient at the open circuit is +1. If the length of the feed line is l, we can use Equation (2.38) to obtain the reflection coefficient at the connector of the open-circuit line:

$$S_{11open} = \Gamma(l)_{open} = e^{-2\gamma l} \tag{7.24}$$

Similarly, the reflection coefficient at the connector of the antenna can be expressed as

$$S_{11ant} = \Gamma(l)_{ant} = \Gamma_0 e^{-2\gamma l} \tag{7.25}$$

Thus, the reflection coefficient of the antenna (Γ_0) is the ratio of the measured S_{11ant} to $S11_{open}$

$$\Gamma_0 = \frac{S_{11ant}}{S_{11open}} \tag{7.26}$$

This is the principle of the impedance de-embedding. This result can also be used to obtain the return loss, VSWR, and input impedance as discussed in Chapter 2. A formula can be yielded from Equation (2.27) to calculate the impedance:

$$Z_a = Z_0 \frac{1 + \Gamma_0}{1 - \Gamma_0} \tag{7.27}$$

As seen in Figure 7.31, an open-circuit strip line is made the same as the feed line of the PIFA antenna. The measured and simulated results are given in Figures 6.32 and 6.33. Good agreement has been obtained. Without using the impedance de-embedding, these results would be different, especially the impedance.

7.9.2   MIMO Over-the-Air Testing

MIMO technology is a method for multiplying the capacity of a radio link using multiple transmitting and receiving antennas to exploit multipath propagation. It has become an essential element for many communications standards. In addition to multiple independent antennas used at both the transmitter and receiver, complex signal processing is utilized in the system. Furthermore, a MIMO system only works well in a multipath environment. A detailed discussion on MIMO antenna design is available in Section 8.3.

In terms of the performance valuation, there are two different levels. From MIMO antenna's point of view, the main parameters of interest are *mutual coupling* (or *isolation*), *envelope correlation coefficient* (ECC), and *diversity gain* (DG). These parameters will be further discussed in Sections 8.2 and 8.3. From MIMO system's point of view, the key performance figure of merit is the *throughput*, which is the data rate passing through the radio channel. The maximum throughput can be considered as the system capacity, and it is measured in bits per second (bit/s).

To obtain the mutual coupling or isolation between two antenna elements, we normally use a VNA to measure their transmission coefficient S_{12} or S_{21} inside an anechoic chamber, which is relatively straightforward. However, to find that the ECC is more complicated, since it is a measure of the correlation between the radiation patterns of MIMO receiving antenna pairs (see Section 8.3). Ideally, we should first get the 3D radiation patterns of these antennas (which is not an easy task unless a near-field measurement system is used) and then calculate this parameter. In practice, many people measure the S-parameters and radiation efficiencies of these antennas (η_1 and η_2) first and then calculate the ECC (or ρ_e) using the equation below [39, 40]:

$$\rho_e = \frac{|S_{11}{}^*S_{12} + S_{21}{}^*S_{22}|^2}{\left|\left(1 - |S_{11}|^2 - |S_{21}|^2\right)\left(1 - |S_{22}|^2 - |S_{12}|^2\right)\eta_1\eta_2\right|} \tag{7.28}$$

where the notations are of the usual meaning. This may not be very accurate but a good estimation for low loss antennas (more can be found from Section 8.3.2). The expected $ECC = \rho_e < 0.3$ for a good MIMO ($\rho_e < 0.5$ is acceptable in some cases).

The diversity gain is another key performance indicator that shows the increase in signal-to-interference ratio due to a diversity scheme and normally gives a direct understanding of how a MIMO/diversity system outperforms a single channel system. To obtain the DG, the cumulative distribution function (CDF) of the received signal must be obtained in a multipath environment (such as a reverberation chamber or a city center) and compare with a single element case at typically 1% CDF level. An example of a two-antenna MIMO is given in Figure 7.32: at 1% CDF level, the combined signal is better and at −11.5 dB, and the theoretical Rayleigh distribution signal is at −20.33 dB while the power at single element a or b is −21.24 dB, thus this MIMO/diversity antenna has an *effective DG* (compared with ideal case, i.e. Rayleigh distribution) of −11.05 − (−20.33) = 9.28 dB and an *apparent DG* (compared with the measured single element) of −11.05 − (−21.24) = 10.19 dB, respectively [28, 41, 42]. How to measure the DG accurately and efficiently has been studied by many people. A reverberation chamber

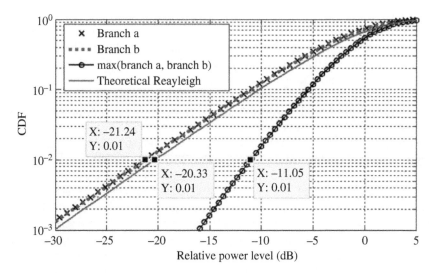

Figure 7.32 CDF plot of both branches/antennas and the combined signal, theoretical Rayleigh distribution

seems to be an ideal environment: it is indoor and multipath with certain controllable parameters. More information on this topic can be found from such as [28, 42, 43].

All these measurements are done using the conventional *conducted testing method* where RF cables and connector are used to link the equipment and AUT. As the number of antenna elements in MIMO systems continues to increase, we now have massive MIMO systems, typically over 100 elements. The conventional conducted testing approach becomes almost impossible. In addition, as the frequency increases, antennas become smaller and the integration of antennas with other devices is set to grow. Antenna-on-chip and antenna-in-package have become reality in some applications. The conventional method is no longer suitable, and an alternative measurement method is required.

Unlike conducted testing, OTA testing removes the cables and connectors linked to the DUT, which should have a transmitter or receiver) to evaluate the device or system performance over the air, in the real-life scenario. Typically, the DUT is placed in a free space environment inside a test chamber where real-life situations are simulated. The device is subjected to different test conditions to check how the device responds in various situations. OTA testing is a way to ensure that the DUT will perform the way it is intended to, by measuring its entire signal path and antenna performance. These tests are important as they should reflect the installed device/system true performance.

While choosing an antenna, data sheets are good indicators of an antenna characteristics and performance in free space. OTA tests are perfect for assessing the performance of an antenna when it has been integrated within a device. To ensure a well-behaved RF environment (i.e. a free space plus elimination of external interference) for repeatable results, the OTA testing is normally managed inside of an anechoic chamber.

For a MIMO system, a main feature is that its performance is closely associated with the propagation channel, which can be divided into urban, suburban, and rural environments. A more detailed classification (such as urban-micro and urban-macro) may be required for some applications. They are different from an anechoic chamber, on-site measurements have been conducted for some mobile systems, but the results are normally site-specific and not general enough for comparison. Research into MIMO OTA for standardization purposes has been ongoing in The Wireless Association (i.e. CTIA, stands for Cellular Telecommunications Industry Association), the Third Generation Partnership Project (3GPP), and the European Cooperation in Science and Technology for many years [44]. This was motivated by the urgent need to develop accurate, realistic, and cost-effective test standards for LTE and 5G systems. Although many MIMO networks are already deployed and some testing methods are proposed, there are not yet any formal international standards for MIMO OTA testing (to year 2020). The development of MIMO OTA test standards has proven to be particularly complex compared with single-input single-output (SISO) OTA. Unlike SISO OTA, which was relatively straightforward, MIMO OTA is highly dependent on the interaction between the propagation characteristics of the radio channel and the receive antennas of the user equipment (UE). Consequently, the existing SISO measurement techniques are unable to test the UE's MIMO properties. There are several test methodologies proposed to test MIMO system performance, each with very specific benefits and challenges. They vary widely in their propagation channel characteristics, size, and cost. For example, seven different test methods have been proposed to 3GPP in Technical Report 37.976 [45] for creating the necessary environment to test MIMO performance. The test methods fall into two main groups: five based on anechoic chambers and two based on reverberation chambers.

- Anechoic chamber candidate methods are:
     A1: multi-probe method (arbitrary number and position),
     A2: ring of probes method (symmetrical),
     A3: two-stage method,
     A4: two-channel method,
     A5: spatial fading emulator method.
- Reverberation chamber candidate methods are as follows:
     R1: basic or cascaded reverberation chamber,
     R2: reverberation chamber with channel emulator.

The anechoic and reverberation methods take fundamentally different approaches toward achieving the same end goal – the creation of a spatially diverse radio channel. In the case of the anechoic chamber, multiple probes are used to launch signals at the DUT in order to create known angles of arrival, which map onto the required channel spatial model. This is a powerful approach although in order to achieve arbitrary channel model flexibility, large numbers of probes are required, which is costly and challenging to calibrate due to issues such as backscatter. For example, the multi-probe method (A1) shown in Figure 7.33 (along with an MVG test system) is to create the desired channel model by positioning an arbitrary number of probe antennas in arbitrary positions within the anechoic chamber equidistant from the DUT (this is hard to achieve in reality when the DUT is large), each antenna being faded by a *channel emulator* to provide the desired temporal component. By careful choice of the number and position of the probe antennas, it is possible to construct an arbitrarily complex radio propagation environment. The method is the most conceptually simple since there is a direct relationship between the required angular spread of the channel and the physical location of the probes.

For the radio channel, the existing set of industry models can be easily adapted for OTA testing. The standard channel models as defined for MIMO testing largely comprise three major channel phenomena: (1) multipath, specified by a power-delay profile; (2) fading, specified by

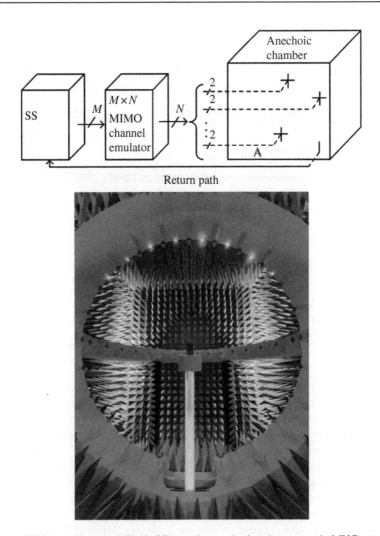

Figure 7.33 Multi-probe MIMO OTA testing method and an example MVG system

a Doppler spectrum; and (3) spatial and directional effects, as specified by MIMO fading correlation. These characteristics will allow a similar set of propagation conditions to be defined for testing a device using an OTA methodology. The channel emulator is the equipment used to recreate the dynamic over the air conditions through propagation models, thus it is expected to provide a wide range of radio channel characteristics (such as the delay spread and even Doppler frequency shifter) and is a very expensive piece of equipment.

For comparison, another measurement setup, two-stage MIMO OTA measurement (A3), is given in Figure 7.34. This method is very different from A1. The first stage involves the measurement of the 3D antenna pattern of the DUT using an anechoic chamber of the size and type used for existing SISO tests. In order to measure the antenna pattern non-intrusively (i.e. without modification of the device or the attachment of cables), a special test function is required, which reports the received power per antenna and relative phase between antennas for

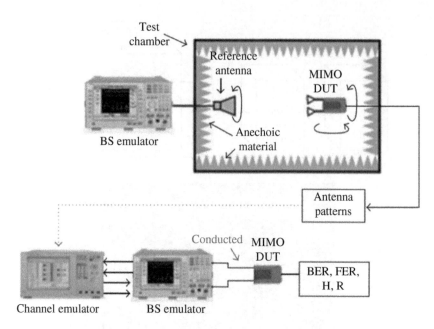

Figure 7.34 Two-stage MIMO OTA testing method. Source: From 3GPP Technical Report 37.976 [45]

a given received signal. The second stage takes the measured antenna pattern and convolves it with the desired channel model using a channel emulator. The output of the channel emulator then represents the faded downlink signal modified by the spatial properties of the DUT's antenna. This signal is then connected to the DUT's temporary antenna connectors as used for traditionally conducted testing. The second stage does not require the use of an anechoic chamber. A typical measurement result is shown in Figure 7.35 where the throughput performances of seven devices for the multi-cluster Uma channel were obtained. Pool 1 Dev 1 performed the best while Pool 4 Dev 1 performed the worst in this case.

In the reverberation chamber method, the spatial richness is provided in 3D by relying on the natural reflections within the chamber, which are further randomized by use of mode stirrers that stir the fields inside the chamber, which over a period of time (one revolution) approaches an averaged isotropic field. However, the instantaneous spatial field is not isotropic, which means that the reverberation chamber can be used to measure spatial multiplexing gain in decorrelated antennas. For further details on these testing methods, refer to [28, 45, 46].

MIMO OTA is still an evolving area. Although multi-probe and two-stage methods have now received wide recognition in the mobile industry. Reverberation-based methods offer some unique features and could be better used for most MIMO OTA testing (but not all aspects). 3GPP is still working on 5G MIMO OTA testing and the Technical Report TR 38.827 is to be finalized [47]. The development of millimeter-wave massive MIMO system and associated measurement technologies is still a hot topic. For example, reference [48] proposed an OTA testing method for millimeter-wave massive MIMO system using cascaded amplitude and phase modulation network and channel emulator. Hopefully in a few years, international standard on MIMO OTA will be made available, maybe not based on one but two methods.

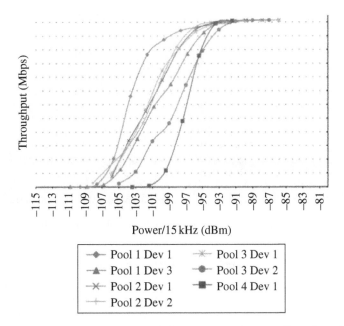

Figure 7.35 Throughput performance of 7 DUTs using multi-cluster Uma channel model [46]

7.9.3 Probe Array in Near Field Systems

Lars Jacob Foged
  Scientific Director, MVG

7.9.3.1 Introduction

Further to the discussion on the near-field chamber in Section 7.4.4, we are going to present a more detailed introduction and examination on the near-field measurement techniques with reference to MVG (former SATIMO) systems in this section.

Spherical near-field measurements are traditionally conducted by sampling two orthogonal field components on a sphere surrounding the AUT. At any point on the spherical measurement surface, the measurement antenna (probe) must point to the center of the sphere. In principle, it does not matter which of the two antennas moves relative to the other: the AUT may be fixed, with all rotations being done by the probe, it may rotate around two axes with the probe rotating around its bore-sight axis, or it may rotate around one axis with the probe rotating around two axes. If the probe is dual-polarized, there is no need for rotating the probe around its bore-sight axis.

Spherical near-field antenna testing exploits the fact that the AUT can be characterized by a finite, discrete set of coefficients, which expresses the radiation and due to reciprocity, the transmitting and receiving properties of the antenna. These coefficients are the weight factors in a truncated expansion of the AUT radiation in spherical vector waves. This expansion satisfies Maxwell's equations and allows the complete characterization of the AUT radiation everywhere outside a sphere centered in the reference coordinate system and completely enclosing the AUT called the "minimum-sphere" [24]. The size of the minimum sphere

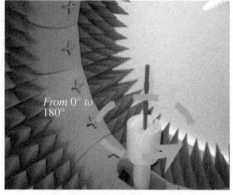

Figure 7.36   Electronic scanning for elevation sampling and AUT rotation for azimuth sampling.

determines how many near-field samples are needed to make a complete characterization of the AUT. From general sampling theory we know that at least one dual-polarized sample is needed for each $(0.5\lambda)^2$ unit area of the minimum sphere. The measurement distance can be anywhere outside the minimum sphere without impacting the sampling requirement. The calculation of the far-field radiation from the spherical wave expansion is often referred to as "near field to far-field transformation," which was discussed in Section 5.2.1 and [24].

7.9.3.2   Multi-Probe Systems

In the traditional approach, described above, the mechanical movement of the probe and/or AUT is a time-consuming task with respect to the overall antenna measurement time. It is evident that an elegant way to reduce the number of mechanical movements, often two or three in traditional systems, to just one is to use several antennas in the form of an array to perform the field sampling in one dimension. This leads to a reduction in mechanical complexity, which significantly reduces the time required to complete a measurement close to a factor equal to the number of antennas in the array. An implementation of a spherical near-field system based on probe arrays is shown in Figure 7.36. The array elements are mounted on a circular arch and embedded in multilayer conformal absorbers. The probe tips protrude through small-crossed slots in the smooth curvature of the absorbers keeping the reflectivity of the probe array at a minimum. The absorbing material also reduces scattering and reflections from the support structure and cabling. The full sphere measurement is performed by electronically scanning the probe array in elevation and rotating the AUT 180° in azimuth as illustrated in Figure 7.36. The array elements operate over more than a decade of bandwidth and are dual-polarized; hence, two orthogonal field components are simultaneously sampled at any measurement point.

7.9.3.3   Probe Array Calibration

Since the array probes have low directivity throughout the operating frequencies and the measurement distance to the AUT is not too close, the probes may be assumed to behave like Hertzian dipoles and the received signal will be proportional to the electric field parallel to the polarization of each probe. This means that traditional probe correction is not required in the near-field to far-field transformation. However, since the probes are manufactured using high-volume industrial production techniques, the probe response cannot be assumed to be equal in amplitude, phase, and polarization orientation. To correct these discrepancies, a probe array calibration must be performed.

MVG's probe arrays are calibrated by pointing a linearly polarized antenna toward each of the probes. Dual-polarized phase and amplitude measurements are taken as the antenna is rotated in polarization in front of each probe. From the data, a set of calibration coefficients are derived for each probe. The calibrated horizontal and vertical components of the electromagnetic field are obtained from the measurements. The process is able to compensate for differences among probes, align the polarization axes, and reduce the cross-polarization components. After calibration, the typically measured probe uniformity is better than ±0.1 dB in amplitude and ±1° in phase.

7.9.3.4 Gain Calibration

The directivity of an AUT is determined by a spherical near-field measurement followed by a near-to-far-field transformation. During the transformation, the TRP is calculated, and the field is normalized to obtain the directivity of the AUT in dBi.

The gain of the AUT is determined by the gain substitution technique already introduced in Section 7.7. Such a technique requires a reference antenna with a known gain. Any calibrated antenna at the desired frequency can be used as reference. Nevertheless, it is a good practice to involve reference antennas as similar as possible to the antenna to be measured (e.g. dipoles for low directive devices and standard gain horn for medium/high directive devices). The far field of the AUT is calculated from the spherical near-field measurement through the application of a near-to-far-field transformation. Another full-sphere measurement of a reference antenna with a well-known gain is carried out at the same frequency and with the same setup as for the AUT and the corresponding far field of the reference antenna is calculated. Comparing the far-fields of the AUT and the reference antenna, the AUT gain can be calculated.

In spherical near-field measurements, there are two practical approaches for gain determination based on the known gain or efficiency performance of the reference. The two methods are often referred to as *Gain Calibration* and *Efficiency Calibration* techniques. Both approaches can determine the gain of the AUT according to the IEEE definition [49].

7.9.3.5 Applications and Performance Validation

The probe array technology is used extensively in many demanding measurement fields including automotive, telecom, aerospace, and defense. There is no theoretical upper limit for the frequency range although spherical near-field systems rarely go beyond 60 GHz since very high frequency antennas tend to be directive due to their physical size, and thus near-field systems based on planar scanning are often preferred due to the simpler mechanics. The lower frequency limit for probe array technology is often imposed by the physical size of the range. Probe arrays have been used down to 70 MHz for automotive testing.

A very efficient way to validate an antenna test range is to measure a known reference antenna or compare measurements with a known reference facility. Alternatively, measurements on a

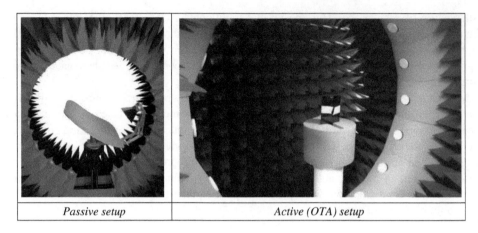

| *Passive setup* | *Active (OTA) setup* |

Figure 7.37   Example of probe array systems: SL50GHz

reference antenna in which calculated performances can be determined with a high confidence level can be used. Double-ridged horns, linear arrays, and reflector antennas have been used to compare the results from different institutes using different facilities. Very good agreements have been obtained [50–52].

Figure 7.37 shows the implementation of a portable spherical multi-probe system (StarLab 50GHz – SL50GHz) with 90 cm diameter working in the 20–50 GHz frequency range.

7.9.3.6   OTA Measurements

A system such as the one shown in Figure 7.37 is capable of measuring both passive and active devices under test. The former are devices that can be connected to an external signal source or receiver (connectorized device), and the latter are devices with an internal signal generator or receiver. For example, antennas of modern wireless devices fall in this second category. Such antennas are in fact strongly integrated with the whole RF system; hence, the users cannot access directly to its outputs/inputs. Measurements of such active devices are typically called OTA testing as discussed in Section 7.9.2 and are studied and standardized by different organizations (e.g. CTIA and 3GPP).

In OTA testing, two spatial power quantities, the *Effective Isotropic Radiated Power* (EIRP) and *Effective Isotropic Sensitivity* (EIS), are typically measured in order to respectively characterize the transmitting and receiving properties of a device. To accomplish this, an external device capable of setting up a communication and control link with the wireless/mobile terminal under test is needed. This slightly complicates such type of testing with respect to traditional antenna testing.

Since the EIRP and EIS are power-related quantities (like the gain), they can be measured directly in a calibrated test setup using the gain substitution method as described earlier. TRP and *Total Isotropic Sensitivity* (TIS), which is also known as *Total Radiated Sensitivity*, are also power-related quantities obtained by integrating the EIRP and the EIS respectively. The TRP and the TIS are the principal figures of merit in the OTA testing, and they represent the average directional line-of-sight (LOS) performance of the transmitter and receiver. Specifically, the

TRP is defined as the integral of the power radiated by the wireless device in all directions over the entire sphere for both polarizations:

$$TRP = \frac{1}{4\pi} \iint EIRP(\theta, \varphi) \sin(\theta) \, d\theta \, d\varphi \tag{7.29}$$

If the EIRP is measured in discrete samples on a regular angular grid around the DUT with N samples in θ and M samples in φ for a total of $N \times M$ measurement points, the TRP can be approximated by a summation:

$$TRP \approx \frac{\pi}{2MN} \sum_{n=0}^{N-1} \sum_{m=0}^{M-1} EIRP(\theta_n, \varphi_m) \sin(\theta_n) \tag{7.30}$$

The TIS refers to the average receiver sensitivity across the entire radiated sphere in order to achieve a specific error rate threshold for both polarizations, which represents the overall receiving capability of the DUT:

$$TIS = 4\pi \iint EIS(\theta, \varphi) \sin(\theta) \, d\theta \, d\varphi \tag{7.31}$$

If the TIS is measured in discrete samples on a regular angular grid around the DUT with N samples in θ and M samples in φ for a total of $N \times M$ measurement points, the TIS can be approximated by a summation:

$$TIS \approx 2MN / \left(\pi \sum_{n=0}^{N-1} \sum_{m=0}^{M-1} \frac{1}{EIS(\theta_n, \varphi_m)} \sin(\theta_n) \right) \tag{7.32}$$

The reported summation formulas for TRP and TIS are valid for regular angular grids. In case an irregular angular grid is used, a normalization factor on each angular term can be used.

As an example, we report below the measurement of the TRP of the wireless device provided by Sony Mobile shown in Figure 7.38. The upper patch array has been measured at 24 GHz in the SL50GHz system as shown in Figure 7.37-right. In particular, the EIRP has been measured by first calibrating the system and then considering different sampling densities ($\Delta\theta$, $\Delta\varphi$).

Results are reported in Table 7.3. The differences in the TRP are approximately 0.1 dB when 5°–20° sampling resolutions are considered. Such difference is considered low enough for this type of application; hence, it can be concluded that for such devices 10°–15° sampling resolution is sufficient in order to measure the TRP.

7.9.3.7 Power Density Measurements

Modern user equipment, such as mobile terminals and tablets, should comply with RF safety guidelines. In RF/microwave frequencies, human exposure to electromagnetic fields is evaluated in terms of incident power density, i.e. free space Poynting vector. The power density is normally measured over rectangular planes at specified distances from the DUT using proper scanners and probes connected to a power meter or a spectrum analyzer. Although it is relatively easy, such measurement procedure can be very time-consuming (due to the high

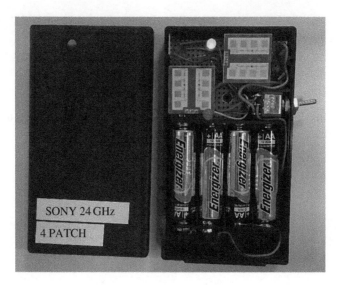

Figure 7.38 Wireless device under test working at 24 GHz. Source: Courtesy of Sony Mobile

Table 7.3 TRP vs sampling grid resolution

$\Delta\theta = \Delta\varphi$ (degrees)	TRP (dBm)
5	4.49
10	4.41
15	4.42
20	4.38
25	3.75
30	4.66

number of points to be measured) and also not very accurate because of the mutual coupling between the DUT and the probe.

An alternative solution is to measure the device in a calibrated system such as the one shown in Figure 7.37 and represents it in terms of equivalent currents (EQC) as described in [53]. The EQC are defined on an equivalent geometry fully enclosing DUT and from them the field can be calculated everywhere outside such a geometry. In other words, starting from the near field measured on the scanning surface, a near-field to near-field transformation on any surface and at any distances can be performed evaluating the power densities according to the safety normative.

It should be noted that the EQC technique requires amplitude and phase data; hence, if an OTA setup is used, a phase recovery technique should also be involved.

In the below example, a 5G mobile terminal mock-up manufactured by Sony Mobile [54] containing a notch element working at 28 GHz is considered. The device has been measured in the multi-probe SL50GHz shown in Figure 7.37. From the measured data the EQC have reconstructed on box conformal to the DUT. Figure 7.39 shows the EQC of the notch element

Figure 7.39 Simulated (left) and reconstructed (right) EQC from measured data of the mobile terminal mock-up

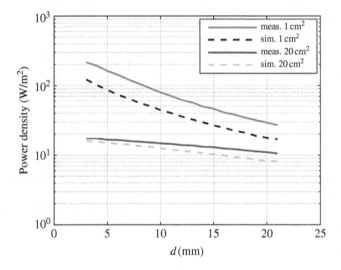

Figure 7.40 Measured (from reconstructed EQCs) and simulated spatial-average power density for the notch element to be compliant with the FCC and ICNIRP power density limits

respectively simulated with a full-wave software (left) and reconstructed from measurements (right). In both cases the EQC has been used to calculate the near field on planes at distance d from the device as also shown in Figure 7.39. From the computed near field, the incident power density has been evaluated. Figure 7.40 shows the maximum spatial average power density versus the distance d. The discrepancy observed between the measured and the simulated power density with 1 cm^2 is probably due to positioning error of the device. The results suggest that the misalignment could be about 3–4 mm. Such error could be improved by better alignment methods. The discrepancies are instead less pronounced considering the maximum spatial average power density with 20 cm^2 meaning that such metric is more robust against possible measurement error.

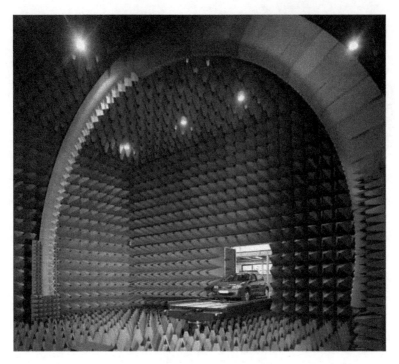

Figure 7.41   Application of probe array systems to automotive testing

The probe array technology is used extensively in many demanding measurement tasks, not limited to mobile phones. Figure 7.41 shows the implementation of an automotive system for the characterization of complete car systems from 70 MHz to 18 GHz. We believe that near-field systems offer many advantages; of course, there are also challenges ahead.

7.10   Summary

In this chapter, we have addressed three practical issues: antenna materials, fabrication, and measurements. The developments on new materials have shown increasing impact on antenna design and performance. The advancements on fabrication and manufacturing technologies have enabled us to make a wide range of novel antennas, which were not possible before. All these areas are interlinked and a breakthrough in one area will result in new development in other areas. The recent development on metamaterials and metasurfaces was discussed in detail. A wide range of antenna fabrication technologies and examples have been presented and discussed. The antenna measurement is a vital part for antenna development. Here we have introduced the essential antenna measurement equipment and facilities and provided an overview and procedures on measuring the most important antenna parameters. The latest development on MIMO OTA testing has also been addressed. The multi-probe near-field system of MVG has been introduced and applied to a selected example.

References

1. D. Kajfez and P. Guillon (Eds.), *Dielectric Resonator*, Artech House, Norwood, MA, 1986.
2. A. A. Kishk, Y. Tin, and A. W. Glisson, "Conical dielectric resonator antennas for wideband applications," *IEEE Trans. Antennas Propag.*, 50, 469–474, 2002.
3. K. M. Luk and K. W. Leung (Ed.), *Dielectric Resonator Antennas*, Research Studies Press Ltd, 2003.
4. G. DeJean, R. Bairavasubramanian, D. Thompson, G. Ponchak, M. Tentzeris, J. Papapolymerou, "Liquid crystal polymer (LCP): a new organic material for the development of multilayer dual-frequency/ dual-polarization flexible antenna arrays," *IEEE Antennas Wireless Propag. Lett.*, 4, 22–26, 2005.
5. J. K. Fink, *Liquid Crystal Polymers in High Performance Polymers*, 2nd Edition, Science Direct, 2014.
6. A. Mehdipour, I. Rosca, Abdel-Razik Sebak, Christopher W. Trueman, and Suong V. Huo "Carbon nanotube composites for wideband millimeter-wave antenna applications," *IEEE Trans. Antennas Propag.*, 59 (10), 3572–3578, 2011.
7. C. Caloz and T. Itoh, *Electromagnetic Metamaterials: Transmission Line Theory and Microwave Applications*, Wiley-IEEE Press, New York, USA, 2005.
8. C. L. Holloway, E. F. Kuester, J. A. Gordon, John O'Hara, Jim Booth, David R. Smith, "An overview of the theory and applications of metasurfaces: the two-dimensional equivalents of metamaterials." *IEEE Antenna Propag. Mag.*, 54(2) 10–35, 2012.
9. H. T. Chen, A. J. Taylor, N. F. Yu, "A review of metasurfaces: physics and applications," *Rep. Prog. Phys.*, 79(7), 076401, 2016. https://doi.org/10.1088/0034-4885/79/7/076401.
10. V. G. Veselago, "The electrodynamics of substances with simultaneously negative values of ε and μ," *Sov. Phys. Uspekhi*, 10(4), 509–514, 1968.
11. J. B. Pendry, "Negative refraction makes a perfect lens," *Phys. Rev. Lett.* 85, 3966, 2000.
12. D. R. Smith, W. J. Padilla, D. C. Vier, S. C. Nemat-Nasser, and S. Schultz, "Composite medium with simultaneously negative permeability and permittivity," *Phys. Rev. Lett.*, 84(18), 4184–4187, 2002.
13. D. Sievenpiper, L. Zhang, R. F. J. Broas, N. G. Alexopolous, E. Yablonovitch, "High-impedance electromagnetic surfaces with a forbidden frequency band," *IEEE Trans. Antennas Propag.*, 47(11), 2059–2074, 1999.
14. A. Li, S. Singh and D. Sievenpiper, "Metasurfaces and their applications," *Nanophotonics*, 7(6), 2018, https://doi.org/10.1515/nanoph-2017-0120.
15. J. Wang, Y. Li, Z. H. Jiang, T. Shi, M. Tang, Z. Zhou, Z. N. Chen, and C. Qiu, "Metantenna: when metasurface meets antenna again," *IEEE Trans. Antennas Propag.*, 68(3), 1332–1347, 2020.
16. A. Syed, K. Demarest and D. D. Deavours, "Effects of antenna material on the performance of UHF RFID tags", IEEE Int Conf. on RFID, 2007.
17. B. K. Tehrani, Jo Bito, Jimmy G. Hester, Wenjing Su, Ryan A. Bahr, Benjamin S. Cook, Manos M. Tentzeris, Advanced antenna fabrication processes (MEMS/LTCC/LCP/Printing) in Chen, Z.N., Liu, D., Nakano, H., Qing, X., Zwick, Th. *Handbook of Antenna Technologies*, Springer, 2016.
18. Poole, I. How to calibrate a VNA. https://www.electronics-notes.com/articles/test-methods/rf-vector-network-analyzer-vna/how-to-calibrate-vna.php.
19. Diamond Engineering: http://www.diamondeng.net/.
20. A. Jernberg, M. Pinkasy, G. Pinchuk, T. Haze, R. Konevky, L. Shmidov, R. Braun, B. Gershkovich, and J. Foged "Short focal length compact antenna test range for large L Ku band antenna measurements", IEEE Indian Conf. on Antennas and Propagation (InCAP), December 2018.
21. G. E. Evans, *Antenna Measurement Techniques*, Artech House, 1990.
22. C. A. Balanis, *Antenna Theory: Analysis and Design*, 2nd edition, John Wiley & Sons, Inc., 1997.
23. IEEE Std 1720-2012, *Recommended Practice for Near-Field Antenna Measurements*, IEEE 2012.

24. J. E. Hansen (Ed.), *Spherical Near-Field Antenna Measurements*, IEE Electromagnetic Waves Series 26, Peter Peregrinus Ltd. 1988.
25. Microwave Vision Group (MVG): https://www.mvg-world.com/.
26. IEC 61000-4-21:2011, *Electromagnetic compatibility (EMC): Testing and Measurement Techniques – Reverberation Chamber Test Methods*. 2nd edition, IEC 2011.
27. S. Boyes and Y. Huang, *Reverberation Chamber: Theory and Applications to EMC and Antenna Measurements*, John Wiley & Sons, 2016.
28. Q. Xu and Y. Huang, *Anechoic and Reverberation Chamber: Theory, Design, and Measurements*. John Wiley & Sons, 2019.
29. Bluetest: http://www.bluetest.se/
30. IEEE Std 149, *IEEE Standard Test Procedures for Antennas*, Revised, IEEE 2021.
31. J. S. Hollis, T. J. Lyno and L. Clayton, *Microwave Antenna Measurements*, Scientific-Atlanta, 1970.
32. W. H. Kummer and E. S. Gillespie, "Antenna measurements – 1978," *Proc. IEEE*, 66(4), 483–507, 1978.
33. H. A. Wheeler, "The radiansphere around a small antenna," *Proc. IRE*, 47, 1325–1331, 1959.
34. E. H. Newman, P. Bohley and C. H. Walter, "Two methods for the measurments of antenna efficiency," *IEEE Trans. Antennas Propag.*, 23, 457–461, July 1975.
35. R. H. Johnston, J. G. McRory, "An improved small antenna radiation-efficiency measurement method", *IEEE Antennas Propag. Mag.*, 40(5), 40–48, 1998.
36. Y. Huang, "Radiation efficiency measurements of small antennas" in: *Handbook of Antenna Technologies* by Chen, Liu, Nakano, Qing, Zwick (Eds.), Springer, Singapore. https://doi.org/10.1007/978-981-4560-44-3_71.
37. S. J. Boyes, P. J. Soh, Y. Huang, G. A. E. Vandenbosch, and N. Khiabani, "Measurement and performance of textile antenna efficiency on a human body in a reverberation chamber," *IEEE Trans. Antennas Propag.*, 61(2), 871 – 881, 2013.
38. Y. Huang, Y. Lu and H. Chattha, "Antenna efficiency measurements," IEEE iWAT 2010, Lisbon, Portugal, March 2010.
39. P. Hallbjorner, "The significance of radiation efficiencies when using S-parameters to calculate the received signal correlation from two antennas," *IEEE Antenna Wireless Propag. Lett.*, 4, 97–100, 2005.
40. R. Cornelius, A. Narbudowicz, M. J. Ammann and D. Heberling, "Calculating the envelope correlation coefficient directly from spherical modes spectrum," 2017 11th European Conf. on Antennas and Propagation (EUCAP), Paris, 2017, pp. 3003–3006. doi: 10.23919/EuCAP.2017.7928132.
41. P.-S. Kildal and K. Rosengren, "Correlation and capacity of MIMO systems and mutual coupling, radiation efficiency, and diversity gain of their antennas: simulations and measurements in a reverberation chamber," *IEEE Commun. Mag.*, 42(12), 104–112, 2004.
42. P. Kildal, C. Orlenius and J. Carlsson, "OTA testing in multipath of antennas and wireless devices with MIMO and OFDM," *Proc. IEEE*, 100(7), 2145–2157, 2012.
43. Q. Xu, Y. Huang, X. Zhu, S. S. Alja'afreh and L. Xing, "A new antenna diversity gain measurement method using a reverberation chamber, "*IEEE Antennas Wireless Propag. Lett.*, 14, 935–938, 2015.
44. M. Rumney, R. Pirkl, M. H. Landmann, D. A. Sanchez-Hernandez, "MIMO over-the-air research, development, and testing", *Int. J. Antennas Propag.*, 2012, Article ID 467695, 8pp, 2012.
45. 3GPP Technical Report 37.976: Measurement of radiated performance for Multiple Input Multiple Output (MIMO) and multi-antenna reception for High Speed Packet Access (HSPA) and LTE terminals. www.3gpp.org.
46. Y. Jing, X. Zhao, H. Kong, S. Duffy, M. Rumney, "Two-stage over-the-air test method for LTE MIMO device performance evaluation", *Int. J. Antennas Propag.*, 2012, Article ID 572419, 6pp, 2012.

47. 3GPP TR 38.827: Study on radiated metrics and test methodology for the verification of multi-antenna reception performance of NR User Equipment (UE). www.3gpp.org.
48. L. Xin, Y. Li, H. Sun and X. Zhang, "OTA testing for massive MIMO devices using cascaded APM networks and channel emulators", *Int. J. Antennas Propag.*, 2019, Article ID 6901383, 14pp., 2019.
49. IEEE Std 145-1993 (1993). *IEEE Standard Definitions of Terms for Antennas.* IEEE.
50. L. J. Foged, Ph. Garreau, O. Breinbjerg, S. Pivnenko, M. Castañer and J. Zackrisson "Facility comparison and evaluation using dual ridge horn," AMTA 2005, Newport RI, USA, 2005.
51. L. J. Foged, B. Bencivenga, O. Breinbjerg, S. Pivnenko, G. Di Massa and M. Sierra-Castañer, "Measurement facility comparisons within the European Centre of Excellence", AMTA 2007, St. Louis, Missouri, USA, 2007.
52. M. A. Saporetti, L. J. Foged, A. A. Alexandridis, F. las Heras, C. López, Y. Kurdi and M. Sierra Castañer "International Facility Comparison Campaign at L/C Band Frequencies," AMTA 2017, 15–20 October, Atlanta, GA, USA, 2017.
53. J. L. Araque Quijano and G. Vecchi, "Improved accuracy source reconstruction on arbitrary 3-D surfaces," *IEEE Antennas Wireless Propag. Lett.*, 8, 1046–1049, 2009.
54. B. Xu, Z. Ying, L. Scialacqua, A. Scannavini, L. J. Foged, T. Bolin, and K. Zhao, "Analysis and diagnosis of 28 GHz antennas in 5G Mobile terminal" *IEEE Access*, 6, 48088–48101, 2018.

Problems

Q7.1. What are the popular materials for making antennas?

Q7.2. What is the popular method for fabrication a planar antenna?

Q7.3. Explain the concept of metamaterials and metasurfaces.

Q7.4. Give one example to show how to use metamaterials/surfaces for antenna design.

Q7.5. Explain which equipment is required to measure antenna input impedance and VSWR.

Q7.6. Can we perform antenna impedance measurements inside a standard office? Justify you answer.

Q7.7. How do you measure the radiation pattern of an antenna?

Q7.8. How do you measure the gain of an antenna?

Q7.9. Explain the impedance de-embedding technique.

Q7.10. What is a network analyzer? What are its major functions for antenna measurements?

Q7.11. Explain the concept of S-parameters.

Q7.12. Explain the concept of antenna near-field measurement. How to obtain the radiation pattern through the near-field measurement?

Q7.13. What are the major antenna ranges (measurement facilities)? Compare them in terms of the advantages and disadvantages.

Q7.14. What is an OATS? Are there reflections in OATS? If yes, how to remove/minimize them?

Q7.15. What is an anechoic chamber? How to determine the thickness of the RAM used for the chamber?

Q7.16. What is a reverberation chamber? What are the antenna parameters that can be measured inside the chamber?

Q7.17. How to measure antenna radiation efficiency? Give two measurement methods and compare their advantages and limitations for antenna efficiency measurement.

Q7.18. Explain the concept of OTA testing. How to conduct the MIMO OTA testing?

8

Special Topics

In this chapter, we will address a few "hot" topics and timely issues in antennas, which include electrically small antennas, mobile terminal and base station antennas, diversity and MIMO antennas, multi-band and ultra-wideband (UWB) antennas, radio frequency identification (RFID) antennas, reconfigurable antennas, automotive antennas, and reflector antennas. The aim is to broaden your knowledge and provide you with the information on latest developments in antennas.

8.1 Electrically Small Antennas

8.1.1 *The Basics and Impedance Bandwidth Limits*

8.1.1.1 Introduction

Whether to reduce wind loading, to accommodate the size specification, and to allow more attractive products or simply to reduce costs, antenna engineers are routinely asked to reduce the size of antennas without significantly compromising their performance. Hence, it is important to know that there are fundamental limits that determine the relationship between antenna size, bandwidth, and efficiency. When the antenna size becomes too small compared with wavelength of interest, either the bandwidth or the efficiency *must* be compromised: this section gives fundamental relations between antenna volume, bandwidth, and efficiency.

The bandwidth of an antenna can be broadened to some degree using passive circuitry. However, once again, there are limits to the improvements that can be realized and the circuits introduce losses of their own. Again, this section quantifies the improvements that can be achieved.

Antennas: From Theory to Practice, Second Edition. Yi Huang.
© 2022 John Wiley & Sons Ltd. Published 2022 by John Wiley & Sons Ltd.
Companion website: www.wiley.com/go/huang_antennas2e

Though somewhat arbitrary, an antenna is often considered to be electrically small when the following condition is met [1]

$$\frac{r}{\lambda} \le \frac{1}{2\pi} \text{ or } \beta r \le 1 \tag{8.1}$$

where r is the radius of a sphere that just contains the antenna and λ is the wavelength. Note that, electrically small antenna could be physically large if the frequency is low. The concept of a sphere that just encloses the antenna is illustrated in Figure 8.1. The feed terminals are not necessarily at the center of the sphere. We will see that this concept is important to the theory and interpretation of minimum antenna Q.

In the remainder of this section on the basics of small antennas, first "slope parameters" are introduced and related to the simple series and parallel resonant circuits of Chapter 2 (where quality factor, Q, is derived). It is shown that quality factor can be expressed in terms of the resistance and the derivative of the reactance with frequency. This applies at any frequency, not just at resonance, allowing basic antenna Q to be determined from the antenna impedance (and variation with frequency), independent of any matching networks.

Second, relations between quality factor and fractional impedance bandwidth are derived, with particular emphasis given to the criteria for maximum bandwidth.

Third, the fundamental limits of antenna size, unloaded quality factor and efficiency are reviewed.

Finally, the limits of bandwidth broadening using passive circuits are summarized. This yields the maximum possible bandwidth available from an antenna of a given size that is matched with an idealized circuit. This circuit requires an infinite number of components. Hence, a more practical analysis is given focusing particularly on single-stage matching circuits often referred to as "*double-tuning*" circuits. This analysis is extended to include the case where

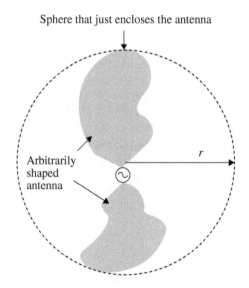

Sphere that just encloses the antenna

Arbitrarily shaped antenna

r

Figure 8.1   A sphere that just encloses the antenna

the bandwidth broadening components have finite losses. A simple, but important relation is derived showing the efficiency of a double-tuning circuit as a function of the antenna and circuit quality factors. This relation puts the concentration on double-tuning in context, indicating that higher order bandwidth broadening (via additional circuitry) is often not worthwhile. One example is given at the end to demonstrate the theory presented.

8.1.1.2   Slope Parameters

Series and parallel slope parameters, *x* and *b*, respectively, are defined [2] as

$$x = \frac{\omega}{2}\frac{dX}{d\omega} \tag{8.2}$$

$$b = \frac{\omega}{2}\frac{dB}{d\omega} \tag{8.3}$$

where X and B are the reactance and susceptance, respectively. These parameters allow distributed circuits such as transmission lines to be represented by series or parallel lumped equivalent circuits.

The slope is normally specified only at resonance or antiresonance (for series and parallel equivalent circuits, respectively). However, here we allow the slope to be calculated at any frequency and show that the use of the slope parameters is quite general.

Using the series resonant circuit of Figure 2.28, the series slope parameter, x, is given by

$$x = \frac{1}{2}\left(\omega L + \frac{1}{\omega C}\right) \tag{8.4}$$

Comparing this with Equation (2.60), the generalized formula for quality factor gives $x = QR$. Thus, the quality factor can immediately be written in terms of the slope parameter

$$Q = \frac{x}{R} = \frac{\omega}{2R}\frac{dX}{d\omega} \tag{8.5}$$

It is not necessary to insist that the circuit is resonant, as is often assumed. It is convenient to apply this description to antenna problems, where the antenna has a series resonant nature but is not resonant: in particular small antennas often have such characteristics. Similarly, the parallel slope parameter can be applied to antennas with substantially parallel resonances. In the parallel case,

$$Q = \frac{b}{G} = \frac{\omega}{2G}\frac{dB}{d\omega} \tag{8.6}$$

where G is the conductance of the circuit.

So far, the Q factor has been defined for series or parallel circuits, both in a fundamental form and in a slope parameter form that is readily applied to distributed components and some antennas. It has also been shown in Chapter 2 that the unloaded Q and fractional bandwidth only obey the commonly used relation $B_F = 1/Q$ for series or parallel circuits. In the text that follows, slope parameters are used for antennas that display predominantly series resonant impedance characteristics. In such cases, Equation (8.5) is used to relate the slope parameters to quality factor. However, it should be understood that the quality factor, when derived in this way, is approximate, with an accuracy dependent on how well the antenna response can be modeled as a series resonant circuit (over a narrow bandwidth). *For antenna problems, the loaded quality factor is required in order to find the impedance (power transfer) bandwidth.*

8.1.1.3   Impedance Bandwidth and Q factor

The unloaded bandwidths of simple circuits are related to their unloaded quality factors in Section 2.3.3. The bandwidths are based on power delivered to the load with *assumed* perfect sources (i.e. a voltage source is assumed for a series circuit and a current source for a parallel circuit, but a power source should be used for receiving antenna as discussed in Section 5.4.1). In practice, power transfer will depend on the relationship between the source and load impedances. Hence, it is important to consider the impedance bandwidth of the resonant circuit and to go on to find the bandwidth of power transfer from the source to the load. The most straightforward way to do this is to consider the bandwidth over which the reflection coefficient, Γ or transmission coefficient, τ is less than some predetermined limit. An example might be to consider the points at which $|\Gamma|^2 = |\tau|^2 = 0.5$, since this corresponds to the 3 dB points of power transmission with a reflection coefficient of 0.707. In practical designs, a reflection of half the power at the antenna is seldom tolerated and the reflection coefficient is normally specified to be lower than this: for example, mobile phone antennas are often designed to a specification of $\Gamma = 0.5$ (i.e. 25% power is reflected back which is still quite high, reflecting the difficulty associated with small antenna design and the variability of the antenna impedance when the phone is used), whereas HF (high frequency) broadcast antennas (dipole or monopole arrays that operate in a number of bands over a total bandwidth of approximately an octave) are often designed for $\Gamma = 0.2$ (the power reflection is now down to 4%).

We assume that a series resonant circuit can be used to represent an antenna and, as such, give the relevant components the subscript A, as shown in Figure 8.2, which is different from Figure 2.28 by considering the source impedance Z_0. A tuning inductor or capacitor is added to make it resonant, thus it is called *single-tuned matching circuit*.

Taking $Z_0 = \kappa R_A$, the antenna power transmission coefficient τ is given by

$$|\tau|^2 = 1 - |\Gamma|^2 = \frac{4\kappa R_A^2}{(1 + \kappa)^2 R_A^2 + X_A^2} \tag{8.7}$$

where R_A and X_A are the antenna resistance and reactance, respectively, Z_0 is the source characteristic impedance, and κ is a constant of proportionality. Rearranging gives

$$X_A = \pm \frac{R_A}{|\tau|^2} \sqrt{4\kappa - (1 + \kappa)^2 |\tau|^2} \tag{8.8}$$

It is related to the frequency response via Equation (2.66). Hence, we need to solve

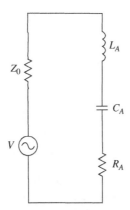

Figure 8.2 Series resonant antenna fed by a resistive source

$$Q_0\left(\frac{\omega}{\omega_0} - \frac{\omega_0}{\omega}\right) = \pm\frac{1}{|\tau|}\sqrt{4\kappa - (1+\kappa)^2|\tau|^2} \tag{8.9}$$

where Q_0 is the Q factor when it is resonant – this is also called *unloaded Q* and can be obtained once the antenna impedance is known as shown in Section 2.3.3. Equation (8.9) is a quadratic equation with two positive solutions for ω. The difference between these two solutions gives the *fractional bandwidth*:

$$B_F = \frac{\omega_2 - \omega_1}{\omega_0} = \frac{1}{Q_0}\left\{\frac{1}{|\tau|}\sqrt{4\kappa - (1+\kappa)^2|\tau|^2}\right\} \tag{8.10}$$

For $\kappa = 1$ (i.e. $Z_0 = R_A$), this simplifies to

$$B_F = \frac{2}{|\tau|Q_0}\sqrt{1 - |\tau|^2} = 2\left|\frac{\Gamma}{\tau}\right|\frac{1}{Q_0} \tag{8.11}$$

Taking $|\Gamma|^2 = 0.5$ (equivalent to the 3 dB power bandwidth) gives

$$B_F = 2\frac{1}{Q_0} \tag{8.12}$$

Compared with Equation (2.66), it is clear that the loaded bandwidth is twice the unloaded bandwidth previously calculated in Section 2.3.3 where Z_0 was not taken into account. The 3 dB fractional bandwidth is equal to the inverse of the loaded Q at resonance which is understandable since the total loss in this case is $2R_A$, twice of the unloaded case. For example, when $R = 50\ \Omega$, $L = 79.5775$ Nh, and $C = 0.3183$ pF gives $Q_0 = 10$, as shown in Section 2.3.3, the fractional bandwidth was 10%, but now it is 20% when the source impedance is taken into account.

To find out what the optimum matching impedance is for maximum bandwidth, we can find the maximum of Equation (8.10) with κ. Differentiating gives

$$\frac{d(B_F)}{d\kappa} = \frac{1}{2|\tau|Q_0} \frac{4 - 2(1 + \kappa)|\tau|^2}{\sqrt{4\kappa - (1 + \kappa)^2 |\tau|^2}} \tag{8.13}$$

Setting this to zero gives

$$\kappa = \frac{2}{|\tau|^2} - 1 = \frac{1 + |\Gamma|^2}{1 - |\Gamma|^2} \tag{8.14}$$

Note that this relation is independent of Q. Since $|\tau|^2$ is always less than 1, since $Z_0 = \kappa R_A$, it can be seen that $Z_0 > R_A$ is required to maximize the bandwidth. For example, if we require that $|\tau|^2$ is greater than 0.75 (equivalent to a 6 dB return loss or better), $Z_0 = 1.67 R_A$.

The relations derived above can also be applied to a parallel circuit by taking $R_A = \kappa Z_0$ (it is not $Z_0 = \kappa R_A$). Here, $Z_0 < R_A$ is required.

Substituting Equation (8.14) in Equation (8.10) gives the *optimal fractional bandwidth* (i.e. the maximum bandwidth), B_{Fopt}:

$$B_{Fopt} = 2 \frac{|\Gamma|}{1 - |\Gamma|^2} \frac{1}{Q_0} \tag{8.15}$$

This is the maximum fractional bandwidth available from a series or parallel resonant antenna with an optimal resistance at resonance (for a given reflection coefficient). It shows how the fractional bandwidth is linked to the reflection coefficient and the unloaded antenna Q factor. When it is perfectly matched, the reflection coefficient is 0 and the fractional bandwidth is also 0, thus a realistic reflection level should be set to achieve a viable bandwidth in practice. The bandwidth and target reflection coefficient are usually known, from which the required antenna Q can be calculated. Alternatively, the system bandwidth and antenna Q may be known, and it may then be desirable to find the minimum possible reflection coefficient that can be achieved over the band. This equation is very useful to aid antenna designs.

8.1.1.4 Fundamental Limits of Antenna Size, Q, and Efficiency

It has long been recognized that, for a given frequency, the size of an antenna cannot be indefinitely reduced without compromising the Q, the efficiency or both. This problem was first explored in the 1940s. In 1947, Wheeler derived a relation between antenna volume and the maximum achievable "power factor" (equal to the inverse of the quality factor) [3]. In 1948, Chu extended Wheeler's analysis by expressing radiated fields in terms of spherical modes [4]. Chu's method is general and can be applied to any antenna; however, to allow this, Chu specified that the antenna should be contained within a sphere (as indicated in Figure 8.1) in which no energy can be stored. This is an idealized, rather optimistic, assumption that makes the underlying mathematics tractable and has often been used in subsequent work. Most antennas will store some energy within this sphere, increasing the Q. Chu used a partial fraction expansion of the wave impedance of spherical modes that exist outside the sphere bounding the antenna to obtain an equivalent (approximate) ladder network from which the Q can be calculated using circuit analysis.

A second phase of work on the limits of small antennas occurred in the 1960s. In 1960, Harrington related antenna size, minimum Q, and gain (including losses) for linearly and circularly polarized waves [5]. Later Collin and Rothschild presented an exact method of finding the minimum Q without using the approximate equivalent network of Chu, for both spherical and cylindrical modes [6]. In 1969, Fante extended these results to include antennas with mixed polarization [7].

Nearly 30 years then passed before a third phase of work was undertaken in the mid-1990s, McLean presented a simple but exact method of determining the minimum antenna Q, based on the observation that the lowest order TM_{01} and TE_{01} spherical modes have fields that correspond to infinitesimal electric and magnetic current elements respectively. Both linear and circular polarization can be treated this way [8, 9]. In 1999, Foltz showed that antennas constrained to shapes other than spheres and cylinders could also be treated [10]. Later, Thiele et al. addressed the question of why the previous theory has been found to be too optimistic in practice [11]. Further related research work has been conducted by many people, especially on how to design electrically small antennas achieving the widest bandwidth.

From all of the efforts of the workers listed above, as well as others, we know that the minimum unloaded Q at resonance of a linearly polarized antenna (when a single lowest order TM mode is implied) is given by the well-known *Zhu's limit* below [4, 9]:

$$Q_0 = \eta_A \left(\frac{1}{(\beta r)^3} + \frac{1}{\beta r} \right) \tag{8.16}$$

where r is the minimum radius of a sphere that just encloses the antenna, β is the wave number $(2\pi/\lambda)$, and η_A is the antenna radiation efficiency. It should be noted here that the antenna may be brought to resonance using an external component. Should this be the case, the Q in Equation (8.16) is that of the antenna and component combined.

A related study on the effect of antenna size on gain, bandwidth, and efficiency was conducted by Harrington who has found that the maximum normal gain of such a small antenna of radius r is given by [5]:

$$G = (\beta r)^2 + 2(\beta r) \tag{8.17}$$

Systems having larger gain than this are called *super-gain antennas*. For large βr, the gain of a uniformly illuminated aperture of radius r is equal to the above-defined normal gain. Therefore, one cannot obtain a gain higher than that of the uniformly illuminated aperture without resorting to a super-gain antenna which is a special topic, and it is hard to realize in reality.

The fundamental limits of antenna size, Q factor, and efficiency are shown graphically in Figure 8.3. For all points on and above the solid black line, the antenna can be 100% efficient. However, for all points below this line it can be seen that small antennas exhibit a trade-off between size, Q factor (bandwidth), and efficiency. Clearly, there is little advantage in trading loss for bandwidth. The only other mechanism that allows these curves to be exceeded is that of interaction, either between the antenna and its supporting structure or via parasitic coupling. In the case of mobile communications equipment, both the antenna and the handset (particularly the PCB) that carries the radio are fed by the source. Radiation from the PCB increases the effective volume of the antenna and hence, the bandwidth.

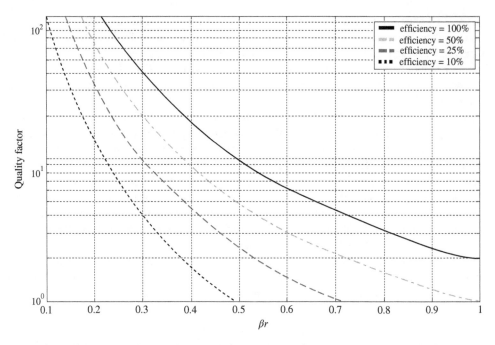

Figure 8.3   The fundamental limits of antenna size, Q factor (bandwidth), and efficiency

8.1.1.5   The Limits of Bandwidth Broadening

Narrowband antennas often exhibit impedance characteristics that can be modeled by series or parallel resonant circuits. It is easy to show that the bandwidth of a series resonant circuit can be enlarged using a shunt connected parallel circuit of the same resonant frequency. Similarly, the bandwidth of a parallel resonant circuit can be enlarged by the series connection of a series resonant circuit. For example, consider the series resonant antenna shown in Figure 8.4 (we could add a source impedance to it as in Figure 8.2, but the results are the same for the derivation to follow), with a corresponding parallel resonant tuning circuit formed by L_M and C_M, thus it is called *double-tuned matching circuit* which was defined directly from the single-tuned matching case.

It is interesting to look at the energy storage, and, therefore, the Q factor of this configuration. The energy stored and dissipated in the series components is given by Equation (2.59), whereas the energy stored in the parallel components is given by Equation (2.71).

$$Q = Q_A + Q_{ME} \tag{8.18}$$

Here Q_A is the Q of the series arm, while Q_{ME} is the equivalent Q of the parallel circuit, with a conductance equal to $1/R_A$. Hence, the series and parallel contributions to Q can be considered to be additive and the total Q is increased. However, if the parallel and series arms of the circuit are resonant at the same frequency, the impedance bandwidth of the circuit is increased (this will be seen later in this section). Thus, we have a seemingly contradictory situation where both the impedance bandwidth and the Q increase simultaneously. This illustrates a basic

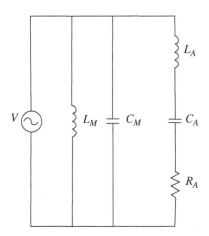

Figure 8.4   A double-tuned series-parallel resonant circuit

problem – there is no strict relationship between antenna Q and bandwidth. The formulas derived previously can only be applied (with care) when the antenna impedance can be closely modeled as either a series or parallel resonant circuit. Fortunately, this limitation can often be satisfied, particularly for narrow bandwidths.

Bandwidth broadening concepts of simple, for example, resonant, circuits were generalized by Bode [12] and in two well-known papers by Fano [13], [14]. Fano's theory states that the maximum achievable fractional bandwidth, $B_{F\infty}$, of a series or parallel resonant circuit in combination with an optimal bandwidth-broadening network, composed of an infinite number of elements, is given by

$$B_{F\infty} = \frac{\pi}{Q_0 \ln \left(\frac{1}{|\Gamma|}\right)} \tag{8.19}$$

where $|\Gamma|$ is a chosen maximum reflection coefficient at the source. The improvement in bandwidth offered by the infinite-order bandwidth-broadening network, F_∞ is given by ratio of Equations (8.19) and (8.15), as follows:

$$F_\infty = \frac{B_{F\infty}}{B_{Fopt}} = \frac{\pi \left(1 - |\Gamma|^2\right)}{2|\Gamma| \ln \left(\frac{1}{|\Gamma|}\right)} \tag{8.20}$$

Interpreting the theory presented is not straightforward, since these relations only strictly apply to antennas that can be represented by series or parallel equivalent circuits. Often, antenna designs that are reported as low Q are designs that incorporate lumped or distributed bandwidth-broadening circuitry on the antenna itself. The widely cited Goubau antenna [15] is believed to be an example of this. When this is the case, some of the available bandwidth-broadening has already been realized on the antenna, such that only reduced improvement is possible from an external circuit.

Assuming that the antenna is well represented by a series or parallel LCR circuit, the formula given previously for the limits of (linearly polarized) antennas and bandwidth broadening can be combined to yield the maximum fractional bandwidth (by combining Equations (8.19) and (8.16)):

$$B_{MAX} = \frac{\pi}{\eta_A \left(\dfrac{1}{(\beta r)^3} + \dfrac{1}{\beta r} \right) \ln \left(\dfrac{1}{|\Gamma|} \right)} \tag{8.21}$$

This relation gives the maximum possible bandwidth available from an antenna of a given size that is matched with an idealized circuit containing an infinite number of components. However, it is rather optimistic, predominantly for two reasons:

- It is assumed that the antenna stores no energy within the enclosing radius, r. In practice, this is never achieved.
- Practical bandwidth broadening circuits have finite losses – this is addressed in more detail in the text that follows.

Quite clearly, a network of infinite order is impractical. Hence, it is useful to consider how much improvement can be gained from a network with just a limited number of stages. In the following part, one stage is considered in detail – i.e. a double-tuning network. The analysis is complete, i.e. losses are included. The results of this analysis are then extrapolated qualitatively to circuits of higher order.

Consider a series resonant antenna that is double-tuned by a parallel resonant LC circuit (as shown in Figure 8.4). This operation is shown graphically in Figure 8.5.

The admittance of the antenna, Y_A, is given by

$$Y_A = G_A + jB_A = \frac{R_A - jX_A}{R_A^2 + X_A^2}. \tag{8.22}$$

Optimum double-tuning occurs when the susceptance of the antenna is tuned to zero at the band edges. In Figure 8.5, this is termed the "cross-over" point. It is not worth double-tuning beyond this point. Doing so would only increase the band-edge mismatch, since the constant conductance circle that passes through the band edges lies outside the constant reflection coefficient circle that passes through the point at which the antenna is resonant, as illustrated in Figure 8.6. When the antenna is optimally double-tuned, it is observed that the maximum reflection coefficient is always along the resistive axis.

The resistance associated with the double-tuning "cross-over" point, R_C, is given by

$$R_C = \frac{R_A^2 + X_{AE}^2}{R_A} \tag{8.23}$$

where X_{AE} is the reactance at the band edges (denoted by f_1 and f_2). The optimum reference impedance is then given by

$$Z_0 = \sqrt{R_A R_C} = \sqrt{R_A^2 + X_{AE}^2}. \tag{8.24}$$

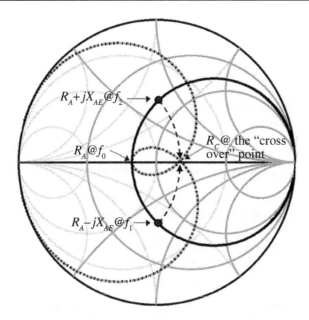

Figure 8.5 Series resonant circuit with parallel double-tuning (solid line is the impedance of the series resonant circuit, dashed line is the impedance after double-tuning, arrows show the movement of the band edges)

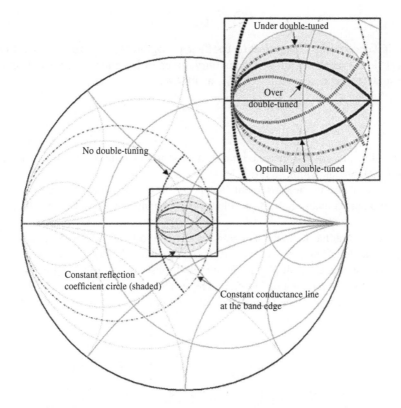

Figure 8.6 Series resonant circuit with different levels of parallel double-tuning

For optimum double-tuning, it can be proved that

$$L_M = \frac{Z_0^2}{\omega_0^2 L_A}$$

(8.25)

Since the double-tuning circuit has the same resonant frequency as the antenna, the required capacitance, C_M, is given by

$$C_M = \frac{1}{\omega_0^2 L_M}$$

(8.26)

The relations above allow the double-tuning components to be chosen based only on the characteristic impedance of the system and the antenna resonant frequency and equivalent inductance.

Having derived the necessary relations, the bandwidth improvement available from a single double-tuning network can be found (for a given reflection coefficient) and related to the maximum possible improvement given by Fano's theory.

We have already seen that, for a particular system or characteristic impedance, the antenna resistance at resonance required for double-tuning is identical to that required for an optimum match without any additional circuitry. Referring to Figure 8.5, the optimum reflection coefficient prior to double-tuning is worst at the band edges.

It is found that the fractional bandwidth with double-tuning, B_{FDT}, can be written

$$B_{FDT} = \frac{1}{Q_{A0}} \frac{2\sqrt{|\Gamma|}}{1 - |\Gamma|}$$

(8.27)

where $|\Gamma|$ is a chosen, or target reflection coefficient magnitude. Comparing this with the optimum bandwidth available without any circuitry, given in Equation (8.15), yields the double-tuning bandwidth improvement factor, F_{DT} as follows:

$$F_{DT} = \frac{1 + |\Gamma|}{\sqrt{|\Gamma|}}$$

(8.28)

The variation of F_{DT} and the maximum possible improvement, F_∞, given previously in Equation (8.20) are shown in Figure 8.7.

It can be seen that the improvement factors are largest for small reflection coefficients. Importantly, approximately half of the total improvement is achieved with a double-tuning network.

The analysis so far has assumed all components to be lossless. Of course, this is not the situation encountered in practice, and the effects of component losses must be considered. Here only the losses associated with double-tuning will be considered. Referring to Figure 8.4, the loss of the double-tuning components can be represented by an additional parallel resistance, R_M. Using Equation (2.72), the Q of the double-tuning network, Q_M will then be given by

$$Q_M = \frac{\left(\omega C_M + 1/\omega L_M\right)}{2G_M} = \frac{1}{\omega L_M} \left\{ \frac{1 + \left(\omega/\omega_0\right)^2}{2G_M} \right\}$$

(8.29)

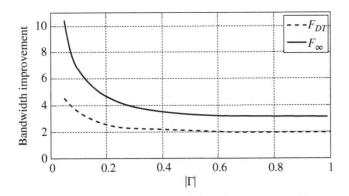

Figure 8.7 Variation of bandwidth improvement factors F_{DT} and F_∞

We assume that the antenna resistance is optimized and that this optimum condition is unaffected by the finite Q of the double-tuning circuit (a reasonable assumption if the conductivity of the double-tuning circuit is considerably less than that of the antenna). Using Equation (8.25) gives

$$L_M = \frac{Z_0^2}{\omega_0 Q_{A0} R_A} \tag{8.30}$$

Substituting Equation (8.30) in Equation (8.29) gives

$$Q_M = \frac{Q_{A0} R_A}{2 Z_0^2 G_M} \left\{ \frac{\omega}{\omega_0} + \frac{\omega_0}{\omega} \right\} \tag{8.31}$$

We have previously seen that the term in $\{..\}$ is approximately equal to 2 over narrow to moderate bandwidths. Using this and rearranging for G_M gives

$$G_M = \frac{Q_{A0} R_A}{Z_0^2 Q_M} \tag{8.32}$$

Using Equations (8.22) and (8.24), the band-edge (i.e. highest or worst case) conductance of the antenna is given by

$$G_{AE} = \frac{R_A}{Z_0^2} \tag{8.33}$$

Hence, the minimum (band-edge) efficiency is given by

$$\eta_{min} = \frac{G_{AE}}{G_{AE} + G_M} = \frac{1}{1 + Q_{A0}/Q_M} \tag{8.34}$$

This is a very simple relation dependent only on the antenna and double-tuning circuit Q factors. There is no bandwidth dependency. Note also that Equation (8.34) does not include the effects of mismatch.

Here we have provided in-depth discussion on limits of bandwidth broadening, especially through double-tuning technique based on Wheeler's original method. Fano established the theoretical limit on the matching bandwidth with respect to the number of tuned circuits in the impedance-matching network [13]. Using the same antenna model (capacitor or inductor with constant radiation resistance), Fano derived an exact relationship for fractional bandwidth B_n, Q, and Γ, where $n = 1, 2, 3, 4, \ldots$ is the number of tuned circuits in the matching network (single tuned, $n = 1$; double tuned, $n = 2$; triple tuned, $n = 3$, etc.). Wheeler's equations for single-tuned edge-band and double-tuned matching are in exact agreement with Fano's equations for single and double-tuning [16]. Fano's solution is a set of three simultaneous transcendental equations that relate B_n, Q, and Γ. We can summarize the major equations linking the fractional bandwidth B, Q-factor and reflection coefficient Γ for different tuning matching circuits (single-tuned, double-tuned, and infinite-tuned) in Table 8.1. The products of B and Q_0 for VSWR = 2 (i.e. $\Gamma = 1/3$) are also given in the table. Having compared the results, it is clear that the single-tuned edge-band matching can provide a small benefit (from 0.707 to 0.750) over the single-tuned mid-band matching case. For a given Q, the double-tuned matching case provides more than twice of the bandwidth of the single-tuned mid-band matching case for a VSWR = 2, which was validated by an interesting example given in reference [17] where a spherical-cap dipole antenna was employed. In theory, an infinite number of tuning circuits will only provide a two-thirds (66%) increase over the double-tuned matching case, thus there is a trade-off between the number of tuning circuits (performance) and complexity/cost, and number of tuned circuits should be carefully selected in reality.

The double-tuned matching case is commonly used for impedance matching since this case provides a good measure of the bandwidth that can be achieved for the impedance matching of a narrow-bandwidth antenna without making the matching circuit too complicated. It has been pointed out [1–7], there is a law of diminishing returns for Fano matching beyond the $n = 2$ case.

Table 8.1 Equations relating bandwidth, Q-factor, and impedance matching

Impedance matching circuit	Equation	BQ_0 for VSWR = 2	Notes
Single tuned mid-band match (non-Fano)	$B_F = \dfrac{2\|\Gamma\|}{\sqrt{1-\|\Gamma\|^2}}\dfrac{1}{Q_0}$	0.707	Equations (8.10) and (8.11)
Single-tuned edge-band match (Fano-Bode-Lopez, $n = 1$)	$B_{Fopt} = \dfrac{2\|\Gamma\|}{1-\|\Gamma\|^2}\dfrac{1}{Q_0}$	0.750	Equation (8.15) The maximum for single-tuned
Double-tuned match (Fano-Bode-Lopez, $n = 2$)	$B_{FDT} = \dfrac{2\sqrt{\|\Gamma\|}}{1-\|\Gamma\|}\dfrac{1}{Q_{A0}}$	1.732	Equation (8.27)
Infinite-tuned match (Fano-Bode-Lopez, $n = \infty$)	$B_{F\infty} = \dfrac{\pi}{\ln\left(\frac{1}{\|\Gamma\|}\right)}\dfrac{1}{Q_0}$	2.860	Equation (8.19) The maximum for all cases

In practice, one should not expect to achieve a bandwidth that is much larger than that achieved with the double-tuned matching case. A simple rule of thumb for the achievable bandwidth for a VSWR > 2 is that it is approximately equal to the VSWR divided by the Q (i.e. $B = VSWR/Q$, the larger the VSWR, the larger the B).

It is clear that, for good efficiency, the double-tuning circuit Q should be much higher than that of the antenna. For example, dual-band mobile phones have typical antenna Q's of approximately 15, while a lumped circuit double-tuning network will have a typical Q of, at best, 50. Using these values gives a band-edge efficiency of 77%. GSM (Global System of Mobiles) and DCS (Digital Cellular System) have fractional bandwidths of approximately 10%. With double-tuning, the band-edge mismatch efficiency is 92%. This gives a combined band-edge efficiency of 71%. The mismatch efficiency without double-tuning is 71%. Hence, in this example, there is no advantage in using double-tuning – the improvement in mismatch efficiency (return loss) is counterbalanced by the mismatch corrected efficiency (insertion loss) of the double-tuning circuit. Of course, there may be other reasons for using double-tuning. For example, the improved antenna return loss will reduce cumulative losses in subsequent parts of the RF chain. Also, the double-tuning circuit provides a degree of filtering. However, it is clear that the use of bandwidth broadening techniques requires the use of high Q resonant circuits (with respect to the antenna Q).

Circuits with higher orders of bandwidth broadening require resonators with increased slope parameters, increasing the losses of each stage. Also, the losses of each stage of the network will be cumulative. Hence, high-order bandwidth broadening circuits will require very high Q resonators. Since approximately half of the maximum bandwidth improvement is provided by a single resonator with only moderate loss, higher order circuits are often not worthwhile.

It should also be recognized that personal devices such as mobile phones are used in a manner such that the driving-point impedance of the antenna is variable within limits defined by the modes of use. For example, the antenna of a mobile phone will present different impedances when in free-space and when held in "talk position". High-order bandwidth broadening circuits are critical and, hence, may become counterproductive when large impedance variations are experienced. Simple double-tuning circuits are thought to be more robust in this respect.

8.1.1.6 Discussions and Conclusions

We have used many equations in this section to introduce and derive some basic and important relations between antenna bandwidth, efficiency, and size. In particular, it is shown that when antennas can be represented as series or parallel resonant circuits, as is often the case for small antennas, quality factor can be found from the resistance and the derivative of the reactance with frequency. This applies at any frequency, not just at resonance, and allows comparison between different designs in a fundamental way.

It is shown that antennas represented by series or parallel resonant circuits have optimum bandwidth when mismatched at resonance. Series resonant circuits are required to have a resistance at resonance that is lower than the system (or characteristic) impedance. The converse applies to parallel resonant circuits. For maximum bandwidth, the level of mismatch depends only on the reflection coefficient magnitude at which the bandwidth is measured. However, for minimum reflection coefficient over a given bandwidth, the level of mismatch is inversely related to the antenna quality factor.

The limits of antenna size, bandwidth, and efficiency are reviewed, as are the limits of passive bandwidth broadening. It is illustrated that there is no universal relationship between bandwidth and quality factor and that some care must be applied when interpreting how well a particular antenna performs. Some small antennas have distributed reactive networks integrated within the antenna structure. Such networks can provide a degree of bandwidth broadening. When this is the case, further bandwidth broadening circuitry will give improvements lower than those stated both in this chapter and in the literature – some of the available improvement will have been "used" on the antenna.

A combined formula is given showing the maximum possible bandwidth available from an antenna of a given size that is matched with an idealized circuit containing an infinite number of components.

A more practical analysis of the bandwidth broadening available from a double-tuning circuit with finite losses is undertaken. A simple new formula for the efficiency of a double-tuning network is derived. This relation indicates the degree to which the Q of any double-tuning circuit must be higher than the Q of the antenna (or any other load) that requires broadbanding. From this result, it is reasoned that high-order bandwidth broadening is only possible if extremely good circuit technologies are used. High-order circuitry is also thought to be too critical for applications such as mobile phones, such that it could become counterproductive if the antenna impedance were to vary due to user interaction. Double-tuning is thought to be less susceptible to changes in the antenna impedance that occur due to user interaction and places less stringent requirements on component quality. Double-tuning can approximately double the bandwidth of an antenna designed for a reflection coefficient of 0.5 (such a typical mobile phone antenna). This is approximately half the improvement available from a network of infinite order. Hence, often only double-tuning is required – higher order bandwidth broadening circuits are unlikely to bring significant additional improvements overall.

To maximize the benefits of bandwidth broadening, high-quality circuit technologies are required. For some applications, such components are available. For example, air wound inductors and vacuum capacitors – with Q factors of around 200 and 1000 respectively – are used for AM radio broadcast applications. For other applications such as mobile phones, size, and technology constraints are such that inductor and capacitor Qs are typically 30 and 100, respectively. Inductors are avoided where possible. In such cases, it is often more desirable to form distributed double-tuning circuits on the antenna itself.

There are some other related research papers. For example, a performance comparison of fundamental small-antenna designs is given in [18]. The work was to determine which basic design approach or configuration, if any, offers the best performance in terms of: achieving an impedance match; the radiation-pattern shape; the radiation efficiency; the half-power bandwidth. The studied antennas included the spherical folded helix, the cylindrical folded helix, the matched disk-loaded dipole, the matched spherical-cap dipole, and the multiarm spherical resonator. Furthermore, the fundamental approaches used in impedance matching and optimizing the bandwidth of these small antennas were described. The study on the fundamental efficiency limits for small metallic antennas was presented in [19], and it was shown that the smaller the antenna, the smaller the efficiency. How to produce an efficient, small but broadband antenna is always a topic of interest, although there are some fundamental limits.

8.1.2   Antenna Size Reduction Techniques

Small is beautiful when talking about antennas. There are many ways of reducing the size of an antenna. Some of the most common are:

- Top Loading
- Matching
- Reactive Loading
- Dielectric Loading

Each of these methods will be dealt with in the sections that follow.

8.1.2.1   Top Loading

Consider the monopole above ground shown in Figure 8.8. The current distribution indicated can be maintained by keeping the length of the antenna approximately constant but having some of the structure located parallel to the ground, as shown for the "T," "Star" and "Disc" configurations. This allows the height of the antenna to be reduced without changing the resonant frequency.

As indicated for the "T" antenna in Figure 8.9, currents on the upper part of the structure cancel with currents induced in the ground. Hence, while the current distribution is maintained, the part of the antenna that is parallel to the ground contributes very little to radiation. Radiation thus emanates from a smaller part of the "T" antenna than for the monopole and the radiation resistance reduces.

However, the radiation resistance is still higher than it would be for a monopole of the same height, h, as the "T" antenna due to the more rectangular shape of the current distribution of the

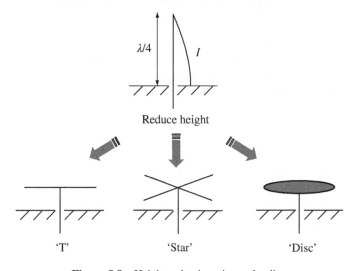

Figure 8.8   Height reduction via top-loading

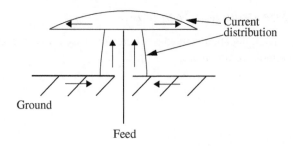

Figure 8.9 Current distribution of a top-loaded "T" antenna

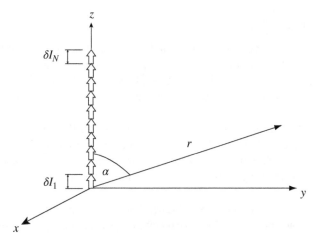

Figure 8.10 Linear antenna with elemental currents

vertical part of the latter. This can be explained by considering the linear antenna shown in Figure 8.10. The current on the antenna is broken up into small elements (we have seen this in Chapter 6 using the MoM for wire antenna simulation), from which it is possible to calculate the radiation resistance for different current distributions.

For any small linear antenna, the radiation resistance, R_r, is given by the real part of Equation (5.13), that is,

$$R_r = 80\pi^2 \left(\frac{\delta l}{\lambda}\right)^2 \left|\sum_{n=1}^{N} \frac{I_n}{I_0}\right|^2 \tag{8.35}$$

where I_0 is the maximum antenna current, δl is the element length, and I_n is the elementary antenna current. With a triangular variation of current, as occurs for very small monopoles, the radiation resistance is given by

$$R_r = 20\pi^2 \left(\frac{N\delta l}{\lambda}\right)^2 \tag{8.36}$$

For $(N\delta l)/\lambda = 0.1$, the radiation resistance is approximately equal to 2 Ω. For a top loaded antenna of the same height, the distribution of the radiating current (the vertical part in Figure 8.9) is close to rectangular, and the radiation resistance is given by

$$R_r = 80\pi^2 \left(\frac{N\delta l}{\lambda}\right)^2 \qquad (8.37)$$

Again taking $(N\delta l)/\lambda = 0.1$, the radiation resistance is approximately equal to 8 Ω. Inspection of Equations (8.36) and (8.37) also shows that a factor of four improvement is possible. Hence, top loading not only allows the resonant frequency to be maintained as the antenna height is reduced but it also improves the radiation resistance for a given height. Hence, it is a very popular and widespread method of achieving a low-profile structure. An early example of the use of top loading is shown in Figure 8.11.

The Titanic famously sank in 1912. Marconi installed transmitter equipment was used to sound the alarm using the (normally not shown) large "T" antenna.

8.1.2.2 Matching

Most electrically small antennas have a low radiation resistance and high, normally capacitive, reactance. A simple approach, illustrated in Figure 8.12, is to match this impedance to 50Ω, or whatever the system impedance might be – using discrete components: for example, two inductors, or an inductor and a capacitor.

Some attention must be paid to the component quality when using this technique; for example, for MF (medium frequency) broadcasting air core inductors and vacuum capacitors are used in order to minimize losses. Litz wire – a special type of multistrand wire or cable designed

Figure 8.11 "T" on the Titanic. Source: Reproduced with the permission of Marconi Corporation plc

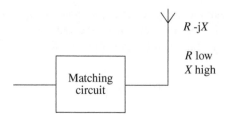

Figure 8.12   Antenna size reduction using impedance matching

to have lower losses due to the skin effect than conventional solid conductors – may also be used. Many strands are used with a conductive coating on each strand. This increases the effective outer area of the conductor for a given circumference. The strands are also twisted in a way that minimizes the induction of opposing electromagnetic fields in other strands.

8.1.2.3   Reactive Loading

A common method of reducing the resonant frequency of a small antenna is to use some form of reactive loading. The aim is to store enough magnetic energy (to create inductance) to counter the electric energy (capacitance) that is associated with most small antennas. This is illustrated for two meandered structures in Figure 8.13.

The bold arrows show, broadly speaking, where radiation occurs. The nonbold arrows can be considered to be in pairs that have opposite directions. Hence, they cancel in the far field and do not produce any radiation. However, they do store energy in the near field, and this can be used to reduce the antenna resonant frequency.

An alternative way of viewing this is that the antenna length is maintained at approximately $\lambda/4$, so that the resonant frequency is the same as that of an equivalent $\lambda/4$ monopole. However, since radiation comes from less of the structure, the radiation resistance must be reduced. In practice, the length of the meander must be somewhat longer than $\lambda/4$ in order to achieve resonance.

Another example of reactive loading – this time applied to planer antennas – is shown in Figure 8.14. Slots and notches are cut in the antenna structure in order to lengthen the path over which current travels (energy is also stored around the slots and notches), reducing the resonant frequency.

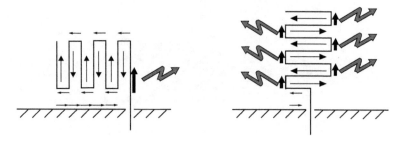

Figure 8.13   Radiation from meandered structures

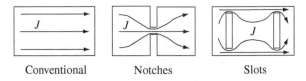

Conventional   Notches   Slots

Figure 8.14   Reactive loading of planar antennas

8.1.2.4   Dielectric Loading

A high dielectric constant material can be used to achieve a slow wave structure (a smaller wavelength in the dielectric) that allows smaller resonators to be realized. Antennas with dielectric loading have the advantage that tuning capacitance can easily built into the structure. For consumer wireless applications, it is also possible to excite a polarization out of the plane of a host PCB. However, dielectric antennas are not immune to the fundamental limits of bandwidth, size, and efficiency: smaller means narrower bandwidth. Also, loss occurs within the dielectric material, and the high Q nature of such antennas tends to increase the losses in conductors. Often these conductors have a conductivity that is compromised somewhat in order to be compatible with the dielectric material from a manufacturing standpoint.

Figure 8.15 shows a meandered, dielectric-loaded monopole antenna manufactured for dual-band mobile phones. The antennas measure 17 mm × 10 mm × 2 mm and the relative permittivity is approximately 20. The total length for the monopole is around $\lambda/4$.

Figure 8.16 shows a half-wave global positioning system (GPS) patch antenna (the frequency is around 1.57 GHz). Since the wavelength is determined by the relative permittivity (dielectric constant) ε_r of the patch substrate, the antenna can be made progressively smaller by increasing ε_r. The bandwidth of GPS is very narrow, so the antenna size can be reduced significantly without necessarily reducing efficiency. However, the measured results shown in Figure 8.17 indicate that as the antenna size is reduced, there is an almost inevitable reduction in antenna efficiency for this antenna (the bandwidth also reduces as expected).

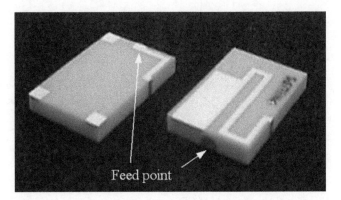

Feed point

Figure 8.15   Meandered, dielectric loaded monopole antenna for dual-band mobile phones (the view on the left shows the bottom of the antenna, intended for connection to the PCB, and the view on the right shows the top of the antenna)

Figure 8.16   A half-wave GPS patch antenna mounted on a finite ground plane

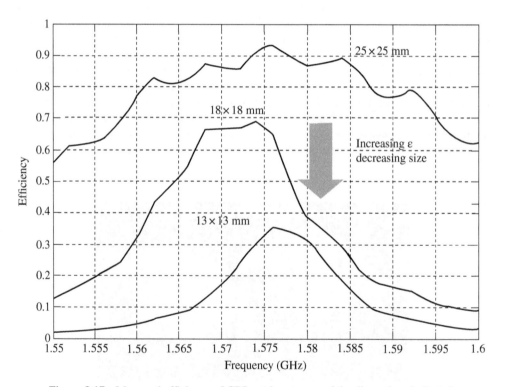

Figure 8.17   Measured efficiency of GPS patch antennas of the dimensions indicated

8.1.3 Summary

In this section, electrically small antennas have been studied in terms of (i) how the antenna Q factor is linked to its size and efficiency (Zhu's limit); (ii) how the antenna bandwidth is related to the Q factor and impedance matching. The relevant theory and equations were derived and discussed. The bandwidth broadening techniques (such as double-tuned) were introduced and evaluated. Four general size reduction techniques were presented, and examples were used to demonstrate the effectiveness of these techniques.

8.2 Mobile Antennas

8.2.1 Introduction

The term "mobile radio" is rather loosely used to describe a radio communication link where at least one end of the link is either at an unknown location or in motion. Marconi's transmission from a fixed site on the Isle of Wight, England, to a tug-boat located at an ill-defined position approximately 18 miles away was probably the first example of such a link in 1888. To be more specific, this was an example of maritime mobile radio. An early example of a land-based mobile transceiver is given in Figure 8.18.

Figure 8.18 An early mobile transceiver: the antenna is incorporated into the bus' chimney.
Source: Courtesy The Marconi Company plc

Here, Marconi and Fleming were pictured with a Thornycroft steam bus outside the Haven Hotel in Poole, England, just before the turn of the twentieth century. Marconi used amplitude modulation (AM) at "low frequencies," so the antenna seen attached to the chimney of the bus was probably electrically small, as defined by Equation (8.1).

Since Marconi's first wireless communication experiments, many other applications have been found which fall into the broad class of mobile radio. The first system to closely resemble the mobile radio networks that we use today was implemented by the Detroit Police Department in 1921 [20]. It used a frequency band centered at approximately 2 MHz for communication between a central controller and police cars. This was an example of a dispatch system, but other modes of communication – such as walkie-talkies, paging, and mobile telephony – soon followed. The first commercial cellular mobile radio system was launched in 1970s, but the widespread use of mobile phones didn't happen until the successful deployment of digital mobile radio systems in 1990s which was represented by the European GSM (Global System for Mobile, viewed as the most successfully second-generation mobile system). Since then, tremendous technology advances have been made and we have witnessed the mobile evolution from the second generation (2G) to the third generation (3G), and then the fourth-generation (4G) mobile system in a span of 30 years. We are now entering the fifth generation (5G) era. In addition to cellular mobile radio systems, there are also many other mobile systems and devices. For example, Bluetooth has been introduced as a wireless technology standard used for exchanging data between fixed and mobile devices over short distances using UHF radio waves in the industrial, scientific and medical (ISM) radio bands, from 2.402 to 2.480 GHz, and building personal area networks (PANs). Wi-Fi is another good example; it is based on the IEEE 802.11 family of standards and provides local area networking of devices and Internet access in indoor environments using the 2.45 GHz ISM radio band and 5 GHz band. A collection of some selected personal mobile devices from 1996 to 2019 are given in Figure 8.19, which clearly shows how they have been evolved over this period: the antenna was changed from external to internal and the screen was from small to large. The function of wearable devices has also become increasingly powerful. Wireless is now available almost everywhere and anytime. The success of mobile radio communications has completely changed the way we live, we shop, we pay, and we work. It is actually changing the world.

This section gives details of the fundamental and design principles of mobile antennas. It also provides a summary of antenna diversity techniques – where two or more receive antennas are used rather than one – and the effect of the human body on antenna performance. Mobile base station antennas are introduced at the end of this section.

8.2.2   Mobile Terminal Antennas

8.2.2.1   The Radio Frequency Bands

Initially mobile phones were designed for only one band. However, dual-band phones – for example, capable of operation in the GSM900 and DCS1800 bands – became the minimum requirement in Europe in the late 1990s due to increased popularity of mobile phones. Tri-band operation in the GSM900, DCS1800, and PCS1900 bands became commonplace for "high-end" phones in the early 2000s, allowing operation in many countries. Tri-band GSM/DCS/PCS antenna operation can be achieved using a fundamentally dual-band design with an extended bandwidth in the higher band. Similarly, four or five bands may also be covered

1996   2000   2007   2014   2019

Figure 8.19 A personal collection of selected mobile devices from 1996 to 2019

by extending the bandwidths of a dual-and tri-band design even further, which is basically the current practice in mobile antenna design.

Cellular radio communications have experienced explosive development over the past 30 years. Every generation change marks a major leap forward in term of technology. A main performance indicator is the data rate which has been increased by at least 10 times from one generation to another. The operational frequency band is also risen from about 800 MHz to millimetre waves in order to gain the required wider bandwidth. The detailed frequency bands vary from one country to another country. The most common 2G, 3G, and 4G bands are GSM850, GSM900, GSM1800, GSM1900, UMTS2100 (Universal Mobile Telecommunications System), LTE2300 (Long Term Evolution), and LTE2500 as shown in Table 8.2. The 5G frequency band plans are much more complex, the frequencies are divided into sub-6GHz bands and millimetre wave bands. The frequency spectrum for sub-6 GHz spans from 600 MHz to 6 GHz while the millimeter-wave (mm-Wave) frequencies span from 24.250 to 52.600 GHz, and also include unlicensed spectrum. Additionally, there may be 5G spectrum in the 5.925–7.150 GHz range and 64 GHz to 86 GHz range in some countries. Therefore, 5G will include all previous cellular spectrum and a large amount spectrum in the sub-6 GHz range and beyond sub-6 GHz is many times current cellular spectrum. The initial 3GPP release of 5G New Radio Non-standalone (5G NR) standards included several sub-6 GHz frequency bands, designated FR1 (frequency range 1 in Table 8.2). The second 3GPP 5G release after IMT-2020 will include FR2 frequency bands (frequency range 2 in Table 8.2) in the millimetre-wave spectrum. It was reported that 64 – 71 GHz may also be allocated to 5G in north America.

Table 8.2 Cellular Radio Frequency Bands and Standards

Generation	Standard	Band	Uplink (MHz)	Downlink (MHz)	Channel bandwidth
2G	GSM/DCS/PCS	GSM850	824–849	869–894	200 kHz
		GSM900	880–914	925–959	
		GSM/DCS1800	1710–1784	1805–1879	
		GSM/PCS1900	1850–1910	1930–1990	
3G	UMTS	UMTS Band 1	1920–1980	2110–2170	5 MHz
		UMTS Band 2	1850–1910	1930–1990	
		UMTS Band 5	824–849	869–894	
4G	LTE	LET Band 17	704–716	734–746	5, 10, 15 MHz
		LET Band 13	777–787	746–756	
		LTE2300	2305–2400	2350–2360	
		LTE2500	2500–2570	2620–2690	
5G	5G	Sub-6 GHz Freq	663–698	617–652	Between 5 and
		range 1	703–748	758–803	100 MHz
		(n71, n28, n78,	3300–3800	3300–3800	
		n79)	4400–5000	4400–5000	
		mm-Wave	**GHz:**	**GHz:**	50, 100, 200,
		Freq range 2	25.50–29.50	25.50–29.50	400 MHz
		(n257, n258,	24.25–27.50	24.25–27.50	
		n260, n261)	27.00–40.00	27.00–40.00	
			27.50–28.35	27.50–28.35	

In addition to cellular radio, there are other radio systems inside a modern mobile phone (normally called a smartphone) which should include:

- Wi-Fi: 2.40–2.49 GHz (up to 600 Mbps) and 5.180–5.825 GHz (up to 1300 Mbps; mainly 5.15–5.35 GHz and 5.725–5.825 GHz).
- GPS: 1575.42 MHz (L1) and 1227.60 (L2) or 1176.45 (L5).
- Bluetooth: 2.40–2.49 GHz.
- NFC (near field coupling): 13.56 MHz.

Some smartphones may even have UWB (ultra-wide-band) for radio positioning, WiMAX (Worldwide Interoperability for Microwave Access) working around 2.6 GHz and wireless charging at 140 kHz.

8.2.2.2 Antenna Design Requirements

It is apparent that many antennas are required to cover all these applications (it was reported that there were 21 antennas in Huawei Mate30 Pro 5G). Some antennas must be broadband or multiband. Thus, the main challenges for the mobile phone antennas are small size, built-in, multiband, and coexistence of a multi-radio system and a multiple-input and multiple-output (MIMO) system (which will be discussed later in this section).

The realization of the compact multiband antenna through excellent design and proper impedance matching is the main design objective of mobile phone antennas. To meet the requirements of mobile phone built-in antennas, monopole antenna, PIFA antenna, and many other types of antennas have been designed. A navigation antenna is specially provided in most phones, which mainly covers the GPS band. LTE provides the essential and far reaching vital solutions for wireless communication systems. It provides high-speed internet along with broader bandwidth and high throughput. During its initial stages in Europe, LTE was assigned 1.8 and 2.6 GHz bands. Due to the increase in the necessity of additional bandwidth, the 790–865 MHz band has been added to the LTE operating system in Europe.

Another design requirement is that mobile antennas must pass various compliances before production, depending on the functionality of the system, its service areas, and the quality and quantity of data to be transmitted. In the cellular frequency bands, mobile phones transmit at relatively high-power levels. GSM, DCS, and UMTS have maximum average output powers of 0.25, 0.125, and 0.125 W (in Class 4 operation) respectively.

In many applications, these systems are only used at low power levels (for example, Bluetooth has a typical output power of 1 mW). Hence, these low power "connectivity" systems have very different design requirements/restrictions to cellular systems (which transmit at relatively high power), both for the antenna and for the RF circuitry.

The type of antenna that is used with a particular type of phone is normally determined by aesthetic considerations and specific absorption rate regulations. This will be covered in more detail later in this chapter. It is a measure of the maximum amount of energy dissipated per unit volume in any part of the phone user's body. When the phone is held in "talk position" (i.e. next to the head), this maximum often occurs close to the antenna.

8.2.2.3 Typical Classical Mobile Antennas

Many mobile phone antennas (some are very innovative and unconventional) have been developed and the number is still growing. It is not possible for us to cover all these antennas. Here we first introduce the following four main types of classical mobile antennas, and then discuss some of the latest smartphone antenna designs to gain a good understanding on this topic. The four classical antennas are:

- monopoles (whips);
- normal mode helices;
- meander line antennas (loaded/unloaded);
- inverted F antennas and planar inverted F antennas.

Monopoles and helical antennas are normally located outside the phone casing, as illustrated in Figure 8.20. There are two broad classes of internal antenna, dependent on the relative positions of the antenna and printed circuit board (PCB):

- adjacent antenna and PCB – where the antenna is installed next to the PCB (in the manner of a monopole);
- coincident antenna and PCB – where the antenna is installed on top of the PCB

These classes are shown diagrammatically in Figure 8.21.

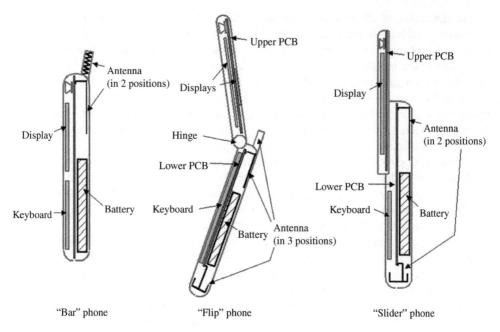

Figure 8.20 Classical mobile phone types: "bar," "flip," and "slider"

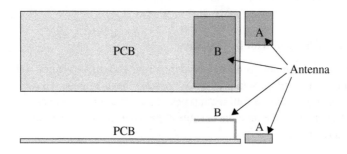

A - Adjacent antenna and PCB
B - Coincident antenna and PCB

Figure 8.21 Basic antenna and PCB arrangements

Monopoles

Monopoles were the antenna of choice for the earliest mobile phones, and they are still used, particularly in countries where coverage is limited. They have the advantage of providing significant clearance between the antenna and the head, which allows low SAR and, perhaps most importantly, high efficiency to be achieved. A simple monopole located over a ground plane is illustrated in Figure 8.22.

Most monopoles are operated at an electrical length that is close to a quarter of a wavelength, when the input resistance is 36 Ω and the input reactance is zero. This impedance matches quite

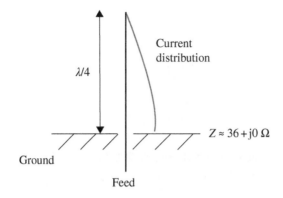

Figure 8.22 Quarter-wave monopole over a ground plane

well with the commonly used system impedance of 50 Ω. When mounted on a typical handset, monopole antennas can be well matched to 50 Ω when slightly shorter than $\lambda/4$, due to the influence of the conductive parts of the handset, which also radiate considerably.

Helical Antennas
A typical helix antenna assembly is shown in Figures 8.23 and 5.16. Helical antennas were widely used in the 1990s, as they occupied considerably less space than the monopoles that preceded them. They were an early example of a compromise between technical performance

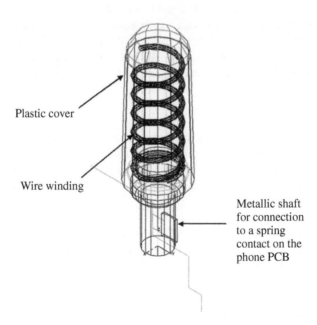

Figure 8.23 A helix antenna assembly (shown as a "wire grid model" as displayed in a typical 3D electromagnetic simulator)

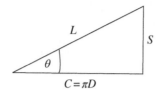

Figure 8.24   Relationship between helix dimensions

and commercial attractiveness. The performance of helical antennas is worse than that of their monopole predecessors; however, they allowed smaller, more esthetically attractive phones to be manufactured. This compromise continued with the later introduction of internal antennas.

The performance is defined by the following parameters (refer to Figure 8.24):

C circumference
S spacing between turns
L unfolded turn length
θ pitch angle: $\tan(\theta)=S/C$

When C/λ is between 0.75 and 1.25, the antenna operates in an *axial mode* as discussed in Chapter 5, when the radiation is along the axis of the antenna (the upward direction as shown in Figure 8.23). The antenna has some useful properties in this mode: the gain is high, the polarization is circular, and the impedance is close to 150 ohms over close to an octave bandwidth. Because of this, axial mode helices are widely used, for example, as satellite antennas. However, they are not normally used for mobile applications.

When C/λ is less than 0.5, the helix is said to operate in a *normal mode*, when the radiation is similar to that of a monopole aligned along the axis of the helix (radiation is predominantly perpendicular to the axis and the polarization is linear, in line with the axis). In this mode, the antenna is resonant with a height that is significantly less than that of a comparable monopole antenna.

Monopole-Like Antennas

Monopole-like antennas operate in a similar fashion to helical antennas, since they are generally both reactively loaded in order to reduce size (as discussed in Section 8.1.2). The distinction made here is that helical antennas are generally external to the phone casing whereas monopole-like antennas are located internally. As such, the latter are often planar or conform to the shape of the phone.

Inverted F Antennas and Planar Inverted F Antennas

The planar inverted F antenna (PIFA), which was used as a major example in previous chapters, became the antenna of choice for mobile phones in the late 1990s due to the requirement to have an internal antenna with a low SAR. They can be seen as evolving from either a monopole or a half-wave patch antenna. Evolution from a monopole is illustrated in Figure 8.25.

Here a quarter-wave monopole is first folded to form an inverted L antenna (ILA), as proposed by King et al. in 1960 [21]. They also proposed the second step in the evolutionary path from monopole to PIFA: that is, the introduction of a shorting pin. At the time, this

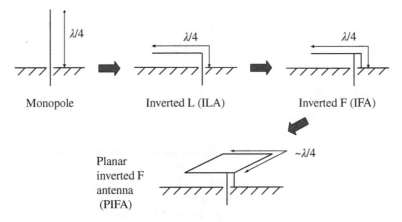

Figure 8.25   Evolution of a PIFA from a monopole antenna

configuration was referred to as a "shunt driven inverted L antenna," though it later became almost universally known as an inverted F antenna (IFA) [22]. The introduction of the shorting pin allows the low impedance of the ILA to be transformed up to a more convenient value; thus, the IFA is widely used in mobile devices. To improve bandwidth (at the expense of antenna size), the wire running parallel to the ground may be replaced by a planar element, giving a PIFA. As shown in Figure 8.25, the sum of two adjacent sides of the PIFA should be approximately quarter of a wavelength long at the desired resonant frequency [23].

Evolution from a half-wave patch antenna is illustrated in Figure 8.26. Here a probe-fed half-wave patch antenna is first short-circuited along a line where the electric field is zero (for the lowest frequency TM_{001} mode). This halves the size of the antenna without changing the resonant mode (though there is a loss of directivity, since radiation now occurs only from one end of the structure rather than two). Such a structure is often referred to as a full short-circuit PIFA (FS-PIFA), since the short circuit runs the full length of one face of the antenna (as indicated by the bold line in Figure 8.26). Replacing this short circuit with a simple shorting pin or tab reduces the resonant frequency further (by increasing the antenna inductance) and yields a

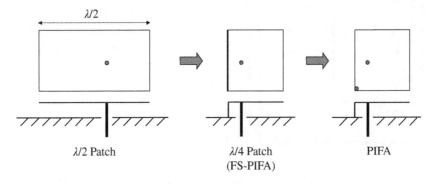

Figure 8.26   Evolution of a PIFA from a half-wave patch antenna

PIFA. There have been a lot of research in PIFA which has many variables for performance optimization. For example, it was shown that properly select the width of the shorting pin and the feed plate, 65% fractional impedance bandwidth could be achieved [24].

8.2.2.4 Typical Smartphone Antennas

A generalized LTE mobile antenna and hardware configuration in smartphones is shown in Figure 8.27. At least 6 antennas are required to cover the essential operation of the phone which has a typical size of 15 cm × 7 cm: three multiband antennas for cellular communications with MIMO functionality; one antenna for GPS navigation and cellular receiving as well; two dual-band antennas for Wi-Fi to ensure the capacity and reliability and one of them is shared with Bluetooth. Most antennas are placed at the top and bottom of the phone and the space available for these antennas is very small and the conventional antennas are too large for it. Thus, in practice, the classical antennas are not suitable and significant modifications have to be made. Some innovative new antennas have utilized all the space available, such as metal frame/rim and the gap between the frame and the PCB which will be discussed further later (e.g. Fig. 8.29).

Figure 8.28 illustrate a real-world example with a number of essential antennas. The one at the bottom is a modified dual-band IFA made using FPC technology as discussed in Section 7.2.6. The feed and shorting pin are clearly marked. The low-band arm is folded to

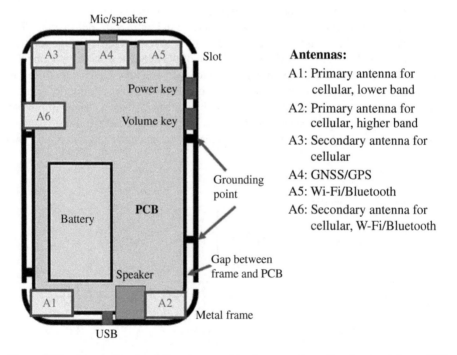

Figure 8.27 A generalized mobile antenna and hardware configuration for smartphones [25]

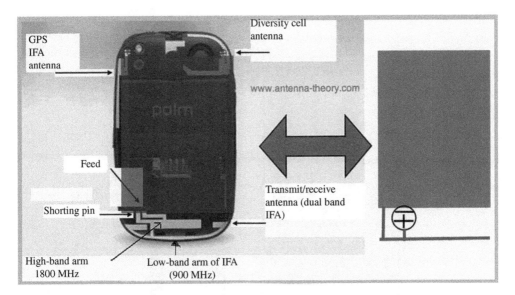

Figure 8.28   An example of smartphone antennas. Source: With permission from Pete at http://www.antenna-theory.com

reduce in order to fit into the space available. An additional open tuning stub is added to improve the impedance matching and bandwidth.

More recently, mental rim antennas become very popular for smartphones since they have utilized the largest size of the smartphone but occupied very little space; furthermore, they can provide multiband and wide bandwidth. One of the early innovative designs is shown in Figure 8.29 [26]. There is a small gap (about 2 mm) between the PCB and the metal frame/rim of the smartphone. The feeding point is at A while the short-circuit point between the PCB and metal rim is at C as shown in Figure 8.29(a), thus the antenna is formed by two loops in parallel: ADEC and AGFC with return paths through the PCB to A. The selection of the feed point and the short-circuit point is used to tune the performance of the antenna. The simulated and measured reflection coefficients are plotted in Figure 8.29(b). They are in good agreement and the antenna has met the design specs to cover GSM850/900 and UMTS2100/LTE 2300/2500 for VSWR = 3 (or S11 < −6 dB). This unbroken metal rim design is attractive for application and better than the well-known broken metal rim design of iPhone 4 which suffered from a performance problem when user's hand/finger touched the rim where two antennas meet – this incident was called "antennagate" in 2010 (https://www.pcmag.com/encyclopedia/term/antennagate).

There are many other related designs. For example, a novel reconfigurable open slot antenna was proposed to cover a wide bandwidth of 698–960 and 1710–2690 MHz [27]. The antenna is located at the bottom portion of the mobile phone and is integrated with metal rim, thereby occupying a small space (7 mm × 68 mm) and providing mechanical stability to the mobile phone. A varactor diode is used to cover the lower band frequencies, so as to achieve a good frequency coverage and antenna miniaturization. It has achieved the desired impedance bandwidth and the total efficiency of minimum 50% throughout the required bands. The antenna

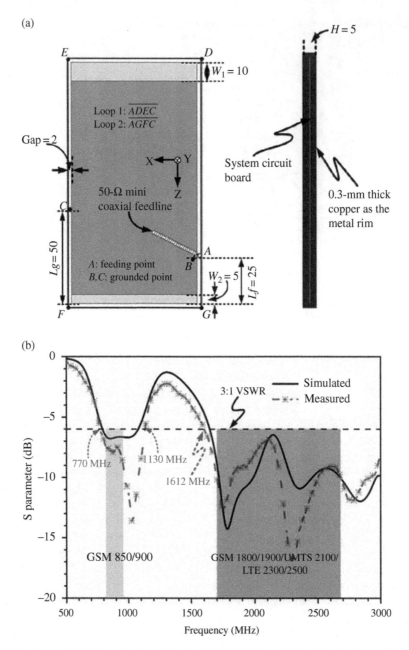

Figure 8.29 A metal rim loop antenna with simulated and measured results. (a) A metal rim antenna with two loops. (b) Simulated and measured reflection coefficient S11 in dB

performance with other components and human hand was also studied. Unlike the design in [26], another identical antenna can be placed at the top edge of the phone as the secondary antenna for diversity/MIMO. Thus, this antenna is an excellent candidate for smartphones and mobile devices.

So far, we have only discussed mobile antennas for cellular communications. There are antennas for other radio services inside the phone. In particular:

- NFC antenna: it is typically an electrically small loop coil antenna to produce magnetic fields at 13.56 MHz for short distance communications. The coil is normally place at the back case of the phone or on top of the battery under a thin plastic cover as shown in Figure 8.30. The number of turns in the coil determines the inductance and the operational distance.
- Wi-Fi/Bluetooth antenna: this antenna has to cover at least three bands now: 2.4, 5.2 and 5.8 GHz. IFA/PIFA and monopole antennas have been widely used for this application [28–30]. This is mainly because their electric length is about a quarter of a wavelength, which is substantially shorter than that of a typical dipole antenna or a patch antenna that requires a half of a wavelength. Figure 8.31 shows one of the most innovative, compact, and simple designs [30], the size is only 23 mm × 5 mm × 4 mm. There is no ground plane clearance (the gap to the rim of the phone) underneath the antenna. The antenna has two metal strips denoted by "Ground" and "Feed," and their widths are 1 and 2 mm, respectively. The thickness of the antenna track is 0.1 mm. As indicated in the figure, there are three resonators: the 2.4-GHz PIFA with a width change from 1.0 to 2.5 mm along the track (for better bandwidth); the 5.2-GHz PIFA, and the 5.8-GHz loop. The loop antenna is formed by the feeding and the grounding strips in conjunction with the antenna track and the PCB between them. The design has been patented and used in many smartphones. It should be pointed out that a slot antenna is also suitable for this application due to its low profile and higher modes could be employed if it is excited properly and the ratio between the higher mode and the fundamental frequencies can be adjusted as demonstrated in [31].
- GPS antenna: GPS signals from satellites are circularly polarized; thus, a GPS antenna should also be circularly polarized to maximize the reception. Many circularly polarized GPS antennas have been developed [32], but they are too large for a slim/low-profile smartphone. Instead, a linearly polarized IFA/PIFA antenna is often applied in practice, as shown in

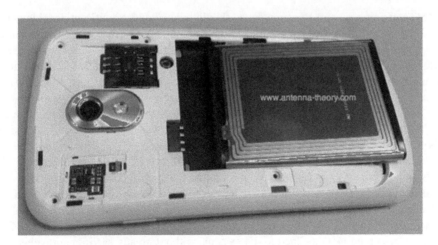

Figure 8.30 A printed NFC antenna inside a smartphone. Source: With permission from Pete at http://www.antenna-theory.com

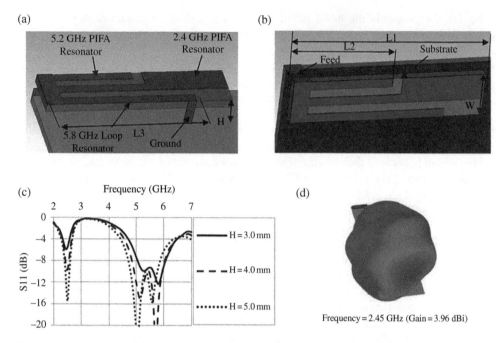

Figure 8.31 A triple-band Wi-Fi antenna for smartphones and its performance. (a) Expended view; (b) top view; (c) simulated S11 for different height H; (d) simulated radiation pattern [30]

Figure 8.28. It works against the ground plane of the main PCB and is printed on a plastic moulding with other electronics. The trade-off for this option is a loss of 3 dB in the received GPS signals due to polarization mismatch.

For 5G smartphones, mm-Wave antennas should be included. Significant development has been taken place over the past a few years. Qualcomm® QTM527 mm-Wave antenna module is the world's first announced, fully-integrated high-power mm-Wave antenna module (https://www.qualcomm.com/products/qtm527-mm-Wave-antenna-module) which is very different from sub-6GHz antennas. Sine at mm-Wave the antenna size is very small, antenna-in-package (antenna module) is the preferred solution and is set to grow over the coming years. The coexistence with other radio system is a major concern. In reference [33], a solution for codesigned mm-Wave and LTE antennas in a metal-rimmed handset was presented. The design showed that both antenna types could be accommodated in a shared volume and be integrated into the same structure for the frequency bands of 700–960 MHz, 1710–2690 MHz, and 25–30 GHz. Simulations and measurements suggested that the system could be designed in such a way that the mm-wave antenna did not hinder the low-band performance. LTE antennas generally reached over 60% total efficiency while the mm-Wave module had a peak gain of 7 dBi with measurement-verified beam-steering capability. The proposed design proves that 5G mm-Wave antennas could be embedded to 4G systems without greatly sacrificing display size or sub-6 GHz antenna performance. Of course, all these individual antenna designs have to take other components and the whole system into account which should at least include the effect of the PCB inside the phone and SAR.

8.2.2.5   The Effect of the PCB

Many small antennas appear to exceed the fundamental limits of antenna size, bandwidth, and efficiency outlined in Section 8.1.1. This is because they do not operate alone: there is a strong interaction with the equipment/PCB that tends to widen the bandwidth. More discussions on this are given later in this section.

8.2.2.6   Specific Absorption Rate (SAR)

SAR is a measure of the rate at which energy is absorbed per unit mass by a human body when exposed to an electromagnetic field. It is defined as the power absorbed per mass of tissue and has units of watts per kilogram (W/kg). SAR is an important parameter in the design of mobile phones because national and international regulations must be met and because consumers may buy a particular model of phone based on the quoted SAR.

SAR is regulated based on the known biological effects thermal heating. The temperature increase of the body cannot be directly correlated with the incident radiation due to the complex thermoregulatory process within the body. Instead, the concept of power absorption per differential unit mass is used as follows:

$$SAR = \frac{dP}{dm} = \frac{dP}{\rho_D \, dV} \text{ (in W/kg)} \tag{8.38}$$

where,

dP power absorbed within a differential volume;
dm mass within the volume;
dV differential volume;
ρ_D mass density.

This can be expressed in terms of the electric field as follows:

$$SAR = \frac{\sigma |E|^2}{\rho_D} \tag{8.39}$$

where

σ conductivity of the body (which will vary within different biological tissues).
E the RMS (root mean square) electric field strength within the body.

SAR is a power relation specified at a particular point in space. To convert this to a temperature rise (and, hence, to a biological effect), it is necessary to average over both time and mass

(a)   (b)

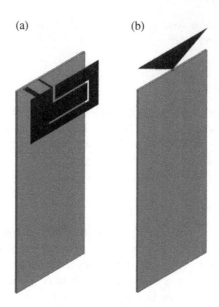

Figure 8.32 PIFA and generic monopole geometries. (a) PIFA (b) monopole

– or equivalently, volume. Typical averaging masses of 10 and 1 g are used in national and international regulations (unfortunately there is no unified world regulation). The value cited is the maximum level measured in the body part studied over the stated volume or mass. Averaging over 1 g gives significantly higher peak SAR values than averaging over a larger mass, since significant SAR variations occur on a small scale. Hence, the lower mass gives the most difficult specification and is used for most designs.

Figure 8.32 shows a PIFA and a generic monopole-like antenna, both mounted on a 100 mm × 40 mm × 1 mm PCB. These represent the two basic classes of internal antenna as described in earlier in this section. The PIFA is a typical dual-band structure, while the monopole-like antenna is triangular – a structure that will have relatively low current density and, therefore, relatively low local field strength.

There have been a lot of studies on SAR produced by mobile phones. For example, the antenna configurations are simulated next to a truncated flat representation of the phone user's head, as shown in Figure 8.33. Such a representation is often referred to as a *phantom*. In both cases, a separation of 5 mm is maintained between the PCB and the phantom surface. The material properties of the head and skin are taken from published databases at 900 and 1800 MHz and presented in Table 8.3.

To minimize reflections at the truncation surfaces of the phantom, these surfaces are defined as impedance boundaries, having the intrinsic impedances of the dielectrics used. Using Equation (3.19), the impedances are as given in Table 8.4

The simulated SAR of the PIFA antenna is shown in Figure 8.34, whereas the SAR of the monopole-like antenna is given in Figure 8.35. The SAR of the PIFA is clearly lower than that of the monopole-like antenna due to the shielding of the PCB and the increased distance of the antenna from the user. This is particularly the case at high frequencies when local fields at the antenna dominate.

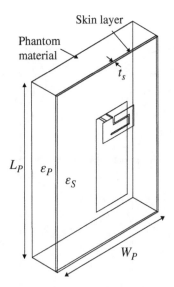

Figure 8.33   Flat phantom

Table 8.3   Relative permittivity and conductivity of phantom and skin layers

Frequency (MHz)	Phantom		Skin	
	Relative dielectric constant, ε_{pr}	Conductivity, σ_p (S/m)	Relative dielectric constant, ε_{sr}	Conductivity, σ_s (S/m)
900	41.5	0.9	4.2	0.0042
1800	40	1.4	4.2	0.0084

Table 8.4   Intrinsic impedance as a function of frequency

Frequency (MHz)	Phantom Impedance (Ω/square)	Skin Impedance (Ω/square)
900	54.35 + j12.06	183.83
1800	57.06 + j9.68	

This example shows that antenna conductors in the plane of the PCB give rise to high local SAR unless they can be moved further from the head. This effectively restricts the use of such antennas to the bottom position of the phone. Only when the phone is long enough that there is sufficient clearance between the antenna and the head, can monopole-like antennas be used with acceptable SAR. This is often the case for "flip" phones.

Measurements of SAR due to a mobile phone have been conducted using a human head "SAR Phantom" in different talk positions on both sides of the head and at different frequencies

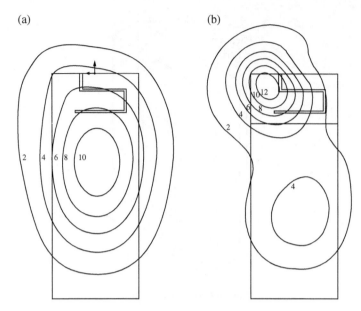

Figure 8.34 SAR in W/kg of a typical conventional PIFA (power normalized to 1W). (a) 900 MHz (b) 1800 MHz

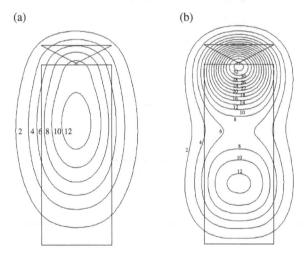

Figure 8.35 SAR in W/kg of a typical triangular monopole (power normalized to 1W). (a) 900 MHz (b) 1800 MHz

at which the device can transmit. Depending on the size and capabilities of the phone, additional testing, e.g. the hearing aid compatibility (HAC) testing, may also be required to represent usage of the device while placed close to the user's body and/or extremities. Various governments have defined maximum SAR levels for RF energy emitted by mobile devices:

- United States: the FCC require that phones sold have a SAR level at or below 1.6 W/kg taken over the volume containing a mass of 1 g of tissue that is absorbing the most signal.
- European Union: the CENELEC specify SAR limits within the EU. For mobile phones and other hand-held devices, the SAR limit is 2 W/kg averaged over the 10 g of tissue absorbing the most signal.

It should be pointed out that there are now accurate head and full human-body models in some software packages (such as CST), most of the smartphone antenna simulations have included the body effects and SAR analyses which requires a lot of computation resources as expected. Simulated SAR distributions on a human head phantom using a smartphone at different frequencies and impedance tuning capacitance C (a common technique used to improve antenna efficiency) are given in Figure 8.36, it is clear that the SAR values have met the FCC limits in all the cases. Further details can be found from reference [34].

Frequency (MHz)		750	825	900	1800	2250	2700
Input power (Watt in dBm)		24	24	24	21	21	21
1-g SAR (W/kg)	C = 0.1 pF	0.23	0.34	0.51	0.36	0.33	0.65
	C = 1.6 pF	0.23	0.54	0.27	0.42	0.39	0.64
	C = 2.4 pF	0.46	0.58	0.10	0.41	0.38	0.64
	C = 3 pF	0.59	0.14	0.04	0.41	0.37	0.64
	C = 3.6 pF	0.34	0.05	0.01	0.41	0.37	0.64

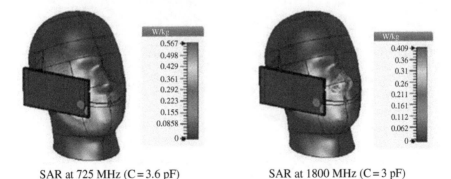

SAR at 725 MHz (C = 3.6 pF)   SAR at 1800 MHz (C = 3 pF)

Figure 8.36 Simulated SAR values and SAR distribution on a head phantom using a smartphone in different scenarios (frequency and tuning capacitance) [34]

Here we have discussed the effects of mobile phones on human body. The effects of human body on mobile phone performance will be discussed later in this section.

8.2.3   Multipath and Antenna Diversity

8.2.3.1   Multipath and Mean Effective Gain

As introduced in Section 3.5, *multipath* is the term used to describe the fact that radio signals often occur at a receiver via several different paths. This is illustrated for a typical mobile radio reception scenario in Figure 8.37. The scenario shown corresponds with that of a macro-cell, where an elevated base-station is free of local scattering (reflections and re-radiation from nearby objects). The mobile, however, is below the height of local clutter and hence receives multipath contributions from many different angles. This is often referred to as a "ring of scatters" model, where typical radii of the scatterers from the mobile are of the order of a few hundred meters.

A summation of the multipath components occurs in the radio receiver, as shown in Figure 8.38. As the mobile moves, each multipath component vector rotates by a different amount, u_i, as indicated. Hence, constructive and destructive interference occurs – a phenomenon that is known as *signal fading*. This reflects the fact that the signal quality is very poor when the vector addition of the multipath components produces a *fade* (a low value due to destructive interference).

Fades tend to occur at intervals of approximately half of a wavelength. Many readers will have experienced this. For example, FM radio has a wavelength of approximately three meters and is subject to fading. If a poor signal is experienced while listening in a car, reception can be improved by moving the car forward by a couple of meters.

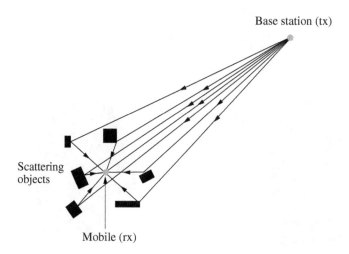

Figure 8.37   Typical macro-cellular multipath scattering scenario

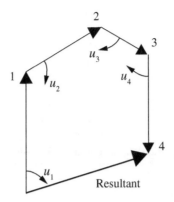

Figure 8.38 The addition of multipath vectors

The performance of an antenna in a multipath environment cannot be quantified simply in terms of its free space gain (radiation pattern) since the gain is only specified in one direction. Thus, we need a more accurate, environment dependent term, the mean effective gain (MEG), given by

$$MEG = \frac{1}{(1 + X_P)} \int\int_{\theta, \varphi} \left[G_\theta(\theta, \varphi) p_\theta(\theta, \varphi) + X G_\varphi(\theta, \varphi) p_\varphi(\theta, \varphi) \right] d\theta d\varphi \qquad (8.40)$$

where

G_θ and G_φ are the antenna power gains in the θ and φ polarizations, respectively.
X is the cross-polar ratio P_φ/P_θ – the ratio of power received by θ and φ polarized isotropic antennas
p_θ and p_φ are the angular density functions of the incoming θ and φ polarized plane waves, respectively

The gain includes all radiation-dependent parameters such as the antenna efficiency, the loss due the proximity of the user, and the radiation pattern (including the effect of the handset and the user). The angular density functions and the cross-polar ratio define the propagation environment.

In any one direction, the gain of the antenna is weighted by the probability of receiving a multipath component from that direction. This weighting is performed for both polarizations. A further weighting is also applied based on the relative powers incident in both polarizations.

For high MEG, the antenna radiation pattern should be well matched to the angular probability distribution and cross-polar ratio of the incoming multipath. The azimuth (φ) distribution is assumed to be uniform, since multipath originates from all angles around the mobile. The elevation (θ, as measured from the ground upward) distributions of some typical environments are given in Table 8.5. The distributions in elevation are considered here to follow a Gaussian distribution, given by

Table 8.5 Typical model parameters for global angle of arrival PDFs

Local environment	Cell type	m_θ	σ_θ	m_φ	σ_φ	X (dB)
Rural	Macrocell	0°	15°	–	–	−15
Suburban	Macrocell	15°	20°	20°	25°	−9
Urban	Macrocell	20°	30°	30°	50°	−7
	Microcell	10°	25°	15°	30°	−7
	Picocell	5°	20°	0°	25°	−7
Indoor	BS Outdoors	0°→5°	15°→20°	0°→5°	20°→25°	−2
	BS Indoors (Picocell)	0°→5°	30°→40°	0°→5°	35°→45°	−2

$$p_\theta(\theta) = \frac{1}{\sqrt{2\pi}\sigma_\theta} \exp\left\{ -\frac{(\theta-m_\theta)^2}{2\sigma_\theta^2} \right\} \tag{8.41}$$

$$p_\varphi(\theta) = \frac{1}{\sqrt{2\pi}\sigma_\varphi} \exp\left\{ -\frac{(\theta-m_\varphi)^2}{2\sigma_\varphi^2} \right\} \tag{8.42}$$

where

m_θ, m_φ, means of the angular probability density functions (PDFs) of the incoming θ and φ polarized plane waves, respectively.
σ_θ, σ_φ. standard deviations of the PDFs of the incoming θ and φ polarized plane waves, respectively.

Distributions other than Gaussian have also been proposed, though the exact form of the distribution is not normally very important to the final MEG. Polarization matching, for example, is much more important.

The shape of a typical θ-polarized urban macro-cell probability distribution is shown in Figure 8.39.

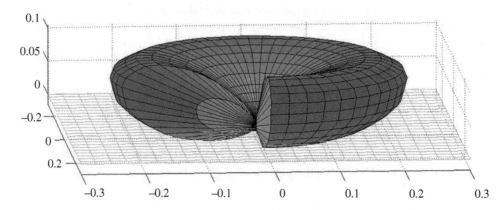

Figure 8.39 Angle of arrival PDF for an urban macro-cell co-polarized with the base station

The antenna MEG is maximized when the antenna gain maximum coincides with the most likely direction of multipath reception. Further information and examples can be found from references such as [35].

8.2.3.2   Antenna Diversity

One of the most effective methods to combat multipath effects is called *antenna diversity* in which more than one antenna is used to overcome the detrimental effects multipath fading to improve receive signals reliability and signal-to-noise ratio (SNR). Figure 8.40 shows the reception from two antennas subjected to the same multipath components. Due to the differences of the antennas (such as the location or polarization), the fading at the received signals from the two antennas may occur at different times/distances (time and distance can be considered to be interchangeable when the mobile moves at a constant velocity), since the antennas process or filter the multipath in different ways. By combining these two signals, the probability of a deep fade is much reduced.

The fades shown in Figure 8.40 are typical of those experienced in narrowband cellular radio systems such as GSM and also those of noncellular systems such as FM radio. Occasionally, fades can cause the signal strength to drop by as much as 30 dB (both antenna 1 and antenna 2 have average signal levels of 0 dB in Figure 8.40). It can be seen that the depth of such fades and the probability that they will occur are reduced when the signals are combined.

To realize antenna diversity there are two requirements:

- Signals must be decorrelated (to some degree) and independent – i.e. two or more signals must be available that fade at different times/positions
- A method of combining the signals must be found.

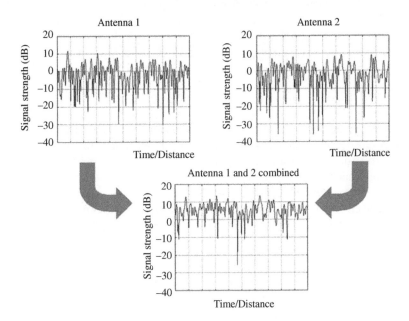

Figure 8.40   Combination of two uncorrelated signals

There are several ways of achieving de-correlated signals: i.e. signals that experience fades at different times or when the mobile is in a different position. These are often referred to as the type of antenna diversity as described below.

Polarization Diversity
Antenna patterns are orthogonally polarized and hence preferentially receive only one of the two incoming polarization states. All multipath components will, to some degree, have both polarization states, the amplitudes and phases of which will be different. Hence, orthogonally polarized antennas will perform different vector summations of the multipath components and they will be de-correlated.

Spatial Diversity
Here multipath components will have different phases at the receive antennas due to their spacing. Unlike for polarization diversity, the amplitudes of the multipath components will be the same at both antennas.

Radiation Pattern Diversity
Here the magnitudes and phases of the antenna radiation patterns are arranged such that they do not overlap. Hence, the antenna directivity causes different multipath components to be added.

All three of the above mechanisms – polarization, spatial, and radiation pattern – are usually present when antennas are closely spaced, such that de-correlated reception can be achieved at smaller spacings than often thought possible. Spacing is often limited by antenna isolation rather than antenna correlation.

The *envelope correlation coefficient* (ECC) which was introduced briefly in Section 7.9.2 to evaluate the correlation between two MIMO antenna elements is now used here for the diversity performance evaluation of two antennas This parameter will be further discussed in Section 8.3.2 and is mathematically defined as

$$\rho_e = \frac{\left| \int_\Omega E_{i\theta} E_{j\theta}^* P_\theta \, d\Omega + X \int_\Omega E_{i\varphi} E_{j\varphi}^* P_\varphi \, d\Omega \right|^2}{\int_\Omega \left[|E_{i\theta}|^2 P_\theta + X |E_{i\varphi}|^2 P_\varphi \right] d\Omega \int_\Omega \left[|E_{j\theta}|^2 P_\theta + X |E_{j\varphi}|^2 P_\varphi \right] d\Omega} \qquad (8.43)$$

where $E_{i\theta}$ and $E_{i\varphi}$ indicate the complex radiation fields of the ith antenna in the θ and φ polarizations, respectively. It is important to note here that all radiation fields have a magnitude and phase in any given direction. Normally only the former is of interest, but the latter must also be considered when calculating correlation.

The ECC is always between zero and unity. When zero, the signals are completely de-correlated (different) and when unity the signals are completely correlated (identical). For two antennas with good isolation, the ECC should be smaller than 0.3 as mentioned earlier.

8.2.3.3   Combining Methods

There are four common diversity combining methods, as follows (in order, with the worst performing method first and the best last).

Switched Combining (SWC)

Here a switch is used to receive from one antenna at a time. When the signal received falls below a certain threshold, the second antenna is used. This does not guarantee better performance, but it does improve the likelihood. As suggested by the method, this is also referred to as threshold diversity.

Selection Combining (SC)

The antenna with the strongest signal is used. This may be implemented with a single receiver or with two receivers. A single receiver is clearly desirable from a cost, size, and power consumption perspective; however, some time must be made available for the antenna signals to be compared. This is relatively straightforward in digital, packet-based systems. For example, DECT (digital enhanced cordless telecommunications) uses selection combining with the antenna selection performed in the preamble that occurs before each data packet is sent.

Equal Gain Combining (EGC)

Here two receivers are required and the signals from the antennas are cophased prior to summation.

Maximal Ratio Combining (MRC)

The signals from the antennas are weighted based on their SNR.

Diversity gain (DG) which was introduced in Section 7.9.2 to evaluation MIMO antenna performance can also be used to evaluate the diversity antenna performance. The DGs for different combining methods are illustrated in Figure 8.41, the differences between these methods are quite small for two completely decorrelated antennas ($\rho_e = 0$). In this figure, the ordinate (y axis) shows the probability that the signal-to-noise ratio is less than the level shown on the abscissa (x axis) – i.e. the degree of certainty that a predefined signal-to-noise ratio will be

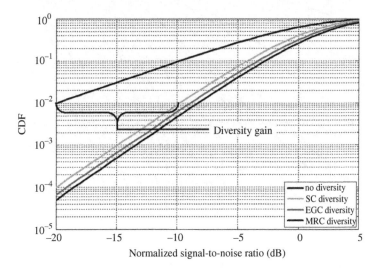

Figure 8.41 Diversity gains of SC, EGC, and MRC with 2 equal power branches and zero correlation (Rayleigh signals)

achieved. For example, a CDF (cumulative distribution function (CDF)) of 10^{-2} indicates a 99% probability that a particular signal level is exceeded.

When the CDF is low enough, the curves with diversity in Figure 8.41 become parallel lines and there is a constant relation between the DGs of the three combining methods. With two branches, EGC performs 0.88 dB better than SC, whereas MRC performs 1.5 dB better than SC. The DG is substantial for all methods; however, the gain depends on the chosen certainty level. For example, for a CDF of 10^{-2}, a SNR of better than –20 dB is always achieved without diversity. With selection combining a SNR of better than –10 dB is always achieved with the same level of certainty. Hence, the DG is 10 dB. For lower levels of certainty (higher CDFs), the DG will be lower than this.

For greater numbers of branches the difference in DG increases, particularly between MRC and SC. The variation in the DG of MRC and SC with the number of branches is shown in Figure 8.42. EGC is not plotted, since an exact formula is not known for more than two branches.

The difference between SC and MRC with large numbers of branches is largely due to the fact that MRC with M branches requires M receivers, whereas SC is limited to one receiver. For all combining methods, the mean power varies with the number of diversity branches. Assuming Rayleigh signal envelope statistics (valid for most multi-path environments), zero correlation, and equal branch powers, this variation is shown in Figure 8.43.

Clearly, both MRC and EGC give a linear improvement with the number of diversity branches. MRC gives the best performance but is only slightly better than EGC. SC is nonlinear

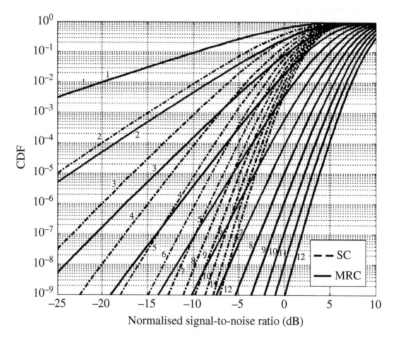

Figure 8.42 Diversity gains of MRC and SC with number of branches (1–12 branches shown from left to right)

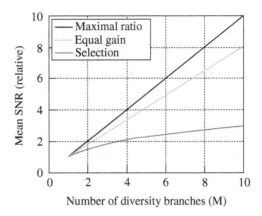

Figure 8.43   Mean SNR of different combining methods with the number of branches

and significantly worse than both MRC and EGC, particularly when the number of diversity branches is high.

8.2.3.4   The Effect of Branch Correlation

The effect of branch correlation for a two-branch MRC system with Rayleigh signals and equal branch powers is shown in Figure 8.44. It shows that there is some DG even with completely correlated signals. This is simply a 3 dB gain that results from having two antennas rather than one (collecting twice as much of the incident energy). An appreciable additional DG occurs at

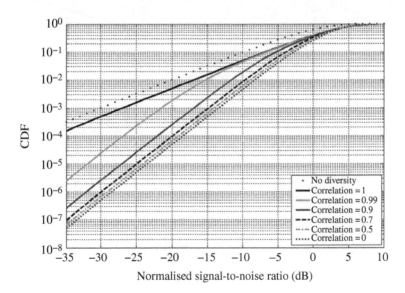

Figure 8.44   The effect of branch correlation

very high levels of correlation. Decreasing the correlation gives diminishing returns, such that it is often considered as unnecessary to aim for a correlation coefficient of less than 0.5–0.7 (a correlation of 0.7 gives a DG of ~1.5 dB less than the optimum).

It can be seen that as the CDF reduces, the curves for different correlation levels reduce at the same rate. Hence, the difference in DG can be considered to be asymptotically constant.

8.2.3.5   The Effect of Unequal Branch Powers

The effect of unequal branch power is shown in Figure 8.45.

For two branches, the DG is nominally reduced by half the SNR difference between the branches.

$$Loss = \frac{SNR_1 - SNR_2}{2} \tag{8.44}$$

The power received by each antenna is given by the MEG.

8.2.3.6   Examples of Diversity Antennas

An example of two very simple diversity antennas within a DECT base-station is shown in Figure 8.46. Here the base-station is relatively large, such that low correlation can be guaranteed using space diversity: the antennas are separated by approximately 2/3 of a wavelength at

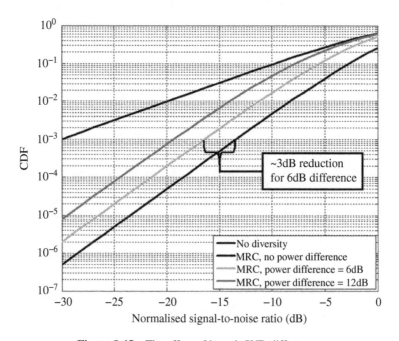

Figure 8.45   The effect of branch SNR differences

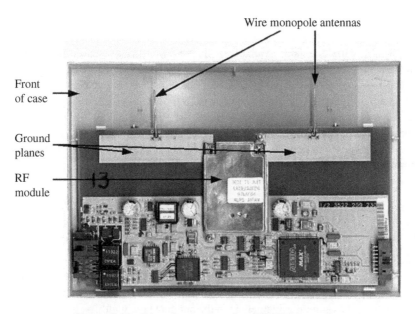

Figure 8.46   Diversity antennas in a DECT base-station

the DECT center frequency of 1890 MHz. Simple quarter-wave wire monopole antennas are used to minimize costs.

An example of two diversity antennas within a DECT handset is shown in Figure 8.47. Here the antennas are arranged orthogonally in an attempt to realize polarization diversity. However, the handset PCB radiates significantly – in the same polarization – for both antennas, so the amount of polarization diversity that can be obtained is limited. Despite this, the differences in the radiation patterns of the two antennas are sufficient to ensure a useably low envelope

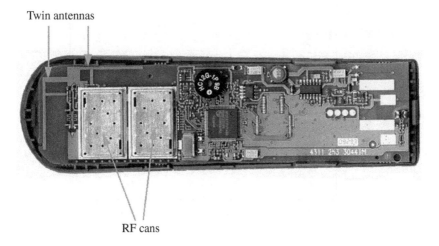

Figure 8.47   Diversity antennas in a DECT handset

correlation coefficient. The antennas used are inverted F antennas that are simply formed by tracks on the main PCB of the handset.

DECT uses SC with time made available in the pre-amble prior to packet transmission. IEEE 802.11b uses Wi-Fi also uses SC in a similar way to DECT. However, UMTS uses time diversity (i.e. there are alternatives to using antennas!). From this classical example, we can clearly see how antenna diversity has been used for both mobile base-station and mobile terminal handset. All modern mobile systems have implemented the antenna diversity technique. It should be pointed out that diversity antennas can be spatially separated, or polarization orthogonal. Some designs use the same radiator, but different excitation ports (hence different modes) to generate antenna diversity, such a compact design can save space [34, 36].

8.2.4   User Interaction

8.2.4.1   Introduction

There is a strong interaction between antennas within portable devices and the user. This interaction causes the antenna resonant frequency to change – often to a lower frequency – and the antenna matching to vary. It also causes the antenna efficiency to reduce, since the body acts as a lossy dielectric. The only improvement is in the antenna bandwidth, as we would expect with reduced efficiency (refer to Section 8.1.1).

8.2.4.2   Body Materials

To consider the effect of user interaction on the performance of mobile phone antennas, we should first examine the material properties of the head. The variation of the relative permittivity of brain material with frequency is shown in Figure 8.48, whereas the loss tangent and skin depth are shown in Figure 8.49 and Figure 8.50, respectively.

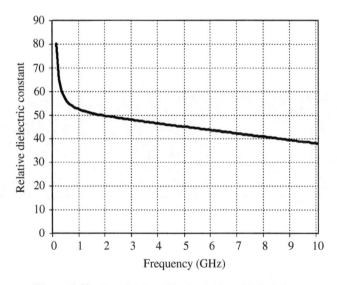

Figure 8.48   Permittivity of brain matter with frequency

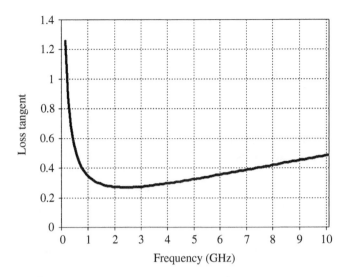

Figure 8.49 Loss tangent of brain matter with frequency

The relative permittivity of brain matter varies slowly from approximately 40 to 50 between 10 and 1 GHz. Below 1 GHz it begins to rise increasingly quickly, reaching a value of approximately 80 at 100 MHz. The loss tangent also rises very quickly as the frequency is reduced below 1 GHz, reaching approximately 1.3 at 100 MHz. However, in contrast to the permittivity, the loss tangent is not monotonic and is a minimum at approximately 2.5 GHz.

Figure 8.50 shows that the skin depth rises sharply at frequencies below 1 GHz. The field penetrates to a depth of approximately 4.2 cm at the lower GSM frequencies at around 900 MHz, whereas at the higher GSM frequencies in the region of 1800 MHz, the penetration is approximately 2.7 cm.

8.2.4.3   Typical Losses

The presence of the user affects the antenna performance in three main ways: the impedance match alters; the radiation efficiency is reduced and the MEG varies. Each of these mechanisms is addressed in the paragraphs that follow.

The radiation efficiency η_R reduces due to the additional losses in the body, particularly the head and hand. It is defined as the ratio of the power radiated (including all loss in the body) to the power accepted by the antenna.

The impedance change that occurs when a phone is held causes a change in the *mismatch loss*. This is expressed as the *match efficiency* (as discussed in Chapter 4 and defined by Equation (4.39)), $\eta_m = |\tau|^2$: the ratio of the power accepted by the antenna to the power supplied (the difference being the reflected power, due to impedance mismatch). The best case is 100% matched.

Figure 8.51 shows the typical match efficiency of a mobile phone antenna in the 900 MHz band and in the presence of 15 different users, all of whom hold the phone in different ways. It can be seen that for 9 users the phone is not significantly detuned. For 3 users, the detuning is such that the resonant frequency is just out-of-band, although the band-edge mismatch

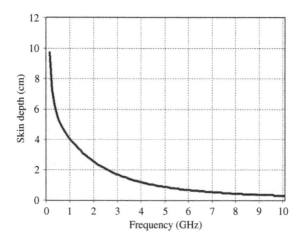

Figure 8.50 Skin depth of brain matter with frequency

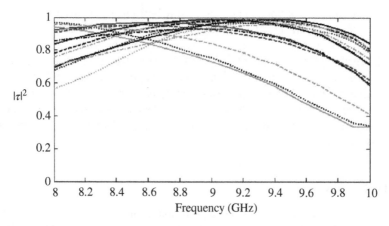

Figure 8.51 Typical match efficiency with users in and around the GSM 900 MHz band

efficiency remains high. For the remaining 3 users, the resonant frequency becomes significantly lower than is required, causing low match efficiency at the upper band edge.

The total efficiency, η_T, is the ratio of the power supplied to the antenna to the integrated
power on a sphere surrounding it (the radiated power). It is equal to the product of the radiation
and match efficiencies.

The total efficiency, match efficiency, and mismatch-corrected efficiency η_C (this is antenna
radiation efficiency in reality, thus $\eta_T = \eta_m \times \eta_C$) of four commercial mobile phone antennas
(made in around 2000) in free space are shown in Table 8.6. The results are averaged over the
GSM (880–960 MHz) and DCS (1710–1880 MHz) bands, using frequency steps of 10 and
25 MHz respectively.

Table 8.6 Average efficiencies of dual-band handsets in free space

Type	880–960 MHz			1710–1880 MHz		
	η_T (%)	$\mid\tau\mid^2$ (%)	η_C (%)	η_T (%)	$\mid\tau\mid^2$ (%)	η_C (%)
Helix	72.8	90.8	80.0	37.7	52.3	72.0
Flip	72.2	97.3	74.2	62.1	93.0	66.9
PIFA-1	70.4	95.2	73.6	64.2	90.9	71.1
PIFA-2	59.1	83.9	69.8	48.9	85.5	57.3

Table 8.7 Average efficiencies of dual-band handsets in the talk position

Type	880–960 MHz			1710–1880 MHz		
	η_T (%)	$\mid\tau\mid^2$ (%)	η_C (%)	η_T (%)	$\mid\tau\mid^2$ (%)	η_C (%)
Helix	5.3	71.7	6.9	4.9	50.8	9.8
Flip	5.7	90.2	6.1	5.8	91.5	6.3
PIFA-1	7.5	83.5	8.7	10.8	91.9	11.7
PIFA-2	3.4	55.2	5.7	10.7	85.0	12.7

It can be seen that the best of these phones has achieved average total efficiencies of the order of 60% at DCS and 70% at GSM in free space. The helical phone has a particularly poor return loss at DCS and, therefore, a match efficiency of approximately 50%. It is assumed that the match is improved (to some extent) using circuitry within the handset.

Typical efficiencies of dual-band phones in the talk position (held next to the head) are shown in Table 8.7. The average total efficiency is in the region of 3–8% in the 900 MHz band and 5–11% in the 1800 MHz band. The average radiation efficiencies are in the region of 6–9% and 6–13% at in the 900 and 1800 MHz bands, respectively. For both frequency bands, approximately half of the reduction in radiation efficiency can be attributed to the losses in the head, with the other half due to losses in the hand.

There is little difference between the average match efficiencies in free space and in the presence of the user at 1800 MHz. However, for 900 MHz, the difference can be appreciable. In general, the resonant frequency reduces and the match at resonance deteriorates. However, the return loss bandwidth generally improves due to the losses induced in the body. In the 1800 MHz band, the net result is that the average stays approximately the same. In the 900 MHz band, the detuning of the resonant frequency generally outweighs any bandwidth enhancement caused by additional body loss (also shown in Figure 8.51.

The mean effective gain, MEG is a function of the total efficiency, the radiation pattern, and the multipath propagation statistics. It is the most representative calculable measure of performance in a real network. Additional loss can result if the radiation pattern changes in the presence of the body such that it becomes less well matched with the likely angles of arrival of multipath components than in free space.

Table 8.8   Mean MEG (%) with 15 users (averages in parentheses are in dB)

	PIFA-1		PIFA-2		Helix		Flip	
	920 MHz	1800 MHz	920 MHz	1800 MHz	920 MHz	1800 MHz	920 MHz	1800 MHz
Rural	4.2	2.9	1.7	2.9	3.0	2.7	3.1	3.6
Suburban	4.2	3.9	1.9	3.8	3.0	2.2	3.3	3.6
Urban Macro	4.4	4.2	2.0	4.1	3.1	2.1	3.4	3.6
Urban Pico	4.4	3.8	2.0	3.8	3.4	2.5	3.6	3.9
Outdoor-to-Indoor	4.8	3.8	2.7	3.8	4.4	2.5	4.6	3.9
Indoor	4.7	4.6	2.5	5.0	4.2	2.4	4.5	4.3
Average	4.5	4.0	2.1	4.1	3.5	2.4	3.7	3.9
	(−13.5)	(−14.0)	(−16.8)	(−13.9)	(−13.5)	(−16.2)	(−14.6)	(−14.1)

Table 8.9   Standard deviation of MEG (%) with 15 users

	PIFA-1		PIFA-2		Helix		Flip	
	920 MHz	1800 MHz	920 MHz	1800 MHz	920 MHz	1800 MHz	920 MHz	1800 MHz
Rural	1.9	1.5	1.5	2.3	2.7	3.0	2.1	3.3
Suburban	1.6	1.7	1.3	2.2	2.3	2.2	2.2	3.1
Urban Macro	1.7	1.7	1.3	2.2	2.3	2.0	2.4	3.0
Urban Pico	2.0	1.7	1.6	2.7	2.9	2.5	2.5	3.2
Outdoor-Indoor	2.6	2.2	2.2	3.5	3.5	2.2	3.1	3.4
Indoor	2.5	2.1	2.1	3.4	3.3	2.1	3.0	3.4
Average	2.0	1.8	1.6	2.6	2.8	2.3	2.5	3.1

The following summary gives the average MEG for 15 users at 920 MHz and 1800 MHz, calculated using the angle of arrival statistics given in Section 8.2.2. Both the mean and the standard deviation are given in Table 8.8 and Table 8.9, respectively.

The MEG is between 2.1 and 4.5%. This is significantly lower than the efficiency, partly because angles close to the horizon are most likely and the phones "waste power" by having high relative gain outside the most likely angles. More importantly, however, the polarization of the phone radiation (reception) is not well matched to the polarization mix of the scattering environment.

The average MEG is −14.6 dB at both GSM and DCS. This represents a reduction of 1.8 dB for GSM and 3.5 dB for DCS when compared to the average efficiency. Clearly, the radiation patterns at GSM must be better matched to the average propagation environment than at DCS.

Figure 8.53 shows the user-averaged radiation pattern of the flip phone, i.e. the gain at each angle is averaged for the 15 users. All users are right-handed, so only one average radiation pattern is required. Clearly, there is more power in the φ polarization than in the θ polarization.

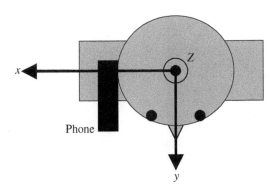

Figure 8.52   Orientation of users with respect to the chamber coordinate system

This is expected, due to the inclination of the phone when in a natural talking position. The blocking region of the head is clearly visible in the negative x, negative y direction. Negative x corresponds to the user's left-hand side while negative y corresponds to the user's back. The left- and right-hand sides of the figure show the radiation pattern viewed from opposite directions in order to better show its 3D nature. Figure 8.52 shows the positioning of the users relative to the coordinate system chosen.

Figure 8.54 shows the user-averaged radiation pattern of "PIFA-1." Again, the phone is predominantly φ polarized, but more so than for the flip phone. Both θ and φ polarizations have significant lobes in a near-zenith direction (along the positive z axis). There is unlikely to be a propagation path supported in this direction, which also contributes to the inferior MEG performance of "PIFA-1." Without this lobe, there is very little energy in the θ polarization. The propagation statistics used assume that this is the dominant polarization coming from the base-station (via a scattering environment), which, again, results in low MEG. Assuming that the dominant polarization is θ directed implies a vertically polarized base-station antenna. It is worth noting that, as polarization diversity with slant 45° polarized antennas is becoming more common at the base-station, this assumption may be questionable.

At 920 MHz, the polarization mix of all phones/antennas is more even than at 1800 MHz, which partly explains the superior relative MEG performance of the phones at GSM. The angular distribution of the patterns is also largely coincident with likely propagation paths. In terms of MEG, all of the phones perform equivalently – within approximately 3 dB of each other at both frequencies.

The strong interaction between antennas within portable devices and the user can also be characterized using commercial EM simulation tools. The main figure of merit is the total efficiency in this case. Figure 8.55a shows the effects of electronic components inside a mobile on the total efficiency of an open slot antenna in Figure 8.55b [34]. Two grip positions are investigated: one-hand grip at the top position of the handset, which is the talk mode and two-hand mode or the data mode which have been used in simulation in CST design studio. The hand models are very similar to CTIA models used by the industry for performance evaluation [37]. It is apparent that the effects of the components are small; but the effects of hands are significant. Generally, the user hand reduces the antenna total efficiency and often degrades the impedance matching. The simulated total efficiency of the antenna for one-hand grip at the top position and two-hand data mode are as shown in Figure 8.55c – the efficiency is reduced

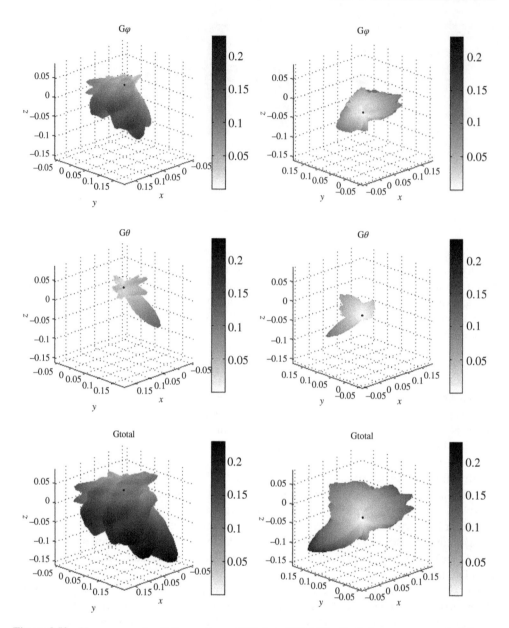

Figure 8.53 User-averaged radiation pattern of "flip" at 1800 MHz. Radiation patterns on the left are viewed from an azimuth angle of 45°, while those on the right are viewed from an azimuth angle of −135°. The origin is marked by a small circle

from about 90 to 40% in the lower band, from 90 to 55% in the higher band for one hand case. The reduction is more significant (90–20%, and 90–40% for the lower and higher bands, respectively) for the two-hand case. But the new antenna performs better than the classical antennas.

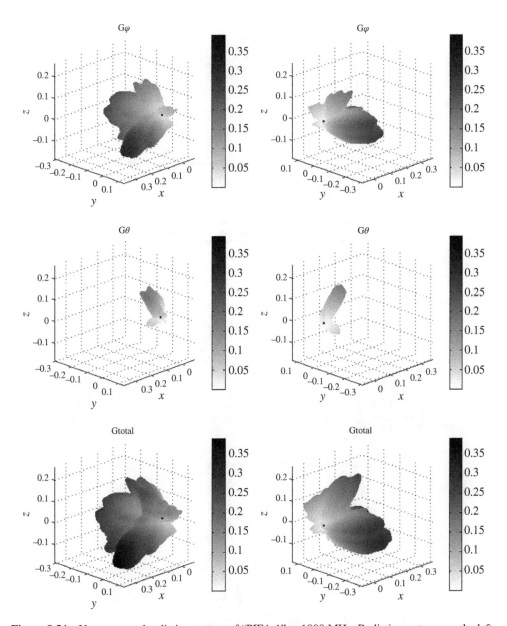

Figure 8.54 User-averaged radiation pattern of "PIFA-1" at 1800 MHz. Radiation patterns on the left are viewed from an azimuth angle of 45°, while those on the right are viewed from an azimuth angle of −135°. The origin is marked by a small circle

8.2.5 Mobile Base-Station Antennas

Compared with mobile hand portables, mobile base stations have more spaces and power for the antennas which have different requirements and specifications. There are basically two types of mobile base stations: indoor and outdoor.

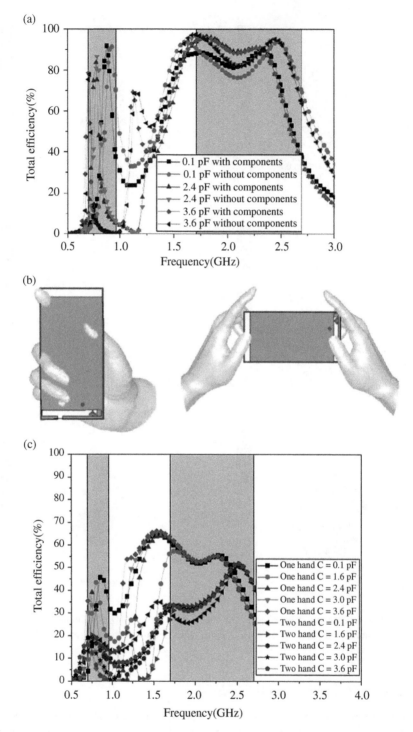

Figure 8.55 Effects of components and hands on total efficiency of the antenna. (a) Total efficiency with and without components near the antenna; (b) one-hand talk mode and two-hand data mode; (c) total efficiency for one-hand and two-hand modes for different matching capacitance

8.2.5.1   Outdoor Base-Station Antennas

For cellular mobile base-station antennas, there two basically two types of antennas: sectorial antennas and omni-directional antennas. A 7-cell configuration (i.e. frequencies are reused after a group of 7 cells) is shown Figure 8.56, the standard antennas (marked as A1) are located at a corner of the cell and they are formed by three directional sectorial antennas and each of the antenna covers 120° with assigned channels (such as 3, 10 or 17). A real-world example is shown in Figure 8.57 which explains why most mobile base stations have three antennas at each site. The main advantage of this arrangement is to minimize cochannel interference. Unlike A1, A2 in Figure 8.56 is an omni-directional antenna, which is normally located at the center of the cell. The main advantage of this arrangement is low cost and normally used in rural environment, not many users around.

Sectorial Antennas
Most mobile base-station antennas are in this category. The typical electrical specifications/performance for such an antenna are given in Table 8.10 where we can see that dual cross polarization (±45°) is required – this is to ensure that the received signal at mobile or base station is not polarization sensitive and also to provide diversity/MIMO. The horizontal half-power beamwidth (HPBW) is about 70° which is relatively broad while the vertical half-power beamwidth is just about 10°. The vertical beam could be titled electrically or mechanically in order to make a good use of the radiated power. The gain of the antenna is 15 dBi which is large to cover a wide area. The required inter-band isolation is over 25 dB which is high. VSWR (return loss) is 1.5 (14 dB) which is also higher than the normal requirements (i.e. VSWR < 2, or return loss >10 dB). Passive inter-modulation (PIM, as discussed in Section 7.1) is −150 dBc which is very

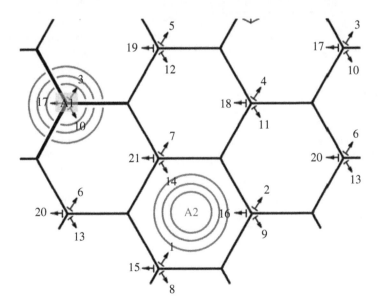

Figure 8.56   7-cell cellular radio configuration with two different antennas

Figure 8.57   Three sectorial antennas to provide 360° cellular radio coverage

Table 8.10   Typical Electrical Specifications of a Mobile Base-station Antenna

Parameter	Specs
Impedance	50 ohms
Frequency band	690–960 MHz
Polarization	±45°
Gain	15 dBi
Horizontal HPBW	70.0°
Vertical HPBW	10.2°
Beam tilt	0–14°
USLS (First lobe)	17 dB
Front to back ratio at 180°	21 dB
Isolation by beam tilt	28 dB, 0°–2°; 30 dB, 3°–14°
Isolation, interband	25 dB
VSWR (or Return loss)	< 1.5 (or > 14 dB)
PIM, 3rd order, 2×20 W	−150 dBc
Input power per port, maximum	300 W

low. USLS is for upper sidelobe suppression and defined to reduced interference (should be at least 17 dB).

Figure 8.58 is a good example of a sectorial antenna which covers both the lower band 698–960 MHz and the higher band 1710–2690 MHz (it was actually divided into penta-band)

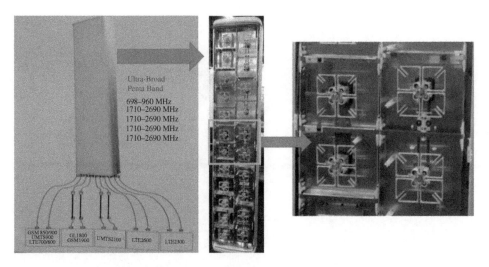

Figure 8.58 A penta-band sectorial antenna with MIMO functionality

and include GSM, UMTS and LET. It has 10 connectors for different wireless standards. From the inside of the antenna, we can see that there are 22 elements with two types of square crossed-dipole antennas to offer MIMO functionality. The large ones are for the lower band while the small ones are for the higher band.

There are plenty of research papers and references about the base station antenna design [38]. For example, a base station antenna with dual-broadband and dual-polarization characteristics is presented in [39]. The proposed antenna contains four parts: a lower-band element, a higher/upper-band element, arc-shaped baffle plates, and a box-shaped reflector. The lower-band element consists of two pairs of dipoles with additional branches for bandwidth enhancement. The upper-band element embraces two crossed hollow dipoles and is nested inside the lower-band element. Four arc-shaped baffle plates are symmetrically arranged on the reflector for isolating the lower- and upper-band elements and improving the radiation performance of upper-band element. As a result, the antenna can achieve a bandwidth of 50.6% for the lower band and 48.2% for the upper band when the return loss is larger than 15 dB, fully covering the frequency ranges 704–960 and 1710–2690 MHz for 2G/3G/4G applications. Measured port isolation larger than 27.5 dB in both the lower and upper bands is also obtained. Some new antenna designs (e.g. antennas integrated to lamps) can be found from [40].

Omni-directional Antennas

The simplest design in this case is a dipole or sleeve dipole. To improve the bandwidth, a biconical antenna could be employed. A main drawback of this type of antenna is the radiation pattern is not titled and the half of the radiated power may be wasted. To solve this problem, we can turn the upper cone of a biconical antenna into a disc, to form a discone antenna as shown in Figure 8.59. The cone length is around quart-wavelength, and the cone angle is used to tune the bandwidth (very wide bandwidth) and radiation pattern. The radiation pattern in the elevation plane is illustrated in Figure 8.59, while the radiation pattern in the horizontal plane is omnidirectional since the antenna is rotational symmetrical.

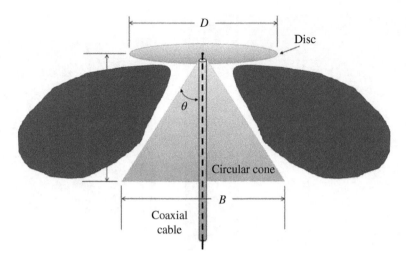

Figure 8.59   A discone antenna and its radiation pattern for mobile base-station

It is interesting to note that the radio coverage at the place very close to an outdoor base-station is normally not very good since the radiation pattern is close to zero radiation.

8.2.5.2   Indoor Base-Station Antennas

For indoor base station antennas, high gain is not of importance as for outdoor base-station antennas. However, because of the limited available spaces in indoor places such as shopping malls, parking garages, and airports, it is favorable to have a single multiband or broadband antenna element rather than multiple narrowband antenna elements to offer multiple services over a certain spot. Moreover, due to the dense uncorrelated multipath reflections in the indoor communications, MIMO antenna systems are vital and efficient. In such cases, colocated antennas with polarization diversity are more favorable than widely separated antennas with spatial diversity for more space saving. Typically, two scenarios are used to install an indoor base-station antenna [38]:

- Wall-mounted antenna: in this scenario, the radiation pattern of an antenna should be unidirectional with a wide horizontal HPBW to ensure wide coverage and an average gain of about 6 dBi. An example of such an antenna is presented in Figure 8.60. The antenna is a quadruple-band antenna for 2G/3G/4G/5G mobile communications [41], which is optimized to cover multiple frequency bands of 0.8–0.96 GHz, 1.7–2.7 GHz, 3.3–3.8 GHz, and 4.8–5.8 GHz and has a compact size with its overall dimensions of $204 \times 175 \times 39$ mm^3. It has an asymmetrical dipole antenna and parasitic patches. A stepped-impedance feeding structure is used to improve the impedance matching of the dipole antenna over these two frequency bands. Meanwhile, the feeding structure also introduces an extra resonant frequency band of 3.3–3.8 GHz. By adding an additional small T-shaped patch, the higher resonant frequency band at 5 GHz is obtained.

Figure 8.60   A broadband wall-mounted antenna

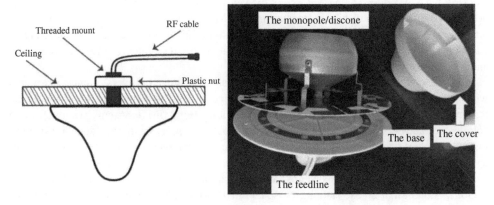

Figure 8.61   A broadband ceiling-mounted antenna

• Ceiling-mounted antenna: in this scenario, the antenna should have a wide bandwidth and a conical radiation pattern, which may be obtained using a discone or monopole antenna as shown in Figure 8.61 which is a bit more complicated than a simple discone/monopole antenna as there are other dipole antennas embedded in the reflector ground plane. The antenna is mounted on the ceiling and radiates downwards toward users. The radiation pattern is very similar to that in Figure 8.59.

For indoor base-station antennas, the radiation pattern is less important due to the multipath feature, the antenna efficiency, bandwidth, and cost are more important in practice.

8.2.6   Summary

In this long section, we have studied mobile terminal and base station antennas in detail. For mobile terminal antennas, a few classical antennas and smartphone antennas were introduced and evaluated. The effects of radiation into human body were examined by investigating the SAR values. The design specifications for these antennas have become very challenging since the antennas must be small and provide multi- or broad-band performance. To combat multipath effects, antenna diversity has been introduced and discussed in detail. Three diversity schemes and four signal combining methods were introduced and discussed. Various examples have been provided to study the user effects on the performance of different mobile antennas by examining the antenna efficiency and MEG values.

For mobile base-station antennas, the typical specifications have been provided and examples have used to study outdoor and indoor antennas. Multiband and multiple antennas are now widely used in base stations to meet the demand for beam steering and MIMO functionality which will be addressed in next section.

8.3   Multiple-Input Multiple-Output (MIMO) Antennas

MIMO (multiple-input multiple-output) antennas have become a popular new technology to increase the capacity (data rate) of a communication system. Many advanced wireless systems like LTE have introduced MIMO as one of the key steps to achieve high data rates without the need for more spectrum or higher transmit power. MIMO employs more than one antenna at both the transmitter and the receiver in a multipath environment to create multiple uncorrelated channels between them. This can be used to provide higher data rates, higher received SNR, and increased robustness to interference [42].

From the antenna point of view, MIMO antennas are similar to diversity antennas which were discussed in Section 8.2.3; they both require uncorrelated antennas and are only suitable for multipath environments where signals are random. But MIMO systems are much more complicated and need significantly more DSP (digital signal processing) than diversity systems [42, 43]. Also, diversity antennas are used for improving the received signal quality and reliability and mainly employed for receiving purpose.

For mobile communications systems, MIMO is applied to both mobile terminals and base stations. Although at a base station, the spatial diversity can be implemented (enough space available) and may perform slightly better than the MIMO based on polarization diversity, the latter offers the advantages of smaller size and no need to meet the spatial diversity requirements on structure designs. Overall, the MIMO system based on polarization diversity is the preferred choice for most applications [38].

8.3.1   MIMO Basics

The MIMO concept as a technique to increase wireless system capacity was first investigated in the 1990s [44, 45]. Since then, it has been attracted intensive research worldwide. Figure 8.62 gives an illustrative diagram of a generic MIMO system [39]. Different from a conventional single antenna system, a MIMO system has multiple transmitting M and multiple receiving antennas N. Each receiving antenna can receive signals from all the transmitting antennas, creating $M \times N$ signal transmission paths. The propagation channel has a channel matrix [H], which is very important for the signal processing.

A MIMO system is not the first time when multiple antennas have been used in wireless communication. Early on, multiple antennas (e.g. antenna array) were used to enhance system SNR through beamforming by concentrating antenna radiation energy to a specific direction. Later on, multi-antennas were used to provide antenna diversity, countering the multipath fading effects. With multiple receiving antennas at a certain distance in between, signals from different propagation paths can be picked up independently by the multiple antennas and then added together constructively. Since the chance of all the signal paths suffering from strong multipath fading is much lower, the system bit error rate (BER) is significantly reduced in this way. However, spatial diversity from multiple antennas has limited benefits. What makes multiple antenna systems powerful and appealing is the extra spatial multiplexing it offers. With multiple independent signal propagation path between the multiple transmitting and multiple receiving antennas, we can transmit multiple independent data streams through those independent paths. Take the example in Figure 8.62, if assuming no signal and noise correlation at both transmitter and receiver sides, the capacity of the MIMO system increases linearly with respect to the minimum number of transmitting/receiving antennas (*min {M, N}*), at no more cost on either power or spectrum resources. Due to this significant advantage in spectral efficiency over conventional single-antenna systems, MIMO has become an essential element in many wireless communication standards, such as 4G LTE, WiMAX and 5G.

The main figure of merit for a MIMO system is the throughput, which is determined by:

- Channel bandwidth
- MIMO antenna configuration
- Modulation coding scheme
- User equipment radio condition (SNR, etc.)
- Duplex mode (e.g. FDD, TDD)

The more elements, the larger the throughput.

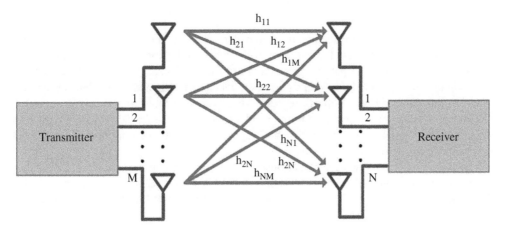

Figure 8.62   A MIMO system

8.3.2   MIMO Antennas and Key Parameters

Since the introduction of MIMO technology, the design of multiple antennas at both the base station and the mobile terminal has received great research attention. In addition to the new design challenges by employing multiple antennas in compact sizes, this new technology has brought in more performance parameters than those of a conventional single antenna system. While single antenna metrics such as total radiation efficiency, radiation patterns, reflection coefficient, and operating bandwidth are still required to characterize MIMO antennas, new performance metrics are added such as mutual coupling/isolation, envelope correlation coefficient, and DG [46].

8.3.2.1   Mutual Coupling/Isolation

The presence of multiple antennas near each other creates an interaction among them known as mutual coupling as discussed in Chapter 5 and [47, 48]. This coupling is due to an induced current flowing on one antenna element from the excitation of the other antenna element. This alters the input impedance of each antenna element in which the input impedance was defined by both self-impedance and mutual impedance. This change produces an impedance matrix att contains both the self-impedances (Z_{ii}) and the mutual impedances (Z_{ij}) as shown in Equation (8.45).

$$Z = \begin{bmatrix} Z_{11} & \cdots & Z_{1N} \\ \vdots & \ddots & \vdots \\ Z_{N1} & \cdots & Z_{NN} \end{bmatrix} \tag{8.45}$$

$$S = \begin{bmatrix} S_{11} & \cdots & S_{1N} \\ \vdots & \ddots & \vdots \\ S_{N1} & \cdots & S_{NN} \end{bmatrix} \tag{8.46}$$

As an inevitable result of the creation of the impedance matrix, antenna BW is no longer governed by only the reflection coefficient. It depends on both reflection coefficients and the transmission coefficients in the S-parameter matrix in Equation (8.46). S_{ii} is the ratio of the reflected wave voltage to the incident one at the port of the ith element when all other elements are terminated with matched loads. While S_{ij} represents the ratio of the transferred voltage from the jth element port to the ith element port to the incident voltage at the jth element port, S_{ij} is also known as the mutual coupling coefficient. Typically, a mutual coupling of −20 to −25 dB is acceptable for base-station antenna.

The *isolation* between two antennas is just the opposite of the mutual coupling, the smaller the coupling, the larger the isolation. It is defined mathematically as −|S_{ij}| and expressed in dB. Thus, an isolation of at least 20 dB is normally required for an MIMO system. It should not be mixed up with the coupling (but many people do in practice).

8.3.2.2   Envelope Correlation Coefficient (ECC)

Up to this point, it has been mentioned many times that in order to achieve a good MIMO and diversity performance, the received signals/antennas should be independent of each other. Highly correlated channels will degrade the MIMO and diversity system performance. This independence can be measured and quantified using the envelope correlation coefficient (ECC) which is mathematically defined by Equation (8.43) and can be rewritten using vector fields as [46]:

$$\rho_e = \frac{\left| \int\int_{4\pi} E_i(\theta,\varphi) \cdot E_j(\theta,\varphi) d\Omega \right|^2}{\int\int_{4\pi} |E_i(\theta,\varphi)|^2 d\Omega \int\int_{4\pi} |E_j(\theta,\varphi)|^2 d\Omega} \tag{8.47a}$$

This requires the (vector) radiated fields E_i and E_j from two concerned antennas (i.e. a field-based method) and gives the most accurate result. However, in practice, this could be very time consuming as it needs a 3D radiation pattern. Therefore, people have tried to calculate the ECC using S-parameters of the antenna as shown below [49]:

$$\rho_e = \frac{\left| S_{ii}^* S_{ij} + S_{ji}^* S_{jj} \right|^2}{\left(1 - |S_{ii}|^2 - |S_{ji}|^2\right)\left(1 - |S_{jj}|^2 - |S_{ij}|^2\right)}, i \neq j \tag{8.47b}$$

Although this S-parameter approach is very simple and fast, it is only accurate for the case of loss-free antennas. Therefore, the S-parameter method has been modified in which the effect of the radiation efficiency has been taken into account [49, 50]. A generalized equation for the calculation of the ECC is given below:

$$\rho_e = \frac{\left| S_{ii}^* S_{ij} + S_{ji}^* S_{jj} \right|^2}{\left(1 - |S_{ii}|^2 - |S_{ji}|^2\right)\left(1 - |S_{jj}|^2 - |S_{ij}|^2\right)\eta_i \eta_j}, i \neq j \tag{8.47c}$$

where η_i and η_j represent the radiation efficacies of the antennas i and j, respectively. Including the antenna efficiencies improves the calculation accuracy of the ECC. However, this accuracy improvement works only for antenna ports with efficiencies larger than 60%; otherwise, it may result in significant calculation errors [50].

It was mentioned that ECC should be less than 0.7 at the base station and 0.5 at the mobile in the past [46]. But now for a good MIMO and diversity system, the ECC between two antennas should satisfy the following condition [48]:

$$ECC = \rho_e \leq 0.3 \tag{8.48}$$

although $ECC < 0.5$ is still acceptable for some systems.

8.3.2.3 Diversity Gain

DG was introduced in Section 7.9.2 to evaluate MIMO antenna performance and also used in Section 8.2.3 when dealing with diversity antennas. As antenna diversity is very effective on combating the multipath fading via sending multiple versions of the same transmitted signal to the receiver through the multipath channel. Therefore, if one of the diversity combining techniques is deployed at the receiver, the receiver can have more freedom in obtaining the best-received signal (best SNR) through a combination scheme, which will improve the reliability of the wireless system. This improvement is quantified by the *DG*, which is the improvement in the received SNR at the output of the diversity combiner relative to the SNR resulting from a reference single antenna at a certain level of performance criterion such as the CDF [46]. The CDF of a diversity antenna is plotted in Figure 8.63 and a DG of 10 dB can be seen when the number of diversity branch is increased from 1 to 2 at 1% probability. The more antennas, the larger the DG.

Although the DG is easily predictable from the measured and calculated CDF, some information such as the frequency range is lost during the post-processing and the generation of the CDFs. In many cases, it is important to know how the DG varies as a function of frequency. To do this, the ECC between measured samples from each diversity branch is calculated using Equation (8.47) and used for the calculation of the DG:

$$DG = 10.48\sqrt{1-\rho_e} \qquad (8.49)$$

There are other related terms related to this DG, such as the *effective DG* and *apparent DG*, one can refer to Section 7.9.2 for more information.

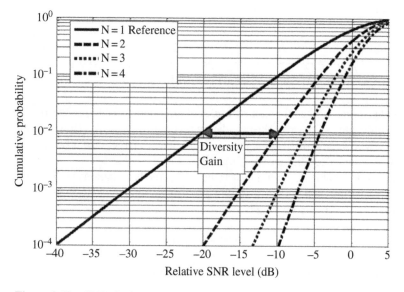

Figure 8.63   CDF of relative SNR threshold for N different diversity branches

8.3.2.4   Total Active Reflection Coefficient (TARC)

In a MIMO system, the presence and activity of adjacent antennas will affect performance of the antenna under consideration if not properly designed. To assess such interactions, the total active reflection coefficient (TARC) was introduced in [46]. It relates the total incident power to the total outgoing power in an N-port system and is mainly used for MIMO antennas and array antennas, where the outgoing power should not be reflected back. The TARC (Γ_T) is the square root of the sum of all outgoing powers at the ports divided by the sum of all incident powers at the ports of an N-port antenna:

$$\Gamma_T = \frac{\sqrt{\sum_{i=1}^{N} |b_i|^2}}{\sqrt{\sum_{i=1}^{N} |a_i|^2}} \tag{8.50}$$

where $[b] = [S][a]$. $[S]$ is the antenna's scattering matrix as discussed in Section 7.3.1, $[a]$ is the excitation vector, $[b]$ represents the scattered/reflected vector. The TARC is a real number between zero and one, although it is typically presented in dB. When the value of the TARC is equal to zero, all the delivered power is accepted by the antenna and when it is equal to one, all the delivered is coming back as outgoing power (thus all power is reflected, but not necessarily in the same port).

Similar to the *active reflection coefficient* (which is the reflection coefficient for a single antenna element in an array antenna, in the presence of mutual coupling), the TARC is a function of frequency and also depends on scan angle and tapering. With this definition, we can obtain TARC directly from the scattering matrix and characterize the MIMO antenna bandwidth and radiation performance.

For the special case of a two port MIMO antenna system, TARC can be evaluated by fixing the angle of the incoming wave of one port, and then sweeping the relative angle with the adjacent port to see the effect of such angle change along with the mutual coupling (*Sij*) on the effective bandwidth and efficiency of the multi-port antenna system.

There are some other parameters which are used for MIMO performance evaluation. For example, the *branch power ratio* (BPR) is another important metric for MIMO antenna systems that shows the ratio between the lowest and highest power levels within the antenna system. In an ideal multi-antenna system, the BPR should be 1 (0 dB), i.e. all ports are receiving the same power (power divided among all ports equally). *Throughput* is another one. But they are more linked to MIMO system performance rather than antenna performance, thus they are not discussed here.

The MIMO antenna measurement is a very challenging issue which has been addressed in Section 7.9.2, as a topic of MIMO OTA testing.

8.3.3   MIMO Antenna Designs

Currently, many base stations mobile terminals use multiple antennas for both diversity and MIMO. The DG obtained increases as the spatial correlation between signals reduces. In a MIMO system, the relationship between spatial correlation and performance is more complex. For example, at the edge of the cell where signal strength is low, it is preferable to use diversity-based single stream transmission strategies that require one of the spatial channels to be significantly stronger than the others. This is possible when the spatial correlation between signals received at the antenna elements is high. In contrast, when the signal strength is high (near the base station), it is beneficial to use spatial multiplexing strategies to increase the capacity. In this situation, spatial correlation among received signals lowers the capacity gains obtained. Thus, the spatial correlation has different implications in diversity-based and MIMO systems. This is the primary reason for the difference in the design principles for diversity-based and MIMO antenna arrays. Spatial correlation might arise due to the antenna arrays employed at the transmitter and receiver (small interelement spacing, mutual coupling, etc.) or due to the channel characteristics [51]. If the wireless propagation environment has sufficient multipath, the channel spatial correlation is generally low. In contrast, when the channel does not have rich multipath or when a strong line of sight exists between the transmitter and receiver, the channel spatial correlation is considerably higher. The impact of high channel spatial correlation can be lowered using effective MIMO antenna design techniques. For example, by using antenna arrays where the elements have orthogonal polarizations or patterns, the spatial correlation of the signals received at the antenna array can be significantly reduced, even when the channel spatial correlation is high.

By spacing antenna elements far apart, or by using elements that have orthogonal radiation patterns or polarizations, it can be ensured that signals received at these antenna elements have undergone independent scattering in the propagation environment and hence, have low correlation. Effective antenna designs can be used to improve MIMO performance by utilizing the following three antenna diversity schemes:

- Spatial diversity – spacing antenna elements far apart;
- Pattern diversity – using antenna elements with orthogonal radiation patterns;
- Polarization diversity – using antenna elements with different (orthogonal) polarizations, typically at ±45°.

Scattering effects might cause a transmitted signal to suffer from a (small) change in its polarization. As an MIMO system takes advantage of multipath, the antenna design at the MIMO front end must be capable of handling these changes in polarization. Further, mobile terminals might be held or placed at different orientations. MIMO antenna designs at the base station must be able to effectively tolerate random orientation effects and still provide good data rates at the receiver. The prohibitive costs of leasing tower space require that base station antennas do not occupy too much space. Further, knowledge of the optimum values of parameters like cross-pol and port-to-port isolation will enable antenna designers to reduce the time and effort needed to meet (possibly) unnecessarily high specifications of these parameters.

To summarize, the requirements of an effective MIMO antenna design are that it should

- exploit the additional spatial dimension by incorporating combinations of the three antenna diversity effects in the design;
- have a structure such that it can cope with random orientation effects at the receive side;
- be able to cope with slight changes in the polarization of the received signals (in the uplink case);
- offer a fair balance between tower space occupied and performance obtained (due to the high costs associated with leasing tower space);
- present a trade-off between design specifications and performance obtained (especially for parameters like cross-pol isolation and port-to-port isolation).

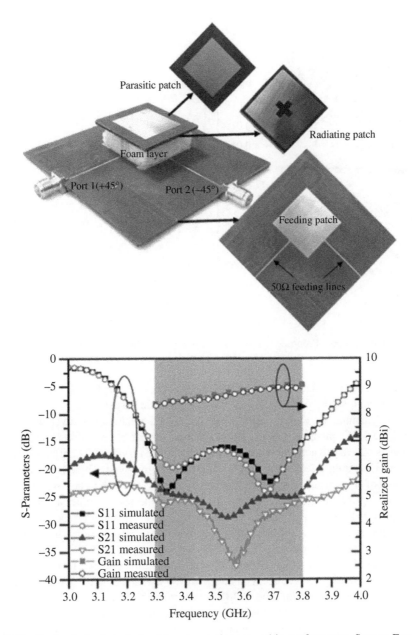

Figure 8.64   Duel-polarized stacked patch antenna element and its performance. Source: From
Alieldin [40]

Figure 8.64 is an example of a MIMO antenna element for 5G band from 3.3 to 3.8 GHz [40].
Its reflection coefficient is < -15 dB and the coupling S21 is < -22.5 dB. It was proposed to
form a 9×3 elements MIMO array for a base station as shown in Figure 8.65 (other configurations may also work). The radiation pattern of the proposed MIMO antenna can be tilted as

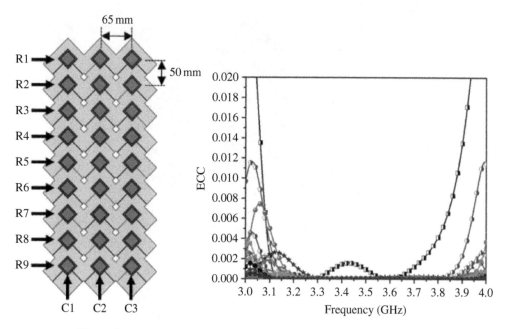

Figure 8.65 Proposed MIMO array (9 × 3) and its ECC performance [40]

required, and the horizontal half-power beamwidth is about 65°. The ECC is much smaller than 0.01 over the band, so the performance is really good in this aspect.

Just as the antenna in Figure 8.58, both spatial and polarization diversities are often employed for MIMO antennas and a main challenging is that the isolation should be at least 25 dB which is not easy to achieve (the antenna in Figure 8.64 has failed marginally to meet this requirement). Decoupling techniques are normally to be employed. There has been a considerable amount of effort made in developing effective decoupling techniques over the years. For examples, decoupling networks, neutralization lines, ground plane modifications, frequency-selective surface (FSS) or metasurface walls, metasurface corrugations or electromagnetic bandgap (EBG) structures, and characteristic modes. More information can be found from [52], a very good review paper.

To implement MIMO at mobile terminals is very challenging due the small space available. A very recent technology to produce a compact MIMO antenna is to use the concept of common mode (CM) and differential mode (DM) sources which can be placed next to each other but maintain a very low coupling between these two orthogonal modes. As seen in Figure 8.66, there are four type antennas: CM wire antenna, DM wire antenna; CM slot antenna, and DM slot antenna. They are orthogonal to each other. There have been a number of mobile antenna designs using this concept. In [53], a highly integrated MIMO antenna unit is proposed for mobile terminals. The inspiration comes from a dipole fed by a differential line which can be considered a DM feed. Some interesting features are provided in this paper: by symmetrically placing one DM antenna and one CM antenna together, a DM/CM antenna can be achieved. Benefitting from the coupling cancellation of antiphase currents and the different distributions of the radiation currents, the DM/CM antenna can obtain high isolation and

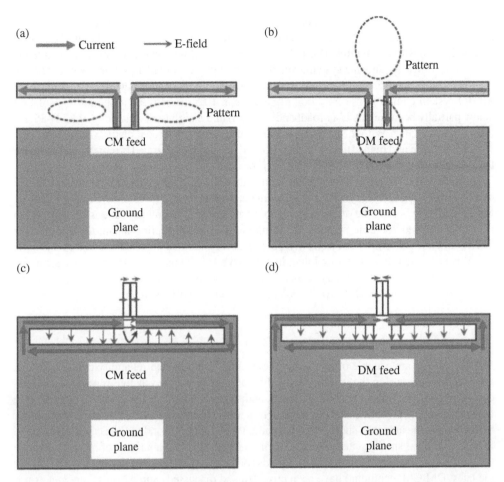

Figure 8.66 Common mode and differential mode for wire-type antennas and slot type antennas. (a) CM feed wire antenna, vertical-pol. (b) DM feed wire antenna, horizontal-pol. (c) CM feed slot antenna, horizontal-pol. (d) DM feed slot antennas, vertical-pol

complementary patterns, even if the radiators of the DM and CM antennas are overlapped similar to Figure 8.66. To validate the concept, a miniaturized DM/CM antenna unit was designed for mobile phones. 24.2 dB isolation and complementary patterns were achieved at 3.5 GHz with small dimensions. One 8 × 8 MIMO antenna array was also constructed by using four DM/CM antenna units and showed good overall performance. The proposed concept of DM/CM design in [53] seems to be an excellent idea for MIMO and diversity antennas, especially when space is a major concern.

It is interesting to note that the same DM/CM concept was used in another recent design [54] where a novel self-decoupled MIMO antenna pair with shared radiator was proposed for 5G smartphones. A radiator was directly excited by two feeding ports which are naturally isolated across a wide bandwidth without using any extra decoupling structures. To offer a deep physical insight of the self-decoupling mechanism, a mode cancellation method based on the synthesis of common and differential modes was shown in the paper. The proposed self-decoupled antenna pair showed an isolation of better than 11.5 dB across the 5G N77 band (3.3~4.2 GHz) with a radiation pattern diversity property. Based on the self-decoupled antenna pair, an 8×8 MIMO antenna system, constituted by four sets of antenna pairs, was simulated, fabricated, and measured. The measurement results demonstrated that, the proposed 8×8 MIMO system could offer an isolation of better than 10.5 dB between all ports (the isolation is low partially because this is a broadband antenna) and a high total efficiency of 63.1~85.1% across 3.3~4.2 GHz. With the advantages of self-decoupling, shared radiator, simple structure, wide bandwidth, and high efficiency, the proposed design demonstrated promising potential for the future highly integrated MIMO antennas for smartphones.

Massive MIMO is the extension of the conventional MIMO technology, which exploits the directivity of a MIMO array with a large element number as one more dimension of freedom. Massive MIMO is also one of the key technologies for the 5G communications system, which is mainly utilized for base stations at the moment. The decoupling techniques in massive MIMO antennas have not been well developed. It is very challenging and literature on this topic is still very limited. It was reported that, in a massive MIMO base station antenna system, the mutual coupling between antenna elements has to be below −30 dB according to the thumb of rules in the industry, which is 10 dB less than current −20 dB requirement!

8.3.4 Summary

MIMO systems are capable of increasing the channel capacity and reliability of wireless systems without increasing the system bandwidth and transmitter power. In this section, the concept of MIMO antenna was first introduced, and the key parameters of a MIMO system were then presented. It has shown that there is a significant overlap with antenna diversity system: for both systems, the antennas should be independent (not correlated and of little coupling) and the ECC should be as small as possible. The DG has been used as a main figure of merit to evaluate the performance of both systems. The methods for effective MIMO antenna designs (including the latest DM/CM technique) have been provided and discussed. Some MIMO antenna examples were employed to illustrate the design considerations and process. Their performances were presented to highlight the challenges. It has shown that how to reduce the coupling and increase the isolation between MIMO antenna elements is a major problem. The measurement of MIMO antennas was discussed in Section 7.9.2.

8.4 Multiband and Wideband Antennas

8.4.1 Introduction

Many modern communication systems have to work well over multiple frequency bands. For example, a well-designed cellular mobile phone is normally expected to operate at least the two GSM bands (900 MHz and 1800 MHz) and the PCS band (1900 MHz). The recent release of the UWB band (from 3.1 to 10.6 GHz) for wireless low-power applications has generated a lot of interest in developing UWB systems [55, 56]. In this section, we are going to deal with these two types of antennas.

8.4.2   Multiband Antennas

8.4.2.1   Techniques

Several basic principles can be employed to realize multiband antennas, as follows:

- Use higher order resonances
- Use resonant traps
- Combine resonant structures
- Use parasitic resonators

Each will be addressed in the sections that follow.

Higher Order Resonances
The use of higher order resonances is illustrated in Figure 8.67, which shows the resonant modes of monopole antennas as their length is increased in $\lambda/4$ increments.

Monopoles often used with a length of $\lambda/4$, when the E field at the feed is a minimum and the current is a maximum. A similar condition exists at the feed when the antenna is $3\lambda/4$. Alternatively, we can say that, for a fixed antenna height, the feed conditions will be similar when the frequency is equivalent to heights of $\lambda/4$ and $3\lambda/4$: the antenna will be resonant at a first frequency and then a second frequency three times the first. Other natural resonances will also exist at higher frequencies.

Higher order resonances are present in many types of resonant antennas such as patches, dipoles, monopoles, slots, and dielectric resonators and are often used (and manipulated) to give multiband operation.

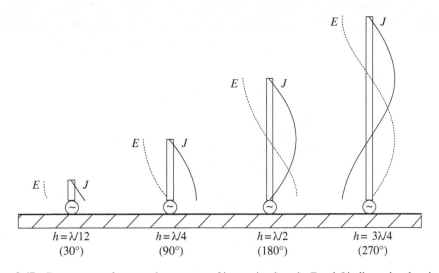

Figure 8.67   Resonances of monopole antennas of increasing length. E and J indicate the electric field and current density magnitudes, respectively

Resonant Traps

Figure 8.68 shows a monopole antenna with a parallel resonant circuit – or "trap" – located approximately halfway along its length. The trap enables the antenna to work at two frequencies: a low frequency f_1 and a higher frequency f_2. At f_2 the trap is tuned to be close to anti-resonance (a high impedance), and current flow is restricted to the lower part of the antenna; resonance is achieved when the lower part of the antenna is close to $\lambda/4$ long. In practice, this resonance is achieved when this part is slightly less than $\lambda/4$ and the trap is antiresonant at a slightly lower frequency than f_2. At f_1 the trap is well below resonance and is inductive, and the antenna is resonant when the full length is close to $\lambda/4$. Again, due to the shortening effect of the inductor, in practice resonance is achieved at a length that is slightly less than $\lambda/4$.

Radio amateurs use traps in dipole antennas located horizontally above ground, for example, to cover the 80- and 40-m bands (3.5–4.0 MHz and 7.0–7.3 MHz, respectively). They are also often used in avionic antennas.

Combined Resonant Structures

Two or more resonant structures can be closely located or even colocated with a single feed point in order to achieve multiband operation. This is illustrated in Figure 8.69(a), which shows two closely spaced, adjacently located monopoles with a common feed point. The larger element is operational at the lower frequency, f_1, whereas the smaller element is operational at the higher frequency, f_2.

Figure 8.69(b) works using the same principle; however, the low-frequency element is realized using a helical element to reduce space (the high-frequency element is within the helix for some of its length).

Parasitic Resonators

In contrast to the previous section, we may add a second element for operation at a second frequency, but without feeding it directly. This is illustrated in Figure 8.70, which shows two monopoles, one of which is fed, whereas the other is parasitically coupled to via the near field

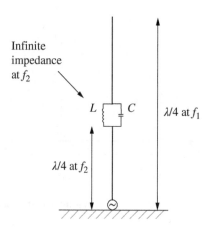

Figure 8.68   A monoopole antenna with a resonant trap

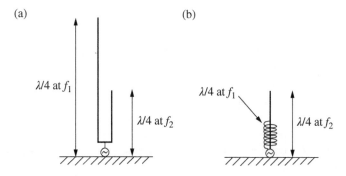

Figure 8.69   Combined resonant structures (a) two monopoles (b) a helical and a monopole

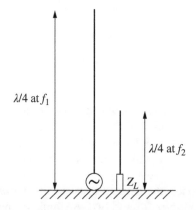

Figure 8.70   Parasitically coupled monopole antennas

of the fed antenna. Here the parasitic antenna is shown with a load, which can be used for tuning. Usually, the load would be reactive in order to maintain high antenna efficiency.

8.4.2.2   Examples

The first example uses the principle of higher order resonances. Figure 8.71 shows a simple normal mode helix antenna mounted on a PCB of typical mobile phone dimensions. The antenna is first resonant at a frequency, f_1, of approximately 0.84 GHz, with a resistance of approximately 27 ohms. At 1.16 GHz, the antenna is antiresonant (when the resistance is very high, and the reactance makes the transition from being positive to negative). The antenna then becomes resonant again at a frequency, f_2, of approximately 2.15 GHz – when the resistance is 10 ohms – before quickly becoming anti-resonant again at approximately 2.35 GHz. The ratio between the first and second resonant frequencies is given by

$$\frac{f_1}{f_2} = 2.6 \tag{8.51}$$

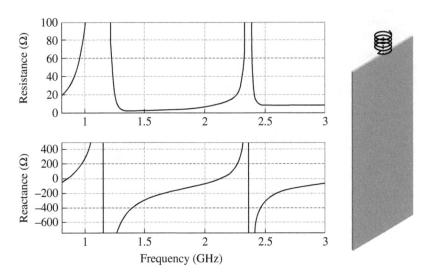

Figure 8.71   A small helical antenna mounted on a mobile phone PCB

This is close to the value of three that we would expect for resonances at electrical lengths of $\lambda/4$ and $3\lambda/4$ (as illustrated in Figure 8.67).

Since the antenna shown in Figure 8.71 is electrically small, the reactance can be seen to vary quickly at the first resonance and even more quickly at the second resonance. This is indicative of narrow bandwidth.

For dual-band mobile phones, the antenna resistance is normally required to be close to 50 ohms and the frequency ratio is close to two. A simple method of altering the frequency ratio is to have different helix pitches in the upper and lower portions of the antenna. This is illustrated in Figure 8.72 for a commercial dual-band helix. The low-frequency resonance is controlled by the overall length of the structure, while the high-frequency resonance is controlled by the pitch of the upper part of the antenna.

Alternatively, Figure 8.73 illustrates a dual-band helical antenna in which two windings of different pitches and lengths are integrated into a single volume with a common feed – an example of combined resonant structures.

8.4.3   Wideband Antennas

There are a wide variety of wideband antennas, many of which are also electrically large. Here we focus on wideband antennas that are also electrically relatively small, since this is often an industrial requirement.

The basic approach to making an electrically small antenna wideband is to make it fat. We saw previously in Section 8.1.1 that the bandwidth of an antenna is related to the size of the sphere that just encloses it. By making the antenna fat, more of the sphere is occupied and the antenna bandwidth can be maximized. Some typical fat monopoles are illustrated in Figure 8.74. These and other similar shapes have formed the basis for UWB antenna designs.

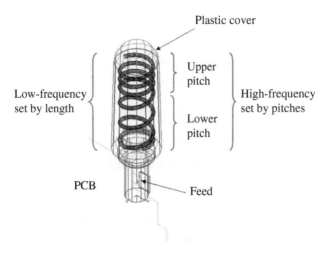

Figure 8.72 A dual-band helical antenna with two pitches for control of the ratio of the resonant frequencies

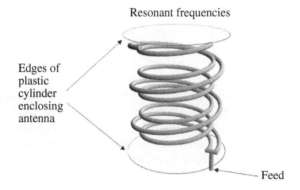

Figure 8.73 A dual-band helical antenna utilizing two different length windings with a common feed point

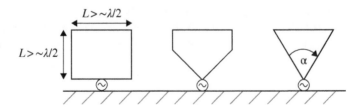

Figure 8.74 Some "fat monopole" antennas

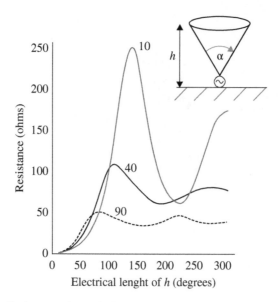

Figure 8.75   Resistance of a conical monopole with electrical length and flare angle

UWB antennas are normally planar. However, as an example of what happens to the feed impedance when an antenna is "fattened", we will concentrate on a three-dimensional conical monopole, which subtends an angle α at the feed. The resistance of such an antenna with height and flare angle is shown in Figure 8.75. When the monopole is thin – i.e. for $\alpha = 10$ – there is a strong resistance peak at approximately 124°. For a very thin monopole, this peak would occur at 180°, when the antenna is half-wave. However, as the flare angle increases, the half-wave peak occurs at progressively lower electrical heights and with a decreasing resistance value. When the flare angle reaches 90° the resistance is close to 50 Ω over a wide range of electrical heights. If the electrical height is greater than approximately 50°, the resistance is close to 50 Ω over a wide frequency range.

The reactance of the conical monopole with height and flare angle is shown in Figure 8.76. As α is increased the reactance variation reduces and is close to zero for heights greater than 90°. For heights greater than 50°, the reactance is within, or capable of being matched to within, acceptable limits.

We can conclude that monopole antennas with wide flare angles can exhibit near-constant resistance with electrical heights greater than approximately 50°. In other words, small, fat antennas can exhibit wideband characteristics. Some UWB antennas have been developed based on this idea [56]. For example, two UWB antennas are shown in Figure 2.46 and the S11 of Antenna 1 is given in Figure 8.77, which has a return loss greater than 10 dB over the UWB band.

8.4.4   Summary

In this short section, we have introduced the basic ideas and techniques to make multiband and UWB antennas. The focus has been on antenna itself to produce multiple resonant current paths. The fatter the antenna, the wider the bandwidth in general. Impedance matching

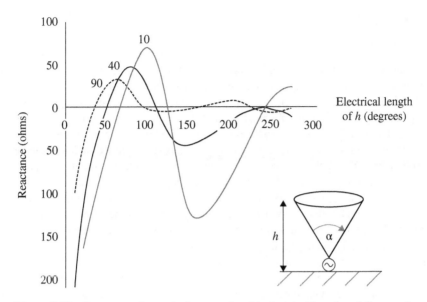

Figure 8.76   Reactance of a conical monopole with electrical length and flare angle

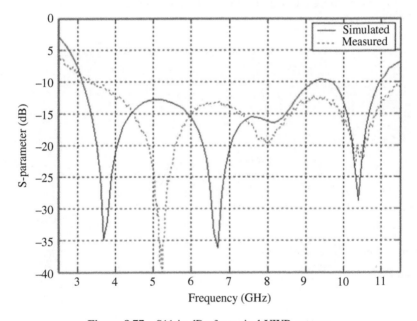

Figure 8.77   S11 in dB of a typical UWB antenna

networks could also be used to broaden the antenna bandwidth as discussed in Section 8.1, but not discussed in this section.

There have been many multiband and wideband antennas developed over the years, they are based on similar principles and the main difference is the design detail and implementation.

8.5 RFID Antennas

8.5.1 Introduction

Radio frequency identification (RFID) systems are short-range – normally digital – wireless systems that, as the name suggests, are used for primarily for identification purposes [57]. Example applications include animal tagging, asset tracking, electronic passports, smartcards, and shop security. A simplified diagram of an RFID system is shown in Figure 8.78.

At one end of the radio link is a small device with limited intelligence: the tag or transponder. Tags are designed for manufacture in very large volumes at very low cost. They can be likened to bar-code labels: they have an Electronic Product Code (EPC), comparable to the Universal Product Code (UPC) format used by bar codes, but they are also programmable and have the capability to store user specific information (i.e. they contain some user memory). Tags contain an integrated circuit (IC) that is connected to a thin planar sheet that serves as a label and as a substrate for the antenna. The tag antenna is normally balanced (to suit the differential inputs of the tag IC), electrically small and linearly polarized.

At the other end of the link is a more expensive and sophisticated device: the reader or interrogator. These commonly used names reflect the fact that the primary jobs of the reader are to prompt communication with the tags (often with several tags simultaneously) and then to receive data from them. However, contrary to the name, the reader often also transfers data, or writes, to the tag. Before prompting the tag to respond, the reader also often supplies it with power: this helps to reduce the cost of the tags, since a battery is expensive to supply, connect and maintain.

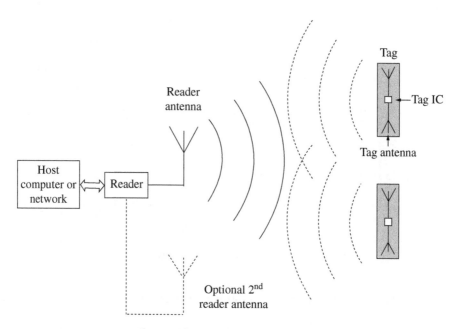

Figure 8.78 A simplified RFID system

There are several classes of tag, dependent on how they generate enough power to communicate, as follows:

– passive
– semi-active
– active

Passive tags are very low cost but tend to be limited in terms of range and data rate. They can only respond once powered by the reader, which may take some time. Semi-active tags have a limited power supply – enough to allow them to "broadcast" their presence to readers, which allows faster operation. Active tags have enough onboard battery power to allow both broadcasting and high data rate communications.

The reader may be either monostatic, where the transmission and reception use the same antenna, or bistatic, where the transmission and reception antennas (and associated circuitry) are separate (as shown in Figure 8.78 with the second antenna used for receiving only). The first and second antennas may also both be used for transmitting and receiving, for example, where the reader antennas are installed on either side of a "gate" through which tags pass (commonly used in shop security systems).

The reader antenna is generally electrically larger than the tag antenna and may be either balanced or unbalanced. Many different types of antenna can be utilized, with radiation patterns and polarization tailored for the particular application. Reader antennas are often circularly polarized when the orientation of the tag is unknown. While this reduces the maximum possible power transfer between the reader and the tag (there is a 3 dB coupling loss between circularly and linearly polarized antennas), it eliminates any likelihood that polarizations could be completely mismatched. This is important in tagging applications, where reliability is required to be very high.

Figure 8.79 shows a typical sequence of events for a passive tag to be identified by a reader.

First, the tag transmits a continuous wave (CW) signal that is received by nearby tags. This signal is rectified to provide enough power to operate the tag. The reader then sends a modulated request for active tags (i.e. tags close enough to the reader to be powered up) to respond. This is followed by a response from the appropriate tag. The tag modulation is achieved by switching the load impedance seen at the terminals of the tag antenna. This causes the tag antenna to alter the field seen at the reader antenna. Although the change is very small, it can be detected and demodulated by the reader. In this way, the tag uses as little power as possible in order to maximize read range: it does not generate a signal, but simply provides enough power to activate a switch.

RFID systems operate at a number of frequency bands that can be grouped as follows:

– low frequency (LF)
– high frequency (HF)
– ultra high frequency (UHF)

There are large differences in the frequencies used in the LF, HF, UHF bands. Because of this, there are also significant differences in the way in which the systems operate and in the applications for which they are appropriate.

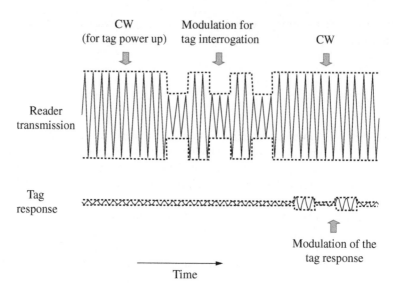

Figure 8.79   A typical sequence of events for a tag to be powered, interrogated and read

LF and HF systems use near field coupling between the tag and the reader antennas: most often the coupling is inductive, but it may also be capacitive. When inductive coupling is used, both the reader and the tag typically use multiturn coils as antennas. In fact, referring to these coils as antennas is a little misleading, since the coils of the reader and tag act more like transformers: the coil of the reader can be considered to be the primary winding and the coil of the tag acts as the secondary. The coupling between the coils (and, therefore the transformation ratio) is dependent on their separation and always occurs within the near field region (where the magnetic field falls with the inverse of distance to the power of three). Near field systems are discussed in detail in Section 8.5.2. They are used particularly in applications where the tag may be attached to a lossy dielectric such a fluid filled bottle, an animal, or a solid object. This is because the tag antenna electrical characteristics tend to change less than for far field (UHF) antennas.

UHF systems use far field coupling, when the antennas of the reader and tag operate in a more conventional way: the transfer of energy is determined by Friis equation (see Section 3.5.1). UHF far field systems typically have greater range than LF and HF systems. As such, they are often used in systems where higher than normal levels of performance are required. They are described in greater detail in Section 8.5.3.

8.5.2   Near Field Systems

For near field systems, loop/coil antennas are normally used for the reader and the tag. Consider the simple reader loop shown in Figure 8.80, where a is the reader loop radius, b is the tag loop radius, and d is the separation between the tag and reader loops. Close to the reader, the magnetic field produced is greatest along the axis of the loop: the z-axis as drawn. Hence, products are often arranged so that the reader and tag loops have the same axis, as shown.

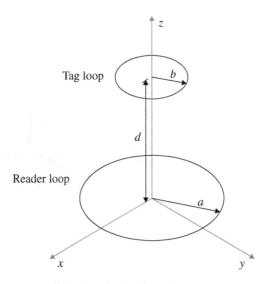

Figure 8.80 Simplified reader and tag loop antennas

With such an alignment, the mutual inductance, M, between the loops is given by [58]

$$M = \mu\sqrt{ab}\left\{\left(\frac{2}{D} - D\right)K(D) - \frac{2}{D}E(D)\right\} \tag{8.52}$$

where

$$D^2 = \frac{4ab}{d^2 + (a+b)^2} \tag{8.53}$$

$$E(D) = \int_0^{\pi/2} \sqrt{1 - D^2\sin^2\theta}\,d\theta \tag{8.54}$$

$$K(D) = \int_0^{\pi/2} \frac{d\theta}{\sqrt{1 - D^2\sin^2\theta}} \tag{8.55}$$

$K(D)$ and $E(D)$ are complete elliptic integrals of the first and second kind, respectively. These formulas indicate that there is an optimum relationship between the sizes of the tag and loop antennas that should be considered at a system design stage.

A typical practical near field system is shown in Figure 8.81. Here the tag antenna size and geometry are normally limited by the application: for example, credit card dimensions are often used for smart cards. Multiple turn loops are also often employed, when the mutual inductance increases linearly with the number of turns.

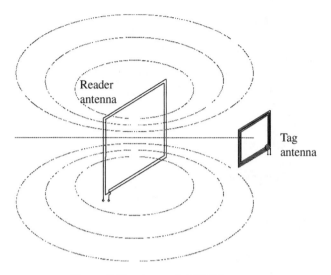

Figure 8.81 A near field RFID system

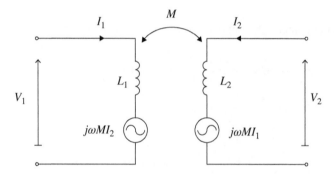

Figure 8.82 Mutual coupling between reader and tag coils

The transfer of energy between the reader and the tag is via the mutual coupling between the two loops. The equivalent circuit of two mutually coupled inductors is shown in Figure 8.82.

The circuit on the left of Figure 8.82 represents the reader, whereas the circuit on the right represents the tag. The transfer of energy that occurs between the two coils is modeled by two current-dependent voltage sources connected in series with the coil inductances. These sources depend on the level of mutual coupling, which varies with the distance between the tag and reader. The mutual coupling is typically low, such that the reader current can be considered to be approximately constant. This allows the tag to be analysed independently of the reader.

In addition to its inductance, the tag loop will have some associated loss resistance and some inter-turn capacitance, which is most simply approximated by a shunt capacitor. The tag IC will have a load resistance and a capacitance that is designed to be appropriate for certain loop dimensions and frequency ranges. The IC is normally modeled as either a series or parallel equivalent circuit. The equivalent circuits of a practical tag inductor and a tag IC are shown

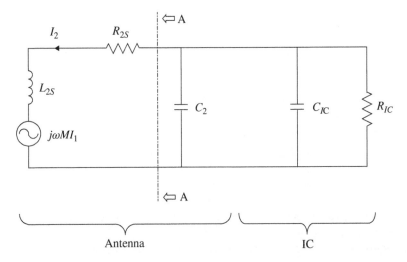

Figure 8.83   Equivalent circuit of a tag antenna and IC

in Figure 8.83. In this example, the IC is modeled as a parallel equivalent circuit, whereas the inductance and resistance of the tag loop are modeled as a series equivalent circuit. To make the system easier to analyze, it is necessary obtain a completely series or parallel equivalent circuit. This can be easily done using Norton or Thevenin equivalents as appropriate. In the example that follows, we will convert components to the left of A-A to a Norton (parallel) equivalent circuit. To do this, we short circuit any independent voltage sources (the source in Figure 8.83 can be considered independent if the mutual coupling is low and I_1 can, therefore, be considered to be constant) and determine the impedance looking into the circuit at A-A. We also short-circuit the terminals at A-A, where the current becomes that of the equivalent Norton current source. The Norton impedance is given by $R_{2S} + j\omega L_{2S}$. To convert this to parallel connected components, we use the equivalent shown in Figure 8.84.

A series connected resistance and reactance, R_S and X_S, respectively, can be represented at a single frequency by a shunt connected resistance and reactance, R_P and X_P, respectively. The relationships between components are

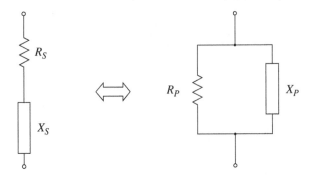

Figure 8.84   Equivalent circuit of a tag antenna and IC

$$R_P = \left(1 + Q^2\right)R_S \tag{8.56}$$

and

$$X_P = \frac{\left(1 + Q^2\right)}{Q^2} X_S \tag{8.57}$$

where

$$Q = \frac{R_P}{X_P} = \frac{X_S}{R_S} \tag{8.58}$$

Using the Norton equivalent of the inductor, its resistance and the voltage induced by the reader give the equivalent circuit shown in Figure 8.85.

With an equivalent circuit consisting of only shunt-connected components, the resonant frequency is simple to calculate. It is given by

$$f_0 = \frac{\omega_0}{2\pi} = \frac{1}{2\pi\sqrt{L_{2P}C_T}} \tag{8.59}$$

where C_T is the combined capacitance

$$C_T = C_{IC} + C_{2P} \tag{8.60}$$

It is also clear that maximum power transfer occurs at the resonant frequency when R_{2P} is equal to R_{IC}. The voltage developed across the load resistor is given by

$$|V_L| = \frac{\sqrt{1 + Q^2}}{Q} \frac{\omega M |I_1|}{\sqrt{\left[1 - \left(\frac{\omega}{\omega_0}\right)^2\right]^2 + \left[\frac{\omega L_{2P}}{R_T}\right]^2}} \tag{8.61}$$

where R_T is the combined resistance of R_{2P} in parallel with R_{IC},

$$R_T = \frac{R_{IC}R_{2P}}{R_{IC} + R_{2P}} \tag{8.62}$$

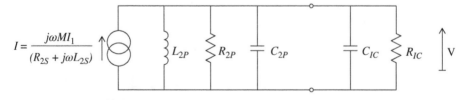

Figure 8.85 Parallel equivalent circuit of a tag antenna and IC

Again, this is a maximum at the resonant frequency given by (8.59). At resonance, and for a given reader current and tag to reader separation, the voltage is predominantly dependent on R_T, which, in turn is usually dominated by the resistance of the tag coil, R_{2P}. This means that the read range (or the range over which there is a voltage great enough to be rectified) is determined by the tag coil quality.

LF and HF tag ICs are designed to have a capacitive input impedance that can be made to resonate with a typical tag coil. Note that tag ICs are normally complementary metal oxide semiconductor (CMOS) devices, which are inherently capacitive. When resonant, the tags produce a high voltage that is then rectified in order to provide a DC power supply to the rest of the tag IC. Once powered, the tag is able to switch its input impedance in order to produce an impedance change at the reader coil. Normally, switching is between the inherent capacitive reactance of the tag IC and a short circuit – an open circuit is seldom used in passive tags due to the high input voltage that could be generated. The impedance change that occurs at the terminals of the reader antenna can be detected by measuring the change in voltage or current that occurs. Hence, digital information can be transferred from the tag to the reader.

A typical rectangular tag antenna is shown in Figure 8.86. The inductance of coils of the type shown in the figure can be approximated by the formulas that follow [59].

$$L = \frac{\mu_0}{\pi} [x_1 + x_2 + x_3 + x_4] N^p \qquad (8.63)$$

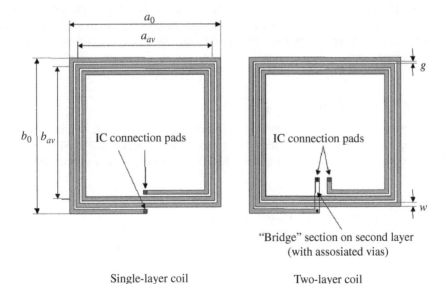

Single-layer coil					Two-layer coil

Figure 8.86   Single- and two-layer tag coils

where

$$x_1 = a_{av} \ln \left[\frac{2a_{av}b_{av}}{d\left(a_{av} + \sqrt{a_{av}^2 + b_{av}^2}\right)} \right] \tag{8.64}$$

$$x_2 = b_{av} \ln \left[\frac{2a_{av}b_{av}}{d\left(b_{av} + \sqrt{a_{av}^2 + b_{av}^2}\right)} \right] \tag{8.65}$$

$$x_3 = 2\left[a_{av} + b_{av} - \sqrt{a_{av}^2 + b_{av}^2} \right] \tag{8.66}$$

$$x_4 = \frac{a_{av} + b_{av}}{4} \tag{8.67}$$

$$d = \frac{2(t + w)}{\pi} \tag{8.68}$$

Here a_0, a_{av}, b_0, b_{av}, g, and w are as shown in Figure 8.86. The track thickness is denoted by t, N is the number of turns, and p is the turn exponent, which is dependent on the manufacturing technology, as indicated in Table 8.11.

These formulas can be used as a starting point for a design. Any final design is likely to be simulated using a 2.5D or 3D electromagnetic simulator, so that the effects of finite conductivity, dielectric materials, etc. can be accounted for.

An example of a commercial coil antenna – used in a European passport – is shown in Figure 8.87. The antenna has 5 turns formed using thin wires. The chip is connected to a metallic strap prior to connection to the coil. The IC is also embedded in a relatively thick layer of insulating material. This prevents the IC from being damaged and also prevents the coil from being excessively bent.

8.5.3 Far-Field Systems

Systems with carrier frequencies above 100 MHz generally operate by transferring power in the far field. The most commonly used frequency bands are in the region of 900 MHz, though the ISM band centered at 2.45 GHz is also used (this band is sometimes termed microwave rather

Table 8.11 Turn exponent with antenna technology

Coil technology	Turn exponent, p
Wires	1.8–1.9
Etched tracks	1.75–1.85
Printed tracks	1.7–1.8

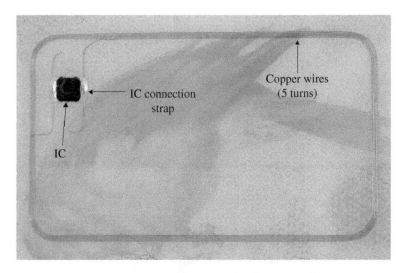

Figure 8.87   Coil antenna used within a passport

Table 8.12   RFID frequencies commonly used within the UHF band

Country	Frequencies (MHz)	Power	Comment
Europe	865.6–867.6	2W ERP	License required in some countries
North America	902–928	4W EIRP	Applies to the United States and Canada
South America	902–928	4W EIRP	
Africa	865.6–867.6	2W ERP	
India	865–867	4W ERP	
China	917–922	2W ERP	Provisional. Temporary license required.
Japan	952–954	4W EIRP	Licence required.
Worldwide	2400–24835		These frequencies apply to the USA and most of Europe. Slightly different frequencies are used in some countries

than UHF). Some typical frequency bands for different parts of the world are indicated in Table 8.12.

Note that for worldwide operation in the region of 900 MHz, a wide frequency range – approximately 860–960 MHz – is required. This means that the antenna must be relatively wideband, with a fractional bandwidth of approximately 11%. Because of this the antenna

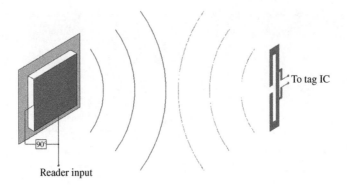

Figure 8.88   Typical far field reader and tag antennas

Q is limited, leading to a limited voltage at the tag IC. Importantly, this indicates that there is a trade-off between bandwidth of operation and range.

A typical far field system is shown in Figure 8.88.

In contrast to near field systems, coil antennas are rarely used in far-field systems. Indeed, a wide variety of antennas are possible. Dipoles, PIFAs, and patches are among the options at the reader. Patches are often used to provide circular polarization (as illustrated in Figure 8.88): this helps to make the communication with an arbitrarily oriented, linearly polarized tag antenna less variable. Tag antennas are rarely circularly polarized (the cost of fabrication is prohibitively high); modified dipoles are most commonly used.

The transfer of energy between far field antennas can be evaluated by consideration of Figure 8.89.

Use Friis' transmission formula as discussed in Section 3.5.1, to give

$$\frac{P_R}{P_T} = \left(\frac{\lambda}{4\pi d}\right)^2 G_T G_R \tag{8.69}$$

This formula applies to perfectly matched antennas with the same polarization. Correction factors, $|\tau|^2$ and χ, must be applied to compensate for the effects of impedance mismatch and polarization misalignment, respectively. Including these effects, the power received by the tag antenna is given by

$$P_R = \left(\frac{\lambda}{4\pi d}\right)^2 P_T G_T G_R |\tau|^2 \chi \tag{8.70}$$

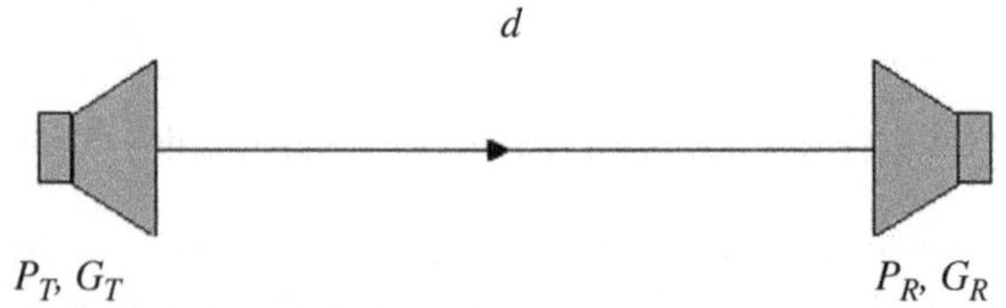

$$d$$

$$P_T, G_T \qquad\qquad\qquad P_R, G_R$$

Figure 8.89   Path loss model

For the tag to operate effectively, it must receive enough power from the tag to "power up" and it must be capable of providing sufficient modulation for the reader to detect. Both depend on the power of the reader, the antenna gains, the polarization matching between the reader and tag antennas, and the mismatch between the tag antenna and the associated IC. The former also depends on the power required to energize the tag, P_{CHIP}, whereas the latter also depends on the effectiveness of the load switching and the sensitivity of the reader. The reader sensitivity is easier to improve than the power required to energize the tag, since the reader is a more elaborate and expensive device. Hence, the range, d, of the system is normally determined by the minimum power required by the tag IC, given by

$$d = \frac{\lambda}{4\pi} \sqrt{\frac{P_T G_T G_R |\tau|^2 \chi}{P_{CHIP}}} \qquad (8.71)$$

To improve range, tags ICs are designed to have the lowest possible P_{CHIP}. The range of a commercial device is calculated in the following example. A tag IC has an equivalent series input resistance of $10 - j245$ ohms. The tag antenna is designed such that its impedance is as close as is possible to the complex conjugate of the chip input impedance (or the impedance at which the tag rectification efficiency is greatest). Hence, an antenna impedance of $10 + j245$ ohms is required. The resistance is low, which conveniently suits electrically small antennas such as dipoles operated well below half of a wavelength. However, the required inductive reactance does not suit small dipoles: such antennas are normally capacitive. Fortunately, it is straightforward to modify the impedance of a dipole in order to make it inductive. A simple technique is to use a shunt feed as illustrated in Figure 8.90. The length of the dipole, l_1, is less than $\lambda/2$, giving a low resistance and a capacitive reactance. The shunt arms provide both an impedance transformation (increasing both the resistance and the capacitive reactance) and a shunt inductance, the combination of which can be used to transform the antenna impedance to the desired value. The impedance transformation is predominantly determined by the ratio of w_2 to w_1 (the higher this ratio the greater the transformation). The shunt inductance is predominantly determined by the length l_2 (though w_1, w_2, and s_1 also have some influence).

Variations of the shunt-fed dipole shown in Figure 8.90 are often used in practice.

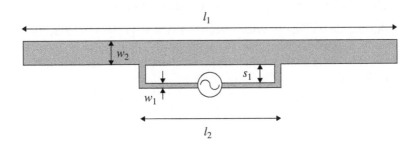

Figure 8.90 Variations of the shunt-fed dipole

8.5.4 Summary

In this section, we have dealt with RFID antennas from theoretical point of view for both near-field system and far-field system. The focus has been on the background and theory behind practical designs. This is an area that has experienced explosive growth for the post 20 years or so. The cost of tags has reduced significantly over the years. It should be pointed out that most tags don't have a ground plane, and their performances may be sensitive to the host medium (e.g. some may not work well for a metal container, or a bottle of water). As IoT is closely linked to RFID, more advanced RFID products will be surely required in the future.

8.6 Reconfigurable Antennas

8.6.1 Introduction

Work has been performed on reconfigurable antennas for many years [60]. As the term suggests, reconfigurable antennas employ some form of variability – often by using switches, variable components, or moving parts – to change the resonant frequency, radiation pattern, and/or impedance match.

For mobile radio, moves toward software-defined radios (SDRs) are driving research on reconfigurable antennas [61]. The Federal Communications Commission (FCC) – an independent United States government agency that regulates radio communications – definition of an SDR is as follows [62]: "a radio that includes a transmitter in which the operating parameters of the transmitter, including the frequency range, modulation type or maximum radiated, or conducted output power can be altered by making a change in software without making any hardware changes."

There are many ways of reconfiguring an antenna, including the following:

- Resonant mode switching/tuning (via shorting, reactive loading)
- Feed network switching/tuning
- Mechanical reconfiguration (with moving parts!)
- Liquid antennas

Each of these methods is considered in the sections that follow. However, it is useful to first consider what switching and variable components are available.

8.6.2 Switch and Variable Component Technologies

Figure 8.91 shows a simplified equivalent circuit of a switch. In the "ON" state, the switch can be represented by a simple resistor, whereas in the "OFF" state, it can be characterized by a series capacitor. For many switches, this capacitor is not lossless, so a parallel resistance is included to indicate that the capacitor has a finite quality factor, Q.

The insertion loss of single pole single throw and series connected single pole dual throw (SPDT) switches is determined primarily by the "ON" resistance, R_{ON}. The isolation is predominantly determined by the "OFF" capacitance, C_{OFF}. Generally, there is a trade-off between R_{ON} and C_{OFF} such that their product, $R_{ON}C_{OFF}$ is constant for a particular technology and DC power supply. Hence, the isolation is usually improved at the expense of insertion loss and vice versa.

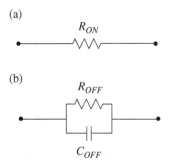

(a)

R_{ON}

(b)

R_{OFF}

C_{OFF}

Figure 8.91   Simple switch equivalent circuit: (a) in the "ON" state, (b) in the "OFF" state

Table 8.13   Comparison of switch technologies

Device	"ON" resistance	Current consumption	Isolation	"OFF" resistance	Linearity
PIN diodes	Low	Low	Reasonable	Appreciable	Reasonable
GaAs & InP FETs	Reasonable	Low (approximately 100µ A)	Good	Reasonable	Good
MEMS	Potentially very low	Very low (a few µA)	Very good	High (high Q)	Very good
Optical switches	Reasonable	Very low	Very good	High	Very good

Table 8.13 gives a comparison of some switch technologies: PIN diodes, Gallium Arsenide (GaAs), Indium Phosphide (InP) field effect transistors (FETS), Micro Electromechanical Systems (MEMS), and optical switches.

PIN diodes have been used with good results in many applications. They have the advantage of potentially low "ON" resistance, but this comes at the cost of high current consumption, which is a disadvantage for low power applications – in particular, for portable devices such as mobile phones. PIN diodes can have a reasonable isolation, but this requires a large negative bias. Again, this is inconvenient in portable devices, where voltages of between around 0 and 3V are available.

GaAs and InP FETs have become increasingly popular in portable devices such as mobile phones because they offer low current consumption. Though the $R_{ON}C_{OFF}$ product of GaAs and InP FETs is approximately the same as for PIN diodes, FETs cannot handle large voltages and are, hence, often used in series. This increases R_{ON} and reduces C_{OFF} and often means that switches made with GaAs and InP FETs have higher "ON" resistances and lower "OFF" capacitances than their PIN-based equivalents.

MEMS devices are in their commercial infancy. They can be likened to miniature relays where metal-to-metal contacts are made and broken by the application (or absence) of an electrostatic field (in contrast to reed relays where actuation is via a magnetic field). Unlike conventional relays, MEMS devices are very small and are fabricated (as far as is possible) using processes that are normally used to manufacture low-cost integrated circuits. Figure 8.92 shows an example. This manufacturing technique, coupled with near lossless "ON" and "OFF" states

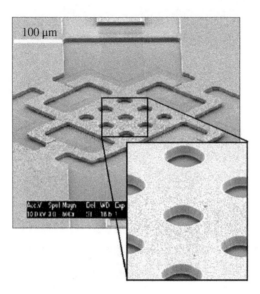

Figure 8.92   Fabrication detail of a typical MEMS switch

– with metal-to-metal contacts and an air gap, respectively – makes MEMS-based switches highly attractive for future applications. The main drawback of MEMS devices is that they typically require high DC actuation voltages.

Switches are not the only components that can be used to make reconfigurable antennas. If large components can be tolerated, mechanically variable inductors and capacitors can be utilized. For example, variable capacitors – where rotary motors provide the variation – are widely used within high power antenna tuning units in the MF and HF bands. On a smaller scale, varactor diodes may be used, though they generally suffer from poor linearity and are therefore limited to low power applications. Active devices may also be used, for example, as variable inductors or capacitors. The main drawbacks of doing so are that the active devices consume power, add noise, and are often nonlinear. In the future, MEMs devices may also be used as variable capacitors, though they tend to have rather nonlinear capacitance curves with DC voltage. MEMS devices may also be used, not as galvanic switches (where a metal-to-metal contact is made), but as capacitive switches with a high "ON" to "OFF" capacitance ratio. This avoids problems associated with the "sticking" and wear of very small metallic contacts.

8.6.3   Resonant Mode Switching/Tuning

Figure 8.93 shows an example of resonant mode tuning, where a PIFA antenna is tuned by a variable capacitor. As the capacitance value is increased, the antenna resonant frequency is reduced. The capacitor is located at the point on the antenna where the electric field is highest, which gives the highest possible resonant frequency tuning ratio for a given capacitance variation.

For physically large (low frequency) applications, the capacitor could be a mechanically variable capacitor; for low power applications or applications where high degrees of linearity are

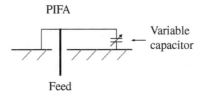

Figure 8.93   A side view of a PIFA antenna that is tuned by a variable capacitance

not required, it could be a varactor diode. Alternatively, discrete variability could be provided by switched capacitors.

8.6.4   Feed Network Switching/Tuning

Antennas can be reconfigured either on the antenna itself (as in the previous section) or by modifying the circuitry feeding the antenna. An obvious example of the later is the phased array. Here, the phases and amplitudes of the signals applied to each element of the array can be modified. In turn, this can alter the antenna radiation pattern in terms of its angle of maximum gain, the value of the gain and the pattern shape.

8.6.5   Mechanical Reconfiguration

Figure 8.94 shows an example of mechanical reconfiguration [63]. A "V" shaped dipole is implemented, and MEMS micro-hinge is used to vary the angle of the "V." This allows the antenna to beam-steer. The largest single mechanical reconfigurable antenna is FAST as we discussed in Chapter 5. When the reflector panels are mechanically reconfigured, the radiation pattern is changed.

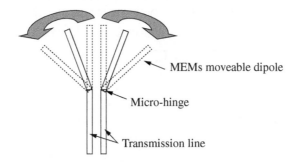

Figure 8.94   A "V" dipole with MEMS-enabled beam-steering

8.6.6   Liquid Reconfigurable Antennas

Liquid antennas are a new type of antennas and can be made reconfigurable. Unlike the conventional metal antennas and solid DRAs, they are made of liquid and a dielectric container/holder is normally required (but not always). Many liquid antennas have been designed, made and tested. They could be categorized in different ways in order to study and summarize their common features in each category. Here we first divide them into liquid metallic and nonmetallic antennas that can be further divided into water-based and non-water-based antennas, as shown Figure 8.95. The most promising candidate seems to be the non-water-based liquid antennas since they offer the most attractive features, such as the flexibility, conformability, fluidity, transparency, and reconfigurability over a large frequency and temperature range. They could be made smaller and cheaper than their metal antenna counterpart.

Most published liquid antennas have been developed to offer the reconfigurability in frequency, radiation pattern, or polarization. Take a monopole antenna as an example, its resonant frequency can be changed by varying the height of the water inside the supporting tube [64, 65] – the water height can be changed using a pump which is the main tuning method up to now for most reconfigurable liquid antennas [66]. Obviously, this tuning speed is much slower than electrical or electronic tuning methods/switches, but the associated circuits and devices are simpler. Furthermore, the tuning range could be much larger than other reconfigurable antennas. One of the reported innovative designs is the passive beam steering antenna that has utilized the special feather of the fluidity and gravity of the liquid to steer the radiation pattern without using a conventional electrical or mechanical steering method [67]. There were some concerns about the radiation efficiency and temperature range (e.g. water becomes ice when temperature is below 0 °C) for liquid antennas which have been addressed by recent research and development in liquid materials for antennas [68]. A very good review paper on liquid reconfigurable antennas is available online [69]. It has reviewed over 100 research papers and provided a very good coverage in this area. Although there are no real-world applications yet, it is believed that the technology is getting mature and will be used in the near future.

8.6.7   Discussion and Summary

Due to the advantages offered, reconfigurable antennas are always on high demand. In [70], the design process of some latest reconfigurable antennas is discussed. Reconfigurable antennas are proposed to cover different wireless services that operate over a wide frequency range. They show significant promise in addressing new system requirements. They exhibit the ability to

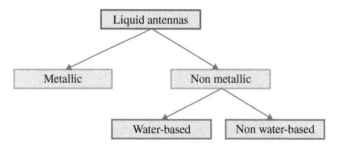

Figure 8.95   Classification of liquid antennas

modify their geometries and behaviour to adapt to changes in surrounding conditions. Reconfigurable antennas can deliver the same throughput as a multiantenna system. They use dynamically variable and adaptable single-antenna geometry without increasing the real estate required to accommodate multiple antennas. The optimization of reconfigurable antenna design and operation by removing unnecessary redundant switches to alleviate biasing issues and improve the system's performance is discussed. Controlling the antenna reconfiguration by software, using Field Programmable Gate Arrays (FPGAs) or microcontrollers is also introduced in the paper. The use of Neural Networks and its integration with graph models on programmable platforms and its effect on the operation of reconfigurable antennas is presented. Finally, the applications of reconfigurable antennas for cognitive radio, MIMO channels, and space applications are highlighted. There are other tunable technologies, such as tunable materials (conductivity, permittivity and permeability), for more information, please refer to [70] and [71].

In this short section, we have briefly introduced switch and variable component technologies for reconfigurable antennas, compared different switches and identified their advantages and drawbacks. The need for reconfigurable antennas is on the rise. Multi-band, cognitive, and software defined radios are the major drivers for this technology.

8.7   Automotive Antennas

Automotive industry is experiencing a revolution: from petrol/diesel to electrical vehicles (EVs), from manual-driving to autonomous vehicles. All these investments and efforts are to ensure that, in the future, the vehicle will be safer and more energy efficient. The function of vehicle will not be limited to transportation from A to B, it may also act as an office or even entertainment center. For a modern car, electronics represent a little more than 30% of the vehicle's total costs, and that percentage is expected to rise. The more electronics a vehicle deploys, the greater the need for electrical power, system components, and need to interconnect those system components. Electronic connections (be it the power supply or control wiring harness), digital communications connections, or radio frequency (RF) interfaces add complexity, cost, and weight. As a consequence, radio systems inside a vehicle have also been increased significantly.

8.7.1   Introduction

For a long period of time, the only radio system inside a vehicle had been the AM/FM radio which was used for the music, news, weather forecast, and traffic update. But over the past 10 years or so, many more RF and microwave systems have been introduced and installed to improve the safety and functionalities of the car. An envisaged example of remote sensing systems in an autonomous car is illustrated in Figure 8.96 where ultrasound, light/optical, radar and wireless technologies have been employed to provide a wide range of services which include but not limited to adaptive cruise control, emergency break, pedestrian detection and collision avoidance, traffic sign recognition, park assist, and wireless communications. Table 8.14 has listed most of the current radio systems inside some current vehicles along with their frequencies and antenna radiation pattern requirements. The frequencies span from 125 kHz to 77 GHz with at least 10 antennas. In this section, we intent to briefly introduce and discuss these radio systems, especially from the antenna design point of view.

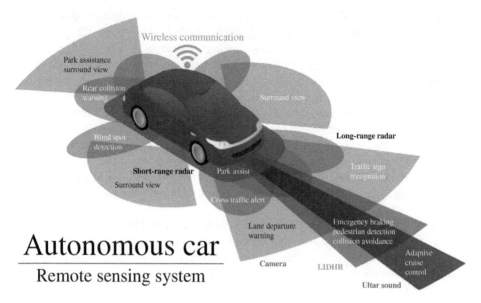

Figure 8.96 An example of remote sensing systems in an autonomous car (innovation-destination.com)

Table 8.14 RF/microwave systems and antenna requirements

No	Application	Frequency	Antenna requirements
1	AM	550–1720 kHz	Omni-directional
2	FM	88–108 MHz	Omni-directional
3	DAB (Digital audio broadcast)	174–240 MHz	Omni-directional
4	GSM, 3G, LET, and 5G mobiles	698–960 MHz	Omni-directional
		1710–2770 MHz	
		3400–3800 MHz	
5	GPS	1.176/1.382/1.575 GHz	Directional, RHCP
6	Wi-Fi/Bluetooth	2.4–2.49 GHz	Omni-directional/ isotropic
7	ETC (Electronic toll collection)	5.8 GHz	Directional
8	Automotive radar	24 and 77 GHz	Directional
9	TPMS (Tire pressure monitor system)	315/413/434 MHz	Isotropic
10	Remote function actuator/smart key fob	125 kHz, 315/434 MHz	Isotropic
11	LCA (Lane change assist)	24 GHz	Directional
12	In-vehicle TV	50–400 MHz	Omni-directional
13	Satellite radio	1.472 GHz	Directional
14	SDARS (Satellite digital audio radio sys)	2.320–2.345 GHz	Directional, RHCP

8.7.2   Antenna Designs

8.7.2.1   AM/FM/DAB Radios

The center frequencies for AM/FM/DAB are around 1, 100, and 200 MHz, respectively, and their corresponding wavelengths are about 300, 3, and 1.5 m, respectively. Thus, it is not possible to install a quarter or half-wavelength antenna for AM, even using size reduction techniques discussed in Section 8.1. Although AM antenna should be large and signal may be of poor quality by current standard in general, AM is still being widely included in a car – the main reason is that the AM radio signal is available in most countries in almost everywhere while FM and DAB are mainly covered in urban and suburban environments. Traditionally, the antennas for AM and FM are co-designed as a whip/monopole antenna, as shown in Figure 8.97. Its length is close to quart-wavelength of FM which is good for FM reception; some designs may be of dual-band and inductive loading, but generally speaking, adaptive impedance tuning is required for AM reception since the antenna is an electrically small monopole for AM in this case.

As the DAB frequencies are now added to the radio specifications while the requirements on the antenna are getting tougher, especially on the low profile, the conventional antenna is therefore out of fashion. Three new antenna designs are presented in Figure 8.98. The "foil antenna" is of low profile and low cost, and it is relatively easy to cover all these three radio frequency bands that could be said the same for the "glass antenna". Many such antennas are based on

Whip/monopole antenna

Figure 8.97   A traditional whip/monopole antenna for car radio

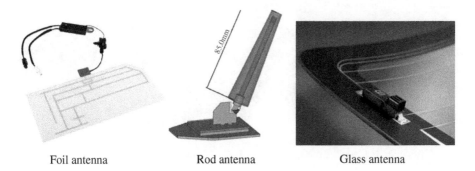

Foil antenna   Rod antenna   Glass antenna

Figure 8.98   Three AM/FM/DAB antennas

inverted-F antenna (IFA) or monopole antennas since it is easy to achieve multi-bands and there are many branches available and their length is typically around quarter-wavelength at the frequency of interest. If it is around 100 MHz, the length should be about 75 cm. A loop antenna may also be selected but it is less flexible. Many glass-mounted antennas are based on wire geometry, although the antenna may or may not be an actual wire. A glass wire antenna is formed by using the wire of a very thin diameter or a silk-screened film that is laminated between layers of glass in the vehicle windows. No additional aerodynamic drag is provided by glass-mounted antennas, and it also creates no wind noise, which is a significant advantage over mast-type designs.

Although the "rod antenna" (also called whip or monopole antenna) is not of low profile, it offers a very good radiation pattern which is the main reason that it is still used although the size has become very small (<10 cm) by employing size reduction techniques. Different types of vehicles may have different strategy to deploy these relatively large antennas. For example, the rear spoiler in a SUV can be used to be fitted with an invisible foil antenna, which may not be a feasible solution for a saloon car.

Another very important aspect is the computer-aided antenna design which has become a routine and the design is now heavily relied on full-wave EM simulation to ensure that the desired impedance bandwidth and radiation are obtained [72]. Figure 8.99 shows how the fields are distributed on a vehicle when a rod antenna is placed at the back of the car for FM and DAB frequencies.

It should be pointed out that many vehicles have more than one AM/FM/DAB antenna in order to introduce antenna diversity and improve the received signal-to-noise ratio. The placement of antenna is a very important aspect of the car system design.

8.7.2.2 Shark-Fin Module

Shark-fin antennas became popular near the turn of the twenty-first century due to its silky design and low file. There are many shark-fin antennas, some are just the replacement of the rod AM/FM radio antenna which could be purchased online at a few dollars. The original equipment manufacturer (OEM) products come with the vehicle are more complicated and expensive. They may consist of several bands with low noise amplifier (LNA) and a matching circuit installed under one or two housings. The communication systems inside the shark fin module include but not limited to:

97 MHz   205 MHz

Figure 8.99   Vehicle-level simulation for radio antenna at different frequencies

Table 8.15 An example of shark fin antenna specifications

	TETRA	GPS ×2	Mobile × 2
Frequency	380–410 MHz	1575 MHz	690–960, 1710–2100 MHz
Input power	1 W	—	1 W
Gain	1.15 dBi	25 dB typical	3 dBi peak
Polarization	Vertical	RHCP	Vertical
VSWR	<2.5	<1.5	<1.5
Impedance	50 Ω	50 Ω	50 Ω
Temperature range	−20 °C–60 °C	20 °C–60 °C	20 °C–60 °C
Dimensions (mm)	105 × 50 × 55	105 × 50 × 55	105 × 50 × 55
Cable/connector	RG316	RG316	RG316

- GNSS (GPS)
- MIMO mobile cellular network
- Wi-Fi and Bluetooth

The specifications of an example shark-fin antenna are listed in Table 8.15. In addition to the standard GPS L1 band and cellular radio bands (690–960 MHz and 1710–2100 MHz), it also covers TETRA (Terrestrial Trunked Radio) band 380–410 MHz for this case, which is specifically designed for use by government agencies, emergency services, (police forces, fire departments, and ambulance) for public safety networks, rail transport staff and services. The GPS antenna is a normally popular dielectric resonant antenna (DRA) with HE11δ mode utilized to cover wideband circular polarization and the module comes with a LNA, thus the total gain of this module is 25 dB (the gain of the antenna is about 5 dBi). Alternatively, a circularly polarized patch antenna can be employed since it has many attractive features such as the low profile, directional radiation pattern, and low cost; furthermore, it can be easily made to cover two or three bands. There are two broadband MIMO antennas (MIMO1 and MIMO2) in the shark-fin module for cellular bands where at least two current paths are clearly shown in Figure 8.100 for each antenna. There is a Wi-Fi/Bluetooth module for communications and controls.

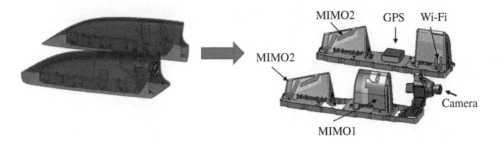

Figure 8.100 Typical shark-fin antennas with and without a camera

8.7.3   24 GHz and 77 GHz Radar

Automotive radars have been on the market since 1999, both in the frequency range around 24 GHz as well as 77 GHz, with a new frequency band ranging from 77 to 81 GHz intended for medium- and short-range sensors. The choice and design of the respective sensor antennas are determined by the requirement for high gain and low loss combined with small size and depth for vehicle integration, the challenges by the millimeter-wave frequency range, and a great cost pressure for this commercial application. Consequently, planar antennas are dominating in the lower frequency range, while lens and reflector antennas had been the first choice at 77 GHz, partly in folded configurations. This is the area that has enjoyed the most attention over the past decade. There are different radar systems in a modern vehicle; some are illustrated in Figure 8.96.

The high-frequency 77-GHz solution offers advantages such as a broader bandwidth and a smaller dimension, but there are more challenges in design and implementation. The 77-GHz band could be divided into two subbands: 76–77 GHz and 77–81 GHz (also called the 79-GHz band). The former has been approved by most countries, while the latter is only available in Europe so far but has been under discussion in other countries. The functions of automotive radar sensors vary with their maximum ranges. Three major classes of radar systems are typically employed in automotive safety systems:

- *Short-range radar (SRR):* this is mainly used for collision proximity warning and safety and to support parking assist features. The distance is up to about 30 m.
- *Medium-range radar (MRR):* this is employed to watch the corners of the vehicle, perform blind spot detection, observe other-vehicle lane crossover, and avoid side/corner collisions. The distance is up to about 150 m.
- *Long-range radar (LRR):* This is for forward-looking sensors, ACC (automatic cruise control), and early collision detection functions. The working distance could be up to 200 m.

LRR is usually mounted in the front grill of the vehicle to measure the distance of objects ahead and it has a narrow beam. SRR offers a broader beam and is used to monitor the vicinity of a vehicle. MRR, which can be installed on the front, the rear, or the side area of the vehicle, is used for different applications, as shown in Figure 8.96. Detailed comparisons of the three radar types are given in Table 8.16 [73].

Table 8.16   Comparison of three radar systems

Radar type	LRR	MRR	SRR
Rang (m)	100–200	60–150	About 30
ERIP	55 dBm	−9 dBm/MHz	−9 dBm/MHz
Frequency band	76–77 GHz	77–81 GHz	77–81 GHz
Distance resolution	0.5 m	0.5 m	0.1 m
Velocity resolution	0.6 m/s	0.6 m/s	0.6 m/s
Velocity accuracy	0.1 m/s	0.1m/s	0.1 m/s
HPBW (Az.)	±15°	±40°	±80°
HPBM (El.)	±5°	±5°	±10°
Applications	ACC	ACC and cross traffic alert	Parking aid, blind spot detection

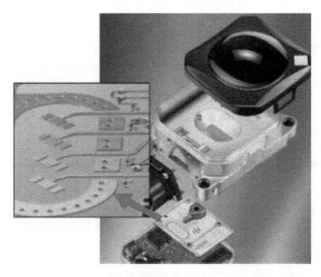

Figure 8.101   Exploded view on Bosch LRR3 and the planar antenna elements feeding the lens

Automotive sensors typically use either a pulse modulation or frequency modulated continuous wave (FMCW) signals with saw tooth or triangular frequency modulation [74]. With increasing requirements toward a much more detailed observation of the scenery in front or around the vehicle, multibeam antennas or scanning antennas have been designed, and solutions based on (digital) beamforming with a number of integrated antennas are in use or under development. General antenna concepts, partly including some system aspects, as well as three realized antenna configurations were described in detail in [74].

A well-known LRR sensor module, Bosch LRR3, is shown in Figure 8.101 which has been available since 2009. Its most prominent feature is the dielectric lens antenna that provides the high gain to achieve the maximum distance of 250 m. It is the only example in this field not scanning its field of view. To achieve angular information in the azimuthal plane, it uses four patch antennas placed in the focal line of the dielectric lens to create four slightly offset beams. All four antennas are receiving antennas with the middle two antennas also simultaneously transmitting. Using phase and amplitude information of all four channels, angular information of detected objects can be determined. The sensor was the first to use SiGe integrated circuits at 77 GHz. The SiGe chip is mounted on a multilayer PCB and wire bonded to an RF substrate on the top side of the PCB.

In many ACC radar modules, it contains a planar antenna array with two transmitting (Tx1 and Tx2) and four receiving antennas. Tx1 is the main transmitting antenna and Tx2 antenna is used for height classification in elevation. MIMO radar modules are also introduced to the automotive industry, more information can be found from such as [75].

TPMS

The tire pressure monitor system (TPMS) is an electronic system designed to monitor the air pressure inside the pneumatic tires on various types of vehicles. A TPMS reports real-time tire-pressure information to the driver of the vehicle, either via a gauge, a pictogram display or a simple low-pressure warning light. TPMS is provided both at an OEM (factory) level as well as an aftermarket solution. They are normally integrated with the valve system. The goal of a TPMS is avoiding traffic accidents, poor fuel economy and increased tire wear due to under-inflated tires through early recognition of a hazardous state of the tires. This functionality first appeared in luxury vehicles in Europe in the 1980s, while mass-market adoption followed the USA passing the 2000 TREAD Act. Mandates for TPMS technology in new cars have contin-ued to proliferate in the twenty-first century in increasing number of countries. Two types of TPMS systems are fitted on cars today:

- **Direct systems:** use radio sensors mounted inside of each wheel to measure the tyre inflation pressures – they "directly" measure the pressure within each tyre and send the data to a con-trol unit
- **Indirect systems:** utilize the vehicle's existing anti-lock braking system (ABS) sensors to measure and compare the rotational speeds of the tyres and vibrations to "indirectly" calcu-late the pressure within the tyres.

As the sensors for the TPMS should be as small as possible and the frequency allocated is around 400 MHz (75 cm), while the wheel is made of steel and conductive, the effective antenna at the tire/wheel is actually larger (since the wheel could act as the antenna) than a quarter-wavelength while the receiving antenna outside the tires is actually smaller than a quar-ter-wavelength, as shown in Figure 8.102 where the field distribution at each tire is also given and receiving antenna is a loop in this case. There are many possible designs for the antenna inside the tire. Since the bandwidth of the system is small and the antenna is next to the metallic wheel rim (near the valve), the simplest design is a whip/monopole antenna as shown in [76]. The most important thing is to ensure the antenna is well matched to achieve the optimum effi-ciency. The computer simulation is vital since it can help us to optimize the design of the antenna and the propagation radio link between the sensors and the receiver by taking car body effects into account.

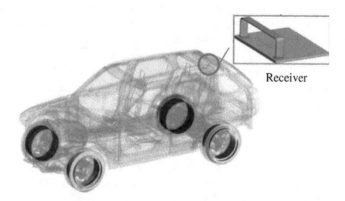

Receiver

Figure 8.102   Receiving antenna and field distribution of a TPMS

8.7.4   Summary

A relatively comprehensive review on RF/microwave systems in a modern vehicle has been provided in this section. It is clear that their frequency coverage is very broad and at least 10 antennas are required for a standard car. We have also looked closely at the antenna designs and specifications for AM/FM/DAB radio, shark fin module, 24/77 GHz radar and TPMS, the main radio systems for a car. There are other RF/microwave systems, including the car key, which are not covered here. There is a growing demand for more radio systems as we move into autonomous car era, more antennas will be required in the future.

Overall, the automotive electronic systems are becoming vital and very complex, the antenna design and placement have to be carefully considered from the system point of view. In addition to aerodynamic and mechanical considerations, the coexistence of different electronic systems and electromagnetic compatibility (EMC) are becoming increasingly challenging. Nevertheless, the antenna design principles are always valid, and the advancement of antenna technology will surely have positive impact on the automotive industry, and vice versa.

8.8   Reflector Antennas

Prof. Anthony K. BROWN
Professor Emeritus, University of Manchester, UK

Reflector antennas are the most common type of structure when electrically large antennas are needed. Applications are widespread and varied from line-of-sight communications to satellite systems and radar (Figure 8.103). As we will see in this Chapter, reflectors are not limited

Figure 8.103   Examples of Reflector Antennas (clockwise from top left): FAST radio telescope, PRC; direct broadcast TV domestic antenna; European Space Agency Artemis satellite; Greenbank radio telescope, USA; air traffic control radar, Easat radar systems

to simple pencil beams but often designed to produce sophisticated beam shapes, giving defined gain coverage over complex angular sectors. An example is shaping the pattern of a satellite system antenna to illuminate a particular country as seen from orbit.

Further to the discussion on reflector antennas in Section 5.2.3, this section commences with basic design criteria for reflectors, developing the principles that underlie their design. Different geometries including offset and multiple reflector designs are discussed. Next, we consider the equations governing the electromagnetics of the reflector antennas and the computational approaches commonly used in their analysis. An introduction to feed design is included. Finally, more advanced reflector designs are introduced, including shaped beam design using single and array feeds.

8.8.1   Fundamentals of Reflector Design

Assume that an antenna has the requirement for a high gain pencil beam such as might be needed in direct broadcast TV or in radio astronomy (Figure 8.103). These two applications indicate the difference in physical size of reflectors that are often used, yet both these are electrically large systems producing narrow, high gain, beams.

From Section 5.2, we know that the relationship between the field across an aperture in an infinite conducting plane and the far field is a Fourier transform. This aperture can be taken as a simple model of the fields across a plane very close to a reflector antenna (Figure 8.104). Hence, the larger the aperture in terms of wavelength, the narrow the beam that can be formed (and so the higher the directivity and gain). The Fourier Transform relationship shows if we taper the amplitude across the aperture, we will increase the beamwidth of the antenna but lower the sidelobes, so a trade-off becomes possible for different applications. The Fourier transform also tells us if we want a beam peak in the direction of the axis of the aperture, then the phase in the aperture plane must be symmetric about its center. If we want to produce a pattern with minimal distortion to the sidelobes and main beam (and thereby maximize the gain) in the far field, then this phase distribution should be uniform across the aperture.

If we now consider an antenna capable of producing this type of well-defined high gain beam, a curved reflector is an suitable candidate. Figure 8.105 illustrates a simple geometry. A curved metallic reflector is illuminated by a much smaller antenna at a defined point. This illuminating antenna, known as a feed, is often a small waveguide horn but could be almost any

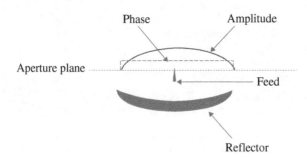

Figure 8.104   A simple reflector antenna model

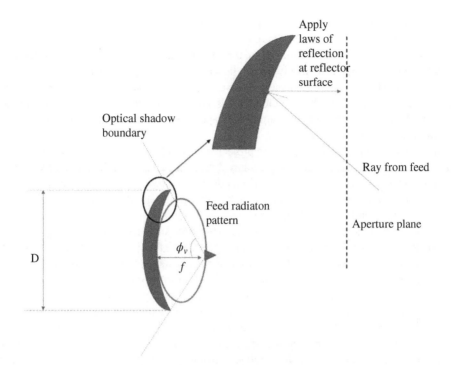

Figure 8.105 Ray tracing approach

small antenna – such as a printed patch, dipole, etc. Of course, the characteristics of this feed play a huge part in the overall performance of the antenna, which will be discussed below.

By exciting the feed input with a suitable source, oscillating currents are induced on the feed structure which therefore produces a radiating wave. This wave impinges on the metallic reflector and in turn excites a current distribution on the reflector surface. This then re-radiates to form the final radiation pattern from the reflector antenna. To get an accurate prediction of this final radiation pattern, we have to compute these various steps, but one can gain significant physical insight by initially considering a ray optic approximation.

Figure 8.105 illustrates the approach. The radiation from the feed is modeled as a set of amplitude weighted rays. At the reflector surface these rays are reflected following the simple laws of reflection and the rays can then be traced to a plane placed close to the reflector aperture (the "aperture plane"). By tracing the rays off the reflector, the aperture plane field distribution is approximated from which a Fourier Transform is used to predict the far field radiation pattern.

Using this approximation, we can undertake a first-order reflector design. As noted above, for a simple well-formed radiation pattern an in-phase aperture distribution is needed. This means the total path length from the feed point to the aperture plane (that is after reflection) must be the same for all points in the aperture plane. Fortunately, there is a simple geometric shape which has this property, a paraboloid with the feed at its focus. This is why this type of antenna is sometimes called a "parabolic reflector" or simply a "parabolic".

The amplitude distribution across the aperture plane is also of course extremely important in forming the radiation pattern. This is normally dominated by the radiation pattern from the feed.

The feed is a small antenna in its own right so it will have its own radiation pattern. This will naturally provide a taper over the reflector (i.e. less power at the edge of the reflector compared to the center) which will be transformed into the aperture plane by the reflector. Therefore, there is a trade-off between the feed size/type (and hence its radiation pattern), the amount of amplitude taper wanted in the aperture plane and the angle between the edge of the reflector and the feed position at the focus of the reflector. This is given by the physical layout of the reflector geometry. Recall the more severe the amplitude taper, the lower the gain but also the lower the sidelobes for a given aperture.

Intuitively one can see that keeping the projected aperture a constant diameter one can select a parabola which will have a focal point at any defined distance from the reflector vertex. The further away the focal point the narrower the angle comes between the edge of the reflector and the focus. So for a given edge taper on the dish as the focal point moves away a narrower feed pattern is required and hence a larger feed.

There is a further factor that can be important. Under the simplifying assumptions made here, the feed radiates a spherical wave from a point aligned to the focus of the reflector. This point on the feed is called the phase center. The power density at distance R from the focus of the reflector of this spherical wave reduces as $1/R^2$. Now the distance from the feed (placed at the focal point) to the center of a parabolic reflector is shorter than the distance to the edge of the reflector, hence the taper (the power at the edge of the reflector compared to that at the center of the dish) is further reduced. After reflection the wave to the aperture plane is essentially a plane wave due to the effect of the parabolic surface and no longer a spherical wave. Therefore, even though the distance from the edge of the reflector to the aperture plane is shorter at the edge than the center, there is no further differential power attenuation across the aperture plane. This first-order explanation allows the inclusion of this effect into the approximate aperture field model. The additional amplitude taper caused by this effect is given by the equation:

$$T\ (\text{dB}) = -20\ \log_{10}\ (r/f) \tag{B.1}$$

where r is the distance from the focal point to any point on the reflector surface and f is the focal length. T must then be used in conjunction with the feed pattern to allow for this differential space taper.

T is heavily dependent of the f/D (focal length to diameter ratio) of the parabolic reflector, which governs the shape of the parabola. The f/D ratio has to be extremely long (greater than about 0.7) for the additional taper caused by this spatial effect to be ignored. For smaller f/D, the reflector is "deeper" with a larger difference in distance between the ray heading to the edge of the reflector and the ray heading toward the center, so the additional spatial taper becomes significant.

Clearly only the power radiated from the feed in angles that hit the reflector can contribute to the antenna pattern in the forward sector. This limit is shown ad the shadow boundary in Figure 8.105. Any power radiating from the feed at wider angles goes past the reflector and appears as far out sidelobes, known as spill over lobes. In reality, radiation in directions close to the edge of the reflector is heavily affected by diffraction at the reflector edge causing rippling and angular spreading in these "spillover lobes".

Not only do these spillover lobes represent a loss of gain in the antenna (which in practice maybe quite small), but they will receive unwanted signals. Most particularly they will receive noise generated by the environment degrading the overall antenna no- temperature.[1] This is a critically important factor in noise-sensitive applications, for example, radio astronomy and some satellite ground stations. Minimization of spill over while still retaining good forward gain from the aperture is often a key design requirement and can be seen to be related both to feed design and reflector geometry.

From these considerations, most applications that are principally interested in forward gain maximization set an edge taper (the amount of power at the edge of the reflector compared to the peak) at about −12 dB to −15 dB, whereas those where side lobe minimization is more important might using −20 dB or lower. These are only rough guidance; each design needs to be assessed on its own merits [71]

Based on these considerations, a first-order design of a reflector can now be undertaken. To simplify matters, we will look initially at a circularly symmetric reflector although a similar approach can be formulated for other shapes of aperture.

Suppose that we require a certain gain G from a reflector with a circularly symmetric shape. We will assume a lossless, perfect, antenna. The gain can then be related to the reflector diameter D via the equation

$$G = \eta \left(\pi D/\lambda\right)^2 \tag{B.2}$$

where η is an efficiency factor (so called *aperture efficiency* as discussed in Section 5.2) due to the tapering of the aperture plus other effects, D is the reflector diameter, and λ is the free space wavelength. If maximizing gain is the key interest typical numbers for η are around 0.55–0.65 though higher efficiencies are possible with sophisticated design. Alternatively, if there is a requirement for low side lobes, requiring much lower power at the reflector edge, the reflector will have lower aperture efficiency.

A first estimate of reflector diameter from the system requirements of gain and operating wavelength is obtained from Equation (B.2).

Next, we need to establish the focal length of the reflector. This is directly related to the feed size and the edge taper required for a given size of reflector. For a simple parabola, the angle from the focal point to the edge of the dish, θ_{max}, is given by

$$\tan\left(0.5\theta_{max}\right) = D/(4f) \tag{B.3}$$

The edge illumination (the power at the edge of the dish compared to the center) due to the space taper effect, T_{max}, can be rewritten (in dB) as

$$T_{max} = -20 \log_{10}\left(1 + (D/(4f))^2\right) \tag{B.4}$$

So if we set the initial edge illumination, we might like and we have estimated the reflector diameter D we can now allow for the space taper T_{max} providing the focal length is set. As a

---

[1] "Antenna noise temperature" relates to the total noise power presented to the receiver by the antenna. This includes both noise received by the antenna and any internally generated noise due to losses. Noise will be received from the ground itself due to thermal emission and to a lesser extent the atmosphere and stellar sources.

starting point a f/D ratio in the range 0.3–0.4 is often used. From this (to first order) the power in dB directed to the edge of the reflector from the feed is calculated. Subtraction from the maximum space taper, T_{max} (in dB), gives the total edge taper. This process can be iterated to give the desired taper. From this the size of feed that might support the design can be estimated. This can then be iterated normally using a more advanced analysis tool to an acceptable design.

To work out the first-order feed aperture size needed for a given feed edge taper, we can use standard equations that relate the beamwidth to size for a particular feed type.

We start by calculating the −10 dB beamwidth of our simple feed. This is related to the edge taper from the feed, E (dB) in [77] as

$$E\,(\text{dB}) = 10\left(\theta_{\text{edge}}/\theta_{10\text{dB}}\right) \tag{B.5}$$

where θ_{edge} is the angle from the feed axis to the edge of the reflector and $\theta_{10\text{dB}}$ is the angle from the feed axis to the angle where the feed pattern is −10 dB down with respect to the feed peak (note the −10 dB beamwidth is conventionally written as the complete angle of the pattern to its −10 dB down points so that the −10 dB beamwidth is $2\theta_{10\text{dB}}$).

A simple feed is often assumed to be a waveguide horn which has a radiation pattern with a −10 dB beamwidth (in degrees) that can be approximated as [78]

$$\text{E}-\text{plane}: 88\lambda b^{-1}\ (\text{valid for } b < 2.5\lambda) \tag{B.6}$$

$$\text{H}-\text{plane}: 31 + 79\lambda a^{-1}\ (\text{valid for } 0.5\lambda < a < 3\lambda) \tag{B.7}$$

where

a is the feed aperture dimension in the H plane
b is the feed aperture dimension in the E plane
λ is the wavelength

From this the size of the (assumed rectangular) feed needed to provide a given edge taper can be found.

As an example, assume that we want gain of 40 dB at 3 GHz. The following steps would apply:

Using Equation (B.2) then yields D to be 4.1 m assuming $\eta = 0.6$.

If we estimate an $f/D = 0.4$, then $f = 0.24$ m.

Equation (B.3) gives an angle to the edge of the dish as 64° and Equation (B.4) a space taper, T_{max}, of −2.9 dB. If we require an overall edge illumination of −15 dB, then the feed needs to provide approximately −12 dB edge illumination contribution at a half angle of 64°. This is equivalent to the feed producing a full beamwidth of 128° at −12 dB.

Equation (B.5) relates this to the 10 dB beamwidth by a factor of 1.1 so the required 10-dB full beamwidth is 116°.

Applying Equations (B.6) and (B.7) gives a rectangular horn of aperture 0.76λ E-plane and 0.92λ in the H-plane. We note these equations are only valid if the horn is tapered at a shallow angle from the aperture to a standard waveguide input. If too sharp an angle is used, then the effect of the taper itself can dominate the patterns [78].

The advantage of these simple equations is that they give a quick trade-off between the various parameters and are amenable to simple spreadsheet or MATLAB calculations. They are only a starting point for optimization. Once a baseline is developed, then computation using one of the techniques discussed in subsection below is used.

8.8.2   Feed Design

From the above the feed and reflector design are closely linked. Feeds themselves are, of course, small antennas which have their own radiation properties.

Overall, the feed radiation pattern must match into the reflector optics. While the emphasis in this Chapter are circular symmetrical antennas, these are by no means the only requirements. Reflectors with elliptical or rectangular rim profiles will have different feed requirements.

In general, however, whatever feed is used its pattern must match into the reflector optics as far as possible. This pattern should also be maintained to a good approximation over the required bandwidth. It must also provide a well-developed phase center (the point on the antenna from which the radiation appears to come) and maintain its phase center position over the bandwidth required. This ensures the reflector remains in focus over the operation bandwidth. Finally, low cross polarization from the feed is normally essential.

For a circularly symmetric dish requiring a pencil beam, then a circularly symmetric feed pattern is ideally needed with low cross polarization. A frequently used approximation is that the feed pattern is circularly symmetric, and Gaussian shaped over the range of angles required to illuminate the dish. Unfortunately, as we will see, in practice, very few feed types actually produce this ideal circularly symmetric beam.

A circular waveguide horn is a commonly used structure due to its simple construction, inherent low loss and reasonable bandwidth (of the order 5–10% is typical). Unfortunately, as is shown below, it has a pattern that is asymmetric in nature.

A circular open-ended waveguide operating in its dominant mode has a field distribution across the feed aperture, as shown in Figure 8.106. Looking first at the E-field we see the field lines are curved. This is a consequence of Maxwells equations, where the E-field must be normal to a conducting wall. Quite clearly the field across the aperture of this type of feed is different in the E and H planes. This means the radiation pattern from a simple circular horn is not circularly symmetric. Depending on the final specification requirements, this may or not be acceptable. It will also, in general, give rise to cross polarization which is unwanted.

If a small feed is needed, appropriate for low f/D and therefore a wide angle from the focal point to the reflector edge, a simple open-ended waveguide may be used. It can be shown that when the feed aperture diameter is approximately 0.8λ the E and H planes are approximately equal at −10 dB beamwidth with low cross polar. Unfortunately, if a larger (or indeed smaller) feed is required, then the feed E and H planes become significantly different and the cross polarization increases [79]. As the f/D grows or if one needs a lower edge illumination on the dish, then a bigger feed is needed.

This will involve some sort of tapered horn as the large horn need to illuminate a large f/D reflector cannot be fed directly by a waveguide (or other transmission line) as this will produce higher order modes in the horn aperture and disrupt the radiation pattern from the feed. Fortunately, if a linear symmetric taper is used from a larger aperture to a smaller one (to allow a standard waveguide at the feed input), it is possible to avoid higher order mode generation

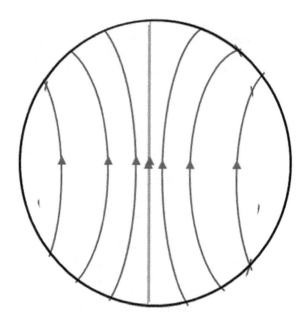

Figure 8.106   Field distribution TE11 mode in circular waveguide – E-field lines shown

so that the pattern narrows as the aperture gets bigger (as expected) but otherwise retains the essential field structure of an open-ended waveguide. However, if the feed horn does not use a perfectly linear taper, or if there are steps or irregularities, then higher order modes may be generated and if they are will affect the radiated field distribution. Even if higher order mode production is eliminated, the asymmetric properties of the radiation patterns from a dominant mode open-ended waveguide will be retained in the larger aperture.

Fortunately, there are a number of well-known feed designs that improve on a simple plane walled tapered horn to produce, at least over a moderate bandwidth, a good symmetric low cross-polarization feed. A good description is provided in [79] and [80] so only an overview is presented here

When a simple open-ended waveguide is not providing adequate bandwidth or incorrect beamwidth for the application, the waveguide can be modified with one or more chokes. Figure 8.107 illustrates the approach. The idea behind the choke is to reduce the current flow on the outside of the waveguide as a quarter wave choke presents a high impedance at the aperture. This in turn reduces back lobes from the feed in addition to lowering cross polarization and providing good pattern symmetry. The approach is only useful for small feeds with a −10 dB beamwidth of 100° over an 8–10% bandwidth [79].

When a narrower beam feed is required another approach to providing beam symmetry is to deliberately introduce some higher order modes into the structure. The idea, originally introduced by Potter [80], is to add to the dominant TE11 mode to one or more higher order modes so that resulting aperture field is approximately symmetric with a cosine type field distribution in both E and H planes. Figure B6 illustrates the idea. Of course, to work the horn aperture has to be big enough to support the higher order modes. In Potters original design the TM01 mode was generated via a step (Figure 8.108) added to the TE11 dominant mode to produce a horn of

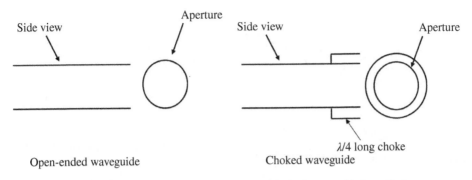

Figure 8.107   Open-ended waveguide and choked waveguide horn feeds

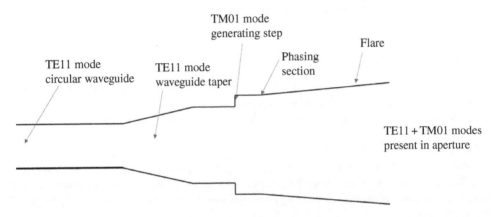

Figure 8.108   Basic Potter Horn design

approximately 4λ diameter with excellent equality of the E and H plane beams and better than −30 dB cross-polarization levels. The −10 dB full beamwidth was approximately 30°. Since then, the idea of a multimode horn has been refined by many authors. These include designs using three rather than two modes, generating the modes by a taper and also incorporating both a choke and a multimode horn. Bhattacharyya and Goyette in [79] give a good overview. Bandwidths of the order 5–10% are typically achievable.

If the potter horn approach is not compatible with the reflector design required a third class of horns have found widespread use. This is the corrugated horn shown in Figure 8.108, where corrugations or radial slots are introduced into the horn. Approximately five slots per wavelength are normally required with each slot less than 0.25 wavelengths – 0.2 being a typical number. In this horn, the intention is to modify the basic wall structure so the propagating wave no longer in a waveguide with a conducting wall but instead the wall becomes essentially reactive due to the slots. Because of this the boundary condition requiring the field to be normal to the conducting surface no longer applies. It will be recalled that it is this boundary condition that causes the curvature of the field lines of the TE11 dominant mode in a circular waveguide which in turn causes the E and H plane beamwidths to be unequal. By introducing this new

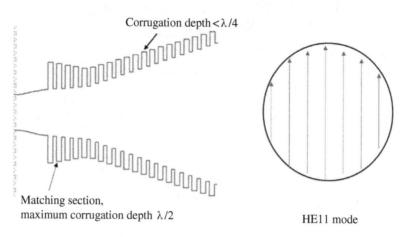

Corrugation depth $< \lambda /4$

Matching section,
maximum corrugation depth $\lambda /2$

HE11 mode

Figure 8.109   Typical Corrugated Horn

boundary condition via the corrugations in the waveguide walls instead of the normal TE11 mode in a waveguide, a hybrid mode propagates termed the HE11 mode. This has a field structure shown in Figure 8.109. In effect, the field lines are "straightened out" giving equal E and H plane radiation patterns from uniform field distribution in both planes. As the horn is metallic the wall impedance is (at least very nearly) purely reactive and the horn has extremely low resistive losses.

One of the major benefits of these horns is the flexibility in the design to encompass a wide range of horn aperture sizes up to many wavelengths in principle. Obviously, they are less attractive than the multimode horn from a manufacturing point of view. Because of their extremely attractive properties corrugated horns have received a lot of attention, [79], [81] present good reviews.

8.8.3   Dual and Multiple Reflector Designs

8.8.3.1   Dual Reflector Designs

As we have seen, we can design a reflector using a parabolic shape and a suitable feed structure. This is a straightforward way of obtaining a high gain aperture, but there are limitations especially for large reflectors. If we use a feed at the focus, we need to get RF energy to that point. That may involve a considerable mechanical structure to support transmit/receive electronics at the focal point. Alternatively, long waveguides or similar may be used to transfer energy from the feed to a ground mounted electronics shelter. Such waveguides may introduce high losses.

One commonly adopted approach to mitigate these and other issues is to adopt a two-reflector approach. This is illustrated in Figure 8.110. The intention here is to move the focal point of the system from the primary focus of the reflector to a point near the vertex of the parabolic reflector by introducing a further reflector. This secondary reflector therefore needs two focal points, one aligned with the main reflector focus with its second focus close to the vertex of the main parabolic reflector. There are two such geometric shapes possible. In the first, a hyperbola is used, in the second an ellipse (Figure 8.110). Both these are adopted from

(a)

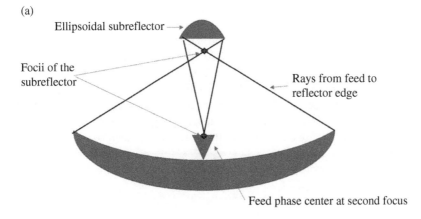

(b)

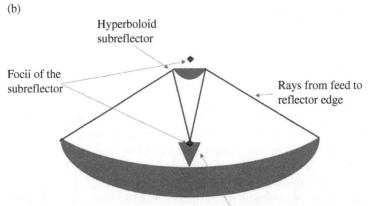

Figure 8.110 Two alternative dual reflector configurations (a) Gregorian geometry; (b) Cassegrain geometry

optical systems and are termed a Cassegrain (for the hyperbola) or a Gregorian (for the ellipse). These secondary reflectors are called "subreflectors".

The feed is now placed at the focus of the subreflector. Radiation from the feed is transformed to give a virtual feed at the focus of the main parabolic reflector which illuminates the main parabola as previously discussed. Of course, depending on the details of the design, the feed design will now be different, and we note it needs to efficiently illuminate the subreflector (else the energy will spill past the subreflector).

The fundamental design equations for a Cassegrain two-reflector system are summarized in [72]. From a first-order design point of view, the addition of the subreflector has produced a system as if it is of a longer focal length. Thus, in two reflector systems, we can undertake an initial feed design based on the feed angle and illumination required as if the feed where placed at a virtual focus at focal length Fe to the main parabola.

While these ray optic approximations give a reasonable starting point to the design, there are second-order effects only a detailed electromagnetic analysis reveals. One example of this is reflector feed interaction. The initial design approach assumes a point source of radiation from the feed. If the distance from reflector to feed is too small, this will not be the case as the sub-reflector will be within the near field of the feed. This can result in phase errors over the main reflector aperture giving both potential gain loss and sidelobe issues. Furthermore, a simple design approach assumes the feed itself is unaffected by the presence of the reflector(s). This may not be true. Radiation leaving the feed which reflects very close to the vertex of the sub- (or main reflector in a single reflector design) will be partially reflected back into the feed causing an increase in VSWR. In extreme cases, some on this energy will be rereflected to the reflector, and a multiple interaction results with a strong frequency dependence and performance degradations.

For these, and other, reasons a full electromagnetic analysis is normally required before finalizing a design.

8.8.3.2 Multiple Reflector Designs

As illustrated above, two reflectors can transfer the focal point of the system to a more convenient point near the main reflector vertex. We can apply the same principles to produce multiple reflector systems to enable the feed to be moved from the reflector vertex to a different point. One use of these systems is in large antenna systems where there may be a requirement to mount the feed closer to the ground. Alternately, it could be a very high-frequency system where using multiple reflectors reduces losses. If use numerically defined reflectors, rather simple geometric shapes, there are other cases where multiple reflector geometries are useful.

As an illustration, let us assume that we have a large reflector system, and we wish to transfer the feed to ground level. One way of achieving this is shown in Figure 8.111. Here two further flat reflectors and two ellipsoidal reflectors are used in a "periscope" type arrangement (note it is also possible to use two paraboloidal rather than ellipsoidal reflectors). In essence, the sub-reflector of a dual reflector system (in the figure a Cassegrain is used) transfers the focal point to near the vertex of the main reflector, as discussed above. This is then aligned to the focus of the upper ellipsoidal reflector which is itself transformed by 45° using a plane reflector. This produces a focus at the second focal point of the ellipsoidal reflector. In turn, this point is aligned to the focal point of the lower reflector which produces a focus at its second focal point. Once again this can be transformed by a 45-degree reflector, so the focus of the system is at ground level where the feed is mounted.

This type of multiple reflector arrangement is known as a beam waveguide. Beam waveguide design needs careful electromagnetic analysis, while a first-order design can be produced using ray optics design the geometry can produce unexpected effects, for example, high cross polarization production, which requires optimization with a full electromagnetic analysis.

8.8.4 Blockage Effects

The simple axially symmetric paraboloidal antenna discussed above is commonly adopted and is usually a cost-effective approach to a narrow pencil beam. There are, however, several limitations to performance. The need for a feed (or a subreflector in a dual reflector design) to illuminate the reflector means it has, by its nature, to shadow (or "block") the reflector aperture (see Figure 8.112). This is also true for any mechanical support structures (or "struts") needed

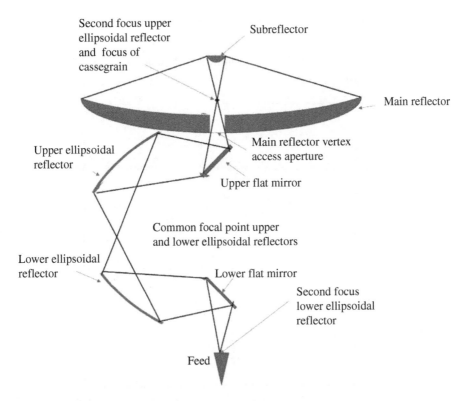

Second focus upper
ellipsoidal reflector
and focus of
cassegrain

Subreflector

Main reflector

Upper ellipsoidal
reflector

Main reflector vertex
access aperture

Upper flat mirror

Common focal point upper
and lower ellipsoidal reflectors

Lower ellipsoidal
reflector

Lower flat mirror

Second focus
lower ellipsoidal
reflector

Feed

Figure 8.111   Beam Waveguide system

to support the feed or subreflector. The effect of this "blockage" region depends on its area and shape. A simple approximation to blockage is to consider the shadow area of the dish and compute the resulting far field via a Fourier Transform of the (complex) unblocked shape. This approach does not allow for diffraction around the blocked area or for complex strut interaction, which can be important, though it does give a good approximation to the blockage effect. Using this simplifying assumption Figure 8.112 gives an approximation to the gain loss. Blockage will not just reduce gain but also increase sidelobe levels. In many applications, this is extremely important as, for example, it increases susceptibility to interference.

8.8.4.1   Offset Reflectors

If we wish to avoid blockage in a reflector system, then we need to remove the feed or subreflector from in front of the reflector aperture. The way this can be readily achieved is through the use of an offset reflector geometry.

Consider initially Figure 8.113. This shows a standard parabolic reflector with the feed point at its focus. Conceptually, we imagine a large feed (with therefore a narrow beamwidth) and rotate the feed, so it illuminates only the top section of the parabola. As it is a narrow beamwidth feed the lower part of the reflector (shown dotted in Figure 8.113) is not illuminated so can be removed. In this way, we now have a geometry where the feed is completely removed from the

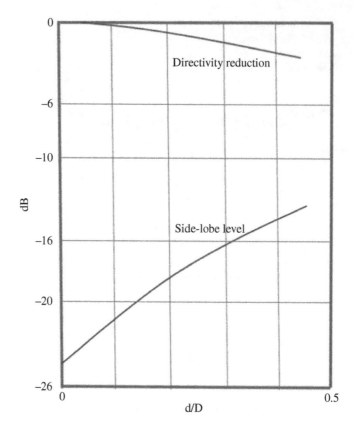

Figure 8.112   Blockage effects in circular symmetric reflectors (d and D are the diameters of the feed and reflector, respectively) [72]

radiating aperture, hence removing the blockage. This is known as an offset reflector geometry. It is important to note the main reflector is still a parabola but no longer rotationally symmetric.

The front fed offset design of Figure 8.113 introduces some issues. Principally, if we operate the antenna in linear polarization and look at the azimuth pattern, then we find a level of cross polarization is produced even with an "ideal" feed. If we use a circularly polarized signal, then there is no cross polar produced but instead the main beam squints slightly away from the mechanical boresight. To calculate these effects, we need to undertake a more rigorous electromagnetic analysis.

This cross-polar effect is due to the way the currents are following on the surface of the reflector. The level of the cross-polar generated is related to the curvature of the main reflector which in turn is related to the f/D ratio of the underlying parabola and the offset angle (θ_0 in Figure 8.112). To gain some insight into this phenomenon, consider a circular symmetric cone of rays leaving the feed hitting the dish and arriving in the aperture plane. In a symmetric front-fed paraboloid, this cone of rays will map into a circle on the aperture plane. However, in an offset geometry such as Figure 8.112, this cone of rays leaving the feed will not map into a

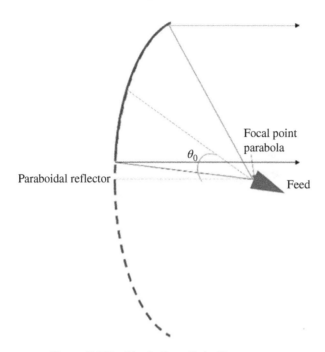

Figure 8.113   Simple Front Fed offset antenna

circle but into an ellipse. This gives a geometric indication that something unusual is happening – in fact cross polar will be generated. Rudge *et al.* [82] give a good summary of this effect.

In an analogous fashion to the symmetric parabolic geometry, it is commonplace to introduce a subreflector to transfer the focal point to a more convenient place near the vertex. Once again, we can use either an ellipsoid or paraboloid. Figure 8.114 illustrates the two geometries, either an offset Cassegrain or an offset Gregorian. Which to choose is a matter of detailed design for a particular application. As can be seen the offset Gregorian requires a longer mechanical support as the subreflector is further away than the main reflector focus, whereas the offset Cassegrain has a shorter mechanical support but is considerably taller that the offset Gregorian. A further point is as shown in Figure 8.114 with the Gregorian geometry the feed points "down," whereas in the Cassegrain the feed point "up". As can be imagined on detailed design, any spill over energy past the subreflector will occur in very different places in the offset plane radiation pattern. Figure 8.103 shows an extremely large, 100 m projected diameter, offset antenna installed at the Greenbank radio astronomy observatory in Virginia, USA.

Whichever dual reflector geometry is chosen, we can now start to cancel out the geometric effects of the main reflector which give rise to the cross polarization. This is achieved by tilting the axis of the subreflector geometry slightly away from the axis system derived from ray optics. In effect, this slight distortion of the axis system causes a cone of rays leaving the feed to cancel the cross polarization induced from the reflector. Referring to Figure 8.115, the cross-polarization cancellation condition can be written [83].

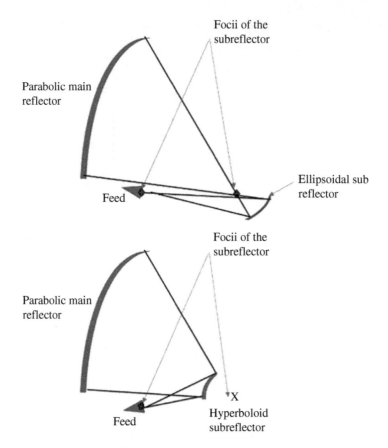

Figure 8.114 Dual offset reflector geometries

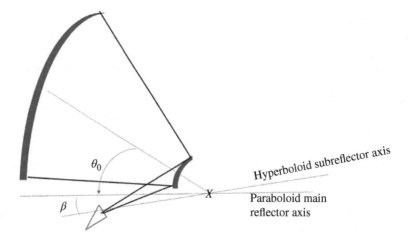

Figure 8.115 Offset Cassegrain geometry showing subreflector axis tilt reflector analysis

$$\tan\left(\theta_0/2\right) = -2e \, \sin\left(\beta\right) \left[e^2 + 1 - 2e \, \cos\left(\beta\right)\right]^{-1} \tag{B.8}$$

where e is the eccentricity of the subreflector, β is the subreflector axis tilt angle, and θ_0 is defined in Figure 8.115. For details, and the equivalent conditions for a Gregorian geometry see [82], [84], [85]. This approach to remove cross-polarization generation was first derived by Mizuguchi [84].

To this point we have been using simple geometric optics to describe some of the design concepts of reflector antennas. To undertake the design in practice one must use a more accurate analysis approach.

There are a number of excellent discussions in the literature on the development of the necessary equations including [86], [84] For this discussion, we will focus on the more widely used techniques for reflector antennas.

8.8.5   Overview of Reflector Analysis

As we know from Maxwell's equations, the radiated field is directly related to the current flowing on the reflector surface which in turn is caused by the incident field from the feed. We could, in principle, attempt a full wave analysis of this problem using the techniques discussed earlier in this book such as Finite Difference techniques. But in practice, most reflectors are electrically large so the concept of using any approach that requires the whole volume enclosing the antenna to be meshed into small cells rapidly becomes impractical. As an example, a realistic reflector could be, say 5 m diameter operating at 14 GHz with a F/D ratio of 0.3. So we would need to define a "solution space" bigger than 5 m × 5 m × 1.5 m or in wavelength terms $233\lambda \times 233\lambda \times 70\lambda$. Using the usual modeling rule of about five cells per wavelength, we have approximately 5×10^8 cells – and this for a relatively modest reflector size. For this reason, surface techniques such as Moment Method approaches (of which there are various useful types in the literature) are attractive where one only needs to model the actual conducting surfaces. Relatively recent advances in modifying the basic moment method using characteristic basis functions are being used for large scale structures [87], [88] and may become appropriate for the reflector problem. One other approach becoming more used is the fast multipole method [89], which is related to moment method type of solutions [90]. Even then problem sizes are large and computationally intensive so structures above perhaps a hundred wavelengths incur very long run times (and potential convergence issues).

Any of these solutions to the analysis of reflector problem do need to be treated with great care when including the feed system explicitly. This is because the electromagnetically important features of the feed are often much less than a wavelength whereas the reflector itself is not normally changing shape that quickly. Accordingly mesh sizing in the feed region is massively smaller than needed in a reflector. As the analysis is already a large electromagnetic problem just how these different mesh regions are handled and still converge to an accurate solution can be extremely difficult to implement. It is still quite common to separate the feed problem away from the reflector analysis, solve the feed separately and input the results as a set of fields, often expressed as different modes in a series expansion of the fields often using spherical wave functions (but not necessarily so). This is a pragmatic solution to the problem but means feed/reflector interactions may be ignored or at best approximated.

These types of solver do have a major benefit in that real-world structures such as complex mechanical feed support mechanisms can be included in the solution allowing second-order scattering to be accurately computed.

8.8.5.1 Physical Optics, GTD and PTD

Given the time, computational source requirements and complexity of moment method or other types of full wave solutions, reflectors are often analysed using an approximate technique known as physical optics (PO) with diffraction at the reflector edge included via either the Physical Theory of Diffraction (PTD) or the Geometric Theory of Diffraction (GTD).

These approaches yield computational efficient solutions to the reflector problem and allow extremely large antennas to be analysed with relatively modest computational memory resources. They do however rely on a number of approximations to reduce the complexity of a solution to Maxwell's equations for large structures. These approximations have been shown to be valid for very many reflector systems, including multiple reflector systems, but are important to be aware of. A good summary of the basis of the technique is given in [79], with an overview provided here.

The PO approximation is based on the fundamental relationship that at the surface of the (assumed conducting) reflector the surface current. J_s can then be derived from the incident magnetic field H at any surface point to be:

$$J_s = 2\hat{n} \times H \tag{B.9}$$

where $\hat{n}$ is the unit normal to the surface. This comes directly from solving Maxwell's equations at a conducting boundary [91]. H is known from the incident fields on the reflector from the feed (multiple reflectors can be analysed by sequentially applying the analysis using an equivalent source for the radiation from the preceding reflector). Once J_s known both far and near fields can be found via the integral [86]

$$E(p) = \frac{-jkZ_0}{4\pi} \int\int_S [J_s \cdot \hat{r}(J_s \cdot \hat{r})] \frac{e^{-jkR}}{R} dS \tag{B.10}$$

$E(p)$ is the required electric field at a point, P, at distance R from the origin of the coordinate system, S is the surface on which the current is running (i.e. the reflector surface in this instance, with zero current assumed on the rear of the reflector), and $\hat{r}$ is the unit vector in the direction of P. In practice, it is this integration that can be computationally demanding. In one approach, the reflector surface is treated as a set of patches, the current established in each patch and the double integral converted to a sum. Other approaches do exist that can be computationally beneficial in some cases [92].

The PO approximation assumes the current on any local region of surface is approximated to the current given by the standard boundary conditions. It is implied that the surface is flat in a local region and infinite in extent. Intuitively, the approximation could get inaccurate near the edge of the reflector as in that case one side of the physical optics current point being computed is metal the other free space. In practice, PO will generally start to get inaccurate about a wavelength away from the reflector edge. Generally, the results of PO alone are sufficient from an engineering viewpoint, but there are a number of circumstances where this is not so. These include calculation very wide-angle sidelobes and spill-over patterns or of on axis front to back ratios for symmetrical antennas.

To improve accuracy in these edge regions, the diffracted field at the edge must be included. To achieve this in the region close to the edge, the analysis is usually modified by using either the GTD or the PTD to include a more accurate model of diffraction. These two techniques provide quite different formulations to cope with the edge diffraction. GTD takes the approach of producing a geometric basis for diffraction to extend the other geometric laws of reflection and refraction. First formulated by Keller [93], the approach considers a ray from the feed incident on the reflector edge. For this discussion, we will assume the edge can be considered as made up of a set of short straight edges each illuminated by a different ray. GTD then provides a cone of output rays within which the diffracted field occurs. The half angle of this cone is related to the incidence angle of the ray to the reflector edge (and the shape of the edge although this is assumed a straight edge in this description). Within this region, the field is given by a diffraction coefficient related to the geometry of the edge and the incident field. The result is a very fast computation of the diffracted part of the field. Unfortunately, the GTD formulation suffers from discontinuities at the optical shadow boundary defined by a ray leaving the feed and just hitting the reflector rim. This theory was later enhanced to produce the Unified Theory of Diffraction (UTD) [94]. While this modification overcomes the issues at the shadow boundary the solution still has a singularity on boresight for a symmetrical reflector. Nonetheless, for a wide variety of problems GTD/UTD provides a good accuracy for the diffracted component of the field. The common alternative to GTD/UTD is called thePTD. As described in [95] PTD approximates the current running around the rim of the reflector based on the incident field and from this to compute the radiating field. A good comparison of these techniques is given in [96].

It can be summarized from the above that PO with an enhanced diffraction accuracy through GTD/UTD or PTD is an effective analysis tool allowing large reflectors to be analyzed. A number of practical approximations are implied in these formulations. These can be summarized as

- Reflector size: the reflector(s) need to be above a few wavelengths as the minimum diameter (as a rule of thumb 5 wavelengths is about the practical limit).
- All reflectors need to be assumed smooth at a wavelength scale.

At least in its most common implementation, the effect of multiple interactions between the feed and a reflector (or sub reflector) is not accurately modeled as the approach has no detailed knowledge of the feed structure. It is assumed that the fields propagated by the field are unaffected by the presence of the reflector.

While the reflector can be in either the near or far field of the feed great care must be taken in describing the field structures from the feed especially if the reflector is in the near field.

Blockage effects can be included in the analysis. Typically, this is achieved by computing the scattering off the subreflector via PO and one of the edge diffraction techniques, zeroing the PO currents on the main reflector below the subreflector and adding in any edge diffraction to the final computed radiated field. A fuller treatment is given in [85]. While it is possible to also include mechanical support struts in the analysis, these are "added in" to the solution as separate sources of scattering. Complex mechanical support structures are therefore difficult to analyse using this approach.

8.8.5.2 Reflector Shaping

So far, we have assumed that the reflectors being used are plane geometric shapes (paraboloidal main reflectors, ellipsoidal or hyperbolical subreflectors). Alternatively, we can use numerically generated shapes, in principle for all the reflectors in a system, to improve performance or radically change the radiation pattern from the reflector system. Some reasons to use beam synthesis include

- Producing a reflector with a shaped beam. Examples include sector-shaped beams, Figure 8.116, elevation beam shaping for a radar system and shaping the beams from a satellite system to more accurately illuminate a required geographic area such as a continent or country.
- Improving overall aperture efficiency and hence gain from a given aperture.

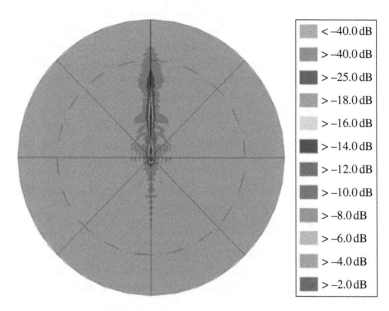

Figure 8.116 Example of a shaped beam given a narrow pencil beam in the azimuth (horizontal) plane and a shaped beam in the elevation (vertical) plane. A single offset reflector was used for a radar system. Looking at the vertical plane we see the beam is extended above the horizon so the radar will be sensitive to targets above the horizon. The pattern falls off quickly below the horizon to minimize illumination of the ground

8.8.5.3   Reducing Sidelobe Levels

There has been a large amount of literature on reflector synthesis techniques [97], [98] provide a good in-depth discussion.

Two alternative starting points are either trying to synthesize a known aperture distribution from which the far field is derived or, alternatively, start directly from the far field and map this through the reflector system. While the detailed implementation is very different mathematically, there are significant similarity in the underlying conceptual thinking.

8.8.5.4   Dual Reflector Synthesis

Initially, consider a two-reflector system. To keep efficiency high, we will assume an offset structure. The feed horn radiates a certain radiation pattern which is transformed by the reflectors into a phase and amplitude distribution across the main reflector aperture plane and hence form the far field radiation pattern. A normal feed horn pattern has more power near its center of the beam. If instead of using a standard subreflector, we distort (or shape), this reflector then we could control the angular amplitude distribution of the fields leaving the subreflector which therefore controls the amplitude distribution across the main reflector. Of course, if we only did this then the system is no longer a focused reflector as the distorted subreflector not only changes the amplitude of the field but its phase. However, if we now look at the phase across the main reflector and distort the main reflector shape appropriately, we can correct for the phase and produce a required amplitude and phase distribution across the main reflector aperture plane. It does not have to be the amplitude that is controlled by the subreflector and the phase by the main reflector, but this helps to visualize the process. More formally by using two reflectors we can synthesis numerically defined shapes which control both amplitude and phase across the resulting aperture plane. In a multireflector system (such as a beam waveguide), it does not have to be the main and subreflectors that are used – it could be any reflectors in the system. Also, it can be as usefully applied to axis-symmetric or offset reflectors. In practice, especially for relatively low degrees of shaping at lower frequencies, the amount of phase correction required by the main reflector can be small in which case successful designs have been produced using the subreflector only.

Of course, there are limitations to this type of reflector shaping. The curvature generated by the numerical process should be smooth and continuous, yet the synthesis procedure will normally represent the surface as patches. The feed pattern itself may be limited in accuracy. The effects of reflectors or shadowing are ignored. Edge diffraction is usually ignored. The synthesis process will not therefore be precise. To complete the design process, an accurate analysis step (as given in Section 6) needs to be included after an initial synthesis. This normally then requires further optimization to meet the design goals. Nonetheless, this approach has been very successfully applied to maximize the gain while minimising spill over (and hence noise temperature), also to produce low sidelobe patterns and other shaped beams [99].

One point to note is that if synthesis is being used in a two-reflector offset system, the pure Mizuguchi conditions are no longer valid to cancel the cross polarization from offset reflector systems. Luh [100] has extended Mizuguchi's work to shaped main reflectors. Minimization of cross-polarization levels remains an important consideration in offset reflector synthesis.

8.8.5.5 Single Reflector Synthesis

There are many applications where use of a two-reflector system is not either economically or mechanically feasible. If we require a shaped beam can this still be achieved with only one reflector? We can indeed achieve such a design, but it is less optimal than a dual or multiple reflector system as a single reflector cannot control both amplitude and phase accurately. One approach is to use a power synthesis. In this we inherently ignore the phase information during the synthesis step, although this is of course included in the analysis step. This then becomes an iterative process.

In power synthesis, the required radiation pattern is considered as a number of small angular sectors. The proportion of power required in each sector compared to the total power radiated in the complete pattern is then calculated. Next a similar calculation is computed for the feed radiation pattern. The main reflector shaping algorithm strives to find a shape such that each of the required radiation pattern sectors is angularly mapped to an appropriate set of angles at the feed horn. The desire is for the total power radiated by the feed to equal the total power radiated in the required far field pattern and for the proportion of that total power in any given angular sector from the feed is mapped into the corresponding far field radiation pattern.

This process is by its nature approximate. At a fundamental level, the far-field radiation pattern is formed by the integration of currents across the reflector where both amplitude and phase are important. This power synthesis approach exerts no control on the phase of the currents flowing on the reflector surface. As such a complete "match" to the far field requirements is never fully achieved. However, by using this power synthesis as a first step, then a fully PO-based analysis step then re-optimization perhaps surprisingly good results can be obtained in certain applications. One such is a surveillance radar reflector (which are installed at many civilian airports) and is illustrated in Figure 8.103. The requirement here is for a narrow-focused beam in the azimuth plane and a shaped beam in elevation. [101] provides results that indicate the type of performance achievable with power only synthesis techniques.

8.8.5.6 Array-Fed Antennas

For some applications, beam synthesis using either single or dual reflector shaping is not appropriate. In these applications, a shaped beam can be provided by using an array to feed a parabolic reflector.

Initially, just consider a single feed in a front fed reflector system. As we move the feed sidewards in the focal plane, then the reflector is no longer in focus. If we plot the phase across the aperture, then the phase distribution is complex depending on the extent of the feed movement. However, Ruze [101] showed that the phase in the aperture plane of a circularly symmetric dish with a feed distanced from the focal point but still in the focal plane can be approximated as a series of terms comprising both linear and nonlinear terms. The nonlinear terms are known as coma. As we move the feed in the aperture plane, the linear term will cause the beam to move off the mechanical boresight of the reflector. Unfortunately, the coma terms will start to degrade the beam, growing the beam width, raising sidelobes and lowering the gain. Provided the movement of the feed is relatively small in terms of both the reflector focal length and diameter the coma can be kept low. The wider the angle of scan is demanded, then the more the feed has to be moved and the higher the coma effects becomes.

If we introduce an array feed, then each element of the array will produce a beam at a different position. Adding these with complex weights can then produce a shaped beam in the far field. Great care has to be taken with this fairly simple approach as the beams from each element may not be independent of each other (i.e. they might not be orthogonal [102]) so that the weight synthesis procedure will be very complex. While this explains the concept behind array fed reflectors there is a better methodology.

If we consider the required shaped beam and map this through the reflector geometry, we can derive the required amplitude and phase distribution in the focal plane to produce that beam (ignoring edge diffraction, etc.). The next step is to produce an array that will establish those focal fields. We start the process by establishing the sampling criteria. In a standard phased array design (as dealt with earlier in this book), an element separation criterion of 0.5–0.7 wavelengths is normally adopted to avoid grating lobe production. Using an array in the focal plane of the reflector is a very different proposition as, depending on the required radiation pattern, the phase and amplitude required in the focal plane may be varying in a much more complex way. This may demand closer element spacing. If the spacing gets too small, the practicality of the array is in doubt and the mutual coupling will be high and limit the performance. So we have a trade-off between required radiation patterns, the geometry of the dish, and the configuration of the array feed.

It is interesting to note that if the requirement is to produce a pencil beams scanned off axis then using a closely spaced array (and not just exciting one element per beam) allows some compensation of the coma effect introduced by off axis feeds so allowing increased performance at wider scan angles [103].

8.8.6 Summary

In this section, we have discussed reflector antennas and their design principles. Fundamentally, reflectors transform patterns from a feed system to the aperture plane of the reflector and hence form a radiation pattern. As we have seen, there are considerable complications in practice, from feed to reflector design from simple approximations to accurate analysis. This chapter has introduced simple parabolic, offset and multiple reflector designs. As will be appreciated, reflector antennas are a large subject, and the interested reader is urged to study the relevant references and indeed other literature for more in-depth work. The reflector antenna remains the most appropriate structure for many applications.

8.9 Chapter Summary

In this chapter, we have discussed eight special topics: electrically small antennas, mobile antennas, MIMO antennas, multi-band and wideband antennas, RFID antennas, reconfigurable antennas, automotive antennas, and reflector antennas. Some topics (such as the electrically small antennas, diversity, and MIMO antennas) were discussed in great depth while some other topics (such as automotive antennas) were more like an overview. It is not possible for us to provide a comprehensive coverage on all these important topics due the page limit. As stated at the beginning, the main objective here is to provide you with some latest developments in antennas so as to broaden your horizon and enable you to apply the theories and design principles you have learned to real-world applications.

References

1. H. Wheeler, 'Small antennas', *IEEE Trans. Antennas Propag.*, AP-23(4), 462–469, 1975.
2. G. L. Matthaei, L. Young, E.M.T. Jones, Microwave Filters, Impedance Matching Networks and Coupling Structures, McGraw-Hill, 1964.
3. H. Wheeler, 'Fundamental limitations of small antennas', *Proc. IRE*, Vol. 35 1479–1484, 1947.
4. L. J. Chu, 'Physical limitations of omni-directional antennas,' *J. Appl. Phys.*, 19, 1163–1175, 1948.
5. R. F. Harrington, 'Effect of antenna size on gain, bandwidth and efficiency,' *J. Res. Natl. Bur. Stand.*, 64D (1), 1–12, 1960.
6. R. E. Collin and S. Rothchild, 'Evaluation of antenna Q,' *IEEE Trans. Antennas Propag.*, AP-12, 23–27, 1964.
7. R. L. Fante, 'Quality factor of general ideal antennas,' *IEEE Trans. Antennas Propag.*, AP-17, 151–157, 1969.
8. J. S. McLean, 'The radiative properties of electrically-small antennas,' IEEE Int. Symp. Electromagn. Compat., 1994, pp. 320–324.
9. J. S. McLean, 'A re-examination of the fundamental limits on the radiation Q of electrically small antennas,' *IEEE Trans. Antennas Propag.*, 44(5), 672–676, 1996.
10. H. D. Foltz and J. S. McLean, 'Limits on the radiation Q of electrically small antennas restricted to oblong bounding regions,' *Antennas Propag. Soc. Int. Symp.*, 4, 2702–2705, 1999.
11. G. A. Thiele, P. L. Detweiler, R. P. Penno, 'On the lower bound of the radiation Q for electrically small antennas,' *IEEE Trans. Antennas Propag.*, 51(6), 1263–1269, 2003.
12. H. W. Bode, 'Network Analysis and Feedback Amplifier Design,' Van Nostrand, 1947.
13. R. M. Fano, 'Theoretical limitations on the broadband matching of arbitrary impedances,' *J. Frankl. Inst.*, 249(1), 57–83, 1950.
14. R. M. Fano, 'Theoretical limitations on the broadband matching of arbitrary impedances,' *J. Frankl. Inst.*, 249(2), 139–154, 1950.
15. G. Goubau, 'Multi-element monopole antennas,' Proceedings of the ECOM-ARO Workshop on Electrically Small Antennas, Fort Monmouth, 1976, pp. 6–7.
16. A. R. Lopez, 'Wheeler and Fano impedance matching', *IEEE Antennas Propag. Mag.*, 49(4), 116–119, 2007.
17. A. R. Lopez, 'Double-tuned impedance matching', *IEEE Antennas Propag. Mag.*, 54(2), 109–116, 2012.
18. S. R. Best and D. L. Hanna, 'a performance comparison of fundamental small-antenna designs,' *IEEE Antennas Propag. Mag.*, 52(1), 47–70, 2010.
19. C. Pfeiffer, 'Fundamental efficiency limits for small metallic antennas,' *IEEE Trans. Antennas Propag.*, 65 (4), 1642–1650, 2017, doi: 10.1109/TAP.2017.2670532.
20. J. D. Parsons, The Mobile Radio Propagation Channel, 2nd Edition, John Wiley & Sons, 2000.
21. R. King, C. W. Harrison, D. H. Denton, 'Transmission line missile antennas,' *IRE Trans. Antennas Propag.*, 8, 88–90, 1960.
22. H. Mishima and T. Taga, 'Mobile antennas and duplexer for 800 MHz band mobile telephone system,' *Antennas Propag. Soc. Int. Symp.*, 18, 508–511, 1980.
23. K. Ogawa and T. Uwano, 'A diversity antenna for very small 800-MHz band portable telephones,' *IEEE Trans. Antennas Propag.*, 42(9), 1342–1345, 1994.
24. H. T. Chattha, Y. Huang and Y. Lu, 'PIFA bandwidth enhancement by changing the widths of feed and shorting plates,' *IEEE Antennas Wirel. Propag. Lett.*, 8, 637–640, 2009, doi: 10.1109/LAWP.2009.2023251.
25. W. Hong, 'Solving the 5G mobile antenna puzzle: assessing future directions for the 5g mobile antenna paradigm shift,' *IEEE Microw. Mag.*, 18(7), 86–102, 2017, doi: 10.1109/MMM.2017.2740538.
26. Y. Ban, Y. Qiang, Z. Chen, K. Kang and J. Guo, 'A dual-loop antenna design for hepta-band WWAN/LTE metal-rimmed smartphone applications,' *IEEE Trans. Antennas Propag.*, 63(1), 48–58, 2015, doi: 10.1109/TAP.2014.2368573.

27. M. Stanley, Y. Huang, H. Wang, H. Zhou, Z. Tian and Q. Xu, 'A novel reconfigurable metal rim integrated open slot antenna for octa-band smartphone applications,' *IEEE Trans. Antennas Propag.*, 65(7), 3352–3363, 2017, doi: https://doi.org/10.1109/TAP.2017.2700084.

28. H. Wang and M. Zheng, 'Triple-band wireless local area network monopole antenna,' *IET Microwaves Antennas Propag.*, 2(4), 367–372, 2008, doi: https://doi.org/10.1049/iet-map:20070120.

29. M. Moosazadeh and S. Kharkovsky, 'Compact and small planar monopole antenna with symmetrical L- and U-shaped slots for WLAN/WiMAX applications,' *IEEE Antennas Wirel. Propag. Lett.*, 13, 388–391, 2014, doi: https://doi.org/10.1109/LAWP.2014.2306962.

30. H. Wang and L. Ying, 'A very compact internal WLAN antenna, *Microw. Opt. Technol. Lett.*, 61:2150–2154, 2019.

31. H. Wang, D. Zhou, L. Xue, S. Gao and H. Xu, 'Modal analysis and excitation of wideband slot antennas,' *IET Microwaves Antennas Propag.*, 11(13), 1887–1891, (2017), doi: https://doi.org/10.1049/iet-map.2017.0084.

32. M. Chen and C. Chen, 'A compact dual-band GPS antenna design,' *IEEE Antennas Wirel. Propag. Lett.*, 12, 245–248, 2013, doi: https://doi.org/10.1109/LAWP.2013.2247972.

33. J. Kurvinen, H. Kähkönen, A. Lehtovuori, J. Ala-Laurinaho and V. Viikari, 'Co-designed mm-wave and LTE handset antennas,' *IEEE Trans. Antennas Propag.*, 67(3), 1545–1553, 2019, doi: https://doi.org/10.1109/TAP.2018.2888823.

34. M. Stanley, Mobile Phone Antennas for MIMO and 5G Millimetre Wave Communications. PhD thesis, the University of Liverpool, 2018

35. A. A. Glazunov, A. F. Molisch, F. Tufvesson, 'Mean effective gain of antennas in a wireless channel,' *IET Microwaves Antennas Propag.* 3(2), 214–227, 2009, doi: https://doi.org/10.1049/iet-map:20080041.

36. H. T. Chattha, Y. Huang, S. J. Boyes and X. Zhu, 'Polarization and pattern diversity-based dual-feed planar inverted-F antenna," *IEEE Trans. Antennas Propag.*, 60(3), 1532–1539, 2012, doi: https://doi.org/10.1109/TAP.2011.2180308.

37. Test Plan for Wireless Device Over the-Air Performance, Method of Measurement for Radiated RF Power and Receiver Performance, CTIA certification Program, (2016), http://www.ctia.org/docs/default-source/defaultdocument-library/ctia-test-plan-for-wireless-device-over-the-air-performancever-3-5-2.pdf

38. Z. N. Chen, and K-M Luk, Antennas for Base Stations in Wireless Communications, McGraw-Hill Professional, 2009.

39. H. Huang, Y. Liu and S. Gong, 'A dual-broadband, dual-polarized base station antenna for 2G/3G/4G applications,' *IEEE Antennas Wirel. Propag. Lett.*, 16, 1111–1114, 2017.

40. A. Alieldin, Smart Base-station Antennas for MIMO and 5G Applications, PhD thesis, The University of Liverpool, 2019.

41. Q. Hua, Y. Huang, et al., 'A novel compact quadruple-band indoor base station antenna for 2G/3G/4G/5G systems,' *IEEE Access*, 7, 151350–151358, 2019, doi: https://doi.org/10.1109/ACCESS.2019.2947778.

42. A. Hottinen, O. Tirkkonen, and R. Wichman, Multi-antenna Transceiver Techniques for 3G and Beyond. John Wiley, 2003.

43. G. Tsoulos (ed.) MIMO System Technology for Wireless Communications, CRC, 2006.

44. G. J. Foschini and M. J. Gans, 'On limits of wireless communications in a fading environment when using multiple antennas,' *Wirel. Pers. Commun.*, 6(3), 311–335, 1998.

45. E. Telatar, 'Capacity of multi-antenna gaussian channels,' *Eur. Trans. Telecommun.*, 10(6), 585–595, 1999.

46. R. G. Vauchan and J. B Anderson, 'Antenna diversity in mobile communications,' *IEEE Trans. Veh. Technol.*, 36(4), 149–172, 1987.

47. S. C. K. Ko and R. D. Murch, 'Compact integrated diversity antenna for wireless communications,' *IEEE Trans. Antennas Propag.*, 49(6), 954–960, 2001.

48. B. K. Lau, 'Multiple antenna terminals,' in MIMO: From Theory to Implementation by A. Sibille, C. Oestges, A. Zanella (Eds.), Academic, San Diego, CA, USA, 2011, pp. 267–298.

49. P. Hallbjorner, 'The significance of radiation efficiencies when using S-parameters to calculate the received signal correlation from two antennas,' *IEEE Antenna Wirel. Propag. Lett.*, 4, 97–100, 2005.

50. M. S. Sharawi, A. T. Hassan, and M. U. Khan, 'Correlation coefficient calculations for MIMO antenna systems: a comparative study,' *Int. J. Microw. Wirel. Technol.*, 9(10), 1991–2004, 2017.

51. J. Yang, S. Pivnenko, T. Laitinen, J. Carlsson, and X. Chen, 'Measurements of diversity gain and radiation efficiency of the Eleven antenna by using different measurement techniques,' Eur. Conf. Antennas and Propag., 2010, pp. 1–5.

52. X. Chen, S. Zhang and Q. Li, 'A review of mutual coupling in MIMO systems,' *IEEE Access*, 6, 24706–24719, 2018.

53. H. Xu, S. S. Gao, H. Zhou, H. Wang and Y. Cheng, 'A highly integrated MIMO antenna unit: differential/common mode design,' *IEEE Trans. Antennas Propag.*, 67(11), 6724–6734, 2019, doi: https://doi.org/10.1109/TAP.2019.2922763.

54. L. Sun, Y. Li, Z. Zhang and H. Wang, 'Self-decoupled MIMO antenna pair with shared radiator for 5G smartphones,' *IEEE Trans. Antennas Propag.*, 68(5), 3423–3432, 2020, doi: https://doi.org/10.1109/TAP.2019.2963664.

55. R. Aiello and A. Batra, Ultra Wideband Systems: Technologies and Applications, Elsevier Inc., 2006.

56. B Allen, M. Dohler, E. Okon, W. Malik, A. Brown, D. Edwards (ed.), Ultra Wideband Antennas and Propagation for Communications, Radar and Imaging, John Wiley, 2006.

57. K. Finkenzeller, RFID Handbook: Fundamentals and Applications in Contactless Smart Cards and Identification, 3rd Edition, John Wiley, 2010.

58. S. Ramo, J. Whinnery, T. Van Duzer, Fields and Waves in Communication Electronics, 3rd Edition, John Wiley & Sons Inc, 1993.

59. I•CODE Coil Design Guide, Philips Semiconductors Application Note, 2002.

60. J. T. Bernhard, Reconfigurable Antennas, Morgan & Claypool, 2006.

61. J. H. Reed, Software Radio: A Modern Approach to Radio Engineering, Prentice Hall, 2002.

62. FCC's Notice of Proposed Rulemaking, ET Docket 00-47, issued December 8, 2000.

63. J. Chiao, Y. Fu, I. Chio, M. DeLisio, L. Lin, 'MEMS reconfigurable Vee antenna,' *IEEE Microwave Symp. Digest*, 4(13-19), 1515–1518, 1999.

64. L. Xing, Investigations on Water-Based Liquid Antennas for Mobile Communications, PhD Thesis, The University of Liverpool, 2015.

65. Y. Huang, L. Xing, C. Song, S. Wang and F. Elhouni, 'Liquid antennas: past, present and future,' *IEEE Open J. Antennas Propag.*, 2 473–487, 2021, doi: https://doi.org/10.1109/OJAP.2021.3069325.

66. L. Xing, Q. Xu, J. Zhu, Y. Zhao. S. Alja'afreh, C. Song, Y. Huang. 'A high efficiency wideband frequency reconfigurable water antenna with a liquid control system: usage for VHF and UHF applications,' *IEEE Antennas Propag. Mag.*, 63, 61–70, 2019.

67. C. Song, E. Bennet, J. Xiao and Y. Huang, 'Passive beam-steering gravitational liquid antennas,' *IEEE Trans. Antennas Propag.*, 68(4), 3207–3212, 2020, doi: https://doi.org/10.1109/TAP.2019.2937362.

68. E. Bennett C. Song, Y. Huang, and J. Xiao, 'Measured relative complex permittivities for multiple series of ionic liquids,' *J. Mol. Liq.*, 24(11571), 2019.

69. E. Motovilova and S. Y. Huang, 'A review on reconfigurable liquid dielectric antennas,' *Materials*, 13, 1863, 2020, doi: https://doi.org/10.3390/ma13081863.

70. J. Costantine, Y. Tawk, S. E. Barbin and C. G. Christodoulou, 'Reconfigurable antennas: design and applications,' *Proc. IEEE*, 103(3), 424–437, 2015, doi: https://doi.org/10.1109/JPROC.2015.2396000.

71. R. L. Haupt and M. Lanagan, 'Reconfigurable antennas,' *IEEE Antennas Propag. Mag.*, 55(1), 49–61, 2013, doi: https://doi.org/10.1109/MAP.2013.6474484.

72. M. Rütschlin and D. Tallini, 'Simulation for antenna design and placement in vehicles,' Antennas, Propagation & RF Technology for Transport and Autonomous Platforms 2017, Birmingham, 2017, pp. 1–5, doi: 10.1049/ic.2017.0021.

73. J. Hasch, E. Topak, R. Schnabel, T. Zwick, R. Weigel and C. Waldschmidt, 'Millimeter-wave technology for automotive radar sensors in the 77 GHz frequency band,' *IEEE Trans. Microwave Theory Tech.*, 60(3), 845–860, 2012, doi: https://doi.org/10.1109/TMTT.2011.2178427.

74. W. Menzel and A. Moebius, 'Antenna concepts for millimeter-wave automotive radar sensors,' *Proc. IEEE*, 100(7), 2372–2379, 2012, doi: https://doi.org/10.1109/JPROC.2012.2184729.

75. S. Sun, A. P. Petropulu and H. V. Poor, 'MIMO radar for advanced driver-assistance systems and autonomous driving: advantages and challenges,' *IEEE Signal Process. Mag.*, 37(4), 98–117, 2020, doi: https://doi.org/10.1109/MSP.2020.2978507.

76. H. Zeng and T. H. Hubing, 'The effect of the vehicle body on em propagation in tire pressure monitoring systems,' *IEEE Trans. Antennas Propag.*, 60(8), 3941–3949, 2012, doi: https://doi.org/10.1109/TAP.2012.2201091.

77. T. S. Saad, The Microwave Engineers Handbook and Buyers Guide, pp. 170, Horizon House 1963.

78. S. Silver, Microwave Antenna Theory and Design, McGraw-Hill Inc, New York 1949.

79. L. Shafai, S. Sharma and S. Rao (ed), Handbook of Reflector Antennas and Feed Systems, Volume 2, Artech House, 2013.

80. P. D. Potter, 'A new horn antenna with suppressed sidelobes and equal beamwidths,' *Microw. J.*, 18(2), 71–78, 1961.

81. P. Clarricoats and A. D. Olver, Corrugated Horns for Microwave Antennas, IEE (now IET), 1984.

82. A. Rudge, K. Milne, A. D. Olver and P. Knight, The Handbook of Antenna Design, Vol. 1, Peter Peregrinus, 1982.

83. W. Rusch, A. Prata, Y. Rahmat-Samii, R. A. Shore 'Derivation and application of the equivalent paraboloid for classical offset cassegrain and gregorian antennas,' *IEEE Trans. Antenna Propag.*, 38 (8), 1141–1149, 1990.

84. Y. Mizugutchi, M. Akagawa and H. Yokoi, 'Offset Dual Reflector Antennas,' 1976 Int. IEEE Antennas and Propagation Soc Symp, pp. 2–5, 1976.

85. S. Sharma, S. Rao and L. Shafai (ed), Handbook of Reflector Antennas and Feed Systems, Volume 1, Artech House, 2013

86. P. J. Wood, Reflector Antenna Analysis and Design, Peter Pereginus, 1986

87. R. Maaskant, R. Mittra and A. G. Tijhuis, 'Fast analysis of large antenna arrays using the characteristic basis function method and the adaptive cross approximation algorithm,' *IEEE Trans. Antennas Propag.*, 56(11), 3440–3451, 2008.

88. V. V. S. Prakash and R. Mittra, 'Characteristic basis function method: a new technique for efficient solution of method of moments matrix equations,' *Microw. Opt. Technol. Lett.*, 36(2), 95–100, 2003.

89. N. Engheta, W. Murphy, V. Rokhlin and M. Vassiliou, 'The fast multipole method for electromagnetic scattering computation,' *IEEE Trans. Antennas Propag.*, 40, 634–641, 1992.

90. D. Davidson, Computational Electromagnetics for RF and Microwave Engineering, Cambridge University Press, 2005.

91. R. E. Collin, Field Theory of Guided Waves, Oxford University Press, 1991.

92. Y. Rahmat-Samii and V. Galindo-Israel, 'Shaped reflector antenna analysis using the Jacobi-Bessel series,' *IEEE Trans. Antennas Propag.*, AP-28, 425–435, 1980.

93. J. B. Keller, 'Geometric theory of diffraction,' *J. Opt. Soc. Am. A*, 52(2), 116–130, 1962.

94. R. G. Kouyoumjian and P. H. Pathak, 'A uniform geometric theory of diffraction of an edge in a perfectly conducting surface,' *Proc. IEEE*, 62, 1448–1461, 1974.

95. P. Y. Ufimisev, 'The Method of Fringe Waves in the Physical Theory of Diffraction.' Sovjetskaye Radio Mscow, 1962.

96. D-W. Duan, Y. Rahmat-Samii and J. Mahon, 'Scattering from a circular disk: a Comparative Study of PTD and GTD techniques,' *Proc. IEEE*, 79(10), 1472–1480, 1991.

97. B. Westcott, Shaped Reflector Antenna Design, Wiley-Blackwell, 1983

98. V. Oliker, 'Development of the Theory and Algorithms for Synthesis of Reflector Antenna Systems,' PN, 1995.

99. A. Cha, 'An offset dual shaped reflector with 84.5 percent efficiency,' *IEEE Trans. Antennas Propag.*, AP-31, 896–902, 1983.

100. H. H. S. Luh, 'Equivalent Hyperboloid (Ellipsoid) and its Applications,' *IEEE Trans. Antennas Propag.*. 48(4), 2000.

101. J. Ruze, 'Lateral -feed displacement in a Paraboloid,' *IEEE Trans. Antennas Propag.*, 13(5), 660–665, 1965.

102. M. Romier, R. Contreres and B. Palacin, 'Overlapping efficiency of Multiple Feed per Beam concepts including orthogonality constraints,' 10th European Conference on Antennas and Propagation (EuCAP), 2016.

103. K. F. Warnick, R. Maaskant and M. V. Ivashina, Phased Arrays for Radio Astronomy, Remote Sensing and Satellite Communications, Cambridge University Press, 2018.

Index

Antennas: From Theory to Practice, Second Edition. Yi Huang.
© 2022 John Wiley & Sons Ltd. Published 2022 by John Wiley & Sons Ltd.
Companion website: www.wiley.com/go/huang_antennas2e